D0321721

Third Edition

Discrete-Event System Simulation

Jerry Banks
AutoSimulations, a Brooks Automation Company

John S. Carson II
AutoSimulations, a Brooks Automation Company

Barry L. Nelson
Northwestern University

David M. Nicol
Dartmouth College

PRENTICE-HALL INTERNATIONAL SERIES
IN INDUSTRIAL AND SYSTEMS ENGINEERING
W. J. Fabrycky and J.H. Mize, Editors

Prentice Hall, Upper Saddle River, NJ 07458

Library of Congress Cataloging-in-Publication Data

Banks, Jerry
Discrete-event system simulation / Jerry Banks, John S. Carson II, Barry L. Nelson, and David Nicol—3rd ed.
p. cm.—(Prentice-Hall international series in industrial and systems engineering)
Includes bibliographical references and index.
ISBN 0–13–088702–1
1. Simulation methods. I. Carson, John S. II. Nelson, Barry L. III. Title. IV. Series.

T57.62.B35 2000 00-031381
003′.83–dc21 CIP

To
Susie, Jay, and Danny
Jonna, Jennifer, and Jonathan
Sharon and LeRoy
Elizabeth, Caitrin, Thomas, and Galen

Vice-President and Editorial Director, ECS: *Marcia Horton*
Acquisitions Editor: *Laura Curless*
Production Supervision: *Harriet Damon Shields and Scott Disanno*
Vice-President of Production and Manufacturing: *David W. Riccardi*
Executive Managing Editor: *Vince O'Brien*
Managing Editor: *David A. George*
Manufacturing Buyer: *Pat Brown*
Manufacturing Manager: *Trudy Pisciotti*
Marketing Manager: *Danny Hoyt*
Cover Director: *Jayne Conte*
Cover: *Bruce Kenselaas*
Composition: *PreTEX, Inc.*

Printed in the United States of America

10 9 8 7 6 5 4 3 2

ISBN 0-13-088702-1

Prentice-Hall International (UK) Limited, *London*
Prentice-Hall of Australia Pty. Limited, *Sydney*
Prentice-Hall Canada Inc., *Toronto*
Prentice-Hall Hispanoamericana, S.A., *Mexico*
Prentice-Hall of India Private Limited, *New Delhi*
Prentice-Hall of Japan, Inc., *Tokyo*
Pearson Education Asia Pte. Ltd.
Editora Prentice-Hall do Brasil, Ltda., *Rio de Janeiro*

Contents

Preface xi

About the Authors xiii

PART ONE: Introduction to Discrete-Event System Simulation

1 Introduction to Simulation **3**

1.1 When Simulation Is the Appropriate Tool 4
1.2 When Simulation Is Not Appropriate 5
1.3 Advantages and Disadvantages of Simulation 6
1.4 Areas of Application 7
1.5 Systems and System Environment 9
1.6 Components of a System 10
1.7 Discrete and Continuous Systems 12
1.8 Model of a System 13
1.9 Types of Models 13
1.10 Discrete-Event System Simulation 14
1.11 Steps in a Simulation Study 15
References 20
Exercises 21

2 Simulation Examples **23**

2.1 Simulation of Queueing Systems 24
2.2 Simulation of Inventory Systems 41
2.3 Other Examples of Simulation 47
2.4 Summary 55
References 55
Exercises 56

3 General Principles **63**

3.1 Concepts in Discrete-Event Simulation 64
3.1.1 The Event-Scheduling/Time-Advance Algorithm 67
3.1.2 World Views 72
3.1.3 Manual Simulation Using Event Scheduling 75
3.2 List Processing 85
3.2.1 Lists: Basic Properties and Operations 86
3.2.2 Using Arrays for List Processing 87

3.2.3 Using Dynamic Allocation and Linked Lists 90
3.2.4 Advanced Techniques 92
3.3 Summary 92
References 93
Exercises 93

4 Simulation Software 95

4.1 History of Simulation Software 96
4.1.1 The Period of Search (1955–60) 97
4.1.2 The Advent (1961–65) 97
4.1.3 The Formative Period (1966–70) 98
4.1.4 The Expansion Period (1971–78) 98
4.1.5 Consolidation and Regeneration (1979–86) 99
4.1.6 The Present Period (1987–present) 99
4.2 Selection of Simulation Software 100
4.3 An Example Simulation 104
4.4 Simulation in C++ 104
4.5 Simulation in GPSS 114
4.6 Simulation in CSIM 119
4.7 Simulation Packages 123
4.7.1 Arena 123
4.7.2 AutoMod 124
4.7.3 Deneb/QUEST 125
4.7.4 Extend 126
4.7.5 Micro Saint 127
4.7.6 ProModel 127
4.7.7 Taylor ED 128
4.7.8 WITNESS 128
4.8 Experimentation and Statistical Analysis Tools 129
4.8.1 Common Features 129
4.8.2 Analysis Tools 129
4.9 Trends in Simulation Software 131
4.9.1 High-Fidelity Simulation 131
4.9.2 Data Exchange Standards 132
4.9.3 The Internet 132
4.9.4 Old Paradigm versus New Paradigm 133
4.9.5 Component Libraries 133
4.9.6 Distributed Manufacturing Simulation/High Level Architecture 133
4.9.7 Embedded Simulation 134
4.9.8 Optimization 134
References 134
Exercises 136

PART TWO: Mathematical and Statistical Models

5 Statistical Models In Simulation 153

5.1 Review of Terminology and Concepts 154
5.2 Useful Statistical Models 160
5.3 Discrete Distributions 165
5.4 Continuous Distributions 170
5.5 Poisson Process 190
5.6 Empirical Distributions 193
5.7 Summary 196
References 196
Exercises 197

6 Queueing Models 204

6.1 Characteristics of Queueing Systems 205
6.1.1 The Calling Population 206
6.1.2 System Capacity 207
6.1.3 The Arrival Process 207
6.1.4 Queue Behavior and Queue Discipline 209
6.1.5 Service Times and the Service Mechanism 209
6.2 Queueing Notation 211
6.3 Long-Run Measures of Performance of Queueing Systems 212
6.3.1 Time-Average Number in System L 213
6.3.2 Average Time Spent in System per Customer, w 215
6.3.3 The Conservation Equation: $L = \lambda w$ 216
6.3.4 Server Utilization 218
6.3.5 Costs in Queueing Problems 223
6.4 Steady-State Behavior of Infinite-Population Markovian Models 224
6.4.1 Single-Server Queues with Poisson Arrivals and Unlimited Capacity: $M/G/1$ 225
6.4.2 Multiserver Queue: $M/M/c/\infty/\infty$ 231
6.4.3 Multiserver Queues with Poisson Arrivals and Limited Capacity: $M/M/c/N/\infty$ 237
6.5 Steady-State Behavior of Finite-Population Models $(M/M/c/K/K)$ 239
6.6 Networks of Queues 243
6.7 Summary 245
References 246
Exercises 247

PART THREE: Random Numbers

7 Random-Number Generation 255

7.1 Properties of Random Numbers 255
7.2 Generation of Pseudo-Random Numbers 256
7.3 Techniques for Generating Random Numbers 258
 7.3.1 Linear Congruential Method 258
 7.3.2 Combined Linear Congruential Generators 262
7.4 Tests for Random Numbers 264
 7.4.1 Frequency Tests 266
 7.4.2 Runs Tests 270
 7.4.3 Tests for Autocorrelation 278
 7.4.4 Gap Test 281
 7.4.5 Poker Test 283
7.5 Summary 284
 References 284
 Exercises 285

8 Random-Variate Generation 289

8.1 Inverse Transform Technique 290
 8.1.1 Exponential Distribution 290
 8.1.2 Uniform Distribution 294
 8.1.3 Weibull Distribution 294
 8.1.4 Triangular Distribution 295
 8.1.5 Empirical Continuous Distributions 296
 8.1.6 Continuous Distributions without a Closed-Form Inverse 300
 8.1.7 Discrete Distributions 301
8.2 Direct Transformation for the Normal and Lognormal Distributions 307
8.3 Convolution Method 309
 8.3.1 Erlang Distribution 309
8.4 Acceptance-Rejection Technique 310
 8.4.1 Poisson Distribution 311
 8.4.2 Gamma Distribution 314
8.5 Summary 315
 References 316
 Exercises 316

PART FOUR: Analysis of Simulation Data

9 Input Modeling 323

9.1 Data Collection 324
9.2 Identifying the Distribution with Data 327
 9.2.1 Histograms 327
 9.2.2 Selecting the Family of Distributions 331

9.2.3 Quantile-Quantile Plots 333
9.3 Parameter Estimation 336
9.3.1 Preliminary Statistics: Sample Mean and Sample Variance 336
9.3.2 Suggested Estimators 338
9.4 Goodness-of-Fit Tests 343
9.4.1 Chi-Square Test 343
9.4.2 Chi-Square Test with Equal Probabilities 346
9.4.3 Kolmogorov-Smirnov Goodness-of-Fit Test 348
9.4.4 p-Values and "Best Fits" 350
9.5 Selecting Input Models without Data 351
9.6 Multivariate and Time-Series Input Models 353
9.6.1 Covariance and Correlation 354
9.6.2 Multivariate Input Models 354
9.6.3 Time-Series Input Models 356
9.7 Summary 358
References 359
Exercises 360

10 Verification and Validation of Simulation Models 367

10.1 Model Building, Verification, and Validation 368
10.2 Verification of Simulation Models 369
10.3 Calibration and Validation of Models 374
10.3.1 Face Validity 376
10.3.2 Validation of Model Assumptions 377
10.3.3 Validating Input-Output Transformations 377
10.3.4 Input-Output Validation: Using Historical Input Data 388
10.3.5 Input-Output Validation: Using a Turing Test 392
10.4 Summary 393
References 394
Exercises 395

11 Output Analysis for a Single Model 398

11.1 Types of Simulations with Respect to Output Analysis 399
11.2 Stochastic Nature of Output Data 402
11.3 Measures of Performance and Their Estimation 407
11.3.1 Point Estimation 407
11.3.2 Interval Estimation 409
11.4 Output Analysis for Terminating Simulations 410
11.4.1 Statistical Background 410
11.4.2 Confidence-Interval Estimation for a Fixed Number of Replications 411
11.4.3 Confidence Intervals with Specified Precision 414
11.4.4 Confidence Intervals for Quantiles 416

11.5 Output Analysis for Steady-State Simulations 418
11.5.1 Initialization Bias in Steady-State Simulations 419
11.5.2 Statistical Background 426
11.5.3 Replication Method for Steady-State Simulations 430
11.5.4 Sample Size in Steady-State Simulations 434
11.5.5 Batch Means for Interval Estimation in Steady-State Simulations 436
11.5.6 Confidence Intervals for Quantiles 440
11.6 Summary 441
References 441
Exercises 442

12 Comparison and Evaluation of Alternative System Designs 450

12.1 Comparison of Two System Designs 451
12.1.1 Independent Sampling with Equal Variances 454
12.1.2 Independent Sampling with Unequal Variances 456
12.1.3 Correlated Sampling, or Common Random Numbers 456
12.1.4 Confidence Intervals with Specified Precision 466
12.2 Comparison of Several System Designs 467
12.2.1 Bonferroni Approach to Multiple Comparisons 468
12.2.2 Bonferroni Approach to Selecting the Best 473
12.3 Metamodeling 476
12.3.1 Simple Linear Regression 477
12.3.2 Testing for Significance of Regression 481
12.3.3 Multiple Linear Regression 484
12.3.4 Random-Number Assignment for Regression 484
12.4 Optimization via Simulation 485
12.4.1 What Does "Optimization via Simulation" Mean? 487
12.4.2 Why Is Optimization via Simulation Difficult? 488
12.4.3 Using Robust Heuristics 489
12.4.4 An Illustration: Random Search 492
12.5 Summary 495
References 496
Exercises 497

13 Simulation of Manufacturing and Material Handling Systems 502

13.1 Manufacturing and Material Handling Simulations 502
13.1.1 Models of Manufacturing Systems 503
13.1.2 Models of Material Handling 505
13.1.3 Some Common Material Handling Equipment 506
13.2 Goals and Performance Measures 507
13.3 Issues in Manufacturing and Material Handling Simulations 508
13.3.1 Modeling Downtimes and Failures 508
13.3.2 Trace-Driven Models 513

13.4 Case Studies of the Simulation of Manufacturing and Material Handling Systems 515
13.5 Summary 517
References 517
Exercises 518

14 Simulation of Computer Systems 528

14.1 Introduction 528
14.2 Simulation Tools 531
14.2.1 Process Orientation 533
14.2.2 Event Orientation 537
14.3 Model Input 542
14.3.1 Modulated Poisson Process 543
14.3.2 Virtual Memory Referencing 547
14.4 High-Level Computer-System Simulation 553
14.5 CPU Simulation 557
14.6 Memory Simulation 563
14.7 Summary 566
References 567
Exercises 568

Appendix Tables 571

A.1 Random Digits 572
A.2 Random Normal Numbers 573
A.3 Cumulative Normal Distribution 574
A.4 Cumulative Poisson Distribution 576
A.5 Percentage Points of the Students t Distribution with v Degrees of Freedom 580
A.6 Percentage Points of the Chi-Square Distribution with v Degrees of Freedom 581
A.7 Percentage Points of the F Distribution with $\alpha = 0.05$ 582
A.8 Kolmogorov–Smirnov Critical Values 583
A.9 Maximum-Likelihood Estimates of the Gamma Distribution 584
A.10 Operating-Characteristic Curves for the Two-Sided t-Test for Different Values of Sample Size n 585
A.11 Operating-Characteristic Curves for the One-Sided t-Test for Different Values of Sample Size n 586

Index 587

The simulation software and products mentioned in this text are distinguished by their trademarks and manufacturers as follows:

GPSS/H™ and Proof Animation™ are trademarks of Wolverine Software Corporation. SIMAN® and Arena® are registered trademarks of Rockwell Software Inc. OptQuest® is a registered trademark of OptTek Systems, Inc. SIMSCRIPT® and SIMSCRIPT II® are registered trademarks of CACI Products Company. SLAM® and SLAM II® are registered trademarks of Symix, Inc. CSIM® is a registered trademark of Mesquite Software, Inc. Microsoft Excel®, Windows®, and Visio® are registered trademarks of Microsoft Corporation. AutoCad® is a registered trademark of Autodesk, Inc. AutoMod™ and AutoStat™ are trademarks of AutoSimulations, a Brooks Automation Company. Deneb/QUEST®, Deneb/IGRIP®, and Deneb/ERGOSim® are registered trademarks of Delmia, Inc. Extend™ is a trademark of Imagine That, Inc. ProModel® and SimRunner® are registered trademarks of PROMODEL Corporation. Taylor ED™ is a trademark of F&H Simulations, Inc. WITNESS® is a registered trademark of Lanner Group. ExpertFit® is a registered trademark of Averill M. Law and Associates, Inc. Stat::Fit® is a registered trademark of Geer Mountain Software Corporation. @Risk's BestFit® is a registered trademark of Palisade Corporation. Maple® is a registered trademark of Waterloo Maple, Inc.

Preface

The objective of the third edition remains the same: to provide a basic treatment of all the important aspects of discrete-event simulation. The manner and level of presentation and much of the content remain the same as the second edition. What has changed is the addition of David Nicol as an author, new and updated material, reorganization of some of the material, the addition of a chapter on the modeling of computer systems, and the addition of the text's own website.

The text now has fourteen chapters. Chapter 1, *Introduction to Simulation*, has been generally updated. New sections have been added on when simulation is and is not the appropriate tool to use.

The text's website has solutions in Excel for the examples in Chapter 2, *Simulation Examples*. Students are encouraged to solve hand simulations using Excel.

Chapter 3 on *General Principles* has been updated to include properties and operations of current programming languages. In Chapter 4 on *Simulation Software*, simulation in C++ replaces FORTRAN. We also have an up-to-date discussion of the features of currently available simulation software, and a new discussion of the features of output analysis software, including optimization. We also have a new section on trends in simulation software.

Chapter 5 on *Statistical Models in Simulation* remains basically unchanged, except that we have added the lognormal distribution. In Chapter 6, *Queueing Models*, we now provide a more concise presentation of the basic models. We also added the $M/M/c/N/\infty$ queue and greater coverage of networks of queues.

Chapter 7 on *Random-Number Generation* is unchanged except for updating. In Chapter 8, *Random-Variate Generation*, we reorganized and simplified the material on empirical continuous distributions. We also added a generator for the lognormal distribution, and provided parameter estimates for the lognormal in Chapter 9, *Input Modeling*. In addition, Chapter 9 discusses p-values and "best fits" as used in input modeling software, and contains an expanded discussion of input modeling without data.

We left Chapter 10 on *Verification and Validation of Simulation Models* largely unchanged. In Chapter 11, *Output Analysis for a Single Model*, we greatly reorganized the material on types of simulation models with respect to output analysis (terminating vs. steady state), and on output analysis for each type. We also added confidence intervals for quantiles. And we enhanced our discussion of the batch means method, including a suggested algorithm. In Chapter 12, *Comparison and Evaluation of Alternative System Designs*, a section on optimization via simulation has been included.

The material in Chapter 13 on the *Simulation of Manufacturing and Material Handling Systems* has been updated. However, the material on simulation packages for manufacturing and material handling applications has been moved to Chapter 4 with the general-purpose simulation packages.

Finally, Chapter 14, *Simulation of Computer Systems*, is new to this edition. It focuses on how discrete-event simulation is used in the design and evaluation of computer systems. The chapter emphasizes the hierarchical nature of computing systems, and points out how simulation techniques vary, depending on the level of abstraction. Topics in model representation and model input are considered, as are examples of simulating a web-server system, a CPU that executes instructions out-of-order, and memory hierarchies.

As mentioned above, the text now has its own website that can be accessed at the URL `http://www.bcnn.net`. At that website, you will find access to our Excel solutions, helpful simulation links, and errata. We hope that the last category will be empty!

Discrete-Event System Simulation can serve as a textbook in the following types of courses:

1. An introductory simulation course in engineering, computer science, or management (Chapters 1–9 and selected parts of Chapters 10–12 when no companion language text is used. If a companion language text is used, skip Chapter 4. Use Chapter 13 or 14, as appropriate to the discipline.)
2. A second course in simulation (All of Chapters 10–12, a companion language text, and an outside project. Add Chapter 13 or 14 if appropriate.)

Jerry Banks
John S. Carson II
Barry L. Nelson
David M. Nicol

About the Authors

Jerry Banks is Senior Simulation Technology Advisor, AutoSimulations, a Brooks Automation Company, in their Atlanta, Georgia, office. He retired in June, 1999 as Professor, School of Industrial and Systems Engineering, Georgia Institute of Technology. He is the author, co-author, or co-editor of ten books, several chapters in texts, and numerous technical papers. He is the editor of the award-winning *Handbook of Simulation*, published in 1998 by John Wiley & Sons, Inc. His latest title, *Getting Started with AutoMod*, was published by AutoSimulations in 2000. He was a founding partner in the simulation consulting firm Carson/Banks & Associates, Inc. located in Atlanta. The firm was purchased by AutoSimulations, Inc. in May of 1994. He served eight years as the IIE representative to the Board of the Winter Simulation Conference, including two years as Board Chair. He was the recipient of the INFORMS College on Simulation Distinguished Service Award for 1999.

John S. Carson II is the East Coast Consulting Manager for AutoSimulations, a Brooks Automation Company. With AutoSimulations since 1994, he has over 22 years experience in simulation in a wide range of application areas, including manufacturing, distribution, warehousing and material handling, transportation and rapid transit systems, port operations and shipping, and medical/health care systems. His current interests center on the simulation of transportation systems, train systems, bulk and liquid processing, and user interface development. He has been an independent simulation consultant, and has taught at the Georgia Institute of Technology, the University of Florida, and the University of Wisconsin-Madison.

Barry L. Nelson is a Professor in the Department of Industrial Engineering and Management Sciences at Northwestern University, and is Director of the Master of Engineering Management Program there. His research centers on the design and analysis of computer simulation experiments on models of stochastic systems, concentrating on multivariate input modeling and output analysis problems. He has published numerous papers and two books. He has served as the Simulation Area Editor of *Operations Research* and President of the INFORMS (then TIMS) College on Simulation, and has held many positions for the annual Winter Simulation Conference, including Program Chair in 1997 and Board Member currently.

David M. Nicol is Professor and Chair of the Department of Computer Science at Dartmouth College. He is a longtime contributor in the field of parallel and distributed discrete-event simulations, having written one of the early Ph.D. theses on the topic. He has also worked in parallel algorithms, algorithms for mapping workload in parallel architectures, performance analysis, and reliability modeling and analysis. His research contributions extend to well over 100 articles in leading computer science journals and conferences. His research is largely driven by problems encountered in industry and government—he has worked closely with researchers at NASA, IBM, AT&T, Bellcore, and Sandia National Laboratories. His current interests lie in modeling and simulation of very large systems, particularly communications and other infrastructure, with applications in evaluating system security. He has served on the editorial board of *ACM Transactions on Modeling and Computer Simulation*, where he is presently Editor-in-Chief.

PART ONE

Introduction to Discrete-Event System Simulation

1

Introduction to Simulation

A *simulation* is the imitation of the operation of a real-world process or system over time. Whether done by hand or on a computer, simulation involves the generation of an artificial history of a system, and the observation of that artificial history to draw inferences concerning the operating characteristics of the real system.

The behavior of a system as it evolves over time is studied by developing a simulation *model*. This model usually takes the form of a set of assumptions concerning the operation of the system. These assumptions are expressed in mathematical, logical, and symbolic relationships between the *entities*, or objects of interest, of the system. Once developed and validated, a model can be used to investigate a wide variety of "what-if" questions about the real-world system. Potential changes to the system can first be simulated in order to predict their impact on system performance. Simulation can also be used to study systems in the design stage, before such systems are built. Thus, simulation modeling can be used both as an analysis tool for predicting the effect of changes to existing systems, and as a design tool to predict the performance of new systems under varying sets of circumstances.

In some instances, a model can be developed which is simple enough to be "solved" by mathematical methods. Such solutions may be found by the use of differential calculus, probability theory, algebraic methods, or other mathematical techniques. The solution usually consists of one or more numerical parameters which are called *measures of performance* of the system. However, many real-world systems are so complex that models of these systems are virtually impossible to solve mathematically. In these instances, numerical, computer-based simulation can be used to imitate the behavior of the system over time. From the simulation, data are collected as if a real system were being observed. This simulation-generated data is used to estimate the measures of performance of the system.

This book provides an introductory treatment of the concepts and methods of one form of simulation modeling—discrete-event simulation modeling. The first chapter initially discusses when to use simulation, its advantages and disadvantages, and actual areas of application. Then the concepts of system and model are explored. Finally, an outline is given of the steps in building and using a simulation model of a system.

1.1 When Simulation Is the Appropriate Tool

The availability of special-purpose simulation languages, massive computing capabilities at a decreasing cost per operation, and advances in simulation methodologies have made simulation one of the most widely used and accepted tools in operations research and systems analysis. Circumstances under which simulation is the appropriate tool to use have been discussed by many authors, from Naylor et al. [1966] to Banks et al. [1996]. Simulation can be used for the following purposes:

1. Simulation enables the study of, and experimentation with, the internal interactions of a complex system, or of a subsystem within a complex system.
2. Informational, organizational, and environmental changes can be simulated, and the effect of these alterations on the model's behavior can be observed.
3. The knowledge gained in designing a simulation model may be of great value toward suggesting improvement in the system under investigation.
4. By changing simulation inputs and observing the resulting outputs, valuable insight may be obtained into which variables are most important and how variables interact.
5. Simulation can be used as a pedagogical device to reinforce analytic solution methodologies.
6. Simulation can be used to experiment with new designs or policies prior to implementation, so as to prepare for what may happen.
7. Simulation can be used to verify analytic solutions.
8. By simulating different capabilities for a machine, requirements can be determined.
9. Simulation models designed for training allow learning without the cost and disruption of on-the-job learning.

10. Animation shows a system in simulated operation so that the plan can be visualized.
11. The modern system (factory, wafer fabrication plant, service organization, etc.) is so complex that the interactions can be treated only through simulation.

1.2 When Simulation Is Not Appropriate

This section is based on an article by Banks and Gibson [1997], who gave ten rules for determining when simulation is not appropriate. The first rule indicates that simulation should not be used when the problem can be solved using common sense. An example is given of an automobile-tag facility serving customers who arrive randomly at an average rate of 100/hour and are served at a mean rate of 12/hour. To determine the minimum number of servers needed, simulation is not necessary. Just compute 100/12 = 8.33, indicating that nine or more servers are needed.

The second rule says that simulation should not be used if the problem can be solved analytically. For example, under certain conditions, the average waiting time in the example above can be determined from curves that were developed by Hillier and Lieberman [1995].

The next rule says that simulation should not be used if it is easier to perform direct experiments. An example of a fast-food drive-in restaurant is given, where it was less expensive to have a person use a hand-held terminal and voice communication to determine the effect of adding another order station on customer waiting time.

The fourth rule says not to use simulation, if the costs exceed the savings. There are many steps in completing a simulation, as discussed in Section 1.11, and these must be done thoroughly. If a simulation study costs $20,000 and the savings might be $10,000, simulation would not be appropriate.

Rules five and six indicate that simulation should not be performed if the resources or time are not available. If the simulation is estimated to cost $20,000 and only $10,000 is available, the suggestion is not to venture into a simulation study. Similarly, if a decision is needed in two weeks and a simulation will take a month, the simulation study is not advised.

Simulation takes data, sometimes lots of data. If no data is available, not even estimates, simulation is not advised.

The next rule concerns the ability to verify and validate the model. If there is not enough time or the personnel are not available, simulation is not appropriate.

If managers have unreasonable expectations—say, too much too soon—or the power of simulation is overestimated, simulation may not be appropriate.

Last, if system behavior is too complex or can't be defined, simulation is not appropriate. Human behavior is sometimes extremely complex to model.

1.3 Advantages and Disadvantages of Simulation

Simulation is intuitively appealing to a client because it mimics what happens in a real system or what is perceived for a system that is in the design stage. The output data from a simulation should directly correspond to the outputs that could be recorded from the real system. Additionally, it is possible to develop a simulation model of a system without dubious assumptions (such as the same statistical distribution for every random variable) of mathematically solvable models. For these, and other reasons, simulation is frequently the technique of choice in problem solving.

In contrast to optimization models, simulation models are "run" rather than solved. Given a particular set of input and model characteristics, the model is run and the simulated behavior is observed. This process of changing inputs and model characteristics results in a set of scenarios that are evaluated. A good solution, either in the analysis of an existing system or the design of a new system, is then recommended for implementation.

Simulation has many advantages, and even some disadvantages. These are listed by Pegden, Shannon, and Sadowski [1995]. The advantages are:

1. New policies, operating procedures, decision rules, information flows, organizational procedures, and so on can be explored without disrupting ongoing operations of the real system.
2. New hardware designs, physical layouts, transportation systems, and so on, can be tested without committing resources for their acquisition.
3. Hypotheses about how or why certain phenomena occur can be tested for feasibility.
4. Time can be compressed or expanded allowing for a speedup or slowdown of the phenomena under investigation.
5. Insight can be obtained about the interaction of variables.
6. Insight can be obtained about the importance of variables to the performance of the system.
7. Bottleneck analysis can be performed indicating where work-in-process, information, materials, and so on are being excessively delayed.
8. A simulation study can help in understanding how the system operates rather than how individuals think the system operates.
9. "What-if" questions can be answered. This is particularly useful in the design of new systems.

The disadvantages are:

1. Model building requires special training. It is an art that is learned over time and through experience. Furthermore, if two models are constructed by two competent individuals, they may have similarities, but it is highly unlikely that they will be the same.

2. Simulation results may be difficult to interpret. Since most simulation outputs are essentially random variables (they are usually based on random inputs), it may be hard to determine whether an observation is a result of system interrelationships or randomness.
3. Simulation modeling and analysis can be time consuming and expensive. Skimping on resources for modeling and analysis may result in a simulation model or analysis that is not sufficient for the task.
4. Simulation is used in some cases when an analytical solution is possible, or even preferable, as discussed in Section 1.2. This might be particularly true in the simulation of some waiting lines where closed-form queueing models are available.

In defense of simulation, these four disadvantages, respectively, can be offset as follows:

1. Vendors of simulation software have been actively developing packages that contain all or part of models that need only input data for their operation. Such models have the generic tag "simulators" or "templates."
2. Many simulation software vendors have developed output analysis capabilities within their packages for performing very thorough analysis.
3. Simulation can be performed faster today than yesterday, and even faster tomorrow. This is attributable to the advances in hardware that permit rapid running of scenarios. It is also attributable to the advances in many simulation packages. For example, some simulation software contains constructs for modeling material handling using transporters such as fork lift trucks, conveyors, automated guided vehicles, and others.
4. Closed-form models are not able to analyze most of the complex systems that are encountered in practice. In nearly eight years of consulting practice by one of the authors, not one problem was encountered that could have been solved by a closed-form solution.

1.4 Areas of Application

The applications of simulation are vast. The Winter Simulation Conference (WSC) is an excellent way to learn more about the latest in simulation applications and theory. There are also numerous tutorials at both the beginning and advanced levels. WSC is sponsored by six technical societies and the National Institute of Standards and Technology (NIST). The technical societies are American Statistical Association (ASA), Association for Computing Machinery/Special Interest Group on Simulation (ACM/SIGSIM), Institute of Electrical and Electronics Engineers: Computer Society (IEEE/CS), Institute of Electrical and Electronics Engineers: Systems, Man and Cybernetics Society (IEEE/SMCS), Institute of Industrial Engineers (IIE), Institute for Operations Research and the Management Sciences: College on Simulation (INFORMS/CS), and The Society for Computer Simulation (SCS). Note that

IEEE is represented by two bodies. Information about the upcoming WSC can be obtained from

`http://www.wintersim.org`

Some of the areas of application at a recent WSC, with the subject matter within those areas, are listed below:

Manufacturing Applications

Analysis of electronics assembly operations
Design and evaluation of a selective assembly station for high-precision scroll compressor shells
Comparison of dispatching rules for semiconductor manufacturing using large-facility models
Evaluation of cluster tool throughput for thin-film head production
Determining optimal lot size for a semiconductor back-end factory
Optimization of cycle time and utilization in semiconductor test manufacturing
Analysis of storage and retrieval strategies in a warehouse
Investigation of dynamics in a service-oriented supply chain
Model for an Army chemical munitions disposal facility

Semiconductor Manufacturing

Comparison of dispatching rules using large-facility models
The corrupting influence of variability
A new lot-release rule for wafer fabs
Assessment of potential gains in productivity due to proactive reticle management
Comparison of a 200-mm and 300-mm X-ray lithography cell
Capacity planning with time constraints between operations
300-mm logistic system risk reduction

Construction Engineering

Construction of a dam embankment
Trenchless renewal of underground urban infrastructures
Activity scheduling in a dynamic, multiproject setting
Investigation of the structural steel erection process
Special-purpose template for utility tunnel construction

Military Applications

Modeling leadership effects and recruit type in an Army recruiting station
Design and test of an intelligent controller for autonomous underwater vehicles
Modeling military requirements for nonwarfighting operations
Multitrajectory performance for varying scenario sizes
Using adaptive agents in U.S. Air Force pilot retention

Logistics, Transportation, and Distribution Applications

Evaluating the potential benefits of a rail-traffic planning algorithm
Evaluating strategies to improve railroad performance
Parametric modeling in rail-capacity planning
Analysis of passenger flows in an airport terminal
Proactive flight-schedule evaluation
Logistics issues in autonomous food production systems for extended-duration space exploration
Sizing industrial rail-car fleets
Product distribution in the newspaper industry
Design of a toll plaza
Choosing between rental-car locations
Quick-response replenishment

Business Process Simulation

Impact of connection bank redesign on airport gate assignment
Product development program planning
Reconciliation of business and systems modeling
Personnel forecasting and strategic workforce planning

Human Systems

Modeling human performance in complex systems
Studying the human element in air traffic control

1.5 Systems and System Environment

To model a system, it is necessary to understand the concept of a system and the system boundary. A *system* is defined as a group of objects that are joined together in some regular interaction or interdependence toward the accomplishment of some purpose. An example is a production system manufacturing automobiles. The machines, component parts, and workers operate jointly along an assembly line to produce a high-quality vehicle.

A system is often affected by changes occurring outside the system. Such changes are said to occur in the *system environment* [Gordon, 1978]. In modeling systems, it is necessary to decide on the *boundary* between the system and its environment. This decision may depend on the purpose of the study.

> In the case of the factory system, for example, the factors controlling the arrival of orders may be considered to be outside the influence of the factory and therefore part of the environment. However, if the effect of supply on demand is to be considered, there will be a relationship between factory output and arrival of orders, and this relationship must be considered an activity of the system. Similarly, in the case of a bank system, there may be a limit on the maximum interest rate that can be paid. For the study of a single bank, this would be regarded as a constraint imposed by the environment. In a study of the effects of monetary laws on the banking industry, however, the setting of the limit would be an activity of the system. [Gordon, 1978]

1.6 Components of a System

In order to understand and analyze a system, a number of terms need to be defined. An *entity* is an object of interest in the system. An *attribute* is a property of an entity. An *activity* represents a time period of specified length. If a bank is being studied, customers might be one of the entities, the balance in their checking accounts might be an attribute, and making deposits might be an activity.

The collection of entities that compose a system for one study might be only a subset of the overall system for another study [Law and Kelton, 2000]. For example, if the bank mentioned above is being studied to determine the number of tellers needed to provide for paying and receiving, the system can be defined as that portion of the bank consisting of the regular tellers and the customers waiting in line. If the purpose of the study is expanded to determine the number of special tellers needed (to prepare cashier's checks, to sell traveler's checks, etc.), the definition of the system must be expanded.

The *state* of a system is defined to be that collection of variables necessary to describe the system at any time, relative to the objectives of the study. In the study of a bank, possible state variables are the number of busy tellers, the number of customers waiting in line or being served, and the arrival time of the next customer. An *event* is defined as an instantaneous occurrence that may change the state of the system. The term *endogenous* is used to describe activities and events occurring within a system, and the term *exogenous* is used to describe activities and events in the environment that affect the system. In the bank study, the arrival of a customer is an exogenous event, and the completion of service of a customer is an endogenous event.

Table 1.1 lists examples of entities, attributes, activities, events, and state variables for several systems. Only a partial listing of the system components is shown. A complete list cannot be developed unless the purpose of the study is known. Depending on the purpose, various aspects of the system will be of interest, and then the listing of components can be completed.

Table 1.1. Examples of Systems and Components

System	*Entities*	*Attributes*	*Activities*	*Events*	*State Variables*
Banking	Customers	Checking account balance	Making deposits	Arrival; departure	Number of busy tellers; number of customers waiting
Rapid rail	Riders	Origination; destination	Traveling	Arrival at station; arrival at destination	Number of riders waiting at each station; number of riders in transit
Production	Machines	Speed; capacity; breakdown rate	Welding; stamping	Breakdown	Status of machines (busy, idle, or down)
Communications	Messages	Length; destination	Transmitting	Arrival at destination	Number waiting to be transmitted
Inventory	Warehouse	Capacity	Withdrawing	Demand	Levels of inventory; backlogged demands

1.7 Discrete and Continuous Systems

Systems can be categorized as discrete or continuous. "Few systems in practice are wholly discrete or continuous, but since one type of change predominates for most systems, it will usually be possible to classify a system as being either discrete or continuous" [Law and Kelton, 2000]. A *discrete system* is one in which the state variable(s) change only at a discrete set of points in time. The bank is an example of a discrete system, since the state variable, the number of customers in the bank, changes only when a customer arrives or when the service provided a customer is completed. Figure 1.1 shows how the number of customers changes only at discrete points in time.

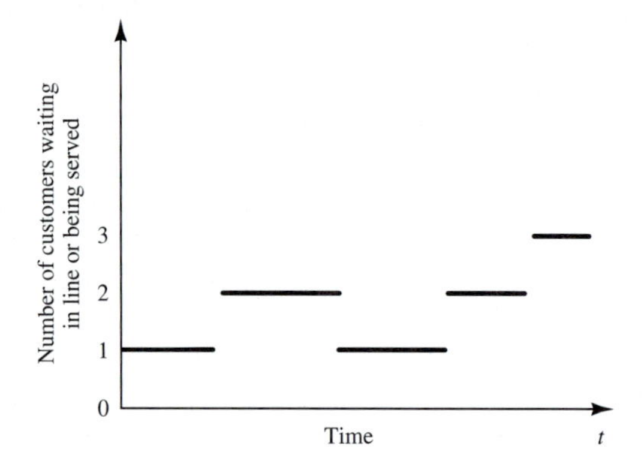

Figure 1.1. Discrete-system state variable.

A *continuous system* is one in which the state variable(s) change continuously over time. An example is the head of water behind a dam. During and for some time after a rain storm, water flows into the lake behind the dam. Water is drawn from the dam for flood control and to make electricity. Evaporation also decreases the water level. Figure 1.2 shows how the state variable, head of water behind the dam, changes for this continuous system.

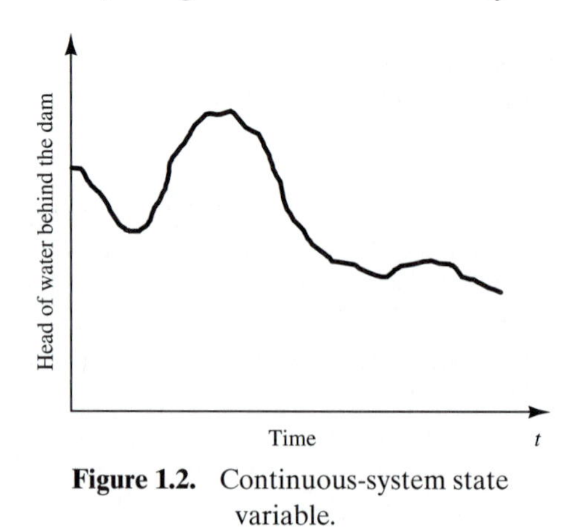

Figure 1.2. Continuous-system state variable.

1.8 Model of a System

Sometimes it is of interest to study a system to understand the relationships between its components or to predict how the system will operate under a new policy. Sometimes it is possible to experiment with the system itself, but not always. A new system may not yet exist; it may be only in hypothetical form or at the design stage. Even if the system exists, it may be impractical to experiment with it. For example, it may not be wise or possible to double the unemployment rate to determine the effect of employment on inflation. In the case of a bank, reducing the numbers of tellers to study the effect on the length of waiting lines may infuriate the customers so greatly that they move their accounts to a competitor. Consequently, studies of systems are often accomplished with a model of a system.

We had a consulting job for the simulation of a redesigned port in western Australia. At $200 million for a loading/unloading berth, it's not advisable to invest that amount only to find that the berth is inadequate for the task.

A *model* is defined as a representation of a system for the purpose of studying the system. For most studies, it is necessary to consider only those aspects of the system that affect the problem under investigation. These aspects are represented in a model of the system, and the model, by definition, is a simplification of the system. On the other hand, the model should be sufficiently detailed to permit valid conclusions to be drawn about the real system. Different models of the same system may be required as the purpose of investigation changes.

Just as the components of a system were entities, attributes, and activities, models are represented similarly. However, the model contains only those components that are relevant to the study. The components of a model are discussed more extensively in Chapter 3.

1.9 Types of Models

Models can be classified as being mathematical or physical. A mathematical model uses symbolic notation and mathematical equations to represent a system. A simulation model is a particular type of mathematical model of a system.

Simulation models may be further classified as being static or dynamic, deterministic or stochastic, and discrete or continuous. A *static* simulation model, sometimes called a Monte Carlo simulation, represents a system at a particular point in time. *Dynamic* simulation models represent systems as they change over time. The simulation of a bank from 9:00 A.M. to 4:00 P.M. is an example of a dynamic simulation.

Simulation models that contain no random variables are classified as *deterministic*. Deterministic models have a known set of inputs which will result in a unique set of outputs. Deterministic arrivals would occur at a dentist's office if all patients arrived at the scheduled appointment time. A *stochastic* simulation model has one or more random variables as inputs. Random inputs lead to random outputs. Since the outputs are random, they can be considered only as estimates of the true characteristics of a model. The simulation of a bank would usually involve random interarrival times and random service times. Thus, in a stochastic simulation, the output measures—the average number of people waiting, the average waiting time of a customer—must be treated as statistical estimates of the true characteristics of the system.

Discrete and continuous systems were defined in Section 1.7. Discrete and continuous models are defined in an analogous manner. However, a discrete simulation model is not always used to model a discrete system, nor is a continuous simulation model always used to model a continuous system. Tanks and pipes are modeled discretely by some software vendors, even though we know that fluid flow is continuous. In addition, simulation models may be mixed, both discrete and continuous. The choice of whether to use a discrete or continuous (or both discrete and continuous) simulation model is a function of the characteristics of the system and the objective of the study. Thus, a communication channel could be modeled discretely if the characteristics and movement of each message were deemed important. Conversely, if the flow of messages in aggregate over the channel were of importance, modeling the system using continuous simulation could be more appropriate. The models considered in this text are discrete, dynamic, and stochastic.

1.10 Discrete-Event System Simulation

This book is about discrete-event system simulation—the modeling of systems in which the state variable changes only at a discrete set of points in time. The simulation models are analyzed by numerical rather than by analytical methods. *Analytical* methods employ the deductive reasoning of mathematics to "solve" the model. For example, differential calculus can be used to determine the minimum-cost policy for some inventory models. *Numerical* methods employ computational procedures to "solve" mathematical models. In the case of simulation models, which employ numerical methods, models are "run" rather than solved; that is, an artificial history of the system is generated based on the model assumptions, and observations are collected to be analyzed and to estimate the true system performance measures. Since real-world simulation models are rather large, and since the amount of data stored and manipulated is so vast, the runs are usually conducted with the aid of a computer. However, much insight can be obtained by simulating small models manually.

In summary, this book is about discrete-event system simulation in which the models of interest are analyzed numerically, usually with the aid of a computer.

1.11 Steps in a Simulation Study

Figure 1.3 shows a set of steps to guide a model builder in a thorough and sound simulation study. Similar figures and discussion of steps can be found in other sources [Shannon, 1975; Gordon, 1978; Law and Kelton, 2000]. The number beside each symbol in Figure 1.3 refers to the more detailed discussion in the text. The steps in a simulation study are as follows:

Problem formulation. Every study should begin with a statement of the problem. If the statement is provided by the policy makers, or those that have the problem, the analyst must ensure that the problem being described is clearly understood. If a problem statement is being developed by the analyst, it is important that the policy makers understand and agree with the formulation. Although not shown in Figure 1.3, there are occasions where the problem must be reformulated as the study progresses. In many instances, policy makers and analysts are aware that there is a problem long before the nature of the problem is known.

Setting of objectives and overall project plan. The objectives indicate the questions to be answered by simulation. At this point a determination should be made concerning whether simulation is the appropriate methodology for the problem as formulated and objectives as stated. Assuming it is decided that simulation is appropriate, the overall project plan should include a statement of the alternative systems to be considered, and a method for evaluating the effectiveness of these alternatives. It should also include the plans for the study in terms of the number of people involved, the cost of the study, and the number of days required to accomplish each phase of the work with the anticipated results at the end of each stage.

Model conceptualization. The construction of a model of a system is probably as much art as science. Pritsker [1998] provides a thorough discussion of this step. "Although it is not possible to provide a set of instructions that will lead to building successful and appropriate models in every instance, there are some general guidelines that can be followed" [Morris, 1967]. The art of modeling is enhanced by an ability to abstract the essential features of a problem, to select and modify basic assumptions that characterize the system, and then to enrich and elaborate the model until a useful approximation results. Thus, it is best to start with a simple model and build toward greater complexity. However, the model complexity need not exceed that required to accomplish the purposes for which the model is intended. Violation of this principle will only add to model-building expenses. It is not necessary to have a one-to-one mapping between the model and the real system. Only the essence of the real system is needed.

It is advisable to involve the model user in model conceptualization. This will both enhance the quality of the resulting model and increase the confidence of the model user in the application of the model. (Chapter 2 describes a

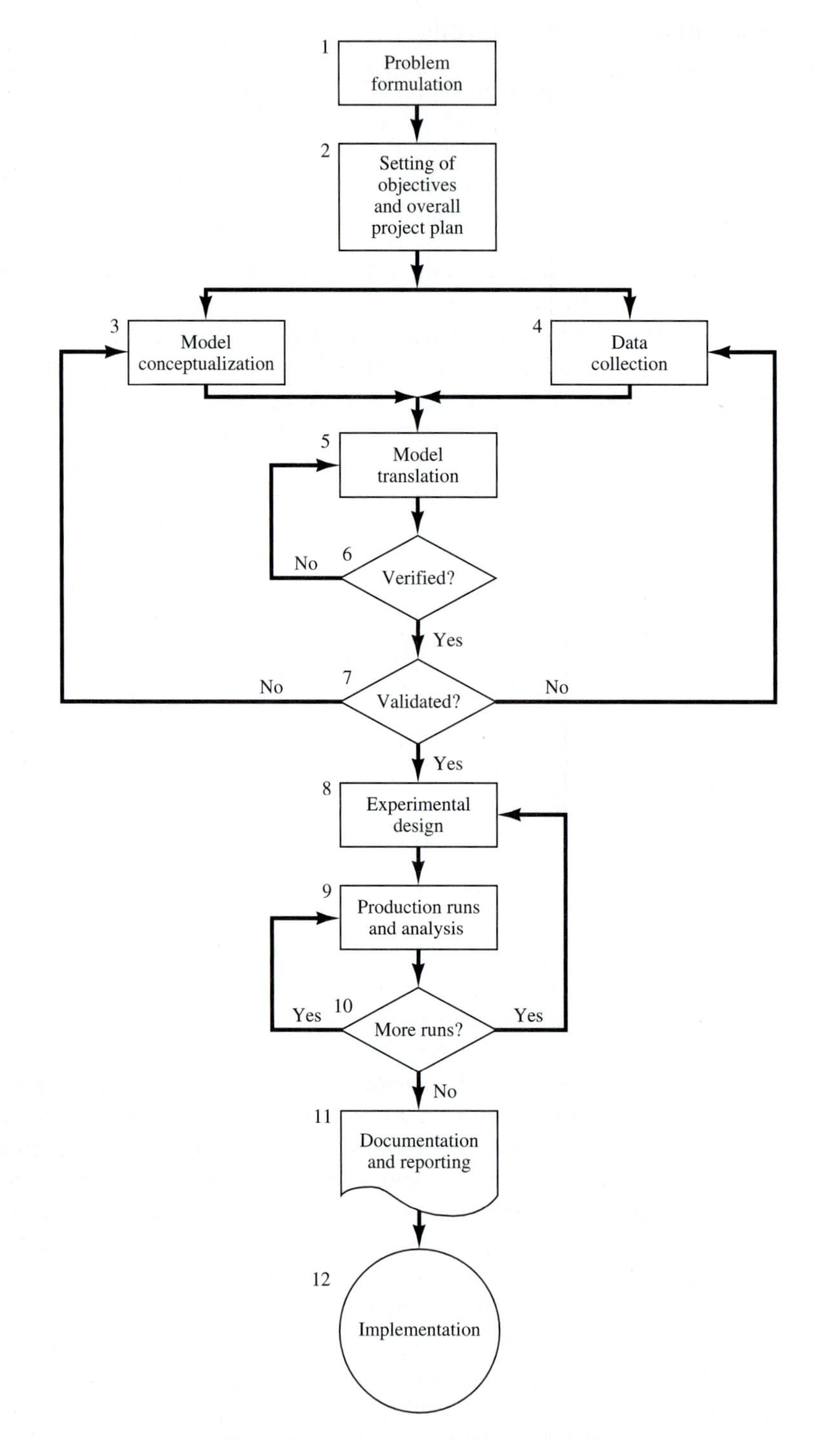

Figure 1.3. Steps in a simulation study.

number of simulation models. Chapter 6 describes queueing models that can be solved analytically. However, only experience with real systems—versus textbook problems—can "teach" the art of model building.)

Data collection. There is a constant interplay between the construction of the model and the collection of the needed input data [Shannon, 1975]. As the complexity of the model changes, the required data elements may also change. Also, since data collection takes such a large portion of the total time required to perform a simulation, it is necessary to begin it as early as possible, usually together with the early stages of model building.

The objectives of the study dictate, in a large way, the kind of data to be collected. In the study of a bank, for example, if the desire is to learn about the length of waiting lines as the number of tellers change, the types of data needed would be the distributions of interarrival times (at different times of the day), the service-time distributions for the tellers, and historic distributions on the lengths of waiting lines under varying conditions. These last data will be used to validate the simulation model. (Chapter 9 discusses data collection and data analysis; Chapter 5 discusses statistical distributions which occur frequently in simulation modeling. See also an excellent discussion by Vincent [1998].)

Model translation. Since most real-world systems result in models that require a great deal of information storage and computation, the model must be entered into a computer-recognizable format. We use the term "program," even though it is possible to accomplish the desired result in many instances with little or no actual coding. The modeler must decide whether to program the model in a simulation language such as GPSS/H™ (discussed in Chapter 4) or to use special-purpose simulation software. For manufacturing and material handling, Chapter 4 discusses Arena®, AutoMod™, CSIM, Extend™, Micro Saint, ProModel®, Deneb/QUEST®, Taylor Enterprise Dynamics (ED), and WITNESS™. Simulation languages are powerful and flexible. However, if the problem is amenable to solution with the simulation software, the model development time is greatly reduced. Furthermore, most of the simulation software packages have added features that enhance their flexibility, although the amount of flexibility varies greatly.

Verified? Verification pertains to the computer program prepared for the simulation model. Is the computer program performing properly? With complex models it is difficult, if not impossible, to translate a model successfully in its entirety without a good deal of debugging. If the input parameters and logical structure of the model are correctly represented in the computer, verification has been completed. For the most part, common sense is used in completing this step. (Chapter 10 discusses verification of simulation models, and Balci [1998] also discusses this topic extensively.)

Validated? Validation is the determination that a model is an accurate representation of the real system. Validation is usually achieved through the calibration of the model, an iterative process of comparing the model to actual system behavior and using the discrepancies between the two, and the insights gained, to improve the model. This process is repeated until model accuracy is judged acceptable. In the example of a bank mentioned above, data were collected concerning the length of waiting lines under current conditions. Does the simulation model replicate this system measure? This is one means of validation. (Chapter 10 discusses the validation of simulation models, and Balci [1998] also discusses this topic extensively.)

Experimental design. The alternatives that are to be simulated must be determined. Often, the decision concerning which alternatives to simulate may be a function of runs that have been completed and analyzed. For each system design that is simulated, decisions need to be made concerning the length of the initialization period, the length of simulation runs, and the number of replications to be made of each run. (Chapters 11 and 12 discuss issues associated with the experimental design, and Kleijnen [1998] discusses this topic extensively.)

Production runs and analysis. Production runs, and their subsequent analysis, are used to estimate measures of performance for the system designs that are being simulated. [Chapters 11 and 12 discuss the analysis of simulation experiments, and Chapter 4 discusses software to aid in this step, including AutoStat (in AutoMod), OptQuest (in several simulation softwares), SimRunner (in ProModel), and the Arena Output Analyzer.]

More runs? Based on the analysis of runs that have been completed, the analyst determines if additional runs are needed and what design those additional experiments should follow.

Documentation and reporting. There are two types of documentation: program and progress. Program documentation is necessary for numerous reasons. If the program is going to be used again by the same or different analysts, it may be necessary to understand how the program operates. This will build confidence in the program, so that model users and policy makers can make decisions based on the analysis. Also, if the program is to be modified by the same or a different analyst, this can be greatly facilitated by adequate documentation. One experience with an inadequately documented program is usually enough to convince an analyst of the necessity of this important step. Another reason for documenting a program is so that model users can change parameters at will in an effort to determine the relationships between input parameters and output measures of performance, or to determine the input parameters that "optimize" some output measure of performance.

Musselman [1998] discusses progress reports that provide the important, written history of a simulation project. Project reports give a chronology of work done and decisions made. This can prove to be of great value in keeping the project on course.

Musselman suggests frequent reports (monthly, at least) so that even those not involved in the day-to-day operation can keep abreast. The awareness of these others can usually enhance the successful completion of the project by surfacing misunderstandings early, when the problem can be solved easily. Musselman also suggests maintaining a project log providing a comprehensive record of accomplishments, change requests, key decisions, and other items of importance.

On the reporting side, Musselman suggests frequent deliverables. These may or may not be the results of major accomplishments. His maxim is that "it is better to work with many intermediate milestones than with one absolute deadline." Possibilities prior to the final report include a model specification, prototype demonstrations, animations, training results, intermediate analyses, program documentation, progress reports, and presentations. He suggests that these deliverables should be timed judiciously over the life of the project.

The result of all the analysis should be reported clearly and concisely in a final report. This will enable the model users (now, the decision makers) to review the final formulation, the alternative systems that were addressed, the criterion by which the alternatives were compared, the results of the experiments, and the recommended solution to the problem. Furthermore, if decisions have to be justified at a higher level, the final report should provide a vehicle of certification for the model user/decision maker and add to the credibility of the model and the model-building process.

Implementation. The success of the implementation phase depends on how well the previous eleven steps have been performed. It is also contingent upon how thoroughly the analyst has involved the ultimate model user during the entire simulation process. If the model user has been thoroughly involved and understands the nature of the model and its outputs, the likelihood of a vigorous implementation is enhanced [Pritsker, 1995]. Conversely, if the model and its underlying assumptions have not been properly communicated, implementation will probably suffer, regardless of the simulation model's validity.

The simulation model-building process shown in Figure 1.3 can be broken down into four phases. The first phase, consisting of steps 1 (Problem Formulation) and 2 (Setting of Objective and Overall Design), is a period of discovery or orientation. The initial statement of the problem is usually quite "fuzzy," the initial objectives will usually have to be reset, and the original project plan will usually have to be fine-tuned. These recalibrations and clarifications may occur in this phase, or perhaps after or during another phase (i.e., the analyst may have to restart the process).

The second phase is related to model building and data collection and includes steps 3 (Model Conceptualization), 4 (Data Collection), 5 (Model Translation), 6 (Verification), and 7 (Validation). A continuing interplay is required among the steps. Exclusion of the model user during this phase can have dire implications at the point of implementation.

The third phase concerns running the model. It involves steps 8 (Experimental Design), 9 (Production Runs and Analysis), and 10 (Additional Runs). This phase must have a thoroughly conceived plan for experimenting with the simulation model. A discrete-event stochastic simulation is in fact a statistical experiment. The output variables are estimates that contain random error, and therefore a proper statistical analysis is required. Such a philosophy differs sharply from that of the analyst who makes a single run and draws an inference from that single data point.

The fourth phase, implementation, involves steps 11 (Documentation and Reporting) and 12 (Implementation). Successful implementation depends on continual involvement of the model user and the successful completion of every step in the process. Perhaps the most crucial point in the entire process is step 7 (Validation), because an invalid model is going to lead to erroneous results, which if implemented could be dangerous, costly, or both.

REFERENCES

BALCI, O. [1998], "Verification, Validation, and Testing," in *Handbook of Simulation*, ed. Jerry Banks, John Wiley, New York.

BANKS, J., AND R. R. GIBSON [1997], "Don't Simulate When: 10 rules for determining when simulation is not appropriate," *IIE Solutions*, September.

BANKS, J., M. SPEARMAN, AND V. NORMAN [1996], "Second Look at Simulation," *OR/MS Today*, Vol. 22, No. 4, August 1996.

GORDON, G. [1978], *System Simulation*, 2d ed., Prentice-Hall, Upper Saddle River, NJ.

HILLIER, F. S., and Gerald J. Lieberman [1995], *Introduction to Operations Research*, 6th ed., McGraw-Hill, New York.

KLEIJNEN, J. P. C. [1998], "Experimental Design for Sensitivity Analysis, Optimization, and Validation of Simulation Models," in *Handbook of Simulation*, ed. Jerry Banks, John Wiley, New York.

LAW, A. M., AND W. D. KELTON [2000], *Simulation Modeling and Analysis*, 3d ed., McGraw-Hill, New York.

MORRIS, W. T. [1967], "On the Art of Modeling," *Management Science*, Vol. 13, No. 12.

MUSSELMAN, K. J. [1998], "Guidelines for Success," in *Handbook of Simulation*, ed. Jerry Banks, John Wiley, New York.

NAYLOR, T. H., J. L. BALINTFY, D. S. BURDICK, AND K. CHU [1966], *Computer Simulation Techniques*, John Wiley, New York.

PEGDEN, C. D., R. E. SHANNON, AND R. P. SADOWSKI [1995], *Introduction to Simulation Using SIMAN*, 2d ed., McGraw-Hill, New York.

PRITSKER, A. A. B. [1995], *Introduction to Simulation and SLAM II*, 4th ed., John Wiley, New York.

PRITSKER, A. A. B. [1998], "Principles of Simulation Modeling," in *Handbook of Simulation*, ed. Jerry Banks, John Wiley, New York.

SHANNON, R. E. [1975], *Systems Simulation: The Art and Science*, Prentice-Hall, Upper Saddle River, NJ.

VINCENT, S. [1998], "Input Data Analysis," in *Handbook of Simulation*, ed. Jerry Banks, John Wiley, New York.

EXERCISES

1. Name several entities, attributes, activities, events, and state variables for the following systems:
   (a) A small appliance repair shop
   (b) A cafeteria
   (c) A grocery store
   (d) A laundromat
   (e) A fast-food restaurant
   (f) A hospital emergency room
   (g) A taxicab company with 10 taxis
   (h) An automobile assembly line
2. Consider the simulation process shown in Figure 1.3.
   (a) Reduce the steps by at least two by combining similar activities. Give your rationale.
   (b) Increase the steps by at least two by separating current steps or enlarging on existing steps. Give your rationale.
3. A simulation of a major traffic intersection is to be conducted with the objective of improving the current traffic flow. Provide three iterations, in increasing order of complexity, of steps 1 and 2 in the simulation process of Figure 1.3.
4. In what ways and at what steps might a personal computer be used to support the simulation process of Figure 1.3?
5. A simulation is to be conducted of cooking a spaghetti dinner to determine what time a person should start in order to have the meal on the table by 7:00 P.M. Read a recipe for preparing a spaghetti dinner (or ask a friend or relative, etc., for the recipe). As best you can, trace what you understand to be needed in the data-collection phase of the simulation process of Figure 1.3 in order to perform a simulation in which the model includes each step in the recipe. What are the events, activities, and state variables in this system?
6. What events and activities are associated with the operation of your checkbook?
7. Read an article in the current *WSC Proceedings* on the application of simulation related to your major area of study or interest, and prepare a report on how the author accomplishes the steps given in Figure 1.3. *WSC Proceedings* are available on-line at

   `http://www.informs-cs.org/wscpapers.html`

8. Get a copy of a recent *WSC Proceedings* and report on the different applications discussed in an area of interest to you. *WSC Proceedings* are available on-line at

   `http://www.informs-cs.org/wscpapers.html`

9. Get a copy of a recent *WSC Proceedings* and report on the most unusual application that you can find. *WSC Proceedings* are available on-line at

   `http://www.informs-cs.org/wscpapers.html`

10. Go to the Simulation Education website at

`http://www.acs.ilstu.edu/faculty/wjyurci/nsfteachsim/indexnew.html`

and address the following:

(a) Use the links there to thoroughly answer the question, "What is simulation?"

(b) What kinds of careers are available in simulation?

(c) What are some recent applications of discrete-event simulation in the news?

(d) What are some simulation education organizations that are currently active?

11. Go to the Winter Simulation Conference website at

`http://www.wintersim.org`

and address the following:

(a) What advanced tutorials were offered at the previous WSC or are planned at the next WSC?

(b) Where and when will the next WSC be held?

(c) What is the history of WSC?

2

Simulation Examples

This chapter presents several examples of simulations that can be performed by devising a simulation table either manually or with a spreadsheet. The simulation table provides a systematic method for tracking system state over time. These examples provide insight into the methodology of discrete system simulation and the descriptive statistics used for predicting system performance.

The simulations in this chapter entail three steps:

1. Determine the characteristics of each of the inputs to the simulation. Quite often, these may be modeled as probability distributions, either continuous or discrete.
2. Construct a simulation table. Each simulation table is different, for each is developed for the problem at hand. An example of a simulation table is shown in Table 2.1. In this example there are p inputs, x_{ij}, $j = 1, 2, \ldots, p$, and one response, y_i, for each of repetitions $i = 1, 2, \ldots, n$. Initialize the table by filling in the data for repetition 1.
3. For each repetition i, generate a value for each of the p inputs, and evaluate the function, calculating a value of the response y_i. The input values may be computed by sampling values from the distributions determined in step 1. A response typically depends on the inputs and one or more previous responses.

This chapter gives a number of simulation examples in queueing, inventory, and reliability. The two queueing examples provide a single-server and

Table 2.1. Simulation Table

	Inputs						*Response*
Repetitions	x_{i1}	x_{i2}	$\cdots$	x_{ij}	$\cdots$	x_{ip}	y_i
1							
2							
3							
.							
.							
.							
n							

two-server system, respectively. (Chapter 6 provides more insight into queueing models.) The first inventory example involves a problem that has a closed-form solution; thus the simulation solution can be compared to the mathematical solution. The second inventory example pertains to the classic order-level model.

Finally, there is an example that introduces the concept of random normal numbers and a model for the determinination of lead-time demand.

2.1 Simulation of Queueing Systems

A queueing system is described by its calling population, the nature of the arrivals, the service mechanism, the system capacity, and the queueing discipline. These attributes of a queueing system are described in detail in Chapter 6. A simple single-channel queueing system is portrayed in Figure 2.1.

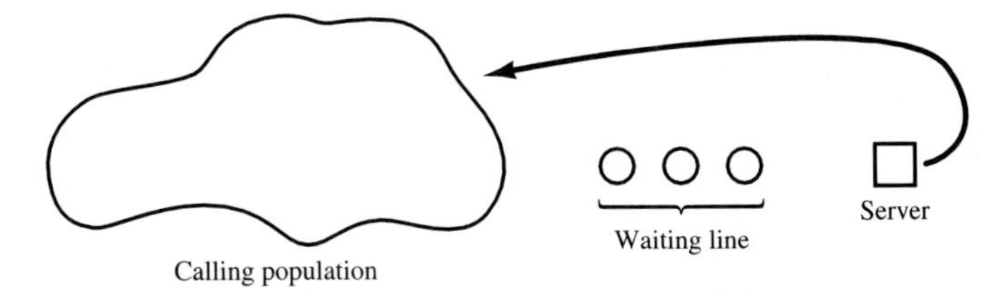

Figure 2.1. Queueing system.

In the single-channel queue, the calling population is infinite; that is, if a unit leaves the calling population and joins the waiting line or enters service, there is no change in the arrival rate of other units that may need service. Arrivals for service occur one at a time in a random fashion; once they join the waiting line, they are eventually served. In addition, service times are of some random length according to a probability distribution which does not change over time. The system capacity has no limit, meaning that any number of units can wait in line. Finally, units are served in the order of their arrival (often called FIFO: first in, first out) by a single server or channel.

Arrivals and services are defined by the distribution of the time between arrivals and the distribution of service times, respectively. For any simple single- or multi-channel queue, the overall effective arrival rate must be less than the total service rate, or the waiting line will grow without bound. When queues grow without bound, they are termed "explosive" or unstable. (In some reentrant queueing networks in which units return a number of times to the same server before finally exiting the system, the condition about arrival rate being less than service rate may not guarantee stability. See Harrison and Nguyen [1995] for more explanation. Interestingly, this type of instability was noticed first, not in theory, but in actual manufacturing in semiconductor plants.) More complex situations may occur—for example, arrival rates that are greater than service rates for short periods of time, or networks of queues with routing. However, this chapter sticks to the simplest, more basic queues.

Prior to introducing several simulations of queueing systems, it is necessary to understand the concepts of system state, events, and simulation clock. (These concepts are studied systematically in Chapter 3.) The *state* of the system is the number of units in the system and the status of the server, busy or idle. An *event* is a set of circumstances that cause an instantaneous change in the state of the system. In a single-channel queueing system there are only two possible events that can affect the state of the system. They are the entry of a unit into the system (the arrival event) or the completion of service on a unit (the departure event). The queueing system includes the server, the unit being serviced (if one is being serviced), and units in the queue (if any are waiting). The *simulation clock* is used to track simulated time.

If a unit has just completed service, the simulation proceeds in the manner shown in the flow diagram of Figure 2.2. Note that the server has only two possible states: it is either busy or idle.

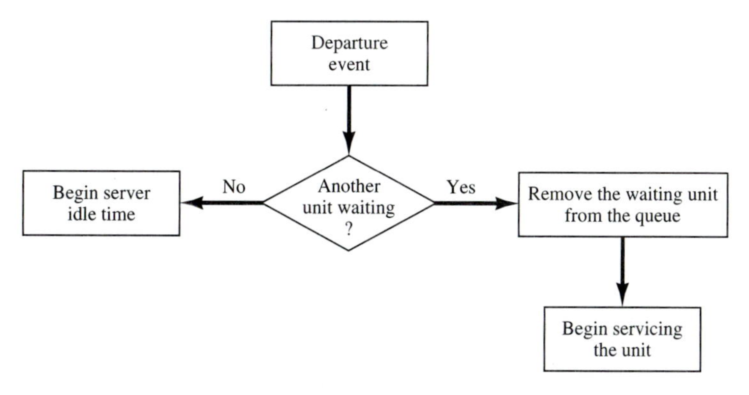

Figure 2.2. Service-just-completed flow diagram.

The arrival event occurs when a unit enters the system. The flow diagram for the arrival event is shown in Figure 2.3. The unit may find the server either idle or busy; therefore, either the unit begins service immediately, or it enters the queue for the server. The unit follows the course of action shown in Figure 2.4.

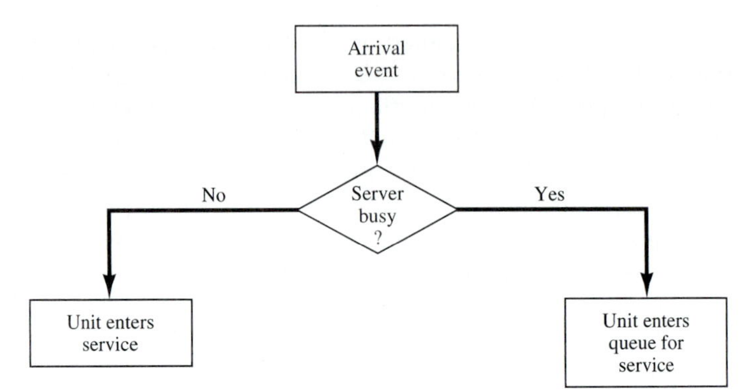

Figure 2.3. Unit-entering-system flow diagram.

If the server is busy, the unit enters the queue. If the server is idle and the queue is empty, the unit begins service. It is not possible for the server to be idle and the queue to be nonempty.

		Queue status	
		Not empty	Empty
Server status	Busy	Enter queue	Enter queue
	Idle	Impossible	Enter service

Figure 2.4. Potential unit actions upon arrival.

After the completion of a service the server may become idle or remain busy with the next unit. The relationship of these two outcomes to the status of the queue is shown in Figure 2.5. If the queue is not empty, another unit will enter the server and it will be busy. If the queue is empty, the server will be idle after a service is completed. These two possibilities are shown as the shaded portions of Figure 2.5. It is impossible for the server to become busy if the queue is empty when a service is completed. Similarly, it is impossible for the server to be idle after a service is completed when the queue is not empty.

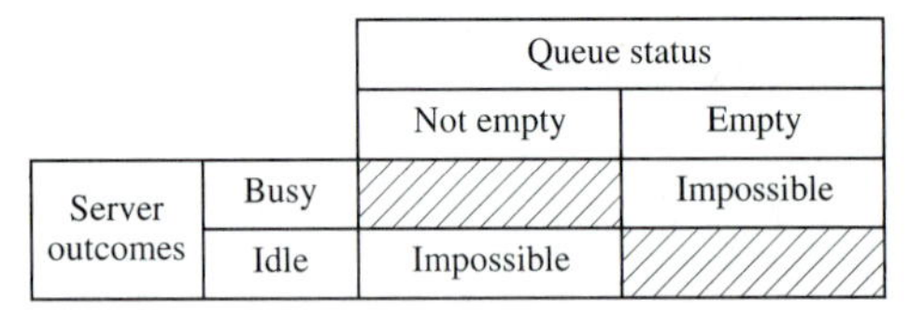

		Queue status	
		Not empty	Empty
Server outcomes	Busy		Impossible
	Idle	Impossible	

Figure 2.5. Server outcomes after service completion.

Now, how can the events described above occur in simulated time? Simulations of queueing systems generally require the maintenance of an event list for determining what happens next. The event list tracks the future times at which the different types of events occur. Simulations using event lists are described in Chapter 3. This chapter simplifies the simulation by tracking each unit explicitly. Simulation clock times for arrivals and departures are computed in a simulation table customized for each problem. In simulation, events usually occur at random times, the randomness imitating uncertainty in real life. For example, it is not known with certainty when the next customer will arrive at a grocery checkout counter, or how long the bank teller will take to complete a transaction. In these cases, a statistical model of the data is developed from either data collected and analyzed, or subjective estimates and assumptions.

The randomness needed to imitate real life is made possible through the use of "random numbers." Random numbers are distributed uniformly and independently on the interval (0, 1). Random digits are uniformly distributed on the set $\{0, 1, 2, \ldots, 9\}$. Random digits can be used to form random numbers by selecting the proper number of digits for each random number and placing a decimal point to the left of the value selected. The proper number of digits is dictated by the accuracy of the data being used for input purposes. If the input distribution has values with two decimal places, two digits are taken from a random-digits table (such as Table A.1) and the decimal point is placed to the left to form a random number.

Random numbers can also be generated in simulation packages and in spreadsheets such as Excel®. For example, Excel has a macro function called RAND() that returns a "random" number between 0 and 1. When numbers are generated using a procedure, they are often referred to as pseudo-random numbers. Since the method is known, it is always possible to know the sequence of numbers that will be generated prior to the simulation. The most commonly used methods for generating random numbers are discussed in Chapter 7.

In a single-channel queueing system, interarrival times and service times are generated from the distributions of these random variables. The examples that follow show how such times are generated. For simplicity, assume that the times between arrivals were generated by rolling a die five times and recording the up face. Table 2.2 contains a set of five interarrival times generated in this manner. These five interarrival times are used to compute the arrival times of six customers at the queueing system.

Table 2.2. Interarrival and Clock Times

Customer	*Interarrival Time*	*Arrival Time on Clock*
1	–	0
2	2	2
3	4	6
4	1	7
5	2	9
6	6	15

Table 2.3. Service Times

Customer	*Service Time*
1	2
2	1
3	3
4	2
5	1
6	4

The first customer is assumed to arrive at clock time 0. This starts the clock in operation. The second customer arrives two time units later, at a clock time of 2. The third customer arrives four time units later, at a clock time of 6; and so on.

The second time of interest is the service time. Table 2.3 contains service times generated at random from a distribution of service times. The only possible service times are one, two, three, and four time units. Assuming that all four values are equally likely to occur, these values could have been generated by placing the numbers one through four on chips and drawing the chips from a hat with replacement, being sure to record the numbers selected. Now, the interarrival times and service times must be meshed to simulate the single-channel queueing system. As shown in Table 2.4, the first customer arrives at clock time 0 and immediately begins service, which requires two minutes. Service is completed at clock time 2. The second customer arrives at clock time 2 and is finished at clock time 3. Note that the fourth customer arrived at clock time 7, but service could not begin until clock time 9. This occurred because customer 3 did not finish service until clock time 9.

Table 2.4 was designed specifically for a single-channel queue which serves customers on a first-in, first-out (FIFO) basis. It keeps track of the clock time

Table 2.4. Simulation Table Emphasizing Clock Times

A *Customer Number*	B *Arrival Time (Clock)*	C *Time Service Begins (Clock)*	D *Service Time (Duration)*	E *Time Service Ends (Clock)*
1	0	0	2	2
2	2	2	1	3
3	6	6	3	9
4	7	9	2	11
5	9	11	1	12
6	15	15	4	19

Table 2.5. Chronological Ordering of Events

Event Type	*Customer Number*	*Clock Time*
Arrival	1	0
Departure	1	2
Arrival	2	2
Departure	2	3
Arrival	3	6
Arrival	4	7
Departure	3	9
Arrival	5	9
Departure	4	11
Departure	5	12
Arrival	6	15
Departure	6	19

at which each event occurs. The second column of Table 2.4 records the clock time of each arrival event, while the last column records the clock time of each departure event. The occurrence of the two types of events in chronological order is shown in Table 2.5 and Figure 2.6.

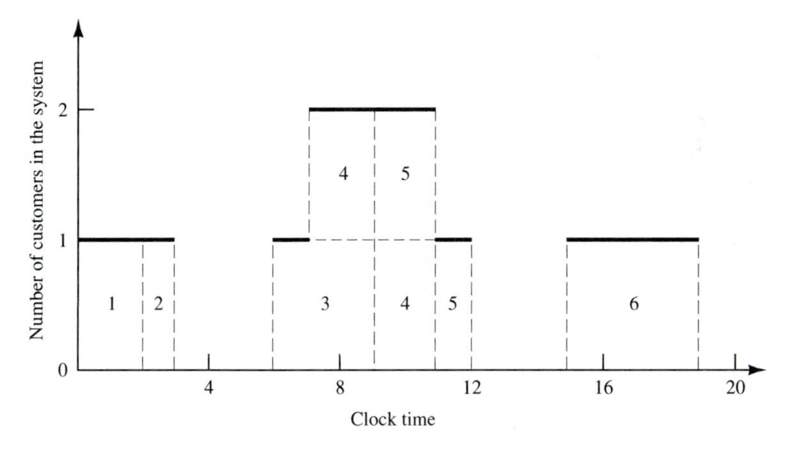

Figure 2.6. Number of customers in the system.

It should be noted that Table 2.5 is ordered by clock time, in which case the events may or may not be ordered by customer number. The chronological ordering of events is the basis of the approach to discrete-event simulation described in Chapter 3.

Figure 2.6 depicts the number of customers in the system at the various clock times. It is a visual image of the event listing of Table 2.5. Customer 1

Table 2.6. Distribution of Time Between Arrivals

Time between Arrivals (Minutes)	*Probability*	*Cumulative Probability*	*Random-Digit Assignment*
1	0.125	0.125	001–125
2	0.125	0.250	126–250
3	0.125	0.375	251–375
4	0.125	0.500	376–500
5	0.125	0.625	501–625
6	0.125	0.750	626–750
7	0.125	0.875	751–875
8	0.125	1.000	876–000

is in the system from clock time 0 to clock time 2. Customer 2 arrives at clock time 2 and departs at clock time 3. No customers are in the system from clock time 3 to clock time 6. During some time periods two customers are in the system, such as at clock time 8, when both customers 3 and 4 are in the system. Also, there are times when events occur simultaneously, such as at clock time 9, when customer 5 arrives and customer 3 departs.

Example 2.1 follows the logic described above while keeping track of a number of attributes of the system. Example 2.2 is concerned with a two-channel queueing system. The flow diagrams for a multichannel queueing system are slightly different from those for a single-channel system. The development and interpretation of these flow diagrams is left as an exercise for the reader.

Example 2.1 Single-Channel Queue

A small grocery store has only one checkout counter. Customers arrive at this checkout counter at random from 1 to 8 minutes apart. Each possible value of interarrival time has the same probability of occurrence, as shown in Table 2.6. The service times vary from 1 to 6 minutes with the probabilities shown in Table 2.7. The problem is to analyze the system by simulating the arrival and service of 20 customers.

Table 2.7. Service-Time Distribution

Service Time (Minutes)	*Probability*	*Cumulative Probability*	*Random-Digit Assignment*
1	0.10	0.10	01–10
2	0.20	0.30	11–30
3	0.30	0.60	31–60
4	0.25	0.85	61–85
5	0.10	0.95	86–95
6	0.05	1.00	96–00

Table 4.4. Experimentation and Analysis Features

Feature	*Description*
Scenario manager	Create user-defined scenarios to simulate
Run manager	Make all runs (scenarios and replications) and save results for future analyses
Warmup capability	For steady-state analysis
Independent replications	Using a different set of random numbers
Optimization	Genetic algorithms, tabu search, etc.
Standardized reports	Summary reports including averages, counts, minimum and maximum, . . .
Customized reports	Tailored presentations for managers
Statistical analysis	Confidence intervals, designed experiments, etc.
Business graphics	Bar charts, pie charts, time lines, etc.
Costing module	Activity-based costing included
File export	To spreadsheet or database, for custom processing and analyses

Table 4.5. Vendor Support and Product Documentation

Feature	*Description*
Training	Regularly scheduled classes of high quality
Documentation	Quality, completeness, online
Help system	General or context sensitive
Tutorials	For learning the package or specific features
Support	Telephone, e-mail, Web
Upgrades, maintenance	Regularity of new versions and maintenance releases that address customer needs
Track record	Stability, history, customer relations

4.3 An Example Simulation

EXAMPLE 4.1 (The Checkout Counter: A Typical Single-Server Queue)

The system, a grocery checkout counter, is modeled as a single-server queue. The simulation will run until 1000 customers have been served. In addition, assume that the interarrival times of customers are exponentially distributed with a mean of 4.5 minutes, and that the service times are (approximately) normally distributed with a mean of 3.2 minutes and a standard deviation of 0.6 minute. (The approximation is that service times are always positive.) When the cashier is busy, a queue forms with no customers turned away. This example was manually simulated in Examples 3.3 and 3.4 using the event scheduling point of view. The model contains two events, the arrival and departure events. Figures 3.5 and 3.6 provide the event logic.

The following three sections illustrate the simulation of this single-server queue in C++, GPSS/H, and CSIM. Although this example is much simpler than models that arise in the study of complex systems, its simulation contains the essential components of all discrete-event simulations. ◀

4.4 Simulation in C++

C++ is a widely used programming language that has been used extensively in simulation. (The definitive book on C++ is its designer's [Stroustup, 1997].) It does not, however, provide any facilities directly aimed at aiding the simulation analyst, who therefore must program all details of the event-scheduling/time-advance algorithm, the statistics-gathering capability, the generation of samples from specified probability distributions, and the report generator. However, the runtime library does provide a random-number generator. Unlike FORTRAN or C, the object-orientedness of C++ does support modular construction of large models. For the most part, the special-purpose simulation languages hide the details of event scheduling, whereas in C++ all the details must be explicitly programmed. However, to a certain extent simulation libraries such as CSIM [Schwetman, 1987] alleviate the development burden by providing access to standardized simulation functionality and hiding low-level scheduling minutiae.

Any discrete-event simulation model written in C++ contains the components discussed in Section 3.1: system state, entities and attributes, sets, events, activities and delays, plus the components listed below. To facilitate development and debugging, it is best to organize the C++ model in a modular fashion using subroutines. The following components are common to almost all models written in C++:

Clock A variable defining simulated time

Initialization subroutine A routine to define the system state at time 0

Min-time event subroutine A routine that identifies the imminent event — that is, the element of the future event list (`FutureEventList`) which has the smallest time-stamp

Event subroutines For each event type, a subroutine to update system state (and cumulative statistics) when that event occurs

Random-variate generators Routines to generate samples from desired probability distributions

Main program Provides overall control of the event-scheduling algorithm

Report generator A routine that computes summary statistics from cumulative statistics and prints a report at the end of the simulation

The overall structure of a C++ simulation program is shown in Figure 4.1. This flow chart is an expansion of the event-scheduling/time-advance algorithm outlined in Figure 3.2. (The steps mentioned in Figure 4.1 refer to the five steps in Figure 3.2.)

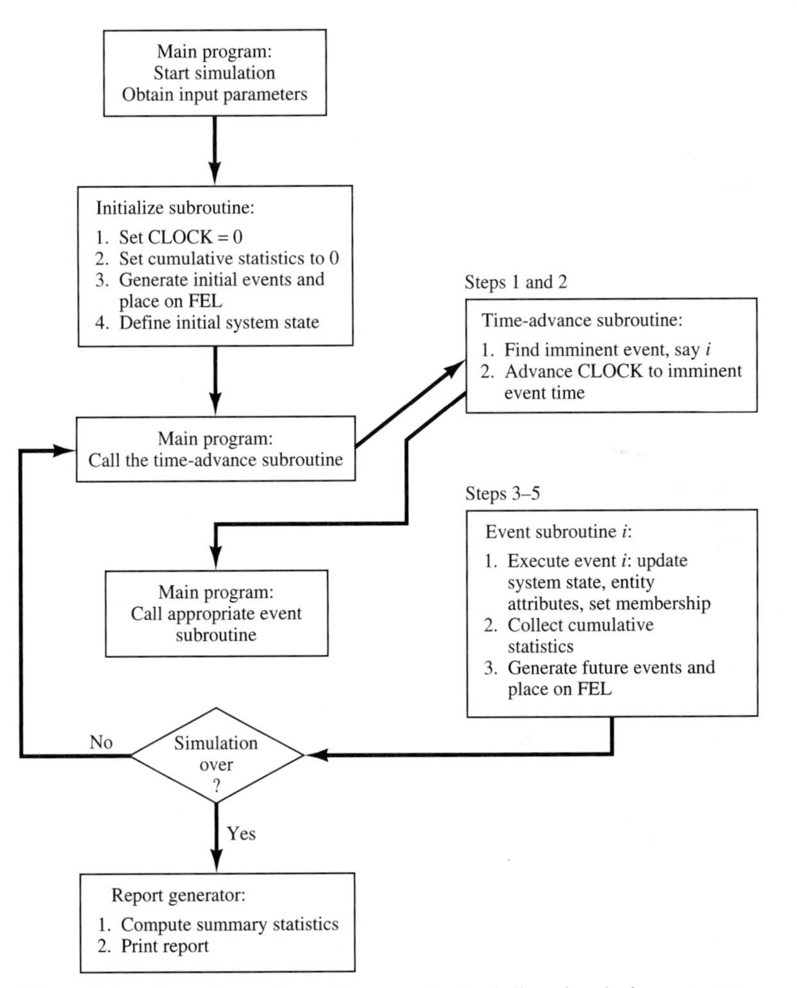

Figure 4.1. Overall structure of an event-scheduling simulation program.

The simulation begins by setting the simulation `Clock` to zero, initializing cumulative statistics to zero, generating any initial events (there will always be at least one) and placing them on the `FutureEventList`, and defining the system state at time 0. The simulation program then cycles, repeatedly passing the current least-time event to the appropriate event subroutines until the simulation is over. At each step, after finding the imminent event but before calling the event subroutine, the simulation `Clock` is advanced to the time of the imminent event. (Recall that during the simulated time between the occurrence of two successive events, the system state and entity attributes do not change in value. Indeed, this is the definition of discrete-event simulation: The system state changes only when an event occurs.) Next, the appropriate event subroutine is called to execute the imminent event, update cumulative statistics, and generate future events (to be placed on the `FutureEventList`). Executing the imminent event means that the system state, entity attributes, and set membership are changed to reflect the fact that the event has occurred. Notice that all actions in an event subroutine take place at one instant of simulated time. The value of the variable `Clock` does not change in an event routine. If the simulation is not over, control passes again to the time-advance subroutine, then to the appropriate event subroutine, and so on. When the simulation is over, control passes to the report generator, which computes the desired summary statistics from the collected cumulative statistics and prints a report.

The efficiency of a simulation model in terms of computer runtime is determined to a large extent by the techniques used to manipulate the `FutureEventList` and other sets. As discussed in Section 3.1, removal of the imminent event and addition of a new event are the two main operations performed on the `FutureEventList`. Virtually every C++ system now includes the "C++ standard libraries," whose priority-queue data structure we will use in this example to implement the event list and whose generic queue we will use to implement the list of waiting customers. The underlying priority-queue organization is efficient, in the sense of having access costs that grow only in the logarithm of the number of elements in the list.

EXAMPLE 4.2 (Single-Server Queue Simulation in C++)

The grocery checkout counter, defined in detail in Example 4.1, is now simulated using C++. A version of this example was simulated manually in Examples 3.3 and 3.4, where the system state, entities and attributes, sets, events, activities, and delays were analyzed and defined.

Class `Event` represents an event. It stores a code for the event type (`arrival` or `departure`), and the event time-stamp. It has associated methods (functions) for creating an event and accessing its data. It also has associated special functions, called operators, which provide semantic meaning to relational operators $<$ and $==$ between events. These are used by the C++ standard library queue implementations. The subroutines for this model and the flow of control are shown in Figure 4.2, which is an adaptation of Figure 4.1

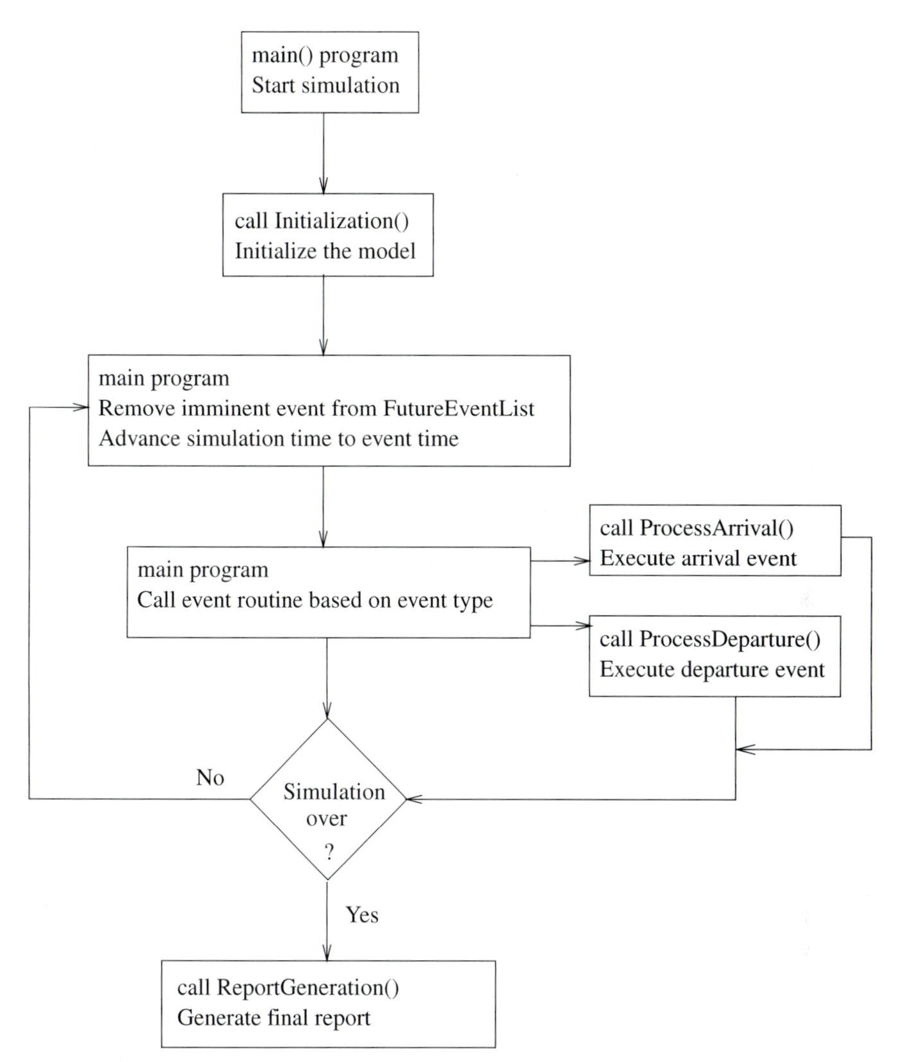

Figure 4.2. Overall structure of C++ simulation of single-server queue.

for this particular problem. Table 4.6 lists the C++ variables used for system state, entity attributes and sets, activity durations, and cumulative and summary statistics; the C++ functions used to generate samples from the exponential and normal distributions; and all the other subroutines needed.

The entry point of the program and the location of the control logic is in routine `main`, shown in Figure 4.3. It controls the overall flow of the event-scheduling/time-advance algorithm. The `#include` statements at the top of the program file identify library functions to be made available to the program—`iostream` for input/output, `math.h` for mathematics, and `queue` for the queue definitions. The class `Event` declaration follows, along with its "friend" func-

Table 4.6. Definitions of Variables, Functions, and Subroutines in the C++ Model of the Single-Server Queue

Variables	*Description*
System state	
`QueueLength`	Number of customers enqueued at current simulated time
`NumberInService`	Number being served at current simulated time
Entity attributes and sets	
`Customers`	FCFS queue of customers in system
Future event list	
`FutureEventList`	Time-ordered list of pending events
Activity durations	
`ServiceTime`	The service time of the most recent customer to begin service
Input parameters	
`MeanInterarrivalTime`	Mean interarrival time (4.5 minutes)
`MeanServiceTime`	Mean service time (3.2) minutes
`SIGMA`	Standard deviation of service time (0.6 minute)
`TotalCustomers`	The stopping criteria—number of customers to be served (1000)
Simulation variables	
`Clock`	The current value of simulated time
Statistical Accumulators	
`LastEventTime`	Time of occurrence of the last event
`TotalBusy`	Total busy time of server (so far)
`MaxQueueLength`	Maximum length of waiting line (so far)
`SumResponseTime`	Sum of customer response times for all customers who have departed (so far)
`NumberOfDepartures`	Number of departures (so far)
`LongService`	Number of customers who spent 4 or more minutes at the checkout counter (so far)
Summary statistics	
`RHO =BusyTime/Clock`	Proportion of time server is busy (here the value of `Clock` is the final value of simulated time)
`AVGR`	Average response time (equal to `SumResponseTime/TotalCustomers`)
`PC4`	Proportion of customers who spent 4 or more minutes at the checkout counter
Functions	*Description*
`exponential(mu)`	Function to generate samples from an exponential distribution with mean `mu`
`normal(xmu,SIGMA)`	Function to generate samples from a normal distribution with mean `xmu` and standard deviation `SIGMA`
Subroutines	*Description*
`Initialization`	Initialization subroutine
`ProcessArrival`	Event subroutine that executes the arrival event
`ProcessDeparture`	Event routine that executes the departure event
`ReportGeneration`	Report generator

```
#include <iostream>
#include <math.h>
#include <queue>

// event representation
class Event {
 friend bool operator<(const Event& e1, const Event& e2);
 friend bool operator==(const Event& e1, const Event& e2);

 public:
  Event() {};
  enum EvtType { arrival, departure };
  Event(EvtType type, double etime) : _type(type), _etime(etime){}
  EvtType get_type() { return _type; }
  double  get_time() { return _etime; }
 protected:
  EvtType _type;
  double  _etime;
};

bool operator <(const Event& e1, const Event& e2) {
        return e2._etime < e1._etime; }

bool operator ==(const Event& e1, const Event& e2) {
        return e2._etime == e1._etime; }

// global variables
double Clock, MeanInterArrivalTime, MeanServiceTime, SIGMA, LastEventTime,
        TotalBusy, MaxQueueLength, SumResponseTime;
long  NumberOfCustomers, QueueLength, NumberInService,
        TotalCustomers, NumberOfDepartures, LongService;
priority_queue<Event> FutureEventList;
queue<Event> Customers;

main(int argc, char* argv[]) {

  MeanInterArrivalTime = 4.5; MeanServiceTime = 3.2;
  SIGMA                = 0.6; TotalCustomers  = 1000;
  long seed = atoi(argv[1]);       // rng seed given as input
  SetSeed(seed);                   // initialize rng stream
  Initialization();

  // Loop until first ''TotalCustomers'' have departed
  while(NumberOfDepartures < TotalCustomers ) {
    Event evt = FutureEventList.top();  // get imminent event
    FutureEventList.pop();              // remove imminent event from queue
    Clock = evt.get_time();             // advance simulation time
    if( evt.get_type() == Event::arrival ) ProcessArrival(evt);
    else  ProcessDeparture(evt);
    }
  ReportGeneration();
}
```

Figure 4.3 C++ main program for single-server queue simulation.

tions that define relational operators on pairs of events. Following this are global variables, whose values are known to all subroutines. The declaration for `FutureEventList` illustrates how the C++ standard libaries work. They contain a "template" for a priority queue, which needs to be told what type of object will be enqueued. That object is the class `Event`, as indicated by the

<Event> portion of the declaration. The relational operators we defined for class Event are used by the standard library code to order instances of class Event.

The main program routine first gives values to variables describing model parameters, and then calls routine Initialization to initialize other variables such as the statistics-gathering variables. Control then enters a loop, which is exited only after TotalCustomers customers have received service. Inside the loop a copy of the imminent event is obtained by calling the top method of the priority queue, and then that event is removed from the event list by a call to pop. The global simulation time Clock is set to the time-stamp contained in the imminent event, and then either ProcessArrival or ProcessDeparture is called, depending on the type of the event. When the simulation is finally over, a call is made to routine ReportGeneration to create and print out the final report.

A listing for subroutine Initialization is given in Figure 4.4. The simulation clock, system state, and other variables are initialized. Note that the first arrival event is created by generating a local Event variable whose constructor accepts the event's type and time. The event time-stamp is generated randomly by a call to function exponential, and the event is inserted into the future event list by calling method push. This logic assumes that the system is empty at simulated time Clock = 0, so that no departure can be scheduled. It is straightforward to modify the code to accommodate alternative starting conditions by adding events to FutureEventList and Customers as needed.

```
void Initialization() {
        // initialize global variables
        Clock = 0.0;
        QueueLength = 0;
        NumberInService = 0;
        LastEventTime = 0.0;
        TotalBusy = 0;
        MaxQueueLength = 0;
        SumResponseTime = 0.0;
        NumberOfDepartures = 0;
        NumNormals = 0;
        LongService = 0;

        // create first arrival event
        Event evt(Event::arrival, exponential(MeanInterArrivalTime));
        FutureEventList.push(evt);
}
```

Figure 4.4 C++ initialization routine for single-server queue simulation.

Figure 4.5 gives a listing of event subroutine ProcessArrival, which executes for each arrival event. The basic logic of the arrival event for a single-server queue was given in Figure 3.5 (where *LQ* corresponds to QueueLength and *LS* corresponds to NumberInService). First, the new arrival is added to the queue Customers of customers in the system. It is worth noting that this data structure (like FutureEventList) makes a *copy* of the event that

```
void ProcessArrival(Event evt) {
        Customers.push(evt);          // push arrival onto the queue
        QueueLength++;                // increment number waiting

        // if the server is idle, fetch the event, do statistics,
        // and put into service
        if( NumberInService == 0 ) ScheduleDeparture();
        else TotalBusy += (Clock - LastEventTime); // server is busy

        // adjust max queue length statistics
        if( MaxQueueLength < QueueLength ) MaxQueueLength = QueueLength;

        // schedule the next arrival
        Event next_arrival(Event::arrival,
              Clock+exponential(MeanInterArrivalTime));
        FutureEventList.push(next_arrival);
        LastEventTime = Clock;
}
```

Figure 4.5 C++ arrival event routine for single-server queue simulation.

is pushed onto it, and gives back a copy when it is extracted from the data structure. The copy we push onto `Customers` at this point will be extracted when the corresponding customer leaves service and will be used to determine the customer's arrival time (and hence, at that point, the customer's total time in the system). Next, if the server is idle (`NumberInService` == 0), then the new customer is to go immediately into service, so subroutine `ScheduleDeparture` is called to do that scheduling. An arrival to an idle queue does not update the cumulative statistics, except possibly the maximum queue length. An arrival to a busy queue does *not* cause the scheduling of its departure, but does increase the total busy time by the amount of simulation time between the current event and the one immediately preceding it (because if the server is busy now, it had to have had at least one customer in service by the end of processing the previous event). In either case, a new arrival is responsible for scheduling the next arrival, one random interarrival time into the future. An arrival event is created with simulation time equal to the current `Clock` value plus an exponential increment, that event is pushed into the future event list, the variable `LastEventTime` recording the time of the last event processed is set to the current time, and control is returned to the main program.

Subroutine `ProcessDeparture`, which executes the departure event, is listed in Figure 4.6, as is subroutine `ScheduleDeparture`. A flow chart for the logic of the departure event was given in Figure 3.6. After removing the event from the queue of all customers, the number in queue is examined. If there are customers waiting, then the departure of the next one to enter service is scheduled. Then, the cumulative statistics recording the sum of all response times, sum of busy time, number of customers who used more than 4 minutes of service time, and number of departures are updated. (Note that the maximum queue length cannot change in value when a departure occurs.) Notice that customers are removed from `Customers` in FIFO order, so that response time `response` of the departing customer can be computed by subtracting the

```
void ProcessDeparture(Event evt) {
        // get the customer description
        Event finished = Customers.front();
        Customers.pop();
        // if there are customers in queue then schedule
        // the departure of the next one
        if( QueueLength ) ScheduleDeparture();
        else NumberInService = 0;
        // measure the response time and add to the sum
        double response  = (Clock - finished.get_time());
        SumResponseTime + = response;
        if( response > 4.0 ) LongService++;     // record long service
        TotalBusy + = (Clock - LastEventTime);  // we were busy
        NumberOfDepartures++;                   // one more gone
        LastEventTime = Clock;
}
void ScheduleDeparture() {
       double ServiceTime;
       // get the job at the head of the queue
       while( (ServiceTime = normal(MeanServiceTime, SIGMA)) < 0 );
       Event depart = Event depart(Event::departure, Clock+ServiceTime);
       FutureEventList.push(depart);
       NumberInService = 1;
       QueueLength--;            // this one isn't waiting anymore
}
```

Figure 4.6 C++ departure event routines for single-server queue simulation.

arrival time of the job leaving service (obtained from the copy of the arrival event removed from the Customers queue) from the current simulation time. After incrementing the total number of departures and saving the time of this event, control is returned to the main program.

Figure 4.6 also gives the logic of routine ScheduleDeparture, called by both ProcessArrival and ProcessDeparture to put the next customer into service. The subroutine normal generating normally distributed service times is called until it produces a nonnegative sample. A new event with type departure is created, with event time equal to the current simulation time plus the service time just sampled. That event is pushed onto FutureEventList, the number in service is set to one, and the number waiting (QueueLength) is decremented to reflect the fact that the customer entering service is waiting no longer.

The report generator, subroutine ReportGeneration, is listed in Figure 4.7. The summary statistics, RHO, AVGR, and PC4, are computed by the formulas in Table 4.6. Then the input parameters are printed, followed by the summary statistics. It is a good idea to print the input parameters at the end of the simulation in order to verify that their values are correct and that they have not been inadvertently changed.

Figure 4.8 provides a listing of functions exponential and normal, used to generate random variates. Both of these functions call function unif, not

```
void ReportGeneration() {
double RHO    = TotalBusy/Clock;
double AVGR   = SumResponseTime/TotalCustomers;
double PC4    = ((double)LongService)/TotalCustomers;

cout << ''SINGLE SERVER QUEUE SIMULATION - GROCERY STORE CHECKOUT COUNTER
\n''
     << endl;
cout << ''\tMEAN INTERARRIVAL TIME                              ''
     << MeanInterArrivalTime << endl;
cout << ''\tMEAN SERVICE TIME                                   ''
     << MeanServiceTime << endl;
cout << ''\tSTANDARD DEVIATION OF SERVICE TIMES                 ''
     << SIGMA << endl;
cout << ''\tNUMBER OF CUSTOMERS SERVED                          ''
     << TotalCustomers << endl << endl ;

cout << ''\tSERVER UTILIZATION                                  ''
     << RHO << endl;
cout << ''\tMAXIMUM LINE LENGTH                                 ''
     << MaxQueueLength << endl;
cout << ''\tAVERAGE RESPONSE TIME                               ''
     << AVGR << '' MINUTES'' << endl;
cout << ''\tPROPORTION WHO SPEND FOUR \n''
     << ''\t MINUTES OR MORE IN SYSTEM                          ''
     << PC4 << endl;
cout << ''\tSIMULATION RUNLENGTH                                ''
     << Clock << '' MINUTES '' << endl;
cout << ''\tNUMBER OF DEPARTURES                                ''
     << TotalCustomers << endl;
}
```

Figure 4.7 C++ report generator for single-server queue simulation.

shown, assumed to generate a random number uniformly distributed on the (0, 1) interval. C++ does have a built-in random-number generator (called `random()`) but superior ones can be built by hand. Such routines are discussed in Chapter 7. The techniques for generating exponentially and normally distributed random variates, discussed in Chapter 8, are based on first generating a $U(0, 1)$ random number. For further explanation, the reader is referred to Chapters 7 and 8.

The output from the grocery checkout-counter simulation is shown in Figure 4.9. It should be emphasized that the output statistics are estimates that contain random error. The values shown are influenced by the particular random numbers that happened to have been used, by the initial conditions at time 0, and by the run length (in this case, 1000 departures). Methods for estimating the standard error of such estimates are discussed in Chapter 11.

In some simulations it is desired to stop the simulation after a fixed length of time, say TE = 12 hours = 720 minutes. In this case, an additional event type, `stop` event, is defined and scheduled to occur by scheduling a stop event as part of simulation initialization. When the stopping event does occur, the cumulative statistics will be updated and the report generator called. The main program and subroutine `Initialization` will require minor changes.

```
double exponential(double mean) {
 return -mean*log( unif() );
}

double SaveNormal;
int    NumNormals = 0;

double normal(double mean, double sigma) {
#define PI 3.1415927
        double ReturnNormal;
        // should we generate two normals?
        if(NumNormals == 0 ) {
          double r1 = unif();
          double r2 = unif();
          ReturnNormal = sqrt(-2*log(r1))*cos(2*PI*r2);
          SaveNormal   = sqrt(-2*log(r1))*sin(2*PI*r2);
          NumNormals = 1;
        } else {
          NumNormals = 0;
          ReturnNormal = SaveNormal;
        }
        return ReturnNormal*sigma + mean ;
}
```

Figure 4.8 Random-variate generators for single-server queue simulation.

Exercise 1 asks the reader to make these changes. Exercise 2 considers the additional change that any customer at the checkout counter at simulated time `Clock` = TE should be allowed to exit the store, but no new arrivals are allowed after time TE. ◄

```
SINGLE SERVER QUEUE SIMULATION - GROCERY STORE CHECKOUT COUNTER
  MEAN INTERARRIVAL TIME                              4.5
  MEAN SERVICE TIME                                   3.2
  STANDARD DEVIATION OF SERVICE TIMES                 0.6
  NUMBER OF CUSTOMERS SERVED                          1000

  SERVER UTILIZATION                                  0.68362
  MAXIMUM LINE LENGTH                                 9
  AVERAGE RESPONSE TIME                               6.73901 MINUTES
  PROPORTION WHO SPEND FOUR
   MINUTES OR MORE IN SYSTEM                          0.663
  SIMULATION RUNLENGTH                                4677.74 MINUTES
  NUMBER OF DEPARTURES                                1000
```

Figure 4.9 Output from the C++ single-server queue simulation.

4.5 Simulation in GPSS

GPSS is a highly structured, special-purpose simulation programming language based on the process-interaction approach and oriented toward queueing systems. A block diagram provides a convenient way to describe the system being simulated. There are over 40 standard blocks in GPSS. Entities called transactions may be viewed as flowing through the block diagram. Blocks represent events, delays, and other actions that affect transaction flow. Thus, GPSS can be used to model any situation where transactions (entities, customers, units of traffic) are flowing through a system (e.g., a network of queues, with the queues preceding scarce resources). The block diagram is converted to block statements, control statements are added, and the result is a GPSS model.

The first version of GPSS was released by IBM about 1961. Since it was the first process-interaction simulation language, and due to its popularity, it has been implemented anew and improved by many parties since 1961, with GPSS/H being the most widely used version in use today. Example 4.3 is based on GPSS/H.

GPSS/H is a product of Wolverine Software Corporation, Annandale, VA (Banks, Carson, and Sy, 1995; Crain and Henriksen, 1999). It is a flexible, yet powerful tool for simulation. Unlike the original IBM implementation, GPSS/H includes built-in file and screen I/O, use of an arithmetic expression as a block operand, an interactive debugger, faster execution, expanded control statements, ordinary variables and arrays, a floating-point clock, built-in math functions, and built-in random-variate generators.

The animator for GPSS/H is Proof Animation™, another product of Wolverine Software Corporation (Henriksen, 1999). Proof Animation provides a 2-D animation, usually based on a scale drawing. It can run in post-processed mode (after the simulation has finished running) or concurrently. In postprocessed mode, the animation is driven by two files: the layout file for the static background, and a trace file that contains commands to make objects move and produce other dynamic events. It can work with any simulation package that can write the ASCII trace file. Alternately, it can run concurrently with the simulation by sending the trace-file commands as messages, or can be controlled directly when using its DLL (dynamic link libary) version.

EXAMPLE 4.3 (Single-Server Queue Simulation in GPSS/H)

Figure 4.10 exhibits the block diagram and Figure 4.11 the GPSS program for the grocery store checkout-counter model described in Example 4.2. Note that the program (Figure 4.11) is a translation of the block diagram together with additional definition and control statements.

In Figure 4.10, the GENERATE block represents the arrival event, with the interarrival times specified by RVEXPO(1, &*IAT*). RVEXPO stands for "random variable, exponentially distributed," the 1 indicates the random-number stream to use, and &*IAT* indicates that the mean time for the exponential distribution comes from a so-called ampervariable &*IAT*. Ampervariable names begin with the "&" character; Wolverine added ampervariables to GPSS because the original IBM implementation had limited support for ordinary global variables, with no user freedom for naming them. (In the discussion that follows, all nonreserved words are shown in italics.)

The next block is a QUEUE with a queue named *SYSTIME*. It should be noted that the QUEUE block is not needed for queues or waiting lines to form in GPSS. The true purpose of the QUEUE block is to work in conjunction with the DEPART block to collect data on queues or any other subsystem.

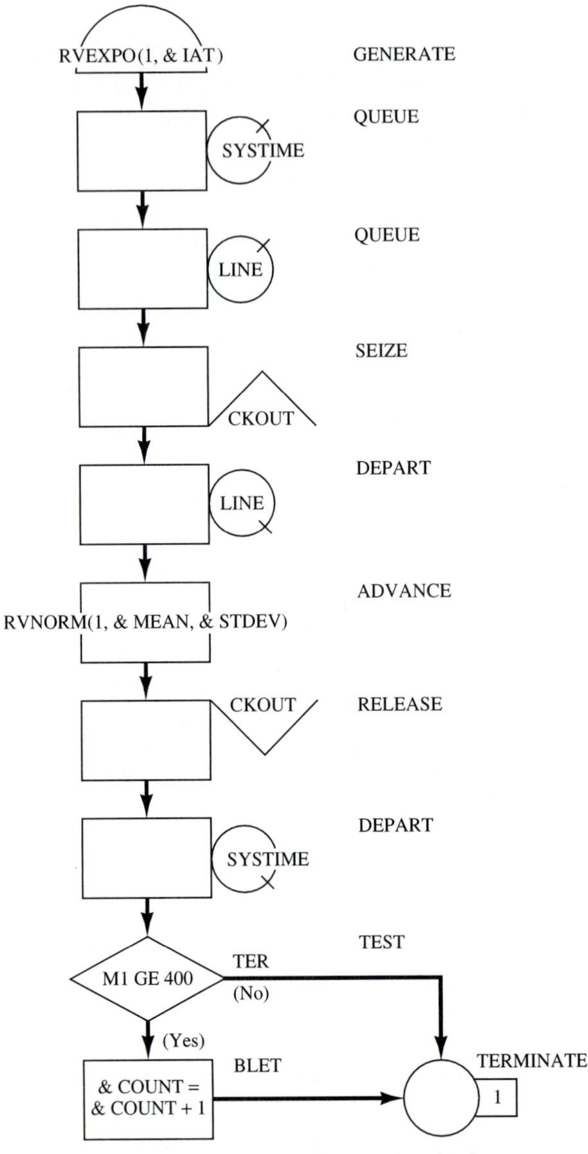

Figure 4.10. GPSS block diagram for single-server queue simulation.

In Example 4.3, we want to measure the system response time — that is, the time a transaction spends in the system. By placing a QUEUE block at the point where transactions enter the system, and the counterpart of the QUEUE block, the DEPART block, at the point where the transactions complete their processing, the response times will be automatically collected. The purpose of the DEPART block is to signal the end of data collection for an individual

```
        SIMULATE
*
*       Define Ampervariables
*
        INTEGER         &LIMIT
        REAL            &IAT,&MEAN,&STDEV,&COUNT
        LET             &IAT=4.5
        LET             &MEAN=3.2
        LET             &STDEV=.6
        LET             &LIMIT=1000
*
*       Write Input Data to File
*
        PUTPIC          FILE=OUT,LINES=5,(&IAT,&MEAN,&STDEV,&LIMIT)
        Mean interarrival time                       **.**  minutes
        Mean service time                            **.**  minutes
        Standard deviation of service time           **.**  minutes
        Number of customers to be served             *****
*
*       GPSS/H Block Section
*
        GENERATE        RVEXPO(1,&IAT)  Exponential arrivals
        QUEUE           SYSTIME         Begin response time data collection
        QUEUE           LINE            Customer joins waiting line
        SEIZE           CHECKOUT        Begin checkout at cash register
        DEPART          LINE            Customer starting service leaves queue
        ADVANCE         RVNORM(1,&MEAN,&STDEV)  Customer's service time
        RELEASE         CHECKOUT        Customer leaves checkout area
        DEPART          SYSTIME         End response time data collection
        TEST GE         M1,4,TER        Is response time GE 4 minutes?
        BLET            &COUNT=&COUNT+1 If so, add 1 to counter
TER     TERMINATE       1
*
        START           &LIMIT          Simulate for required number
*
*       Write Customized Output Data to File
*
        PUTPIC          FILE=OUT,LINES=7,(FR(CHECKOUT)/1000,QM(LINE),_
        QT(SYSTIME),&COUNT/N(TER),AC1,N(TER))
        Server utilization                         .***
        Maximum line length                       **
        Average response time                     **.**    minutes
        Proportion who spend four minutes          .***
             or more in the system
        Simulation runlength                       ****.** minutes
        Number of departures                       ****
*
        END
```

Figure 4.11 GPSS/H program for single-server queue simulation.

transaction. The QUEUE and DEPART block combination is not necessary for queues to be modeled, but rather is used for statistical data collection.

The next QUEUE block (with name *LINE*) begins data collection for the waiting-line before the cashier. The customers may or may not have to wait for the cashier. Upon arrival to an idle checkout counter or after advancing to the head of the waiting line, a customer captures the cashier as represented by the SEIZE block with the resource named *CHECKOUT*. Once the transaction representing a customer captures the cashier represented by the resource CHECKOUT, the data collection for the waiting line statistics ends, as represented by the DEPART block for the queue named LINE. The transaction's service time at the cashier is represented by an ADVANCE block. RVNORM indicates "random variable, normally distributed." Again, random-number stream 1 is being used, the mean time for the normal distribution is given by ampervariable *&MEAN*, and its standard deviation is given by ampervariable *&STDEV*. Next, the customer gives up the use of the facility *CHECKOUT* with a RELEASE block. The end of the data collection for response times is indicated by the DEPART block for the queue *SYSTIME*.

Next, there is a TEST block that checks to see if the time in the system, M1, is greater than or equal to 4 minutes. (Note that M1 is a reserved word in GPSS/H; it automatically tracks transaction total time in system.) In GPSS/H, the maxim is "if true, pass through." Thus, if the customer has been in the system 4 minutes or longer, the next BLET block (for block LET) adds one to the counter *&COUNT*. If not true, the escape route is to the block labeled *TER*. That label appears before the TERMINATE block, whose purpose is the removal of the transaction from the system. The TERMINATE block has a value "1," indicating that one more transaction is added toward the limiting value, or "transactions to go."

The control statements in this example are all of those lines in Figure 4.11 that precede or follow the block section. (There are eleven blocks in the model from the GENERATE block to the TERMINATE block.) The control statements that begin with an asterisk are comments, some of which are used for spacing purposes. The control statement SIMULATE tells GPSS/H to conduct a simulation; if omitted, GPSS/H compiles the model and checks for errors only. The ampervariables are defined as integer or real by control statements INTEGER and REAL. It seems that the ampervariable *&COUNT* should be defined as an integer; however, it will be divided later by the number of customers to obtain a proportion. If it were integer, the result of an integer divided by an integer would be truncation, and that is not desired in this case. The four assignment statements (LET) provide data for the simulation. These four values could have been placed directly in the program; however, the preferred practice is to place them in ampervariables at the top of the program, so that changes can be made more easily, or the model modified to read them from a data file.

To insure that the model data is correct, and for the purpose of managing different scenarios simulated, it is good practice to echo the input data. This is accomplished with a PUTPIC (for "put picture") control statement. The five lines following PUTPIC provide formatting information, with the asterisks being markers (called picture formatting) in which the values of the four ampervariables replace the asterisks when PUTPIC is executed. Thus, "**.**" indicates a value that may have two digits following the decimal point.

```
Mean interarrival time                   4.50  minutes
Mean service time                        3.20  minutes
Standard deviation of service time       0.60  minutes
Number of customers to be served         1000

Server utilization                       0.676
Maximum line length                      7
Average response time                    6.33    minutes
Proportion who spend four minutes        0.646
     or more in the system
Simulation runlength                     4767.27 minutes
Number of departures                     1000
```

Figure 4.12 Customized GPSS/H output report for single-server queue simulation.

The START control statement controls simulation execution. It starts the simulation, sets up a "termination to-go" counter with initial value its operand (*&LIMIT*), and controls the length of the simulation.

After the simulation completes, a second PUTPIC control statement is used to write the desired output data to the same file *OUT*. The printed statistics are all automatically gathered by GPSS. The first output in the parenthesized list is the server utilization. FR*(CHECKOUT)*/1000 indicates that the fractional utilization of the facility *CHECKOUT* is printed. Since FR*(CHECKOUT)* is in parts per thousand, the denominator is provided to compute fractional utilization. QM(*LINE*) is the maximum value in the queue *LINE* during the simulation. QT*(SYSTIME)* is the average time in the queue *SYSTIME*. *&COUNT*/N*(TER)* is the number of customers who had a response time of four or more minutes divided by the number of customers that went through the block with label *TER*, or N*(TER)*. AC1 is the clock time, whose last value gives the length of the simulation.

The contents of the custom output file *OUT* are shown in Figure 4.12. The standard GPSS/H output file is displayed in Figure 4.13. Although much of the same data shown in the file *OUT* can be found in the standard GPSS/H output, the custom file is more compact and uses the language of the problem rather than GPSS jargon. There are many other reasons that customized output files are useful. For example, if 50 replications of the model are to be made and the lowest, highest, and average value of a response are desired, this can be accomplished using control statements with the results in a very compact form, rather than extracting the desired values from 50 standard output reports. ◀

4.6 Simulation in CSIM

CSIM is a system for using the C or C++ language for modeling, along with a rich library of predefined objects to support process-oriented simulation modeling. It is an inexpensive commercial product sold and maintained by Mesquite Software [Schwetman, 1996]; it has a wide user base in industry and academia, principally to model computer and communication systems. CSIM is fast, owing to careful implementation and being a compiled language.

```
RELATIVE CLOCK: 4767.2740  ABSOLUTE CLOCK: 4767.2740

BLOCK CURRENT     TOTAL BLOCK CURRENT     TOTAL
1                  1003 TER                1000
2                  1003
3          3       1003
4                  1000
5                  1000
6                  1000
7                  1000
8                  1000
9                  1000
10                  646

         --AVG-UTIL-DURING--
FACILITY TOTAL AVAIL UNAVL  ENTRIES    AVERAGE   CURRENT  PERCENT  SEIZING   PREEMPTING
          TIME  TIME  TIME            TIME/XACT   STATUS   AVAIL     XACT       XACT
CHECKOUT 0.676                  1000      3.224    AVAIL

  QUEUE   MAXIMUM   AVERAGE   TOTAL     ZERO   PERCENT   AVERAGE   $AVERAGE  QTABLE   CURRENT
         CONTENTS  CONTENTS  ENTRIES  ENTRIES   ZEROS   TIME/UNIT  TIME/UNIT  NUMBER  CONTENTS
SYSTIME         8    1.331    1003        0               6.325
6.235                3
   LINE         7    0.655    1003      334     33.3      3.111
4.665                3

RANDOM  ANTITHETIC   INITIAL   CURRENT   SAMPLE  CHI-SQUARE
STREAM    VARIATES  POSITION  POSITION    COUNT  UNIFORMITY
     1         OFF    100000    103004     3004      0.83
```

Figure 4.13 Standard GPSS/H output report for single-server queue simulation.

A key built-in object in CSIM is the *facility*, which represents a multiserver queue. One can associate different queuing disciplines with a facility, and CSIM internally collects statistics on facility usage. Another important object is the *event*, which is a signaling mechanism for use between processes. A process can call a suspension function that names an event, suspending the caller until some other process explicitly "signals" that event. If multiple processes are suspended on an event, one or all suspended processes are released on the occasion of that event being signaled, depending on the precise signaling call being made. Processes can also communicate data to each other, through CSIM *mailboxes*. A process can retrieve previously sent messages from a mailbox, and can choose to suspend on an empty mailbox until some other process deposits a message there. CSIM provides *storages*, abstract units of allocation used to model process contention for limited resources. A storage is created with a given number of units: a process can attempt to acquire some number of units and block if the requested number are not available. A process that is successful in its request releases the resources when finished with them. This release may unblock one or more suspended processes. CSIM also provides *table* objects, which are used to gather statistics and obtain from them summary information.

EXAMPLE 4.4 (Single-Server Queue Simulation in CSIM)

Source code given in Figure 4.14 expresses the single-server queue example in CSIM. The #include <cpp.h> statement makes the CSIM object definitions available to the C++ program. The first two objects in the global variable area, done and f, are CSIM event and facility objects, respectively. f represents the single queue, with one server and FCFS scheduling as defaults, and event is used to signal when the last customer has finished service. After declaring some variables for gathering statistics, routine sim gives the entry point into the CSIM model.

While a CSIM model uses the C++ language as its basic framework, the runtime behavior of a CSIM model is somewhat different from an ordinary C++ program. This is because CSIM implements asynchronous threads of control with its process constructs, whereas C++ has only one thread of control. Chapter 14 discusses this at more length.

A subroutine used as the entry point of a CSIM process always has a create statement as its first line. In Chapter 14 we examine CSIM's process implementation strategy and see what actually happens by executing this statement. Here we need only recognize that both sim and customer are CSIM processes. The model semantics are expressed from the point of view of a customers passing through the system. When a customer is created (by calling customer), it first determines how many customers are currently in the system by calling the qlength and num_busy functions associated with the server. The former gives the number of customers waiting in queue, the latter the number of servers currently busy. From the number in the system one can determine whether, after this customer joins the queue, the maximum number of customers in queue increases. Following this a random service time, drawn from a normal distribution (constrained to be positive), is sampled. The sampled value (service) is passed to the facility function use, which causes the customer to queue up for use of the server. The process is suspended until the customer acquires the server and the simulation time advances to the customer's departure time. All the details of customer queuing and service-time allocation are completely hidden from the modeler. So, when the statement following the call to use executes, the customer has completed service. At this point we see whether the service time is one of the "long" ones we find interesting. Finally, the customer increments the number of customers completely served, and if that count equals the number of customers we wish to simulate, it "sets" the CSIM event done. Departure from routine customer destroys the process.

Customer generation is the job given to process sim. After initializing various model parameters, it enters a customer-generation loop. Each pass through the loop first suspends the process for a randomly sampled interarrival

```
// C++/CSIM MODEL OF SINGLE SERVER QUEUE
#include <cpp.h>

event done(''done'');               // the event named done
facility f(''f'');                  // the facility named f
long TotalCustomers;               // number of customers
long NumberOfCustomers=0;          // number of customers served
long LongService=0;                // number of long service times
long MaxQueueLength=0;             // maximum number of customers waiting
double MeanServiceTime;
double SIGMA;
double MeanInterArrivalTime;

void customer();                   // forward definition
extern ''C'' void sim()              // main process
{
    create(''sim'');              // procedure is a process
    MeanServiceTime      = 3.2;
    SIGMA                = 0.6;
    MeanInterArrivalTime = 4.5;
    TotalCustomers       = 1000;
    MeanInterArrivalTime = 4.5;

    for(long i=1; i<=TotalCustomers; i++) {
     // hold for an interarrival period
          hold(exponential(MeanInterArrivalTime));
          customer();          // generate new customer process
    }
    done.wait();             // wait for last customer
    ReportGeneration();      // generate a report
}

void customer()                    // arriving customer process
{
       create(''customer'');     // procedure is a process

       // update MaxQueueLength
       long length = f.qlength() + f.num_busy();
       if( length && MaxQueueLength < length)
       MaxQueueLength = length;  // include self

       // sample service time, update LongService
       double service;
       while( (service  = normal(MeanServiceTime, SIGMA) ) < 0);

       f.use(service);          // queue up for service
       if( service > 4.0) LongService++;
       if(++NumberOfCustomers == TotalCustomers)  // are we the last?
          done.set();           // yes, signal main
}
```

Figure 4.14 CSIM Model of single-server queue.

time. The sample is drawn from CSIM function `exponential` and is given to CSIM function `hold`, which accomplishes the suspension. The statement following `hold` is executed at the simulation time equal to the sum of the clock at the time `hold` was first called with the randomly sampled interarrival time. That statement is just a function call to `customer`. This, as we have noted, creates and runs one further customer process. Once `sim` has generated all required customers, it calls the `wait` function associated with CSIM event `done`. This causes it to suspend until `done` is set by the last customer process of interest. The call to `done.set()` in turn releases `sim` to call `ReportGeneration` to give its report. Our CSIM implementation of `ReportGeneration` can be essentially that illustrated in Figure 4.7, with the exception of calling CSIM-provided facility routine `util` to get server utilization, and `resp` to get the average response time :

```
double RHO   = f.util();
double AVGR  = f.resp()
```

CSIM bridges the gap between models developed in pure C++ and models developed in languages specifically designed for simulation. It provides the flexibility offered by a general programming language, with essential support for simulation. ◀

4.7 Simulation Packages

All the simulation packages described in later subsections run on a PC under Microsoft Windows 95/98 and/or NT. Some also run on various UNIX workstations and a few on Apple machines. Although in terms of specifics the packages all differ, generally they have many things in common.

Common characteristics include a graphical user interface, animation, and automatically collected outputs to measure system performance. In virtually all packages, simulation results may be displayed in tabular or graphical form in standard reports and interactively while running a simulation. Outputs from different scenarios can be compared graphically or in tabular form. Most provide statistical analyses that include confidence intervals for performance measures and comparisons, plus a variety of other analysis methods. Some of the statistical analysis modules are described in Section 4.8.

All the packages described here take the process-interaction world view. A few also allow event scheduling and mixed discrete-continuous models. For animation, some emphasize scale drawings in 2-D or 3-D, while others emphasize iconic-type animations based on schematic drawings or process flow diagrams. A few offer both scale-drawing and schematic-type animations. Almost all offer dynamic business graphing in the form of timelines, bar charts, and pie charts.

4.7.1 Arena

Arena Business, Standard, and Professional Editions are offered by Systems Modeling Corporation [Sadowski and Bapat, 1999]. Arena can be used for simulating discrete and continuous systems. A recent addition to the Arena family of products is OptQuest for Arena, an optimization software package (discussed in Section 4.8.2).

The Arena Business Edition is targeted at modeling business processes and other systems in support of high-level analysis needs. It represents process dynamics in a hierarchical flow chart and stores system information in data spreadsheets. It has built-in activity-based costing and is closely integrated with the flow-charting software Visio®.

The Arena Standard Edition is designed for more detailed models of discrete and continuous systems. First released in 1993, Arena employs an object-based design for entirely graphical model development. Simulation models are built using graphical objects called modules to define system logic and physical components such as machines, operators, and clerks. Modules are represented by icons plus associated data entered in a dialog window. These icons are connected to represent entity flow. Modules are organized into collections called templates. The Arena template is the core collection of modules providing general-purpose features for modeling all types of applications. In addition to standard features, such as resources, queues, process logic, and system data, the Arena template includes modules focused on specific aspects of manufacturing and material handling systems. Arena SE can also be used to model combined discrete/continuous systems, such as pharmaceutical and chemical production, through its built-in continuous modeling capabilities.

The Arena Professional Edition enhances Arena SE with the capability to craft custom simulation objects that mirror components of the real system, including terminology, process logic, data, performance metrics, and animation. The Arena family also includes products designed specifically to model call centers and high-speed production lines, namely Arena Call Center and Arena Packaging.

At the heart of Arena is the SIMAN simulation language. For animating simulation models, Arena's core modeling constructs are accompanied by standard graphics for showing queues, resource status, and entity flow. Arena's 2-D animations are created using Arena's built-in drawing tools and by incorporating clipart, AutoCAD, Visio, and other graphics.

Arena's Input Analyzer automates the process of selecting the proper distribution and its parameters for representing existing data, such as process and interarrival times. The Output Analyzer and Process Analyzer (discussed in Section 4.8.2) automate comparison of different design alternatives.

4.7.2 AutoMod

The AutoMod Product Suite is offered by AutoSimulations [Rohrer, 1999]. It includes the AutoMod simulation package, AutoStat for experimentation and analysis, and AutoView for making AVI movies of the built-in 3-D animation. The main focus of the AutoMod simulation product is manufacturing and material handling systems.

AutoMod has built-in templates for most common material handling systems, including vehicle systems, conveyors, automated storage and retrieval systems, bridge cranes, power and free conveyors, and kinematics for robotics. With its Tanks and Pipes module, it also supports continuous modeling of fluid and bulk material flow.

The Path Mover vehicle system can be used to model lift trucks, humans walking or pushing carts, automated guided vehicles, trucks and cars. All the movement templates are based on a 3-D scale drawing (drawn or imported from CAD as 2-D or 3-D). All the components of a template are highly parameterized. For example, the conveyor template contains conveyor sections, stations for load induction or removal, motors, and photo-eyes. Sections are defined by length, width, speed, acceleration and type (accumulating or nonaccumulating), plus other specialized parameters. Photo-eyes have blocked and cleared timeouts that facilitate modeling of detailed conveyor control logic.

In addition to the material handling templates, AutoMod contains a full simulation programming language. Its 3-D animation can be viewed from any angle or perspective in real time. The user can freely zoom, pan, or rotate the 3-D world.

An AutoMod model consists of one or more systems. A system can be either a process system, in which flow and control logic are defined, or a movement system based on one of the material handling templates. Each model must contain one process system and may contain any number of movement systems. Processes can contain complex logic to control the flow of either manufacturing materials or control messages, to contend for resources, or to wait for user-specified times. Loads can move between processes with or without using movement systems.

In the AutoMod world view, loads (products, parts, etc.) move from process to process and compete for resources (equipment, operators, vehicles and queues). The load is the active entity, executing action statements in each process. To move between processes, loads may use a conveyor or vehicle in a movement system.

AutoStat, described in Section 4.8.2, works with AutoMod models to provide a complete environment for the user to define scenarios, conduct experimentation, and perform analyses. It offers optimization based on an evolutionary strategies algorithm.

4.7.3 Deneb/QUEST

QUEST is offered by Deneb Robotics [Donald, 1998]. Deneb also offers a number of workcell simulators including IGRIP for robotic simulation and programming and ERGOSim for ergonomics analyses. QUEST (Queuing Event Simulation Tool) is a manufacturing-oriented simulation package. Robots and workcells simulated in IGRIP and ERGOSim can be imported into QUEST models.

QUEST models are based on 3-D CAD geometry. QUEST combines a simulation environment with graphical user interface and material flow modules for labor, conveyors, automated guided vehicles, kinematics, power and free conveyors, and automated storage and retrieval systems.

For Deneb/IGRIP users, existing robot workcell models and robot or other complex machinery cycle times can be directly imported to Deneb/

QUEST. Similarly, models of human operators in a workcell developed in Deneb/ERGOSim can be incorporated into a QUEST process model both visually and numerically.

A QUEST model consists of elements from a number of element classes. Built-in element classes include AGV and transporters, buffer, conveyor, labor, machine, parts and process. Each element has associated geometric data and parameters that define its behavior. Parts may have a route and control rules to govern part flow. Commonly needed behavior logic is selected from comprehensive menus, many parameter-driven. For unique problems, Deneb/QUEST's Simulation Control Language can be used. This structured simulation programming language provides distributed processing with access to all system variables. SCL allows expert users to define custom behaviors and to gain control over the simulation.

Deneb/QUEST's open architecture allows the advanced user to perform batch simulation runs to automatically collect and tabulate data using the Batch Control Language (BCL). Replications and parameter optimization are controlled with batch command files.

4.7.4 Extend

Extend is offered by Imagine That, Inc. [Krahl, 1999]. Extend combines a block diagram approach to model building along with an authoring environment for creating new blocks or complete vertical market simulators. It is process oriented but is also capable of continuous and combined modeling. It provides iconic animation of the block diagram plus an interface to Proof Animation [Henriksen, 1999] from Wolverine Software for 2-D scale-drawing animation.

Elemental blocks in Extend include Generator (for arrivals), Queue, Activity, Resource Pool, and Exit. In Extend, the active entities, called items, are created at Generator blocks and move from block to block by way of connectors. Connectors are also used for "value flows" — that is, for attaching a calculation to a block parameter or retrieving statistical information from a block for reporting purposes.

Each block has an icon and encapsulates code, parameters, user interface, and online help. Extend comes with a large set of elemental blocks; in addition, libraries of blocks for general and specific application areas, such as manufacturing and BPR (business process reengineering), are available. Third-party developers have created Extend libraries for vertical market applications, including supply chain analysis, chemical processing, pulp and paper processing, and radio-frequency analysis.

End users can build models by placing and connecting blocks and filling in the parameters on the dialog window associated with a block. Collections of blocks can be grouped into a hierarchical block representing a submodel such as a particular machine, a workstation, or a subassembly line. In this way, a higher-level block diagram can be displayed to the end user, hiding many of the details. Parameters from the submodel network of blocks that an end user desires to control can be grouped and displayed at the level of the hierarchical block. Using this capability, a firm could build a set of customized blocks representing various components of their manufacturing facility.

For creating new blocks, Extend comes with a compiled C-like simulation programming language called ModL. It contains simulation support, as well as support for custom user interfaces and message communication. Extend has an open architecture, so that in most cases the source code for blocks is available to advanced model builders.

The statistics library of blocks is used to collect and analyze output data, such as computing confidence intervals.

4.7.5 Micro Saint

Micro Saint is offered by Micro Analysis and Design, Inc. [Bloechle and Laughery, 1999]. With Micro Saint, a model builder develops a model by creating a flow-chart diagram to describe a network of tasks. The animation is iconic and is based on the flow-chart representation. Through menus and parameterized icons, the network can have branching logic, sorted queues, and conditional task execution. Micro Saint is used in manufacturing, health care, retail, and military applications. It supports ergonomics and human performance modeling. It works with OptQuest to provide optimization capability.

4.7.6 ProModel

ProModel is offered by PROMODEL Corporation [Price and Harrell, 1999]. It is a simulation and animation tool designed to model manufacturing systems. The company also offers MedModel for healthcare systems, ServiceModel for service systems, and ProcessModel, a flow-chart-based simulation product for business processes. ProModel offers 2-D animation with an optional 3-D-like perspective view.

ProModel has manufacturing-oriented modeling elements and rule-based decision logic. Some systems can be modeled by selecting from ProModel's set of highly parameterized modeling elements. In addition, its simulation programming language provides for modeling special situations not covered by the built-in choices.

The modeling elements in ProModel are parts or entities, locations, resources, path networks, routing and processing logic, and arrivals. Parts arrive and follow the routing and processing logic from location to location. Resources are used to represent people, tools or vehicles that transport parts between locations, perform an operation on a part at a location, or perform maintenance on a location or other resource that is down. Resources may travel on path networks with given speeds, accelerations, and pickup and setdown times. The routing and processing element allows user-defined procedural logic in ProModel's simulation programming language.

ProModel's runtime interface allows a user to define multiple scenarios for experimentation. SimRunner (discussed in Section 4.8.2) adds the capability to perform an optimization; it is based on an evolutionary strategy algorithm, a variant of the genetic algorithm approach.

4.7.7 Taylor ED

Taylor Enterprise Dynamics (ED) is offered by F&H Simulations [Hullinger, 1999]. It is built on different concepts than its predecessor package, Taylor II. Taylor ED is used to model manufacturing, warehousing, and material handling processes plus service and data flow processes. It can also be used to monitor flow processes in real time. It offers both 2-D and 3-D animation.

Taylor ED is based on the concept of an atom. Atoms are Taylor ED's smart objects and model-building resources. Model builders can access standard libraries of atoms as well as create new atoms using Taylor ED's Atom Editor. Atoms are used to represent the products or entities flowing through the system as well as the resources acting on these entities. In fact, everything in Taylor ED is an atom, whether it is a resource, a product, a model (or submodel), a table, a report, a graph, a record, or a library.

To build a model, a user first creates a model layout by selecting atoms from the atom library and placing them on the screen. Next the user connects the atoms and defines the routing of products or entities (also atoms) flowing through the model. Next the user assigns logic, including routing logic, to each atom by editing its parameter fields.

An atom is an object with four dimensions (x, y, z, and time). Each atom can have a location, speed, and rotation and dynamic behavior over time. Atoms can inherit their behavior from other atoms and contain other atoms; atoms can be created and destroyed. Atoms can be viewed simultaneously in 2-D and 3-D animation.

One of the hallmark characteristics of the atom is its reusability. For example, a workcell can be created by using the Atom Editor or by combining several existing atoms. The workcell can be saved as an atom, included in a standard atom library, then reused in a new model or shared with other Taylor ED modelers.

Taylor ED includes both an experiment module and the OptQuest Optimizer (discussed in Section 4.8.2).

4.7.8 WITNESS

WITNESS is offered by The Lanner Group [Mehta and Rawles, 1999]. WITNESS is strongly machine oriented and contains many elements for discrete-part manufacturing. It also contains elements for continuous processing such as the flow of fluids through processors, tanks and pipes.

WITNESS models are based on template elements. Elements may be combined into a designer element module to be reused. The machine element can be single, batch, production, assembly, multistation or multicycle. The behavior of each element is described on a tabbed detail form in the WITNESS user interface.

It offers a 2-D layout animation plus a process flow view. The 2-D graphics offers a walk-through capability controlled by a mouse or a camera attached to an entity.

4.8 Experimentation and Statistical Analysis Tools

4.8.1 Common Features

Virtually all simulation packages offer various degrees of support for statistical analysis of simulation outputs. In recent years, many packages have added optimization as one of the analysis tools. To support analysis, most packages provide scenario definition, run management capabilities, as well as data export to spreadsheets and other external applications.

Optimization is used to find a "near-optimal" solution. The user must define an objective or fitness function, usually a cost or costlike function that incorporates the trade-off between additional throughput and additional resources. Until recently, the methods available for optimizing a system had difficulty coping with the random and nonlinear nature of most simulation outputs. Advances in the field of metaheuristics have offered new approaches to simulation optimization, based on artificial intelligence, neural networks, genetic algorithms, evolutionary strategies, tabu search, and scatter search.

4.8.2 Analysis Tools

This section briefly discusses the Arena Output and Process Analyzers AutoStat for AutoMod, SimRunner for ProModel, and OptQuest, which is used in a number of simulation products.

Arena's Output and Process Analyzer

Arena comes with the Output Analyzer and Process Analyzer. In addition, Arena uses OptQuest for optimization.

The Output Analyzer provides confidence intervals, comparison of multiple systems, and warmup determination to reduce initial-condition biases. It creates various plots, charts, and histograms, smoothes responses, and performs correlation analysis. To compute accurate confidence intervals, it does internal batching (both within and across replications, with no user intervention) and data truncation to provide stationary, independent and normally distributed data sets.

The Process Analyzer adds sophisticated scenario-management capabilities to Arena for comprehensive design of experiments. It allows a user to define scenarios, make the desired runs, and analyze the results. It allows an arbitrary number of controls and responses. Responses can be added after runs have been completed. It will rank scenarios by any response and provide summaries and statistical measures of the responses. A user can view 2-D and 3-D charts of response values across either replications or scenarios.

AutoStat

AutoStat is the run manager and statistical analysis product in the AutoMod product family [Rohrer, 1999]. AutoStat provides a number of analyses, including warmup determination for steady-state analyses, absolute and comparison confidence intervals, design of experiments, sensitivity analysis, and optimization using an evolutionary strategy. The evolutionary strategies algorithm used by AutoStat is well suited to finding a near-optimal solution without getting trapped at a local optimum.

With AutoStat, an end user can define any number of scenarios by defining factors and their range of values. Factors include single parameters such as resource capacity or vehicle speed, single cells in a data file, and complete data files. By allowing a data file to be a factor, a user can experiment with, for example, alternate production schedules, customer orders for different days, different labor schedules, or any other numerical inputs typically specified in a data file. Any standard or custom output can be designated as a response. For each defined response, AutoStat computes descriptive statistics (average, standard deviation, minimum and maximum) and confidence intervals. New responses can be defined after runs are made, because AutoStat archives and compresses the standard and custom outputs from all runs. Various charts and plots are available to provide graphical comparisons.

AutoStat is capable of distributing simulation runs across a local area network and pulling back all results to the user's machine. Support for multiple machines and CPUs gives users the ability to make many more runs of the simulation than would otherwise be possible by using idle machines during off hours. This is especially useful in multifactor analysis and optimization, both of which may require large numbers of runs. AutoStat also has a diagnostics capability that automatically detects "unusual" runs, where the definition of "unusual" is user-definable.

AutoStat also works with two other products from AutoSimulations: the AutoMod Simulator, a spreadsheet-based job-shop simulator, and AutoSched AP, a rule-based simulation package for finite-capacity scheduling in the semiconductor industry.

OptQuest

OptQuest was developed by Fred Glover, University of Colorado and co-founder of Optimization Technologies, Inc. [Glover et al., 1999]. It is available for Arena, CSIM, Micro Saint, QUEST, and Taylor ED.

OptQuest is based on a combination of methods: scatter search, tabu search with adaptive memory, and a neural network component. Scatter search is a population-based approach, in some ways similar to genetic algorithms; existing solutions are combined to make new solutions. Tabu search is then superimposed to prohibit the search from reinvestigating previous solutions and neural networks to screen out solutions likely to be poor.

When using OptQuest, a user defines an objective function to minimize or maximize. Simulation examples include throughput or a cost function. OptQuest allows a user to specify the range of variable parameters as well as constraints on parameters, responses, and combinations of parameters — in fact, any set of conditions that can be represented by a mixed-integer programming formulation.

SimRunner

SimRunner works with ProModel [Price and Harrell, 1999]. Its optimization is based on an evolutionary strategy, a variant of the genetic algorithm approach. A user specifies an objective function (a response to minimize or maximize) and the input factors to vary. In its Stage One Optimization Analysis, SimRunner performs a factorial design of experiments to determine which input factors have a significant effect on the objective function. Those factors that have no significant effect can be eliminated from the search for an optimum. The Stage Two Simulation Optimization uses the evolutionary strategy algorithm to conduct a multivariable optimization search.

4.9 Trends in Simulation Software

This section is based on a plenary address given by Banks [2000] to the European Simulation Multi-Conference 2000. Many of the remarks are reproduced here with the permission of the Society for Computer Simulation International, European Council, Ghent, Belgium. Since it is difficult to predict software developments for even one year, the trends predicted here may be little more than mere speculation!

4.9.1 High-Fidelity Simulation

High-fidelity simulation is exceptional graphics quality and very accurate emulation. The term "virtual" is often used, implying that the emulation is nearly the same as that which is being modeled, i.e., it looks and acts like the real thing. However, the model is in the computer and the actual object is real. In terms of manufacturing and material handling, consider a virtual machine, a virtual factory, and a virtual warehouse. If you are looking at a virtual machine on your computer monitor, it looks very much like that machine. If the machine is updated, it is only necessary to update the simulation model. You don't have to spend $1,000,000 for a modern centerless grinder, you just simulate it in the computer to study its operation. Of course, if you want to make bearings, you will need to buy the grinder.

Another dimension of high fidelity simulation is to make it selective. You could have varying levels of detail. If you wanted the model to run fast and the animation and solution to be approximate, you could set the dial on low. Or, you could set the dial on high and get the opposite result. You could also set the dial between the two extremes.

4.9.2 Data Exchange Standards

Engineering Animation Incorporated, Inc. (EAI), has developed a simulation interface called SDX for Simulation Data Exchange. David Sly (1999) of EAI says:

> The main function of the tool is to communicate to the simulation package all that is known about the layout in an effort to reduce the redundant tasks of redefining equipment in the layout importation to the simulation tool and that equipment's interconnectivity and attributes. This interface adds value in that it can cut considerable time from the process of creating detailed layouts in the simulation tool and defining equipment properties (such as conveyor speeds, etc.). The interface is best applied to applications that have a significant amount of automated material handling equipment such as conveyors, AGVs, fork trucks and cranes, all typically found in automotive plants and distribution centers.

Several simulation software firms have worked with EAI to accept SDX. This data exchange format could emerge as a standard. Rather than redraw the conveyor system with all of its parameters, it is just imported through SDX.

SDX is not the only data exchange format vying for attention. The World-wide Web Consortium (W3C) is developing XML for extensible markup language. XML is a set of rules, guidelines, or conventions for designing text formats for data such that the files are easy to generate and read, they are unambiguous, and they are extensible. It aids in putting information on the Web and in the retrieval of that information.

XML has a large and growing community. A project coordinated through the Manufacturing Engineering Laboratory at the U.S. National Institute of Standards and Technology is to "explore the use of XML as a standard for structured document interchange on the Web, for exchanging complex data objects between tasks in a distributed workflow application."

So, groups are developing XML standards for all types of documentation. Eventually, some group will want to develop XML standards for representation of manufacturing and material handling systems.

4.9.3 The Internet

The Internet can deliver modeling capability to the user. The interface can be at the client (local computer), with the models run on a server. However, this server is not in a local area network (LAN). It is at the software vendor's or some other host's site. The server can always have the latest version of the model. The interface can be in Java, which is independent of the computer and has applets for displaying results. Currently, the speed of Java is not sufficient to serve as the simulation engine.

Another way that we can use the Internet is to have models run on many computers. Thus, if the model is extremely large or if many replications are needed, the computing power can be greatly extended via the Internet by distributing the computational needs.

4.9.4 Old Paradigm versus New Paradigm

In the past, models were always constructed with a limited purpose in mind and to answer specific questions. That would be called the old paradigm. But, it takes time to build models in response to questions.

The new paradigm is to have models built at different levels of detail, but without knowing all of the questions to be asked. This will speed up the response time, since the model already exists. Under the new paradigm, the models become an asset. The firm with more prebuilt models will be able to answer questions faster — but may have more out-of-date models. Indeed, model maintenance becomes a big issue.

4.9.5 Component Libraries

Consider component libraries stored in a neutral file format. This requires that a format can be designed and accepted by numerous software vendors, which is in itself a challenge. System providers could be contributors of modules. For example, a robot manufacturer could contribute a module for a specific type of robot. The result of having module libraries is more rapid development of models. The modules could be prepared so that all that they require is parameterization.

4.9.6 Distributed Manufacturing Simulation/High Level Architecture

Many years ago, the U.S. Department of Defense (DoD) began a development process to enable various simulations located throughout the world to exchange data dynamically during simulation of large-scale military exercises. This is known as distributed simulation. There can be aircraft maneuvers simulated in one computer model, tanks simulated in another, ground troops in yet another, and the supply chain in still another. Also, the models may be at different levels of fidelity. The computers hosting these models are geographically dispersed, but interacting over the Internet.

The technology has matured to the point where it can now be tried in other domains. This technology is collectively called High Level Architecture (HLA). HLA has been designed so that the minimum amount of information is exchanged between the various simulations sufficient to effect clock synchronization and the exchange of data objects.

4.9.7 Embedded Simulation

Embedded simulation occurs when simulation models are placed in other software. This already happens in simulation-based scheduling. It is anticipated that there will be many other cases in which the simulation becomes invisible. For example, in maintenance management software, a policy could be entered, and a very short time later, the user would see the anticipated result of that policy. The user might not know that the output resulted from a simulation.

4.9.8 Optimization

The main problem with optimization is that of computer time. It takes a lot of it when running an optimization on a complex model. This is related to the number of variables, the types of variables (discrete or continuous), the number of constraints on the solution space, the type of constraints (simple bound, linear, or nonlinear), and the type of objective function (linear, quadratic, or nonlinear). Assume that a simulation takes 10 minutes per replication and that 5 replications are conducted at each combination of factor values. Further, assume that the optimization algorithm calls for 1000 scenarios (not unusual). To obtain a near-optimal solution would take 50,000 minutes. That is approximately one month of elapsed time. Decision makers usually do not have the luxury of this amount of time.

The time can be decreased by using fractional factorial experiments to reduce the number of combinations. Another technique, currently in use by at least one software vendor, is to distribute the simulation runs over the computers on a local area network. For example, if there are ten computers available on your network, roughly ten times as many runs can be made overnight or over the weekend. The results are fed back to the user's computer.

As far as simulation research is concerned, optimization is an area that will continue to receive attention.

REFERENCES

BANKS, J., J. S. CARSON, AND J. N. SY [1995], *Getting Started with GPSS/H*, 2d ed., Wolverine Software Corporation, Annandale, VA.

BANKS, J. [1996], "Interpreting Software Checklists," *OR/MS Today*, June.

BANKS, J. [1998], "The Future of Simulation Software: A Panel Discussion," *Proceedings of the 1998 Winter Simulation Conference*, D. J. Medeiros, E. Watson, J. Carson, S. Manivannan, eds., Washington, DC, Dec. 13–16.

BANKS, J. [2000], "The Future of Simulation," *Proceedings of the European Simulation Multi-Conference*, Ghent, Belgium, May 23–26, to appear.

BLOECHLE, W. K., AND K. R. LAUGHERY, JR. [1999], "Simulation Interoperability Using Micro Saint Simulation Software," *Proceedings of the 1999 Winter Simulation Conference*, P. A. Farrington, H. B. Nembhard, D. T. Sturrock, G. W. Evans, eds., Phoenix, AZ, Dec. 5–8, pp. 286–288.

CRAIN, R. C., AND J. O. HENRIKSEN [1992], "Simulation Using GPSS/H," *Proceedings of the 1999 Winter Simulation Conference*, P. A. Farrington, H. B. Nembhard, D. T. Sturrock, G. W. Evans, eds., Phoenix, AZ, Dec. 5–8, pp. 182–187.

DONALD, D. L. [1998], " Tutorial on Ergonomic and Process Modeling Using Quest and IGRIP," *Proceedings of the 1998 Winter Simulation Conference*, D. J. Medeiros, E. Watson, J. Carson, S. Manivannan, eds., Washington, DC, Dec. 13–16, pp. 297–302.

GLOVER, F., J. P. KELLY, AND M. LAGUNA [1999], "New Advances for Wedding Optimization and Simulation," *Proceedings of the 1999 Winter Simulation Conference*, P. A. Farrington, H. B. Nembhard, D. T. Sturrock, G. W. Evans, eds., Phoenix, AZ, Dec. 5–8, pp. 255–260.

HENRIKSEN, J. O. [1999], "General-Purpose Concurrent and Post-Processed Animation with Proof," *Proceedings of the 1999 Winter Simulation Conference*, P. A. Farrington, H. B. Nembhard, D. T. Sturrock, G. W. Evans, eds., Phoenix, AZ, Dec. 5–8, pp. 176–181.

HULLINGER, D. R. [1999], "Taylor Enterprise Dynamics," *Proceedings of the 1999 Winter Simulation Conference*, P. A. Farrington, H. B. Nembhard, D. T. Sturrock, G. W. Evans, eds., Phoenix, AZ, Dec. 5–8, pp. 227–229.

KRAHL, D. [1999], "Modeling with Extend™," *Proceedings of the 1999 Winter Simulation Conference*, P. A. Farrington, H. B. Nembhard, D. T. Sturrock, G. W. Evans, eds., Phoenix, AZ, Dec. 5–8, pp. 188–195.

MEHTA, A., AND I. RAWLES [1999], "Business Solutions Using WITNESS," *Proceedings of the 1999 Winter Simulation Conference*, P. A. Farrington, H. B. Nembhard, D. T. Sturrock, G. W. Evans, eds., Phoenix, AZ, Dec. 5–8, pp. 230–233.

Mesquite Software, Inc. [1997], *User's Guide, CSIM18, Simulation Engine*. Austin, TX.

NANCE, R. E. [1995], "Simulation Programming Languages: An Abridged History," *Proceedings of the 1995 Winter Simulation Conference*, ed. C. Alexopoulos, K. Kang, W. R. Lilegdon, and D. Goldsman, Arlington, VA, Dec 13–16, pp. 1307–1313.

PRICE, R. N., AND C. R. HARRELL [1999], "Simulation Modeling and Optimization Using ProModel," *Proceedings of the 1999 Winter Simulation Conference*, P. A. Farrington, H. B. Nembhard, D. T. Sturrock, G. W. Evans, eds., Phoenix, AZ, Dec. 5–8, pp. 208–214.

PRITSKER, A. A. B., AND C. D. PEGDEN [1979], *Introduction to Simulation and SLAM*, John Wiley, New York.

ROHRER, M. [1999], "AutoMod Product Suite Tutorial (AutoMod, Simulator, AutoStat) by AutoSimulations," *Proceedings of the 1999 Winter Simulation Conference*, P. A. Farrington, H. B. Nembhard, D. T. Sturrock, G. W. Evans, eds., Phoenix, AZ, Dec. 5–8, pp. 220–226.

SADOWSKI, D., AND V. BAPAT [1999], "The Arena Product Family: Enterprise Modeling Solutions," *Proceedings of the 1999 Winter Simulation Conference*, P. A. Farrington, H. B. Nembhard, D. T. Sturrock, G. W. Evans, eds., Phoenix, AZ, Dec. 5–8, pp. 159–166.

SCHWETMAN, H. [1987], "CSIM: A C-based, Process Oriented Simulation Language," *Proceedings of the 1986 Winter Simulation Conference*, J. R. Wilson, J. O. Henriksen, S. D. Roberts, eds., pp. 387–396.

SCHWETMAN, H. [1996], "CSIM18 — The Simulation Engine," *Proceedings of the 1996 Winter Simulation Conference*, J. Charnes, D. Morrice, D. Brunner, and J. Swain, eds., San Diego, CA, pp. 517–521.

"Simulation Software Buyer's Guide," *IIE Solutions*, May 1999.

SLY, D. et al. [1999], "Increasing the Power and Value of Manufacturing Simulation via Collaboration with Other Analytical Tools: A Panel Discussion," *Proceedings of the 1999 Winter Simulation Conference*, P. A. Farrington, H. B. Nembhard, D. T. Sturrock, G. W. Evans, eds., Phoenix, AZ, Dec. 5–8, pp. 749–753.

STROUSTRUP, B. [1997], The C++ Programming Language, 3d ed., Addison-Wesley, Reading, MA.

SWAIN, J. J. [1999], "Simulation Software Survey," *OR/MS Today*, February, Vol. 26, No. 3.

TOCHER, K. D., AND D. G. OWEN [1960], "The Automatic Programming of Simulations," *Proceedings of the Second International Conference on Operational Research*, J. Banbury and J. Maitland, eds., pp. 50–68.

WILSON, J. R. et al. [1992], "The Winter Simulation Conference: Perspectives of the Founding Fathers" *Proceedings of the 1992 Winter Simulation Conference*, J. J. Swain, D. Goldsman, R. C. Crain, and J. R. Wilson, eds., Arlington, VA, Dec. 13–16, pp. 37–62.

EXERCISES

For the exercises below, the reader should code the model in a general-purpose language (such as C or C++), a special-purpose simulation language (such as GPSS/H), or any desired simulation package.

Most problems contain activities that are uniformly distributed over an interval $[a, b]$. Assume that all values between a and b are possible; that is, the activity time is a *continuous* random variable.

The uniform distribution is denoted by $U(a, b)$, where a and b are the endpoints of the interval, or by $m \pm h$, where m is the mean and h is the "spread" of the distribution. These four parameters are related by the equations

$$m = \frac{a+b}{2}, \qquad h = \frac{b-a}{2}$$

$$a = m - h, \qquad b = m + h$$

Some of the uniform random-variate generators available require specification of a and b; others require m and h.

Some problems have activities that are assumed to be normally distributed, which is denoted by $N(\mu, \sigma^2)$, where μ is the mean and σ^2 the variance. (Since activity times are nonnegative, the normal distribution is appropriate only if $\mu \geq k\sigma$, where k is at least 4 and preferably 5 or larger. If a negative value is generated, it is discarded.) Other problems use the exponential distribution with some rate λ, or mean $1/\lambda$. Chapter 5 reviews these distributions; Chapter 8 covers the generation of random variables having these distributions. All of the languages have a facility to easily generate samples from these distributions. For C or C++ simulations, the student may use the functions given in Section 4.4 (Figure 4.8) for generating samples from the normal and exponential distributions.

1. Make the necessary modifications to the C++ model of the checkout counter (Example 4.2) so that the simulation will run for exactly 60 hours. [*Note:* The $U(0, 1)$ random-number generator unif($\cdot$) may have to be changed, depending on the software being used.]

2. In addition to the change in Exercise 1, also assume that any customers still at the counter at time 60 hours will be served, but no arrivals after time 60 hours are allowed. Make the necessary changes to the C++ code and run the model.

3. Implement the changes in Exercises 1 and 2 in any of the simulation packages.

4. Ambulances are dispatched at a rate of one every 15 ± 10 minutes in a large metropolitan area. Fifteen percent of the calls are false alarms which require 12 ± 2 minutes to complete. All other calls can be one of two kinds. The first kind are classified as serious. They constitute 15% of the nonfalse-alarm calls and take 25 ± 5 minutes to complete. The remaining calls take 20 ± 10 minutes to complete. Assume that there are a very large number of available ambulances, and that any number can be on call at any time. Simulate the system for 500 calls to be completed.

5. In Exercise 4, estimate the mean time that an ambulance takes to complete a call.

6. **(a)** In Exercise 4, suppose that there is only one ambulance available. Any calls that arrive while the ambulance is out must wait. Can one ambulance handle the work load?

   (b) Simulate with x ambulances, where $x = 1, 2, 3$, or 4, and compare the alternatives on the basis of length of time a call must wait, percentage of calls that must wait, and percentage of time the ambulance is out on call.

7. A parent volunteers to remind other parents to come to a school meeting next week. The volunteer is given a list with 100 names. It takes 5 ± 2 seconds to find the next number to call, 7 ± 2 seconds to place the call, and 30 ± 5 seconds to give the message. For each parent on the list, the chance of reaching that parent is 35%. How many parents were reached out of the 100 names? How long does it take?

8. A superhighway connects one large metropolitan area to another. A vehicle leaves the first city every 20 ± 15 seconds. Twenty percent of the vehicles have 1 passenger, 30% have 2 passengers, 10% have 3 passengers, and 10% have 4 passengers. The remaining 30% of the vehicles are buses, which carry 40 people. It takes 60 ± 10 minutes for a vehicle to travel between the two metropolitan areas. How long does it take for 5000 people to arrive in the second city?

9. People arrive at a meat counter at a rate of one every 25 ± 10 seconds. There are two sections: one for beef and one for pork. People want goods from them in the following proportion: beef only, 50%; pork only, 30%; beef and pork, 20%. It takes 45 ± 20 seconds for a butcher to serve one customer for one order. All customers place one order, except that "beef and pork" customers place two orders. Assume that enough butchers are available to handle all customers present at any time. How long does it take for 200 customers to be served?

10. In Exercise 9, what is the maximum number of butchers needed during the course of simulation? Would this number always be sufficient to guarantee that no customer ever had to wait?

11. In Exercise 9, simulate with x butchers, where $x = 1, 2, 3, 4$. When all butchers are busy, a line forms. For each value of x, estimate the mean number of busy butchers.

12. A one-chair unisex hair shop has arrivals at the rate of one every 20 ± 15 minutes. One-half of the arriving customers want a dry cut, 30% want a style, and 20% want a trim only. A dry cut takes 15 ± 5 minutes, a style cut takes 25 ± 10 minutes, and a trim takes 10 ± 3 minutes. Simulate 400 customers coming through the hair shop. Compare the given proportion of service requests of each type with the simulated outcome. Are the results reasonable?

13. An airport has two concourses. Concourse 1 passengers arrive at a rate of one every 15 ± 2 seconds. Concourse 2 passengers arrive at a rate of one every 10 ± 5 seconds. It takes 30 ± 5 seconds to walk down concourse 1 and 35 ± 10 seconds to walk down concourse 2. Both concourses empty into the main lobby, adjacent to the baggage claim. It takes 10 ± 3 seconds to reach the baggage claim area from the main lobby. Only 60% of the passengers go to the baggage claim area. Simulate the passage of 500 passengers through the airport system. How many of these passengers went through the baggage claim area? In this problem, the expected number through the baggage claim area can be computed by $0.60(500) = 300$. How close is the simulation estimate to the expected number? Why the difference?

14. In a multiphasic screening clinic, patients arrive at a rate of one every 5 ± 2 minutes to enter the audiology section. The examination takes 3 ± 1 minutes. Eighty percent of the patients were passed on to the next test with no problems. Of the remaining 20%, one-half require simple procedures which take 2 ± 1 minutes and are then sent for reexamination with the same probability of failure. The other half are sent home with medication. Simulate the system to determine how long it takes to screen and pass 200 patients. (*Note:* Persons sent home with medication are not considered "passed.")

15. Consider a bank with four tellers. Tellers 3 and 4 deal only with business acounts, while Tellers 1 and 2 deal with general accounts. Clients arrive at the bank at a rate of one every 3 ± 1 minutes. Of the clients, 33% are business accounts. Clients randomly choose between the two tellers available for each type of account. (Assume that a customer chooses a line without regard to its length and does not change lines.) Business accounts take 15 ± 10 minutes to complete and general accounts 6 ± 5. Simulate the system for 500 transactions to be completed. What percentage of time is each type of teller busy? What is the average time that each type of customer spends in the bank?

16. Repeat Exercise 15 assuming that customers join the shortest line for the teller handling their type of account.

17. In Exercises 15 and 16, estimate the mean delay of business customers and of regular customers. (Delay is time spent in the waiting line and is exclusive of service time.) Also estimate the mean length of the waiting line and the mean proportion of customers who are delayed longer than 1 minute.

18. Three different machines are available for machining a special type of part for 1 hour of each day. The processing-time data is as follows:

Machine	*Time to Machine One Part*
1	20 ± 4 seconds
2	10 ± 3 seconds
3	15 ± 5 seconds

Assume that parts arrive by conveyor at a rate of one every 15 ± 5 seconds for the first 3 hours of the day. Machine 1 is available for the first hour, machine 2 for the second hour, and machine 3 for the third hour of each day. How many parts are produced in a day? How large a storage area is needed for parts waiting for a machine? Do parts "pile up" at any particular time? Why?

19. People arrive at a self-service cafeteria at the rate of one every 30 ± 20 seconds. Forty percent go to the sandwich counter, where one worker makes a sandwich in 60 ± 30 seconds. The rest go to the main counter, where one server spoons the prepared meal onto a plate in 45 ± 30 seconds. All customers must pay a single cashier, which takes 25 ± 10 seconds. For all customers, eating takes 20 ± 10 minutes. After eating, 10% of the people go back for dessert, spending an additional 10 ± 2 minutes altogether in the cafeteria. Simulate until 100 people have left the cafeteria. How many people are left in the cafeteria, and what are they doing, at the time the simulation stops?

20. Thirty trucks carrying bits and pieces of a C–5N cargo plane leave Atlanta at the same time for the Port of Savannah. From past experience, it is known that it takes 6 ± 2 hours for a truck to make the trip. Forty percent of the drivers stop for coffee, which takes an additional 15 ± 5 minutes. (a) Model the situation as follows: For each driver, there is a 40% chance of stopping for coffee. When will the last truck reach Savannah? (b) Model it so that exactly 40% of the drivers stop for coffee. When will the last truck reach Savannah?

21. Customers arrive at the Last National Bank every 40 ± 35 seconds. Currently, the customers pick one of two tellers at random. A teller services a customer in 75 ± 25 seconds. Once a customer joins a line, the customer stays in that line until the transaction is complete. Some customers want the bank to change to the single-line method used by the Lotta Trust Bank. Which is the faster method for the customer? Simulate for one hour before collecting any summary statistics, then simulate for an 8-hour period. Compare the two queueing disciplines on the basis of teller utilization (percentage of busy time), mean delay of customers, and proportion of customers who must wait (before service begins) more than 1 minute, and more than 3 minutes.

22. Loana Tool Company rents chain saws. Customers arrive to rent chain saws at the rate of one every 30 ± 30 minutes. Dave and Betty handle these customers. Dave can rent a chain saw in 14 ± 4 minutes. Betty takes 10 ± 5 minutes. Customers returning chain saws arrive at the same rate as those renting chain saws. Dave and Betty spend 2 minutes with a customer to check in the returned chain saw. Service is first-come, first-served. When no customers are present, or Betty alone is busy, Dave gets these returned saws ready for rerenting. For each saw, maintenance and cleanup take him 6 ± 4 minutes and 10 ± 6 minutes, respectively. Whenever Dave is idle, he begins the next maintenance or cleanup. Upon finishing a maintenance or cleanup, Dave begins serving customers if one or more is waiting. Betty is always available for serving customers. Simulate the operation of the system starting with an empty shop at 8:00 A.M., closing the doors at 6:00 P.M., and getting chain saws ready for rerenting until 7:00 P.M. From 6:00 until 7:00 P.M., both Dave and Betty do maintenance and cleanup. Estimate the mean delay of customers who are renting chain saws.

23. In Exercise 22, change the shop rule regarding maintenance and cleanup to get a chain saw ready for rerental. Now Betty does all this work. Upon finishing a cleanup on a saw, she helps Dave if a line is present. (That is, Dave and Betty both serve new customers and check in returned saws until Dave alone is busy, or the shop is empty.) Then she returns to her maintenance and cleanup duties. (a) Estimate the mean delay of customers who are renting chain saws. Compare the two shop rules on this basis. (b) Estimate the proportion of customers who must wait more than 5 minutes. Compare the two shop rules on this basis. (c) Discuss the pros and cons of the two criteria in parts (a) and (b) for comparing the two shop rules. Suggest other criteria.

24. U. of L.A. (University of Lower Altoona) has only one color printer for student use. Students arrive at the printer every 15 ± 10 minutes to use it for 12 ± 6 minutes. If the printer is busy, 60% will come back in 10 minutes to use it. If it is still busy, 50% (of the 60%) will return in 15 minutes. How many students fail to use the printer, compared to 500 that actually finish? Demand and service occur 24 hours a day.

25. A warehouse holds 1000 cubic meters of cartons. These cartons come in three sizes: little (1 cubic meter), medium (2 cubic meters), and large (3 cubic meters). The cartons arrive at the following rates: little, every 10 ± 10 minutes; medium, every 15 minutes; and large, every 8 ± 8 minutes. If no cartons are removed, how long will it take to fill an empty warehouse?

26. Go Ape! buys a Banana II computer to handle all of its simulation needs. Jobs arrive every 10 ± 10 minutes to be batch-processed one at a time. Processing takes 7 ± 7 minutes. The monkeys that run the computer cause a system failure every 60 $\pm$ 60 minutes. The failure lasts for 8 ± 4 minutes. When a failure occurs, the job that was being run resumes processing where it was left off. Simulate the operation of this system for 24 hours. Estimate the mean system response time. (A system response time is the length of time from arrival until processing is completed.) Also estimate the mean delay for those jobs that are in service when a computer system failure occurs.

27. Able, Baker, and Charlie are three carhops at the Sonic Drive-In (service at the speed of sound!). Cars arrive every 5 ± 5 minutes. The carhops service customers at the rate of one every 10 ± 6 minutes. However, the customers prefer Able over Baker, and Baker over Charlie. If the carhop of choice is busy, the customers choose the first available carhop. Simulate the system for 1000 service completions. Estimate Able's, Baker's, and Charlie's utilization (percentage of time busy).

28. Jiffy Car Wash is a five-stage operation that takes 2 ± 1 minutes for each stage. There is room for 6 cars to wait to begin the car wash. The car wash facility holds 5 cars, which move through the system in order, one car not being able to move until the car ahead of it moves. Cars arrive every 2.5 ± 2 minutes for a wash. If the car cannot get into the system, it drives across the street to Speedy Car Wash. Estimate the balking rate per hour. That is, how many cars drive off per hour? Simulate for one 12-hour day.

29. Consider the three machines A, B, and C pictured below. Arrivals of parts and processing times are as indicated (times in minutes).

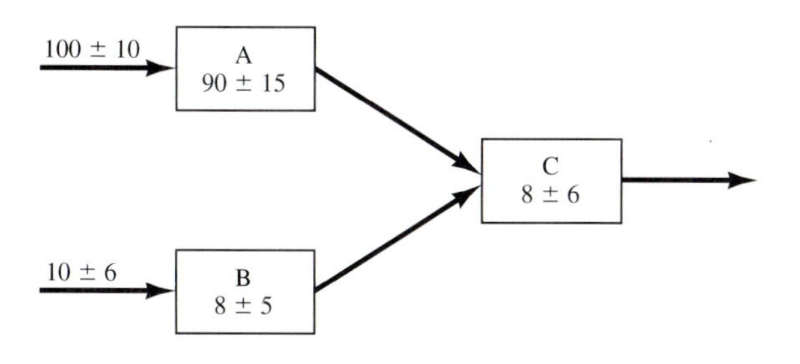

Machine A processes type I parts, machine B processes type II parts, and machine C processes both types of parts. All machines are subject to random breakdown: machine A every 400 ± 350 minutes with a downtime of 15 ± 14 minutes, machine B every 200 ± 150 minutes with a downtime of 10 ± 8 minutes, and machine C almost never, so its downtime is ignored. Parts from machine A are processed at machine C as soon as possible, ahead of any type II parts from machine B. When machine A breaks down, any part in it is sent to machine B and processed as soon as B becomes free, but processing begins over again, taking 100 ± 20 minutes. Again, type I parts from machine A are processed ahead of any parts waiting at B, but after any part currently being processed. When machine B breaks down, any part being processed resumes processing as soon as B becomes available. All machines handle one part at a time. Make two independent replications of the simulation. Each replication will consist of an 8-hour initialization phase to load the system with parts, followed by a 40-hour steady-state run. (Independent replications means that each run uses a different stream of random numbers.) Management is interested in the long-run throughput [i.e., the number of parts of each type (I and II) produced per 8-hour day], the long-run utilization of each machine, and the existence of bottlenecks (long "lines" of waiting parts, as measured by the queue length at each machine). Report the output data in a table similar to the following:

	Run 1	*Run 2*	*Average of 2 Runs*
Utilization A			
Utilization B			
Etc.			

Include a brief statement summarizing the important results.

30. Workers come to a supply store at the rate of one every 10 ± 4 minutes. Their requisitions are handled by one of three clerks; a clerk takes 22 ± 10 minutes to handle a requisition. All requisitions are then passed to a single cashier, who spends 7 ± 6 minutes per requisition. Simulate the system for 120 hours.

(a) Estimate the utilization of each clerk, based on the 120-hour simulation.

(b) How many workers are completely served? How many do the three clerks serve? How many workers arrive? Are all three clerks ever busy at the same time? What is the average number of busy clerks?

31. People arrive at a barbershop at the rate of one every 4.5 minutes. If the shop is full (it can hold five people altogether), 30% of the potential customers leave and come back in 60 ± 20 minutes. The others leave and do not return. One barber gives a haircut in 8 ± 2 minutes, whereas the second talks a lot and it takes 12 ± 4 minutes. If both barbers are idle, a customer prefers the first barber. (Treat customers trying to reenter the shop as if they are new customers.) Simulate this system until 300 customers have received a haircut.

(a) Estimate the balking rate, that is, the number turned away per minute.

(b) Estimate the number turned away per minute who do not try again.

(c) What is the average time spent in the shop?

(d) What is the average time spent getting a haircut (not including delay)?

(e) What is the average number of customers in the shop?

32. People arrive at a microscope exhibit at a rate of one every 8 ± 2 minutes. Only one person can see the exhibit at a time. It takes 5 ± 2 minutes to see the exhibit. A person can buy a "privilege" ticket for $1 which gives him or her priority in line over those who are too cheap to spend the buck. Some 50% of the viewers do this, but they make their decision to do so only if one or more people are in line when they arrive. The exhibit is open continuously from 10:00 A.M. to 4:00 P.M. Simulate the operation of the system for one complete day. How much money is generated from the sale of privilege tickets?

33. Calls are made from the emergency room of Northside Hospital to central supply. Replies are sent back to the emergency room. Messages are created every 6 ± 3 minutes. They go through a pneumatic tube which takes 2 minutes to traverse. Only one message can be carried through the tube at a time. Some 70% of these messages require a reply. To prepare a reply takes 2 ± 1 minute. Simulate until 100 replies have been received at the emergency room. Estimate the utilization of the pneumatic tube. How many messages (and replies) are waiting to be sent, on the average?

34. Two machines, the Dewey and the Truman, are available for drilling parts (A-type and B-type). A-type parts at a rate of one every 10 ± 3 minutes and B-type parts arrive at a rate of one every 3 ± 2 minutes. For B-type parts, workers choose an idle machine, or if both drills are busy, they choose a machine at random and stay with their choice. A-type parts must be drilled as soon as possible; therefore, if a machine is available, preferably the Dewey, it is used; otherwise the part goes to the head of the line for the Dewey drill. All jobs take 4 ± 3 minutes to complete. Simulate the completion of 100 A-type parts. Estimate the mean number of A-type parts waiting to be drilled.

35. A telephone in a police precinct is used for both emergency calls and personal calls. Personal calls are on a first-come, first-served basis and are made at a rate of one every 5 ± 1 minutes. Emergency calls have priority and can preempt other calls.

They arrive at a rate of one every 15 ± 5 minutes. Emergency calls take 3 ± 1 minutes to complete while personal calls take 2 ± 2 minutes. Twenty percent of the people using the phone on a nonemergency basis wish to make another call as soon as possible, but they are given the lowest priority for their second call. Simulate until 200 calls of all types have been completed. Estimate the phone utilization.

36. A computer center has two color printers. Students arrive at a rate of one every 8 ± 2 minutes to use the color printer. They can be interrupted by professors, who arrive at a rate of one every 12 ± 2 minutes. There is one systems analyst who can interrupt anyone, but students are interrupted before professors. The systems analyst spends 6 ± 4 minutes on the color printer and then returns in 20 ± 5 minutes. Professors and students spend 4 ± 2 minutes on the color printer. If a person is interrupted, that person joins the head of the queue and resumes service as soon as possible. Simulate for 50 professor or analyst jobs. Estimate the interruption rate per hour, and the mean length of the waiting line of students.

37. Parts are machined on a drill press. They arrive at a rate of one every 5 ± 3 minutes and it takes 3 ± 2 minutes to machine them. Every 60 ± 60 minutes, a rush job arrives which takes 12 ± 3 minutes to complete. The rush job interrupts the present job. When the regular job returns to the machine, it stays only for its remaining process time. Simulate the machining of 10 rush jobs. Estimate the mean system response time for each type of part. (A response time is the total time that a part spends in the system.)

38. Messages are generated at a rate of one every 35 ± 10 seconds for transmission one at a time. Transmission takes 20 ± 5 seconds. At intervals of 6 ± 3 minutes, urgent messages lasting 10 ± 3 seconds take over the transmission line. Any message in progress must be reprocessed for 2 minutes before it can be resubmitted for transmission. When resubmitted, it goes to the head of the line. Simulate for 24 hours. Estimate the percentage of time the line is busy with ordinary messages.

39. A worker packs boxes that arrive at a rate of one every 15 ± 3 minutes. It takes 10 ± 3 minutes to pack a box. Once every hour the worker is interrupted to wrap specialty orders that take 16 ± 3 minutes to pack. On completing the speciality order, the worker then completes the interrupted order. Simulate for 40 hours. Estimate the mean proportion of time that the number of boxes waiting to be packed is more than five, and the average number of boxes waiting to be packed.

40. A patient arrives at the Emergency Room at Hello-Hospital about every 40 ± 19 minutes. They will be treated by either Doctor Slipup or Doctor Gutcut. Twenty percent of the patients are classified as NIA (need immediate attention) and the rest as CW (can wait). NIA patients are given the highest priority, 3, see a doctor as soon as possible for 40 ± 37 minutes, then have their priority reduced to 2 and wait until a doctor is free again, when they receive further treatment for 30 ± 25 minutes and are discharged. CW patients initially receive a priority of 1 and are treated (when their turn comes) for 15 ± 14 minutes; their priority is then increased to 2, they wait again until a doctor is free, receive 10 ± 8 minutes of final treatment, and are discharged. Simulate for 20 days of continuous operation, 24 hours per day. Precede this by a 2-day initialization period to load the system with patients.

Report conditions at times 0 days, 2 days, and 22 days. Does a 2-day initialization appear long enough to load the system to a level reasonably close to steady-state conditions? (a) Measure the average and maximum queue length of NIA patients from arrival to first seeing a doctor. What percent do not have to wait at all? Also tabulate and plot the distribution of this initial waiting time for NIA patients. What percent wait less than 5 minutes before seeing a doctor? (b) Tabulate and plot the distribution of total time in system for all patients. Estimate the 90% quantile. That is, 90% of the patients spend less than x amount of time in the system. Estimate x. (c) Tabulate and plot the distribution of remaining time in system from after the first treatment to discharge, for all patients. Estimate the 90% quantile. (*Note:* Most simulation packages provide the facility to automatically tabulate the distribution of any specified variable.)

41. People arrive at a newspaper stand with an interarrival time that is exponentially distributed with a mean of 0.5 minute. Fifty-five percent of the people buy just the morning paper, while 25% buy the morning paper and a *Wall Street Journal.* The remainder buy only the *Wall Street Journal.* One clerk handles *Wall Street Journal* sales, while another clerk handles morning paper sales. A person buying both goes to the *Wall Street Journal* clerk. The time it takes to serve a customer is normally distributed with a mean of 40 seconds and a standard deviation of 4 seconds for all transactions. Collect statistics on queues for each type of transaction. Suggest ways for making the system more efficient. Simulate for 4 hours.

42. Bernie remodels houses and makes room additions. The time it takes to finish a job is normally distributed with a mean of 17 elapsed days and a standard deviation of 3 days. Homeowners sign contracts for jobs at exponentially distributed intervals having a mean of 20 days. Bernie has only one crew. Estimate the mean waiting time (from signing the contract until work begins) for those jobs where a positive wait occurs. Also estimate the percentage of time the crew is idle. Simulate until 100 jobs have been completed.

43. Parts arrive at a machine in random fashion with exponential interrival times having a mean of 60 seconds. All parts require 5 seconds to prepare and align for machining. There are three different types of parts, in the proportions shown below. The times to machine each type of part are normally distributed with mean and standard deviation as follows:

Part Type	*Percent*	*Mean (Seconds)*	σ *(Seconds)*
1	50	48	8
2	30	55	9
3	20	85	12

Find the distribution of total time to complete processing for all types of parts. What proportion of parts take more than 60 seconds for complete processing? How long do parts have to wait, on the average? Simulate for one 8-hour day.

44. Shopping times at a department store have been found to have the following distribution:

Shopping Time (Minutes)	*Number of Shoppers*
0–10	90
10–20	120
20–30	270
30–40	145
40–50	88
50–60	28

After shopping, the customers choose one of six checkout counters. Checkout times are normally distributed with a mean of 5.1 minutes and a standard deviation of 0.7 minutes. Interarrival times are exponentially distributed with a mean of 1 minute. Gather statistics for each checkout counter (including the time waiting for checkout). Tabulate the distribution of time to complete shopping and the distribution of time to complete both shopping and checkout procedures. What proportion of customers spend more than 45 minutes in the store? Simulate for one 16-hour day.

45. The interarrival time for parts needing processing is given as follows:

Interarrival Time (Seconds)	*Proportion*
10–20	0.20
20–30	0.30
30–40	0.50

There are three types of parts: A, B, and C. The proportion of each part, and the mean and standard deviation of the normally distributed processing times, are as follows:

Part Type	*Proportion*	*Mean*	*Standard Deviation*
A	0.5	30 seconds	3 seconds
B	0.3	40 seconds	4 seconds
C	0.2	50 seconds	7 seconds

Each machine processes any type of part, one part at a time. Use simulation to compare one to two to three machines working in parallel. What criteria would be appropriate for such a comparision?

46. Orders are received for one of four types of parts. The interarrival time between orders is exponentially distributed with a mean of 10 minutes. The table that follows shows the proportion of the parts by type and the time to fill each type of order by the single clerk.

Part Type	*Percentage*	*Service Time (Minutes)*
A	40	$N(6.1, 1.3)$
B	30	$N(9.1,2.9)$
C	20	$N(11.8,4.1)$
D	10	$N(15.1,4.5)$

Orders of types A and B are picked up immediately after they are filled, but orders of types C and D must wait 10 ± 5 minutes to be picked up. Tabulate the distribution of time to complete delivery for all orders combined. What proportion take less than 15 minutes? What proportion take less than 25 minutes? Simulate for an 8-hour initialization period, followed by a 40-hour run. Do not use any data collected in the 8-hour initialization period.

47. Three independent widget-producing machines all require the same type of vital part, which needs frequent maintenance. To increase production it is decided to keep two spare parts on hand (for a total of $2 + 3 = 5$ parts). After 2 hours of use, the part is removed from the machine and taken to a single technician, who can perform the required maintenance in 30 ± 20 minutes. After maintenance, the part is placed in the pool of spare parts, to be put into the first machine that requires it. The technician has other duties, namely, repairing other items which have a higher priority and which arrive every 60 ± 20 minutes, requiring 15 ± 15 minutes to repair. Also, the technician takes a 15-minute break in each 2-hour time period. That is, the technician works 1 hour 45 minutes, takes off 15 minutes, works 1 hour 45 minutes, takes off 15 minutes, and so on.

- **(a)** What are the model's initial conditions? That is, where are the parts at time 0 and what is their condition? Are these conditions typical of "steady state"?
- **(b)** Make each replication of this experiment consist of an 8-hour initialization phase followed by a 40-hour data-collection phase. Make four statistically independent replications of the experiment, all in one computer run. (That is, make four runs with each, using a different set of random numbers.)
- **(c)** Estimate the mean number of busy machines and the proportion of time the technician is busy.
- **(d)** Parts are estimated to cost the company \$50 per part per 8-hour day (regardless of how much they are in use). The cost of the technician is \$20 per hour. A working machine produces widgets worth \$100 for each hour of production. Develop an expression to represent total cost per hour which can be attributed to widget production (i.e., not all of the technician's time is due to widget production). Evaluate this expression based on the results of the simulation.

48. The Wee Willy Widget Shop overhauls and repairs all types of widgets. The shop consists of five work stations, and the flow of jobs through the shop is as depicted here:

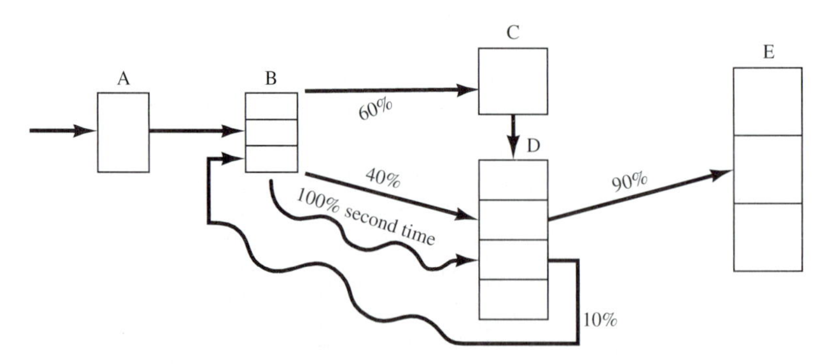

Regular jobs arrive at station A at the rate of one every 15 ± 13 minutes. Rush jobs arrive every 4 ± 3 hours and are given a higher priority except at station C, where they are put on a conveyor and sent through a cleaning and degreasing operation along with all other jobs. For jobs the first time through a station, processing and repair times are as follows:

Station	*Number of Machines or Workers*	*Processing and/or Repair Times (Minutes)*	*Description*
A	1	12 ± 2	Receiving clerk
B	3	40 ± 20	Disassembly and parts replacement
C	1	20	Degreaser
D	4	50 ± 40	Reassembly and adjustments
E	3	40 ± 5	Packing and shipping

The times listed above hold for all jobs that follow one of the two sequences $A \longrightarrow B \longrightarrow C \longrightarrow D \longrightarrow E$ or $A \longrightarrow B \longrightarrow D \longrightarrow E$. However, about 10% of the jobs coming out of station D are sent back to B for further work (which takes 30 ± 10 minutes) and then are sent to D and finally to E. The path of these jobs is as follows:

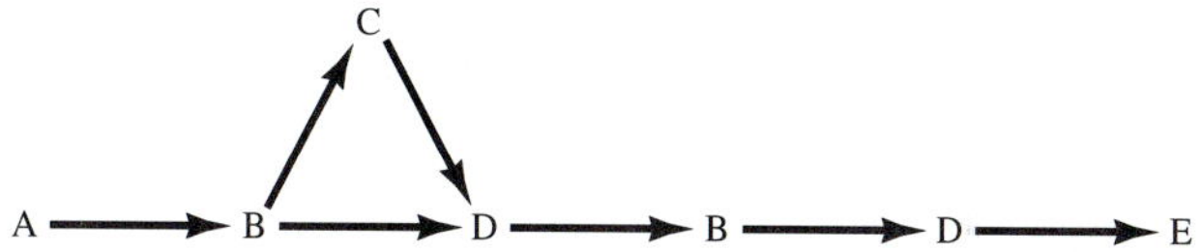

Every 2 hours, beginning 1 hour after opening, the degreasing station C shuts down for routine maintenance, which takes 10 ± 1 minute. However, this routine maintenance does not begin until the current widget, if any, has completed its processing. (a) Make three independent replications of the simulation model, where one replication equals an 8-hour simulation run, preceded by a 2-hour initialization run. The three sets of outputs represent three typical days. The main performance measure of interest is mean response time per job, where a response time is the total time a job spends in the shop. The shop is never empty in the morning, but the model will be empty without the initialization phase. So run the model for a 2-hour initialization period and collect statistics from time 2 hours to time 10 hours. This "warmup" period will reduce the downward bias in the estimate of mean response time. Note that the 2-hour warm-up is a device to load a simulation model to some more realistic level than empty. From each of the three independent replications, obtain an estimate of mean response time. Also obtain an overall estimate, the sample average of the three estimates. (b) Management is considering putting one additional worker at the busiest station (A, B, D, or E). Would this significantly improve mean response time? (c) As an alternative to part (b), management is considering replacing machine C with a faster one that processes a widget in only 14 minutes. Would this significantly improve mean response time?

49. A building materials firm loads trucks with two payloader tractors. The distribution of truck loading times has been determined to be exponential with a mean loading time of 6 minutes. The truck interarrival time is exponentially distributed with an arrival rate of 16 per hour. The waiting time of a truck and driver is estimated to cost $50 per hour. How much (if any) could the firm save (per 10-hour day) if an overhead hopper system that would fill any truck in a constant time of 2 minutes were installed? (Assume that the present tractors could and would adequately service the conveyors loading the hoppers.)

50. A milling-machine department has 10 machines. The runtime until failure occurs on a machine is exponentially distributed with a mean of 20 hours. Repair times are uniformly distributed between 3 and 7 hours. Select an appropriate run length and appropriate initial conditions.

(a) How many repairpersons are needed to ensure that the mean number of machines running is greater than eight?

(b) If there are two repairpersons, estimate the expected number of machines that are either running or being served.

51. Forty people are waiting to pass through a turnstile that takes 2.5 ± 1.0 seconds to traverse. Simulate this system 10 times, each one independent of the others, and determine the range and the average time for all 40 people to traverse.

52. People borrow *Gone with the Wind* from the local library and keep it for 21 ± 10 days. There is only one copy of the book in the library. You are the sixth person on the reservation list (five are ahead of you). Simulate 50 independent cycles of book borrowing to determine the probability that you will receive the book within 100 days.

53. Jobs arrive every 300 ± 30 seconds to be processed through a process that consists of four operations. OP10 requires 50 ± 20 seconds, OP20 requires 70 ± 25 seconds, OP30 requires 60 ± 15 seconds, and OP40 requires 90 ± 30 seconds. Simulate this process until 250 jobs are completed. Then, combine the four operations of the job into one with the distribution 240 ± 100 seconds and simulate the process with this distribution. Does the average time in the system change for the two alternatives?

54. Jobs arrive every $X \pm 0.5X$ seconds to be processed on a machine that requires 80 ± 80 seconds for processing. Based on a simulation of 500 job completions, determine a value of X so that the utilization of the machine is 0.9.

55. Two types of jobs arrive to be processed on the same machine. Type 1 jobs arrive every 50 ± 30 seconds and require 35 ± 20 seconds for processing. Type 2 jobs arrive every 100 ± 40 seconds and require 20 ± 15 seconds for processing. Based on an 8-hour simulation, what is the average number of jobs waiting to be processed?

56. Two types of jobs arrive to be processed on the same machine. Type 1 jobs arrive every 80 ± 30 seconds and require 35 ± 20 seconds for processing. Type 2 jobs arrive every 100 ± 40 seconds and require 20 ± 15 seconds for processing. Engineering has determined that there is excess capacity on the machine. Based on a simulation of 8 hours of operation of the system, find X for Type 3 jobs that arrive every $X \pm 0.4X$ seconds and require a time of 30 seconds on the machine, so that the average number of jobs waiting to be processed is two or less.

57. Using spreadsheet software, generate 1000 uniformly distributed random values with mean 10 and spread 2. Divide the interval from 8 to 12 into 8 subintervals of width 0.5 and count the number of generated values in each subinterval. Plot these frequencies in a bar chart, thus making a histogram. How close did the simulated set of values come to the expected number in each interval ($1/8 \cdot 1000 = 125$)?

58. Using a spreadsheet, generate 1000 exponentially distributed random values with a mean of 10. What is the maximum of the simulated values? What fraction of the generated values is less than the mean of 10? Plot a histogram of the generated values by counting the frequency of observations in subintervals of width 0.5 over a suitable range. [*Hint:* If you cannot find an exponential generator in the spreadsheet you use, use the formula –10*LOG(1 – R), where R is a uniformly distributed random number from 0 to 1 and LOG is the natural logarithm. The rationale for this formula is explained in Chapter 8 on random-variate generators.]

PART TWO

Mathematical and Statistical Models

5

Statistical Models in Simulation

In modeling real-world phenomena there are few situations where the actions of the entities within the system under study can be completely predicted in advance. The world the model builder sees is probabilistic rather than deterministic. There are many causes of variation. The time it takes a repairperson to fix a broken machine is a function of the complexity of the breakdown, whether the repairperson brought the proper replacement parts and tools to the site, whether another repairperson asks for assistance during the course of the repair, whether the machine operator receives a lesson in preventive maintenance, and so on. To the model builder, these variations appear to occur by chance and cannot be predicted. However, some statistical model may well describe the time to make a repair.

An appropriate model can be developed by sampling the phenomenon of interest. Then, through educated guesses (or using software for the purpose), the model builder would select a known distribution form, make an estimate of the parameter(s) of this distribution, and then test to see how good a fit has been obtained. Through continued efforts in the selection of an appropriate distribution form, a postulated model may be accepted. This multistep process is described in Chapter 9.

Section 5.1 contains a review of probability terminology and concepts. Some typical applications of statistical models, or distribution forms, are given in Section 5.2. Then, a number of selected discrete and continuous distributions are discussed in Sections 5.3 and 5.4. The selected distributions are those which describe a wide variety of probabilistic events and, further, appear in different contexts in other chapters of this text. Additional discussion about the distribution forms appearing in this chapter, and about distribution forms mentioned but not described, is available from a number of sources [Hines and Montgomery, 1990; Ross, 1997; Papoulis, 1990; Devore, 1999; Walpole and Myers, 1998; Fishman, 1973]. Section 5.5 describes the Poisson process and its relationship to the exponential distribution. Section 5.6 discusses empirical distributions.

5.1 Review of Terminology and Concepts

1. *Discrete random variables.* Let X be a random variable. If the number of possible values of X is finite, or countably infinite, X is called a discrete random variable. The possible values of X may be listed as $x_1, x_2, \ldots$. In the finite case the list terminates. In the countably infinite case the list continues indefinitely.

EXAMPLE 5.1

The number of jobs arriving each week at a job shop is observed. The random variable of interest is X, where

$$X = \text{number of jobs arriving each week}$$

The possible values of X are given by the range space of X, which is denoted by R_X. Here $R_X = \{0, 1, 2, \ldots\}$.

Let X be a discrete random variable. With each possible outcome x_i in R_X, a number $p(x_i) = P(X = x_i)$ gives the probability that the random variable equals the value of x_i. The numbers $p(x_i), i = 1, 2, \ldots$, must satisfy the following two conditions:

1. $p(x_i) \geq 0$ for all i.

2. $\sum_{i=1}^{\infty} p(x_i) = 1$.

The collection of pairs $(x_i, p(x_i)), i = 1, 2, \ldots$, is called the *probability distribution* of X, and $p(x_i)$ is called the *probability mass function* (*pmf*) of X. ◄

EXAMPLE 5.2

Consider the experiment of tossing a single die. Define X as the number of spots on the up face of the die after a toss. Then $R_X = \{1, 2, 3, 4, 5, 6\}$. Assume the die is loaded so that the probability that a given face lands up is proportional to the number of spots showing. The discrete probability distribution for this random experiment is given by

x_i	1	2	3	4	5	6
$p(x_i)$	1/21	2/21	3/21	4/21	5/21	6/21

The earlier stated conditions are satisfied. That is, $p(x_i) \geq 0$ for $i = 1, \ldots, 6$ and $\sum_{i=1}^{\infty} p(x_i) = 1/21 + \cdots + 6/21 = 1$. The distribution is shown graphically in Figure 5.1.

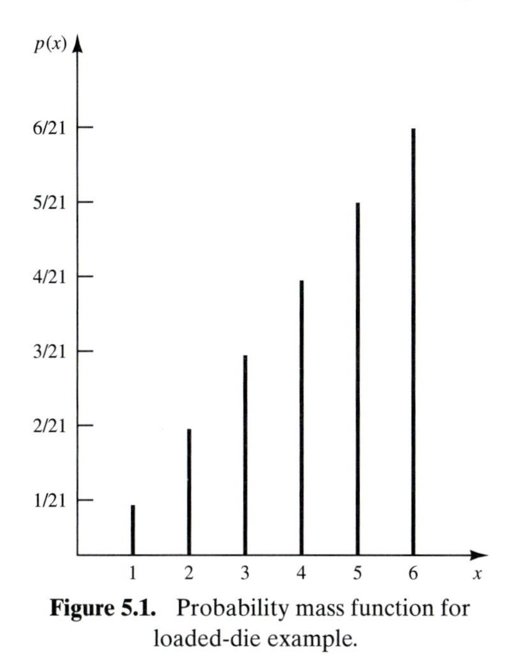

Figure 5.1. Probability mass function for loaded-die example.

2. *Continuous random variables.* If the range space R_X of the random variable X is an interval or a collection of intervals, X is called a continuous random variable. For a continuous random variable X, the probability that X lies in the interval $[a, b]$ is given by

$$P(a \leq X \leq b) = \int_a^b f(x)dx \tag{5.1}$$

The function $f(x)$ is called the *probability density function* (*pdf*) of the random variable X. The pdf satisfies the following conditions:

(a) $f(x) \geq 0$ for all x in R_X.

(b) $\int_{R_X} f(x)dx = 1$.

(c) $f(x) = 0$ if x is not in R_X.

As a result of Equation (5.1), for any specified value x_0, $P(X = x_0) = 0$, since

$$\int_{x_0}^{x_0} f(x)dx = 0$$

Since $P(X = x_0) = 0$, the following equations also hold:

$$P(a \leq X \leq b) = P(a < X \leq b) = P(a \leq X < b) = P(a < X < b) \tag{5.2}$$

The graphical interpretation of Equation (5.1) is shown in Figure 5.2. The shaded area represents the probability that X lies in the interval $[a, b]$. ◄

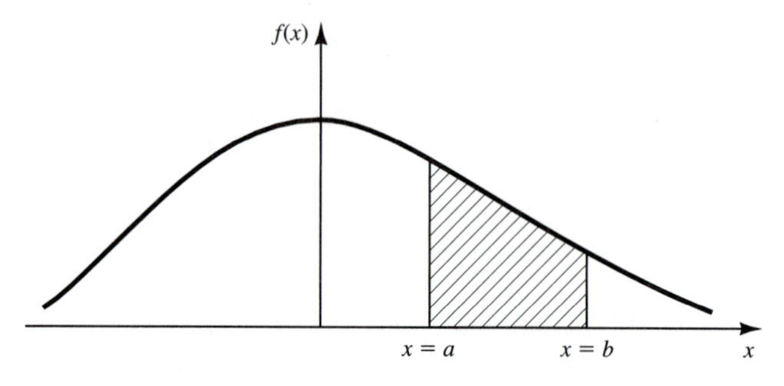

Figure 5.2. Graphical interpretation of $P(a < X < b)$.

EXAMPLE 5.3

The life of a laser-ray device used to inspect cracks in aircraft wings is given by X, a continuous random variable assuming all values in the range $x \geq 0$. The pdf of the lifetime, in years, is as follows:

$$f(x) = \begin{cases} \frac{1}{2}e^{-x/2}, & x \geq 0 \\ 0, & \text{otherwise} \end{cases}$$

This pdf is shown graphically in Figure 5.3. The random variable X is said to have an exponential distribution with mean 2 years.

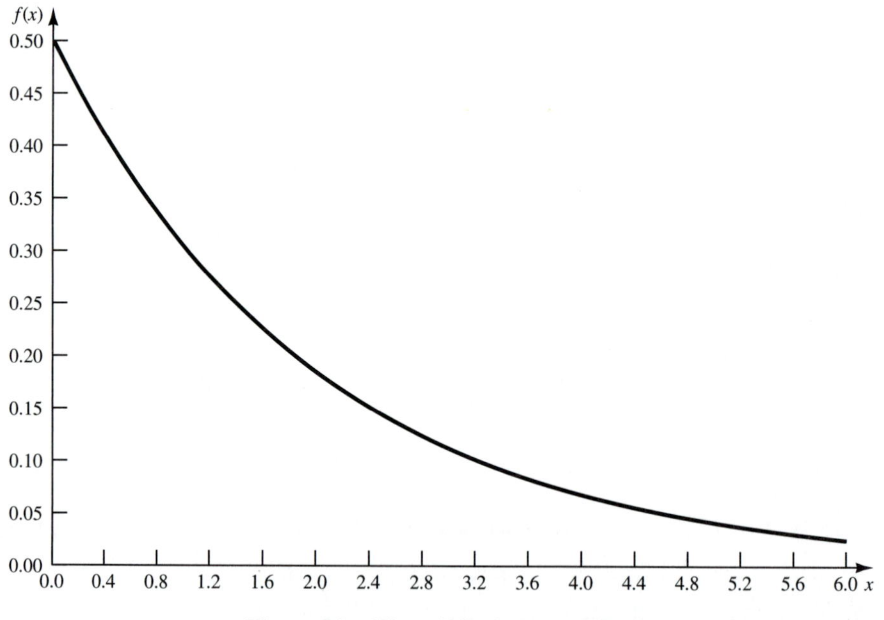

Figure 5.3. The pdf for laser-ray life.

The probability that the life of the laser-ray device is between 2 and 3 years is determined from

$$P(2 \le X \le 3) = \frac{1}{2}\int_2^3 e^{-x/2}dx$$
$$= -e^{-3/2} + e^{-1} = -0.223 + 0.368 = 0.145.$$

3. *Cumulative distribution function.* The cumulative distribution function (cdf), denoted by $F(x)$, measures the probability that the random variable X assumes a value less than or equal to x, that is, $F(x) = P(X \le x)$.

If X is discrete, then

$$F(x) = \sum_{\substack{\text{all} \\ x_i \le x}} p(x_i) \tag{5.3}$$

If X is continuous, then

$$F(x) = \int_{-\infty}^{x} f(t)dt \tag{5.4}$$

Some properties of the cdf are listed here:

(a) F is a nondecreasing function. If $a < b$, then $F(a) \le F(b)$.

(b) $\lim_{x \longrightarrow \infty} F(x) = 1$.

(c) $\lim_{x \longrightarrow -\infty} F(x) = 0$.

All probability questions about X can be answered in terms of the cdf. For example,

$$P(a < X \le b) = F(b) - F(a) \quad \text{for all } a < b \tag{5.5}$$

For continuous distributions not only does Equation (5.5) hold, but also the probabilities in Equation (5.2) are equal to $F(b) - F(a)$. ◀

Example 5.4

The die-tossing experiment described in Example 5.2 has a cdf given as follows:

x	$(-\infty, 1)$	$[1, 2)$	$[2, 3)$	$[3, 4)$	$[4, 5)$	$[5, 6)$	$[6, \infty)$
$F(x)$	0	1/21	3/21	6/21	10/21	15/21	21/21

where $[a, b) = \{a \le x < b\}$. The cdf for this example is shown graphically in Figure 5.4.

If X is a discrete random variable with possible values $x_1, x_2, \ldots$, where $x_1 < x_2 < \ldots$, the cdf is a step function. The value of the cdf is constant in the interval $[x_{i-1}, x_i)$ and then takes a step, or jump, of size $p(x_i)$ at x_i. Thus, in Example 5.2, $p(3) = 3/21$, which is the size of the step when $x = 3$. ◀

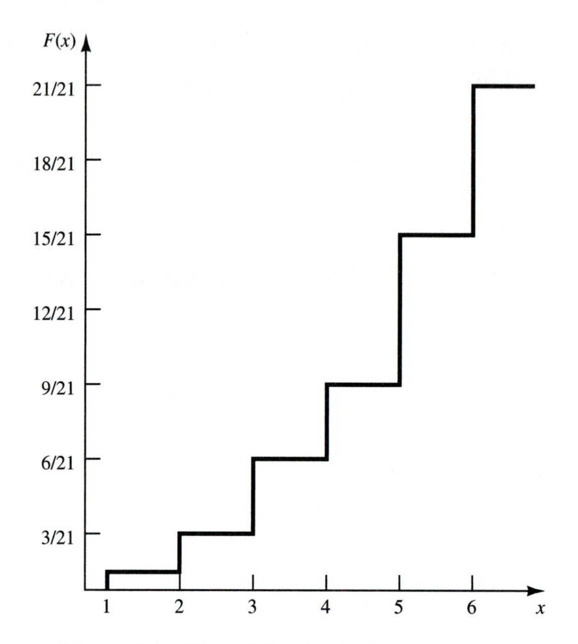

Figure 5.4. The cdf for loaded-die example.

EXAMPLE 5.5

The cdf for the laser-ray device described in Example 5.3 is given by

$$F(x) = \frac{1}{2}\int_0^x e^{-t/2}dt = 1 - e^{-x/2}$$

The probability that the laser-ray device will last for less than 2 years is given by

$$P(0 \leq X \leq 2) = F(2) - F(0) = F(2) = 1 - e^{-1} = 0.632$$

The probability that the life of the laser-ray device is between 2 and 3 years is determined from

$$P(2 \leq X \leq 3) = F(3) - F(2) = (1 - e^{-3/2}) - (1 - e^{-1})$$
$$= -e^{-3/2} + e^{-1} = -0.223 + 0.368 = 0.145$$

as found in Example 5.3.

4. *Expectation.* An important concept in probability theory is that of the expectation of a random variable. If X is a random variable, the expected value of X, denoted by $E(X)$, for discrete and continuous variables is defined as follows:

$$E(X) = \sum_{\text{all } i} x_i p(x_i) \qquad \text{if } X \text{ is discrete} \tag{5.6}$$

and

$$E(X) = \int_{-\infty}^{\infty} xf(x)dx \qquad \text{if } X \text{ is continuous} \tag{5.7}$$

The expected value $E(X)$ of a random variable X is also referred to as the mean, μ, or the first moment of X. The quantity $E(X^n), n \geq 1$, is called the nth moment of X and is computed as follows:

$$E(X^n) = \sum_{\text{all i}} x_i^n p(x_i) \qquad \text{if } X \text{ is discrete} \tag{5.8}$$

and

$$E(X^n) = \int_{-\infty}^{\infty} x^n f(x)dx \qquad \text{if } X \text{ is continuous} \tag{5.9}$$

The variance of a random variable, X, denoted by $V(X)$ or $\text{var}(X)$ or σ^2, is defined by

$$V(X) = E[(X - E[X])^2]$$

A useful identity in computing $V(X)$ is given by

$$V(X) = E(X^2) - [E(X)]^2 \tag{5.10}$$

The mean $E(X)$ is a measure of the central tendency of a random variable. The variance of X measures the expected value of the squared difference between the random variable and its mean. Thus, the variance, $V(X)$, is a measure of the spread or variation of the possible values of X around the mean $E(X)$. The standard deviation, σ, is defined to be the square root of the variance, σ^2. The mean, $E(X)$, and the standard deviation, $\sigma = \sqrt{V(X)}$, are expressed in the same units. ◀

EXAMPLE 5.6

The mean and variance of the die-tossing experiment described in Example 5.2 are determined as follows:

$$E(X) = 1\left(\frac{1}{21}\right) + 2\left(\frac{2}{21}\right) + \cdots + 6\left(\frac{6}{21}\right) = \frac{91}{21} = 4.33$$

To compute $V(X)$ using Equation (5.10), first compute $E(X^2)$ from Equation (5.8) as follows:

$$E(X^2) = 1^2\left(\frac{1}{21}\right) + 2^2\left(\frac{2}{21}\right) + \cdots + 6^2\left(\frac{6}{21}\right) = 21$$

Thus,

$$V(X) = 21 - \left(\frac{91}{21}\right)^2 = 21 - 18.78 = 2.22$$

and

$$\sigma = \sqrt{V(X)} = 1.49$$

◀

EXAMPLE 5.7

The mean and variance of the life of the laser-ray device described in Example 5.3 are determined as follows:

$$E(X) = \frac{1}{2}\int_0^\infty xe^{-x/2}dx = -xe^{-x/2}\Big|_0^\infty + \int_0^\infty e^{-x/2}dx$$

$$= 0 + \frac{1}{1/2}e^{-x/2}\Big|_0^\infty = 2 \text{ years}$$

To compute $V(X)$ using Equation (5.10), first compute $E(X^2)$ from Equation (5.9) as follows:

$$E(X^2) = \frac{1}{2}\int_0^\infty x^2e^{-x/2}dx$$

Thus,

$$E(X^2) = -x^2e^{-x/2}|_0^\infty + 2\int_0^\infty xe^{-x/2}dx = 8$$

Thus,

$$V(X) = 8 - 2^2 = 4 \text{ years}^2$$

and

$$\sigma = \sqrt{V(X)} = 2 \text{ years}$$

With a mean life of 2 years and a standard deviation of 2 years, most analysts would conclude that actual lifetimes, X, have a fairly large variability.

5. *The mode.* The mode is used in describing several statistical models which appear in this chapter. In the discrete case, the mode is the value of the random variable that occurs most frequently. In the continuous case, the mode is the value at which the pdf is maximized. The mode may not be unique; if the modal value occurs at two values of the random variable, the distribution is said to be bimodal. ◀

5.2 Useful Statistical Models

Numerous situations arise in the conduct of a simulation where an investigator may choose to introduce probabilistic events. In Chapter 2, queueing, inventory, and reliability examples were given. In a queueing system, interarrival and service times are often probabilistic. In an inventory model, the time between demands and the lead times (time between placing and receiving an order) may be probabilistic. In a reliability model, the time to failure may be probabilistic. In each of these instances the simulation analyst desires to generate random events and to use a known statistical model if the underlying distribution can be found. In the following paragraphs statistical models appropriate to these application areas will be discussed. Additionally, statistical models useful in the case of limited data are mentioned.

1. *Queueing systems.* In Chapter 2, examples of waiting-line problems were given. In Chapters 2, 3, and 4 these problems were solved using simulation. In the queueing examples, interarrival- and service-time patterns were given. The times between arrivals and the service times were always probabilistic, which is usually the case. However, it is possible to have a constant interarrival time (as in the case of a line moving at a constant speed in the assembly of an automobile), or a constant service time (as in the case of robotized spot welding on the same assembly line). The following example illustrates how probabilistic interarrival times might occur.

Example 5.8

Mechanics arrive at a centralized tool crib as shown in Table 5.1. Attendants check in and check out the requested tools to the mechanics. The collection of data begins at 10:00 A.M. and continues until 20 different interarrival times are recorded. Rather than record the actual time of day, the absolute time from a given origin could have been computed. Thus, the first mechanic could have arrived at time zero, the second mechanic at time 7:13 (7 minutes, 13 seconds), and so on. ◄

Table 5.1. Arrival Data

Arrival Number	*Arrival (Hour:Minutes::Seconds)*	*Interarrival Time (Minutes::Seconds)*
1	10:05::03	—
2	10:12::16	7::13
3	10:15::48	3::32
4	10:24::27	8::39
5	10:32::19	7::52
6	10:35::43	3::24
7	10:39::51	4::08
8	10:40::30	0::39
9	10:41::17	0::47
10	10:44::12	2::55
11	10:45::47	1::35
12	10:50::47	5::00
13	11:00::05	9::18
14	11:04::58	4::53
15	11:06::12	1::14
16	11:11::23	5::11
17	11:16::31	5::08
18	11:17::18	0::47
19	11:21::26	4::08
20	11:24::43	3::17
21	11:31::19	6::36

EXAMPLE 5.9

Another way of presenting interarrival data is to determine the number of arrivals per time period. Since these arrivals occur over approximately $1\frac{1}{2}$ hours, it is convenient to look at 10-minute time intervals for the first 20 mechanics. That is, in the first 10-minute time period, one arrival occurred at 10:05::03. In the second time period, two mechanics arrived, and so on. The results are summarized in Table 5.2. This data could then be plotted in a histogram as shown in Figure 5.5.

Table 5.2. Arrivals in Successive Time Periods

Time Period	*Number of Arrivals*	*Time Period*	*Number of Arrivals*
1	1	6	1
2	2	7	3
3	1	8	3
4	3	9	2
5	4	—	—

The distribution of time between arrivals and the distribution of the number of arrivals per time period are important in the simulation of waiting-line systems. "Arrivals" occur in numerous ways: as machine breakdowns, as jobs coming into a jobshop, as units being assembled on a line, as orders to a warehouse, and so on.

Service times may be constant or probabilistic. If service times are completely random, the exponential distribution is often used for simulation purposes. However, there are several other possibilities. It may be possible that the service times are constant, but some random variability causes fluctuations in either a positive or negative way. For example, the time it takes for a lathe

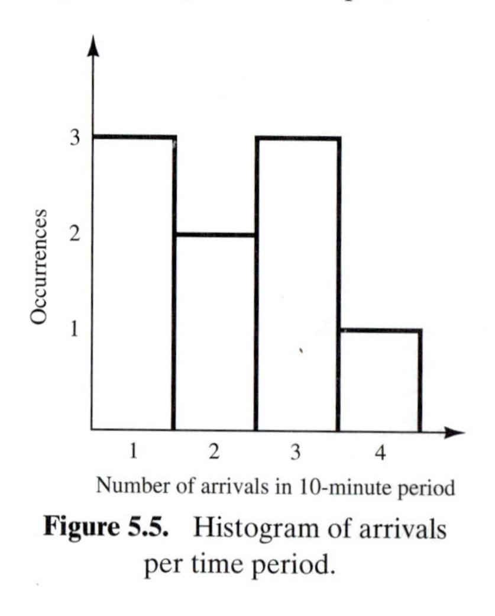

Figure 5.5. Histogram of arrivals per time period.

to traverse a 10-centimeter shaft should always be the same. However, the material may have slight differences in hardness or the tool may wear, causing different processing times. In these cases the normal distribution may describe the service time.

A special case occurs when the phenomenon of interest seems to follow the normal probability distribution, but the random variable is restricted to be greater than or less than a certain value. In this case, the truncated normal distribution can be utilized.

The gamma and Weibull distributions are also used to model interarrival and service times. (Actually, the exponential distribution is a special case of both the gamma and the Weibull distributions.) The differences between the exponential, gamma, and Weibull distributions involve the location of the modes of the pdf's and the shapes of their tails for large and small times [Fishman, 1973]. The exponential distribution has its mode at the origin, but the gamma and Weibull distributions have their modes at some point (≥ 0) which is a function of the parameter values selected. The tail of the gamma distribution is long, like an exponential distribution, while the tail of the Weibull distribution may decline more rapidly or less rapidly than that of an exponential distribution. In practice, this means that if there are more large service times than an exponential distribution can account for, a Weibull distribution may provide a better model of these service times.

2. *Inventory systems.* In realistic inventory systems there are three random variables: (1) the number of units demanded per order or per time period, (2) the time between demands, and (3) the lead time. (The lead time is defined as the time between placing an order for stocking the inventory system and the receipt of that order.) In very simple mathematical models of inventory systems demand is a constant over time, and lead time is zero, or a constant. However, in most realistic cases, and hence, in simulation models, demand occurs randomly in time, and the number of units demanded each time a demand occurs is also random, as illustrated by Figure 5.6.

Distributional assumptions for demand and lead time in inventory-theory texts are usually based on mathematical tractability, but those assumptions may not be valid in a realistic context. In practice, the lead-time distribution can often be fitted fairly well by a gamma distribution [Hadley and Whitin, 1963]. Unlike analytic models, simulation models can accommodate whatever assumptions appear most reasonable.

The geometric, Poisson, and negative binomial distributions provide a range of distribution shapes that satisfy a variety of demand patterns [Fishman, 1973]. The geometric distribution, which is a special case of the negative binomial, has its mode at unity, given that at least one demand has occurred. If demand data are characterized by a long tail, the negative binomial distribution may be appropriate. The Poisson distribution is often used to model demand because it is simple, it is extensively tabulated, and it is well known. The tail of the Poisson distribution is generally shorter than that of the negative binomial, which means that fewer large demands will occur if a Poisson model is used than

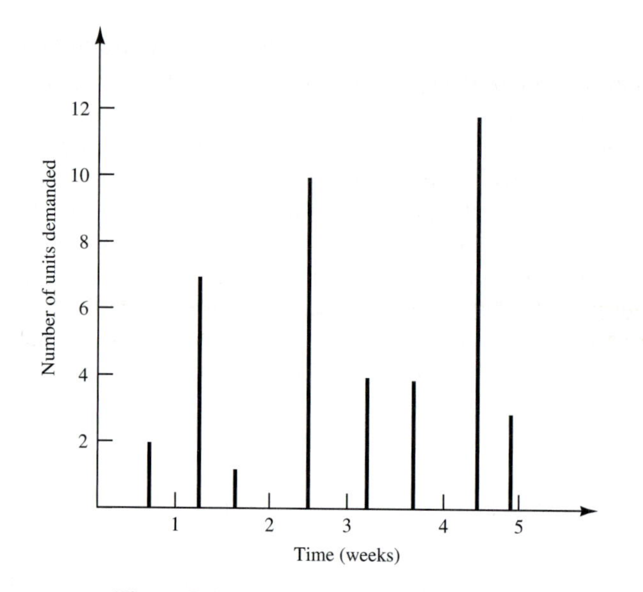

Figure 5.6. Random demands in time.

if a negative binomial distribution is used (assuming that both models have the same mean demand).

3. *Reliability and maintainability.* Time to failure has been modeled with numerous distributions, including the exponential, gamma and Weilbull. If only random failures occur, the time-to-failure distribution may be modeled as exponential. The gamma distribution arises from modeling standby redundancy where each component has an exponential time to failure. The Weibull distribution has been extensively used to represent time to failure, and its nature is such that it may be made to approximate many observed phenomena [Hines and Montgomery, 1990]. When there are a number of components in a system and failure is due to the most serious of a large number of defects, or possible defects, the Weibull distribution seems to do particularly well as a model. In situations where most failures are due to wear, the normal distribution may very well be appropriate [Hines and Montgomery, 1990]. The lognormal distribution has been found to be applicable in describing time to failure for some types of components, and the literature seems to indicate increased use of this distribution in reliability models.

4. *Limited data.* In many instances simulations begin before data collection has been completed. Three distributions have application to incomplete or limited data. These are the uniform, triangular, and beta distributions. The *uniform distribution* can be used when an interarrival or service time is known to be random, but no information is immediately available about the distribution [Gordon, 1975]. However, there are those who do not favor using the uniform distribution, calling it the "distribution of maximum ignorance." It is only necessary to specify the continuous interval in which the random variable

may occur. The *triangular distribution* can be used when assumptions are made about the minimum, maximum, and modal values of the random variable. Finally, the *beta distribution* provides a variety of distributional forms on the unit interval, which, with appropriate modification, can be shifted to any desired interval. The uniform distribution is a special case of the beta distribution. Pegden, Shannon, and Sadowski [1995] discuss the subject of limited data in some detail, and we include further discussion in Chapter 9.

5. *Other distributions.* Several other distributions may be useful in discrete system simulation. The Bernoulli and binomial distributions are two discrete distributions which may describe phenomena of interest. The hyperexponential distribution is similar to the exponential distribution, but its greater variability may make it useful in certain instances. ◀

5.3 Discrete Distributions

Discrete random variables are used to describe random phenomena in which only integer values can occur. Numerous examples were given in Section 5.2 — for example, demands for inventory items. Four distributions are described in the following subsections.

1. *Bernoulli trials and the Bernoulli distribution.* Consider an experiment consisting of n trials, each of which can be a success or a failure. Let $X_j = 1$ if the jth experiment resulted in a success, and let $X_j = 0$ if the jth experiment resulted in a failure. The n Bernoulli trials are called a Bernoulli process if the trials are independent, each trial has only two possible outcomes (success or failure), and the probability of a success remains constant from trial to trial. Thus,

$$p(x_1, x_2, \ldots, x_n) = p_1(x_1) \cdot p_2(x_2) \cdots p_n(x_n)$$

and

$$p_j(x_j) = p(x_j) = \begin{cases} p, & x_j = 1, j = 1, 2, \ldots, n \\ 1 - p = q, & x_j = 0, j = 1, 2, \ldots, n \\ 0, & \text{otherwise} \end{cases} \tag{5.11}$$

For one trial, the distribution given in Equation (5.11) is called the Bernoulli distribution. The mean and variance of X_j are calculated as follows:

$$E(X_j) = 0 \cdot q + 1 \cdot p = p$$

and

$$V(X_j) = [(0^2 \cdot q) + (1^2 \cdot p)] - p^2 = p(1 - p)$$

2. *Binomial distribution.* The random variable X that denotes the number of successes in n Bernoulli trials has a binomial distribution given by $p(x)$, where

$$p(x) = \begin{cases} \binom{n}{x} p^x q^{n-x}, & x = 0, 1, 2, \ldots, n \\ 0, & \text{otherwise} \end{cases} \tag{5.12}$$

Equation (5.12) is motivated by determining the probability of a particular outcome with all the successes, each denoted by S, occurring in the first x trials, followed by the $n - x$ failures, each denoted by an F. That is,

$$P(\overbrace{SSS \ldots\ldots\ldots SS}^{x \text{ of these}}\ \overbrace{FF \ldots\ldots\ldots FF}^{n-x \text{ of these}}) = p^x q^{n-x}$$

where $q = 1 - p$. There are

$$\binom{n}{x} = \frac{n!}{x!(n-x)!}$$

outcomes having the required number of S's and F's. Therefore, Equation (5.12) results. An easy approach to determining the mean and variance of the binomial distribution is to consider X as a sum of n independent Bernoulli random variables, each with mean p and variance $p(1-p) = pq$. Then,

$$X = X_1 + X_2 + \cdots + X_n$$

and the mean, $E(X)$, is given by

$$E(X) = p + p + \cdots + p = np \tag{5.13}$$

and the variance, $V(X)$, is given by

$$V(X) = pq + pq + \cdots + pq = npq \tag{5.14}$$

EXAMPLE 5.10

A production process manufactures computer chips on the average at 2% nonconforming. Every day a random sample of size 50 is taken from the process. If the sample contains more than two nonconforming chips, the process will be stopped. Determine the probability that the process is stopped by the sampling scheme.

Considering the sampling process as $n = 50$ Bernoulli trials, each with $p = 0.02$, the total number of nonconforming chips in the sample, X, would have a binomial distribution given by

$$p(x) = \begin{cases} \binom{50}{x} (0.02)^x (0.98)^{50-x}, & x = 0, 1, 2, \ldots, 50 \\ 0, & \text{otherwise} \end{cases}$$

It is much easier to compute the right-hand side of the following identity to determine the probability that more than two nonconforming chips are found in a sample:

$$P(X > 2) = 1 - P(X \leq 2)$$

The probability $P(X \leq 2)$ is calculated from

$$
\begin{aligned}
P(X \leq 2) &= \sum_{x=0}^{2} \binom{50}{x} (0.02)^x (0.98)^{50-x} \\
&= (0.98)^{50} + 50(0.02)(0.98)^{49} + 1225(0.02)^2(0.98)^{48} \\
&\doteq 0.92
\end{aligned}
$$

Thus, the probability that the production process is stopped on any day, based on the sampling process, is approximately 0.08. The mean number of nonconforming chips in a random sample of size 50 is given by

$$E(X) = np = 50(0.02) = 1$$

and the variance is given by

$$V(X) = npq = 50(0.02)(0.98) = 0.98$$

The cdf for the binomial distribution has been tabulated by Banks and Heikes [1984] and others. The tables decrease the effort considerably for computing probabilities such as $P(a < X \leq b)$. Under certain conditions on n and p, both the Poisson distribution and the normal distribution may be used to approximate the binomial distribution [Hines and Montgomery, 1990].

3. *Geometric distribution.* The geometric distribution is related to a sequence of Bernoulli trials; the random variable of interest, X, is defined to be the number of trials to achieve the first success. The distribution of X is given by

$$p(x) = \begin{cases} q^{x-1}p, & x = 1, 2, \ldots \\ 0, & \text{otherwise} \end{cases} \tag{5.15}$$

The event $\{X = x\}$ occurs when there are $x - 1$ failures followed by a success. Each of the failures has an associated probability of $q = 1 - p$, and each success has probability p. Thus,

$$P(FFF \cdots FS) = q^{x-1}p$$

The mean and variance are given by

$$E(X) = \frac{1}{p} \tag{5.16}$$

and

$$V(X) = \frac{q}{p^2} \tag{5.17}$$

◀

EXAMPLE 5.11

Forty percent of the assembled bubble-jet printers are rejected at the inspection station. Find the probability that the first acceptable bubble-jet printer is the third one inspected. Considering each inspection as a Bernoulli trial with $q = 0.4$ and $p = 0.6$,

$$p(3) = 0.4^2(0.6) = 0.096$$

Thus, in only about 10% of the cases is the first acceptable printer the third one from any arbitrary starting point.

4. *Poisson distribution.* The Poisson distribution describes many random processes quite well and is mathematically quite simple. It was introduced in 1837 by S. D. Poisson in a book concerning criminal and civil justice matters. The title of this rather old text is *Recherches sur la probabilite des jugements en matiere criminelle et en matiere civile.* Evidently, the rumor handed down through generations of probability-theory professors concerning the origin of the Poisson distribution is just not true. Rumor has it that the Poisson distribution was first used to model deaths from the kicks of horses in the Prussian Army.

The Poisson probability mass function is given by

$$p(x) = \begin{cases} \frac{e^{-\alpha}\alpha^x}{x!}, & x = 0, 1, \ldots \\ 0, & \text{otherwise} \end{cases} \tag{5.18}$$

where $\alpha > 0$. One of the important properties of the Poisson distribution is that the mean and variance are both equal to α — that is,

$$E(X) = \alpha = V(X)$$

The cumulative distribution function is given by

$$F(x) = \sum_{i=0}^{x} \frac{e^{-\alpha}\alpha^i}{i!} \tag{5.19}$$

The pmf and cdf for a Poisson distribution with $\alpha = 2$ are shown in Figure 5.7. A tabulation of the cdf is given in Table A.4. ◀

EXAMPLE 5.12

A computer-terminal repairperson is "beeped" each time there is a call for service. The number of beeps per hour is known to occur in accordance with a Poisson distribution with a mean of $\alpha = 2$ per hour. The probability of three beeps in the next hour is given by Equation (5.18) with $x = 3$, as follows:

$$p(3) = \frac{e^{-2}2^3}{3!} = \frac{(0.135)(8)}{6} = 0.18$$

This same result can be read from the left side of Figure 5.7 or from Table A.4 by computing

$$F(3) - F(2) = 0.857 - 0.677 = 0.18$$ ◀

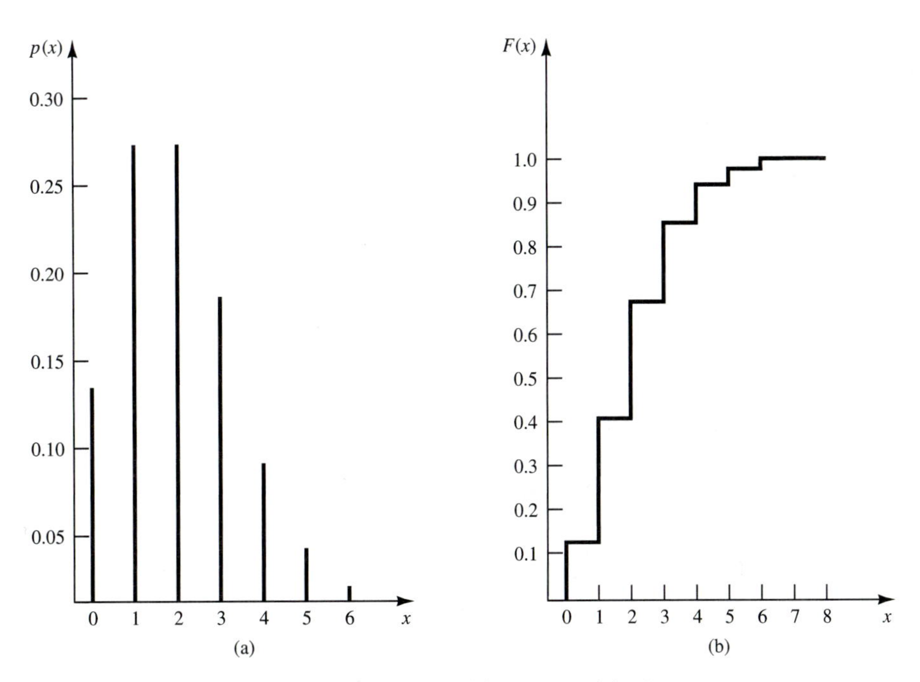

Figure 5.7. Poisson (a) pmf and (b) cdf.

Example 5.13

In Example 5.12 determine the probability of two or more beeps in a 1-hour period.

$$P(2 \text{ or more}) = 1 - p(0) - p(1) = 1 - F(1)$$
$$= 1 - 0.406 = 0.594$$

The cumulative probability, $F(1)$, can be read from the right side of Figure 5.7 or from Table A.4. ◀

Example 5.14

The lead-time demand in an inventory system is the accumulation of demand for an item from the point at which an order is placed until the order is received. That is,

$$L = \sum_{i=1}^{T} D_i \tag{5.20}$$

where L is the lead-time demand, D_i is the demand during the ith time period, and T is the number of time periods during the lead time. Both D_i and T may be random variables.

An inventory manager desires that the probability of a stockout not exceed a certain fraction during the lead time. For example, it may be stated that the probability of a shortage during the lead time not exceed 5%.

If the lead-time demand is Poisson distributed, the determination of the reorder point is greatly facilitated. The reorder point is the level of inventory at which a new order is placed.

Assume that the lead-time demand is Poisson distributed with a mean of $\alpha = 10$ units and that 95% protection from a stockout is desired. Thus, it is desired to find the smallest value of x such that the probability that the lead-time demand does not exceed x is greater than or equal to 0.95. Using Equation (5.19) requires finding the smallest x such that

$$F(x) = \sum_{i=0}^{x} \frac{e^{-10}10^i}{i!} \geq 0.95$$

The desired result occurs at $x = 15$, which can be found by using Table A.4, or by computation of $p(0), p(1), \ldots$. ◀

5.4 Continuous Distributions

Continuous random variables can be used to describe random phenomena in which the variable of interest can take on any value in some interval — for example, the time to failure or the length of a rod. Eight distributions are described in the following subsections.

1. *Uniform distribution.* A random variable X is uniformly distributed on the interval (a, b) if its pdf is given by

$$f(x) = \begin{cases} \frac{1}{b-a}, & a \leq x \leq b \\ 0, & \text{otherwise} \end{cases} \tag{5.21}$$

The cdf is given by

$$F(x) = \begin{cases} 0, & x < a \\ \frac{x-a}{b-a}, & a \leq x < b \\ 1, & x \geq b \end{cases} \tag{5.22}$$

Note that

$$P(x_1 < X < x_2) = F(x_2) - F(x_1) = \frac{x_2 - x_1}{b - a}$$

is proportional to the length of the interval, for all x_1, and x_2 satisfying $a \leq x_1 < x_2 \leq b$. The mean and variance of the distribution are given by

$$E(X) = \frac{a + b}{2} \tag{5.23}$$

and

$$V(X) = \frac{(b - a)^2}{12} \tag{5.24}$$

The pdf and cdf when $a = 1$ and $b = 6$ are shown in Figure 5.8.

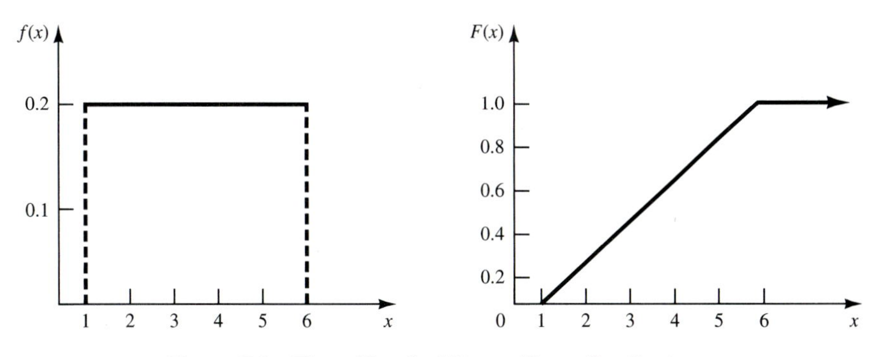

Figure 5.8. The pdf and cdf for uniform distribution.

The uniform distribution plays a vital role in simulation. Random numbers, uniformly distributed between zero and 1, provide the means to generate random events. Numerous methods for generating uniformly distributed random numbers have been devised, two of which are discussed in Chapter 7. Uniformly distributed random numbers are then used to generate samples of random variates from all other distributions, as discussed in Chapter 8.

Example 5.15

A simulation of a warehouse operation is being developed. About every 3 minutes, a call comes for a forklift truck operator to proceed to a certain location. An initial assumption is made that the time between calls (arrivals) is uniformly distributed with a mean of 3 minutes. By Equation (5.24), the uniform distribution with a mean of 3 and the greatest possible variability would have parameter values of $a = 0$ and $b = 6$ minutes. With very limited data (such as a mean of approximately 3 minutes) plus the knowledge that the quantity of interest is variable in a random fashion, the uniform distribution with greatest variance can be assumed, at least until more data are available. ◀

Example 5.16

A bus arrives every 20 minutes at a specified stop beginning at 6:40 A.M. and continuing until 8:40 A.M. A certain passenger does not know the schedule, but arrives randomly (uniformly distributed) between 7:00 A.M. and 7:30 A.M. every morning. What is the probability that the passenger waits more than 5 minutes for a bus?

The passenger has to wait more than 5 minutes only if her arrival time is between 7:00 A.M. and 7:15 A.M. or between 7:20 A.M. and 7:30 A.M. If X is a random variable that denotes the number of minutes past 7:00 A.M. that the passenger arrives, the desired probability is

$$P(0 < X < 15) + P(20 < X < 30)$$

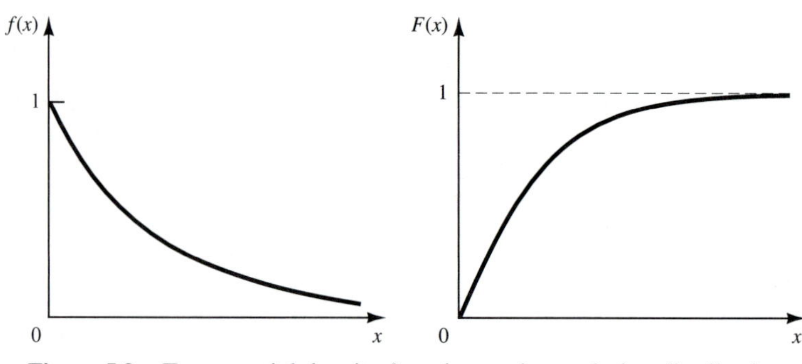

Figure 5.9. Exponential density function and cumulative distribution function.

Now, X is a uniform random variable on $(0, 30)$. Therefore, the desired probability is given by

$$F(15) + F(30) - F(20) = \frac{15}{30} + 1 - \frac{20}{30} = \frac{5}{6}$$

2. *Exponential distribution*. A random variable X is said to be exponentially distributed with parameter $\lambda > 0$ if its pdf is given by

$$f(x) = \begin{cases} \lambda e^{-\lambda x}, & x \geq 0 \\ 0, & \text{elsewhere} \end{cases} \tag{5.25}$$

The density function is shown in Figures 5.9 and 5.3. Figure 5.9 also shows the cdf.

The exponential distribution has been used to model interarrival times when arrivals are completely random and to model service times which are highly variable. In these instances, λ is a rate: arrivals per hour or services per minute. The exponential distribution has also been used to model the lifetime of a component that fails catastrophically (instantaneously), such as a light bulb. Then, λ is the failure rate.

Several different exponential pdf's are shown in Figure 5.10. The value of the intercept on the vertical axis is always equal to the value of λ. Note also that all pdf's eventually intersect. (Why?)

The exponential distribution has mean and variance given by

$$E(X) = \frac{1}{\lambda} \quad \text{and} \quad V(X) = \frac{1}{\lambda^2} \tag{5.26}$$

Thus, the mean and standard deviation are equal. The cdf can be determined by integrating Equation (5.25) to obtain

$$F(x) = \begin{cases} 0, & x < 0 \\ \int_0^x \lambda e^{-\lambda t}\,dt = 1 - e^{-\lambda x}, & x \geq 0 \end{cases} \tag{5.27}$$

◀

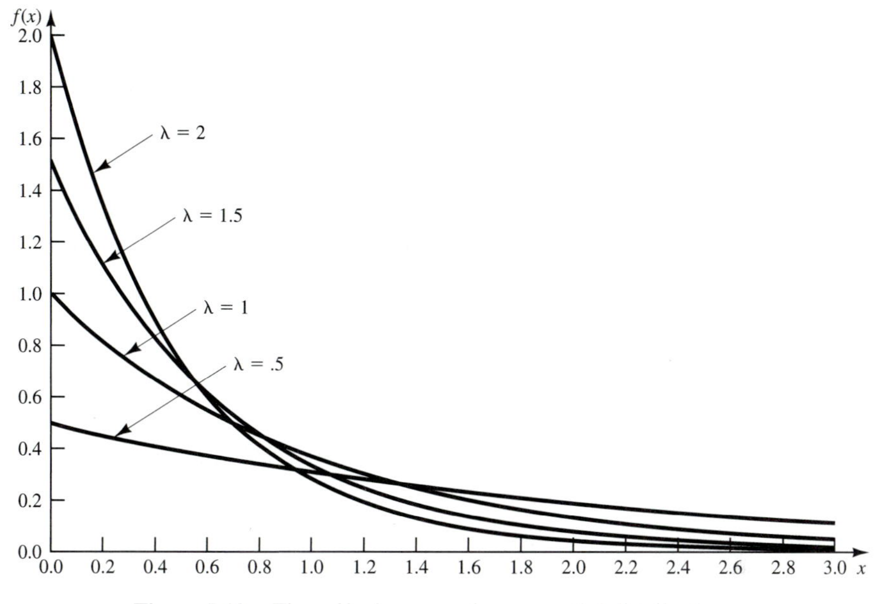

Figure 5.10. The pdf's for several exponential distributions.

EXAMPLE 5.17

Suppose that the life of an industrial lamp, in thousands of hours, is exponentially distributed with failure rate $\lambda = 1/3$ (one failure every 3000 hours, on the average). The probability that the lamp will last longer than its mean life of 3000 hours is given by $P(X > 3) = 1 - P(X \leq 3) = 1 - F(3)$. Equation (5.27) is used to compute $F(3)$, obtaining

$$P(X > 3) = 1 - (1 - e^{-3/3}) = e^{-1} = 0.368$$

Regardless of the value of λ, this result will always be the same! That is, the probability that the random variable is greater than its mean is 0.368, for any value of λ.

The probability that the industrial lamp will last between 2000 and 3000 hours is determined by

$$P(2 \leq X \leq 3) = F(3) - F(2)$$

Again, using the cdf given by Equation (5.27),

$$F(3) - F(2) = (1 - e^{-3/3}) - (1 - e^{-2/3})$$
$$= -0.368 + 0.513 = 0.145$$

One of the most important properties of the exponential distribution is that it is "memoryless," which means that for all $s \geq 0$ and $t \geq 0$,

$$P(X > s + t | X > s) = P(X > t) \tag{5.28}$$

Let X represent the life of a component (a battery, light bulb, computer chip, laser ray, etc.) and assume that X is exponentially distributed. Equation (5.28) states that the probability that the component lives for at least $s + t$ hours, given that it has survived s hours, is the same as the initial probability that it lives for at least t hours. If the component is alive at time s (if $X > s$), then the distribution of the remaining amount of time that it survives, namely $X - s$, is the same as the original distribution of a new component. That is, the component does not "remember" that it has already been in use for a time s. A used component is as good as new.

That Equation (5.28) holds is shown by examining the conditional probability

$$P(X > s + t|X > s) = \frac{P(X > s + t)}{P(X > s)}. \tag{5.29}$$

Equation (5.27) can be used to determine the numerator and denominator of Equation (5.29), yielding

$$\begin{aligned} P(X > s + t|X > s) &= \frac{e^{-\lambda(s+t)}}{e^{-\lambda s}} = e^{-\lambda t} \\ &= P(X > t) \end{aligned}$$

◀

EXAMPLE 5.18

Find the probability that the industrial lamp in Example 5.17 will last for another 1000 hours, given that it is operating after 2500 hours. This determination can be found using Equations (5.28) and (5.27), as follows:

$$P(X > 3.5|X > 2.5) = P(X > 1) = e^{-1/3} = 0.717$$

Example 5.18 illustrates the memoryless property, namely that a used component which follows an exponential distribution is as good as a new component. The probability that a new component will have a life greater than 1000 hours is also equal to 0.717. Stated in general, suppose that a component which has a lifetime that follows the exponential distribution with parameter λ is observed and found to be operating at an arbitrary time. Then, the distribution of the remaining lifetime is also exponential with parameter λ. The exponential distribution is the only continuous distribution which has the memoryless property. (The geometric distribution is the only discrete distribution that possesses the memoryless property.)

3. *Gamma distribution.* A function used in defining the gamma distribution is the gamma function, which is defined for all $\beta > 0$ as

$$\Gamma(\beta) = \int_0^\infty x^{\beta-1} e^{-x} dx \tag{5.30}$$

By integrating Equation (5.30) by parts, it can be shown that

$$\Gamma(\beta) = (\beta - 1)\Gamma(\beta - 1) \tag{5.31}$$

If β is an integer, then using $\Gamma(1) = 1$ and Equation (5.31) it can be seen that

$$\Gamma(\beta) = (\beta - 1)! \tag{5.32}$$

The gamma function can be thought of as a generalization of the factorial notion which applies to all positive numbers, not just integers.

A random variable X is gamma distributed with parameters β and θ if its pdf is given by

$$f(x) = \begin{cases} \frac{\beta\theta}{\Gamma(\beta)}(\beta\theta x)^{\beta-1}e^{-\beta\theta x}, & x > 0 \\ 0, & \text{otherwise} \end{cases} \tag{5.33}$$

The parameter β is called the *shape parameter* and θ is called the *scale parameter*. Several gamma distributions for $\theta = 1$ and various values of β are shown in Figure 5.11.

The mean and variance of the gamma distribution are given by

$$E(X) = \frac{1}{\theta} \tag{5.34}$$

and

$$V(X) = \frac{1}{\beta\theta^2} \tag{5.35}$$

The cdf of X is given by

$$F(x) = \begin{cases} 1 - \int_x^\infty \frac{\beta\theta}{\Gamma(\beta)}(\beta\theta t)^{\beta-1}e^{-\beta\theta t}\,dt, & x > 0 \\ 0, & x \le 0 \end{cases} \tag{5.36}$$

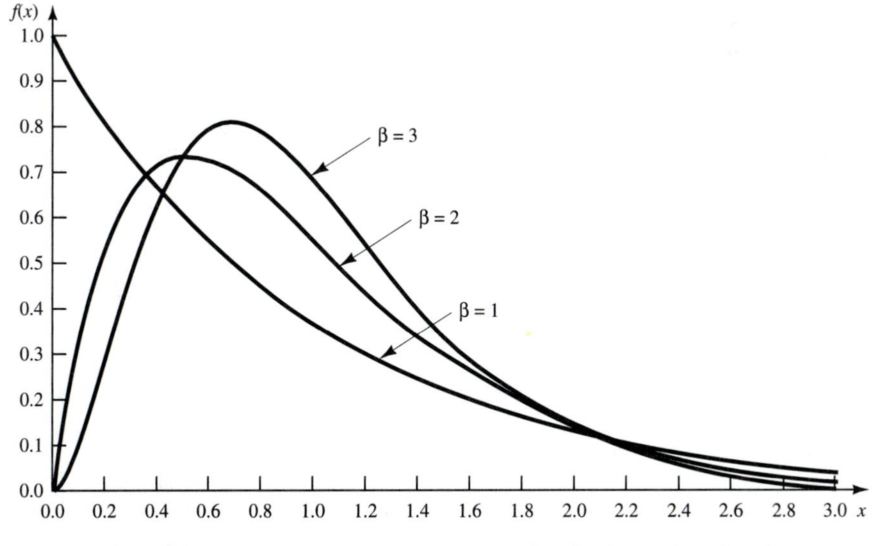

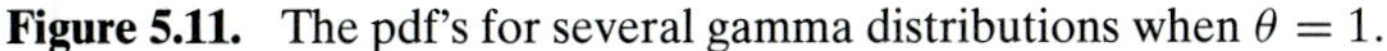

Figure 5.11. The pdf's for several gamma distributions when $\theta = 1$.

When β is an integer, the gamma distribution is related to the exponential distribution in the following manner: If the random variable, X, is the sum of β independent, exponentially distributed random variables, each with parameter $\beta\theta$, then X has a gamma distribution with parameters β and θ. Thus, if

$$X = X_1 + X_2 + \cdots + X_\beta \tag{5.37}$$

where the pdf of X_j is given by

$$g(x_j) = \begin{cases} (\beta\theta)e^{-\beta\theta x_j}, & x \geq 0 \\ 0, & \text{otherwise} \end{cases}$$

and the X_j are mutually independent, X has the pdf given in Equation (5.33). Note that when $\beta = 1$, an exponential distribution results. This result follows from Equation (5.37) or from letting $\beta = 1$ in Equation (5.33).

4. *Erlang distribution.* The pdf given by Equation (5.33) is often referred to as the Erlang distribution of order k when $\beta = k$, an integer. Erlang was a Danish telephone engineer who was an early developer of queueing theory. The Erlang distribution could arise in the following context: Consider a series of k stations that must be passed through in order to complete the servicing of a customer. An additional customer cannot enter the first station until the customer in process has negotiated all the stations. Each station has an exponential distribution of service time with parameter $k\theta$. Equations (5.34) and (5.35), to determine the mean and variance of a gamma distribution, are valid regardless of the value of β. However, when $\beta = k$, an integer, Equation (5.37) may be used to derive the mean of the distribution in a fairly straightforward manner. The expected value of the sum of random variables is the sum of the expected value of each random variable. Thus,

$$E(X) = E(X_1) + E(X_2) + \cdots + E(X_k)$$

The expected values of the exponentially distributed X_j are each given by $1/k\theta$. Thus,

$$E(X) = \frac{1}{k\theta} + \frac{1}{k\theta} + \cdots + \frac{1}{k\theta} = \frac{1}{\theta}$$

If the random variables X_j are independent, the variance of their sum is the sum of the variances, or

$$V(X) = \frac{1}{(k\theta)^2} + \frac{1}{(k\theta)^2} + \cdots + \frac{1}{(k\theta)^2} = \frac{1}{k\theta^2}$$

When $\beta = k$, a positive integer, the cdf given by Equation (5.36) may be integrated by parts, giving

$$F(x) = \begin{cases} 1 - \displaystyle\sum_{i=0}^{k-1} \frac{e^{-k\theta x}(k\theta x)^i}{i!}, & x > 0 \\ 0, & x \leq 0 \end{cases} \tag{5.38}$$

which is the sum of Poisson terms with mean $\alpha = k\theta x$. Tables of the cumulative Poisson distribution may be used to evaluate the cdf when the shape parameter is an integer. ◀

EXAMPLE 5.19

A college professor is leaving home for the summer but would like to have a light burning at all times to discourage burglars. The professor rigs up a device that will hold two light bulbs. The device will switch the current to the second bulb if the first bulb fails. The box in which the light bulbs are packaged says, "Average life 1000 hours, exponentially distributed." The professor will be gone 90 days (2160 hours). What is the probability that a light will be burning when the summer is over and the professor returns?

The probability that the system will operate at least x hours is called the reliability function $R(x)$, where

$$R(x) = 1 - F(x)$$

In this case, the total system lifetime is given by Equation (5.37) with $\beta = k = 2$ bulbs and $k\theta = 1/1000$ per hour, so $\theta = 1/2000$ per hour. Thus, $F(2160)$ can be determined from Equation (5.38) as follows:

$$F(2160) = 1 - \sum_{i=0}^{1} \frac{e^{-(2)(1/2000)(2160)}[(2)(1/2000)(2160)]^i}{i!}$$

$$= 1 - e^{-2.16} \sum_{i=0}^{1} \frac{(2.16)^i}{i!} = 0.636$$

Therefore, the chances are about 36% that the light will be burning when the professor returns. ◀

EXAMPLE 5.20

A medical examination is given in three stages by a physician. Each stage is exponentially distributed with a mean service time of 20 minutes. Find the probability that the exam will take 50 minutes or less. Also, determine the expected length of the exam. In this case, $k = 3$ stages and $k\theta = 1/20$, so that $\theta = 1/60$ per minute. Thus, $F(50)$ can be determined from Equation (5.38) as follows:

$$F(50) = 1 - \sum_{i=0}^{2} \frac{e^{-(3)(1/60)(50)}[(3)(1/60)(50)]^i}{i!}$$

$$= 1 - \sum_{i=0}^{2} \frac{e^{-5/2}(5/2)^i}{i!}$$

The cumulative Poisson distribution, shown in Table A.4, can be used to determine that

$$F(50) = 1 - 0.543 = 0.457$$

The probability is 0.457 that the exam will take 50 minutes or less. The expected length of the exam is determined from Equation (5.34) as

$$E(X) = \frac{1}{\theta} = \frac{1}{1/60} = 60 \text{ minutes}$$

In addition, the variance of X is $V(X) = 1/\beta\theta^2 = 1200$ minutes^2. In general, a gamma distribution is less variable than an exponential distribution with the same mean. Incidentally, the mode of the Erlang distribution is given by

$$\text{mode} = \frac{k-1}{k\theta} \tag{5.39}$$

Thus, the modal value in this example is

$$\text{mode} = \frac{3-1}{3(1/60)} = 40 \text{ minutes}$$

5. *Normal distribution.* A random variable X with mean $\mu(-\infty < \mu < \infty)$ and variance $\sigma^2 > 0$ has a normal distribution if it has the pdf

$$f(x) = \frac{1}{\sigma\sqrt{2\pi}} \exp\left[-\frac{1}{2}\left(\frac{x-\mu}{\sigma}\right)^2\right], \quad -\infty < x < \infty \tag{5.40}$$

The normal distribution is used so often that the notation $X \sim N(\mu, \sigma^2)$ has been adopted by many authors to indicate that the random variable X is normally distributed with mean μ and variance σ^2. The normal pdf is shown in Figure 5.12.

Some of the special properties of the normal distribution are listed here:

1. $\lim_{x\to-\infty} f(x) = 0$ and $\lim_{x\to\infty} f(x) = 0$; the value of $f(x)$ approaches zero as x approaches negative infinity and, similarly, as x approaches positive infinity.
2. $f(\mu - x) = f(\mu + x)$; the pdf is symmetric about μ.
3. The maximum value of the pdf occurs at $x = \mu$. (Thus, the mean and mode are equal.)

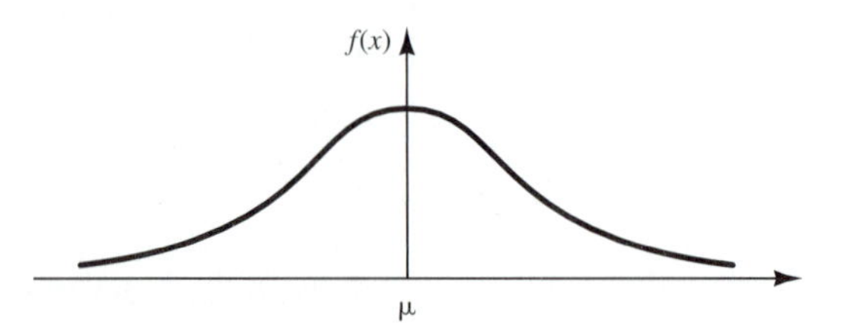

Figure 5.12. The pdf of the normal distribution.

The cdf for the normal distribution is given by

$$F(x) = P(X \leq x) = \int_{-\infty}^{x} \frac{1}{\sigma\sqrt{2\pi}} \exp\left[\frac{-1}{2}\left(\frac{t-\mu}{\sigma}\right)^2\right] dt \qquad (5.41)$$

It is not possible to evaluate Equation (5.41) in closed form. Numerical methods could be used, but it appears that it would be necessary to evaluate the integral for each pair (μ, σ^2). However, a transformation of variables, $z = (t - \mu)/\sigma$, allows the evaluation to be independent of μ and σ. If $X \sim N(\mu, \sigma^2)$, let $Z = (X - \mu)/\sigma$ to obtain

$$\begin{aligned} F(x) &= P(X \leq x) = P\left(Z \leq \frac{x-\mu}{\sigma}\right) \\ &= \int_{-\infty}^{(x-\mu)/\sigma} \frac{1}{\sqrt{2\pi}} e^{-z^2/2} dz \\ &= \int_{-\infty}^{(x-\mu)/\sigma} \phi(z) dz = \Phi\left(\frac{x-\mu}{\sigma}\right) \end{aligned} \qquad (5.42)$$

The pdf

$$\phi(z) = \frac{1}{\sqrt{2\pi}} e^{-z^2/2}, \quad -\infty < z < \infty \qquad (5.43)$$

is the pdf of a normal distribution with mean zero and variance 1. Thus, $Z \sim N(0, 1)$, and it is said that Z has a standard normal distribution. The standard normal distribution is shown in Figure 5.13. The cdf for the standard normal distribution is given by

$$\Phi(z) = \int_{-\infty}^{z} \frac{1}{\sqrt{2\pi}} e^{-t^2/2} dt \qquad (5.44)$$

Equation (5.44) has been widely tabulated. The probabilities $\Phi(z)$ for $Z \geq 0$ are given in Table A.3. Several examples are now given that indicate how Equation (5.42) and Table A.3 are used. ◀

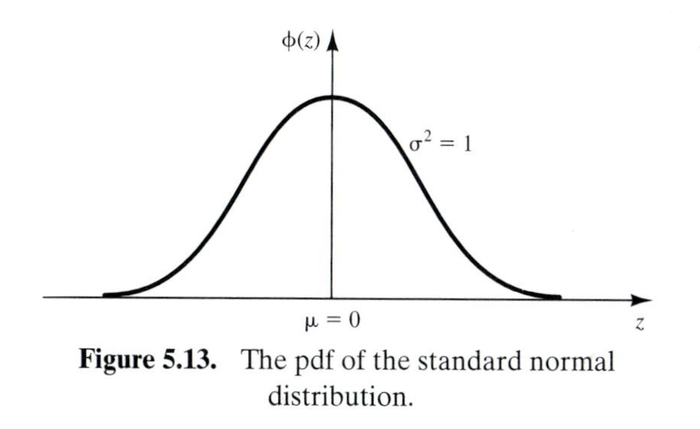

Figure 5.13. The pdf of the standard normal distribution.

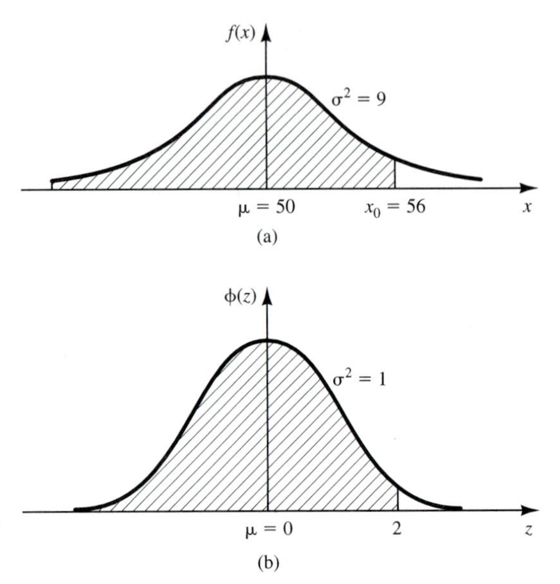

Figure 5.14. Transforming to the standard normal distribution.

EXAMPLE 5.21

It is known that $X \sim N(50, 9)$. Determine $F(56) = P(X \leq 56)$. Using Equation (5.42), we get

$$F(56) = \Phi\left(\frac{56 - 50}{3}\right) = \Phi(2) = 0.9772$$

from Table A.3. The intuitive interpretation is shown in Figure 5.14. Figure 5.14(a) shows the pdf of $X \sim N(50, 9)$ with the specific value, $x_0 = 56$, marked. The shaded portion is the desired probability. Figure 5.14(b) shows the standard normal distribution or $Z \sim N(0, 1)$ with the value 2 marked, since $x_0 = 56$ is 2σ (where $\sigma = 3$) greater than the mean. It is helpful to make both sketches such as those in Figure 5.14 to avoid confusion in determining required probabilities. ◀

EXAMPLE 5.22

The time required to load an oceangoing vessel, X, is distributed $N(12, 4)$. The probability that the vessel will be loaded in less than 10 hours is given by $F(10)$, where

$$F(10) = \Phi\left(\frac{10 - 12}{2}\right) = \Phi(-1) = 0.1587$$

The value of $\Phi(-1) = 0.1587$ is determined from Table A.3 using the symmetry

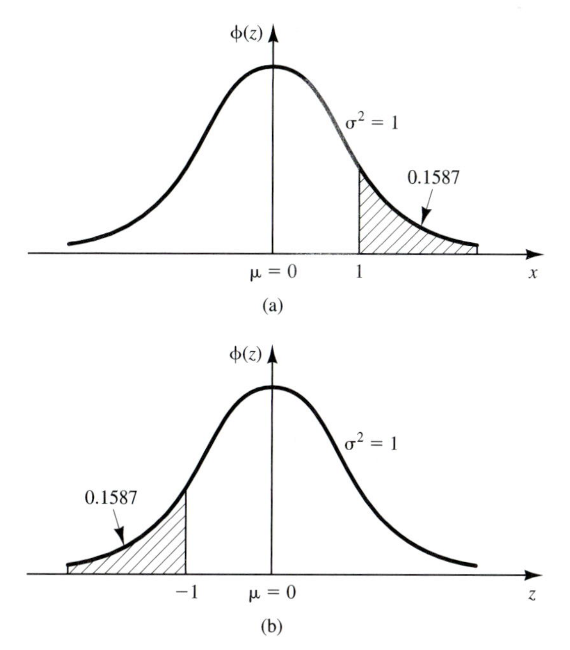

Figure 5.15. Using the symmetry property of the normal distribution.

property of the normal distribution. Note that $\Phi(1) = 0.8413$. The complement of 0.8413, or 0.1587, is contained in the tail, the shaded portion of the standard normal distribution shown in Figure 5.15(a). In Figure 5.15(b) the symmetry property is used to determine the shaded region to be $\Phi(-1) = 1 - \Phi(1) = 0.1587$. [Using this logic, it can be seen that $\Phi(2) = 0.9772$ and $\Phi(-2) = 1 - \Phi(2) = 0.0228$. In general, $\Phi(-x) = 1 - \Phi(x)$.]

The probability that 12 or more hours will be required to load the ship can also be determined by inspection, using the symmetry property of the normal pdf and the mean as shown by Figure 5.16. The shaded portion of Figure 5.16(a) shows the problem as originally stated [i.e., determine $P(X < 12)$]. Now, $P(X > 12) = 1 - F(12)$. The standardized normal in Figure 5.16(b) is used to determine $F(12) = \Phi(0) = 0.50$. Thus, $P(X > 12) = 1 - 0.50 = 0.50$. [The shaded portions in both Figure 5.16(a) and (b) contain 0.50 of the area under the normal pdf.]

The probability that between 10 and 12 hours will be required to load a ship is given by

$$P(10 \leq X \leq 12) = F(12) - F(10) = 0.5000 - 0.1587 = 0.3413$$

using earlier results presented in this example. The desired area is shown in the shaded portion of Figure 5.17(a). The equivalent problem in terms of the

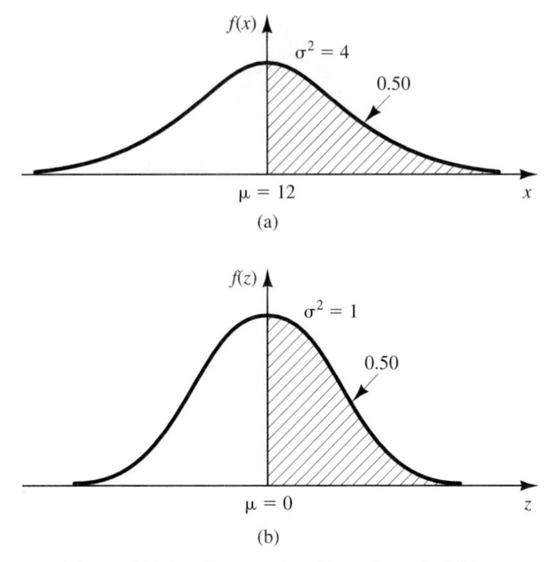

Figure 5.16. Determination of probability by inspection.

standardized normal distribution is shown in Figure 5.17(b). The probability statement $F(12) - F(10) = \Phi(0) - \Phi(-1) = 0.5000 - 0.1587 = 0.3413$, using Table A.3. ◀

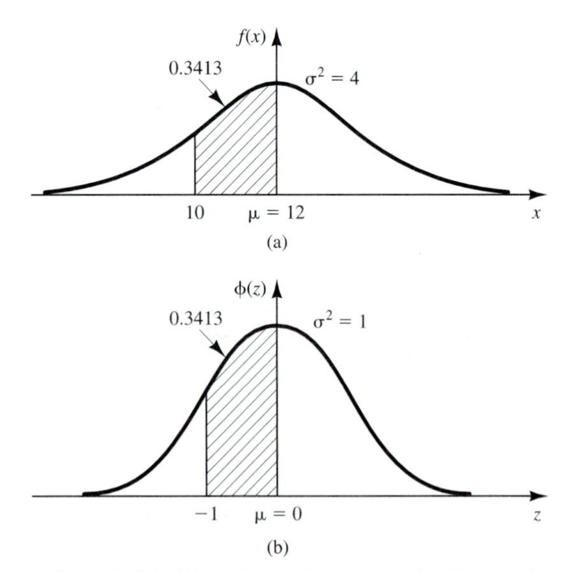

Figure 5.17. Transformation to standard normal for vessel loading problem.

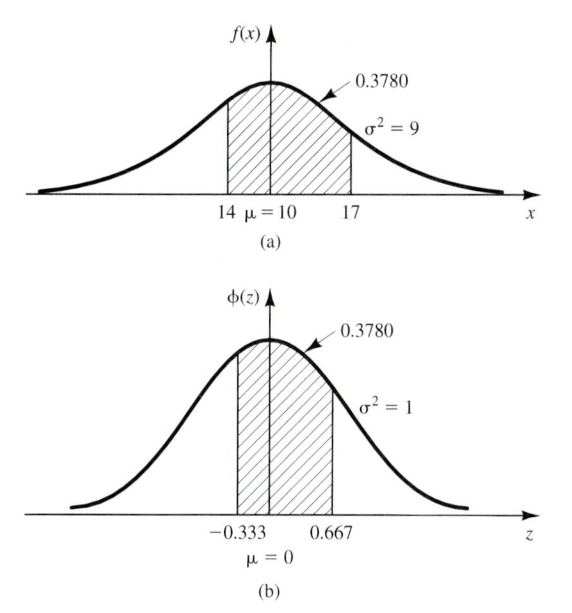

Figure 5.18. Transformation to standard normal for cafeteria problem.

Example 5.23

The time to pass through a queue to begin self-service at a cafeteria has been found to be $N(15, 9)$. The probability that an arriving customer waits between 14 and 17 minutes is determined as follows:

$$P(14 \leq X \leq 17) = F(17) - F(14) = \Phi\left(\frac{17 - 15}{3}\right) - \Phi\left(\frac{14 - 15}{3}\right)$$

$$= \Phi(0.667) - \Phi(-0.333)$$

The shaded area shown in Figure 5.18(a) represents the probability $F(17) - F(14)$. The shaded area shown in Figure 5.18(b) represents the equivalent probability, $\Phi(0.667) - \Phi(-0.333)$, for the standardized normal distribution. Using Table A.3, $\Phi(0.667) = 0.7476$. Now, $\Phi(-0.333) = 1 - \Phi(0.333) = 1 - 0.6304 = 0.3696$. Thus, $\Phi(0.667) - \Phi(-0.333) = 0.3780$. The probability is 0.3780 that the customer will pass through the queue in a time between 14 and 17 minutes. ◀

Example 5.24

Lead-time demand, X, for an item is approximated by a normal distribution with a mean of 25 and a variance of 9. It is desired to determine a value for lead time that will be exceeded only 5% of the time. Thus, the problem is to find x_0 such that $P(X > x_0) = 0.05$, as shown by the shaded area in Figure 5.19(a).

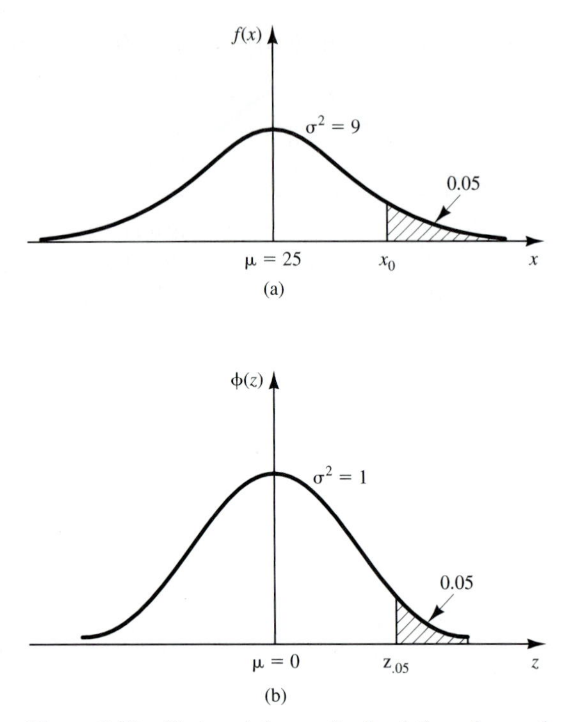

Figure 5.19. Determining x_0 for lead-time demand problem.

The equivalent problem is shown as the shaded area in Figure 5.19(b). Now,

$$P(X > x_0) = P\left(Z > \frac{x_0 - 25}{3}\right) = 1 - \Phi\left(\frac{x_0 - 25}{3}\right) = 0.05$$

or, equivalently,

$$\Phi\left(\frac{x_0 - 25}{3}\right) = 0.95$$

From Table A.3 it can be seen that $\Phi(1.645) = 0.95$. Thus, x_0 can be determined by solving

$$\frac{x_0 - 25}{3} = 1.645$$

or

$$x_0 = 29.935$$

Therefore, in only 5% of the cases will demand during lead time exceed available inventory if an order to purchase is made when the stock level reaches 30. ◀

6. *Weibull distribution.* The random variable X has a Weibull distribution if its pdf has the form

$$f(x) = \begin{cases} \frac{\beta}{\alpha}\left(\frac{x-\nu}{\alpha}\right)^{\beta-1} \exp\left[-\left(\frac{x-\nu}{\alpha}\right)^{\beta}\right], & x \geq \nu \\ 0, & \text{otherwise} \end{cases} \tag{5.45}$$

The three parameters of the Weibull distribution are $\nu(-\infty < \nu < \infty)$, which is the location parameter; $\alpha(\alpha > 0)$, which is the scale parameter; and $\beta(\beta > 0)$, which is the shape parameter. When $\nu = 0$, the Weibull pdf becomes

$$f(x) = \begin{cases} \frac{\beta}{\alpha}\left(\frac{x}{\alpha}\right)^{\beta-1} \exp\left[-\left(\frac{x}{\alpha}\right)^{\beta}\right], & x \geq 0 \\ 0, & \text{otherwise} \end{cases} \tag{5.46}$$

Figure 5.20 shows several Weibull densities when $\nu = 0$ and $\alpha = 1$. Letting $\beta = 1$, the Weibull distribution is reduced to

$$f(x) = \begin{cases} \frac{1}{\alpha}e^{-x/\alpha}, & x \geq 0 \\ 0, & \text{otherwise} \end{cases}$$

which is an exponential distribution with parameter $\lambda = 1/\alpha$.

The mean and variance of the Weibull distribution are given by the following expressions:

$$E(X) = \nu + \alpha\Gamma\left(\frac{1}{\beta} + 1\right) \tag{5.47}$$

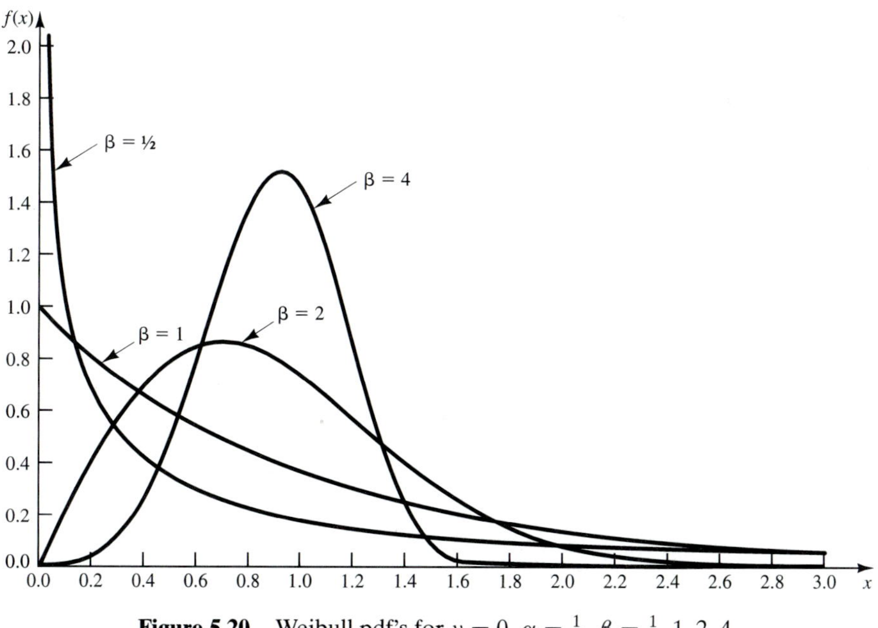

Figure 5.20. Weibull pdf's for $\nu = 0, \alpha = \frac{1}{2}, \beta = \frac{1}{2}, 1, 2, 4$.

$$V(X) = \alpha^2 \left[\Gamma\left(\frac{2}{\beta} + 1\right) - \left[\Gamma\left(\frac{1}{\beta} + 1\right)\right]^2 \right] \tag{5.48}$$

where $\Gamma(\cdot)$ is defined by Equation (5.30). Thus, the location parameter, ν, has no effect on the variance; however, the mean is increased or decreased by ν. The cdf of the Weibull distribution is given by

$$F(x) = \begin{cases} 0, & x < \nu \\ 1 - \exp\left[-\left(\frac{x-\nu}{\alpha}\right)^\beta\right], & x \geq \nu \end{cases} \tag{5.49}$$

EXAMPLE 5.25

The time to failure for a flat panel screen is known to have a Weibull distribution with $\nu = 0$, $\beta = 1/3$, and $\alpha = 200$ hours. The mean time to failure is given by Equation (5.47) as

$$E(X) = 200\Gamma(3 + 1) = 200(3!) = 1200 \text{ hours}$$

The probability that a unit fails before 2000 hours is determined from Equation (5.49) as

$$\begin{aligned} F(2000) &= 1 - \exp\left[-\left(\frac{2000}{200}\right)^{1/3}\right] \\ &= 1 - e^{-\sqrt[3]{10}} = 1 - e^{-2.15} = 0.884 \end{aligned}$$

◀

EXAMPLE 5.26

The time it takes for an aircraft to land and clear the runway at a major international airport has a Weibull distribution with $\nu = 1.34$ minutes, $\beta = 0.5$, and $\alpha = 0.04$ minute. Determine the probability that an incoming airplane will take more than 1.5 minutes to land and clear the runway. In this case $P(X > 1.5)$ is determined as follows:

$$\begin{aligned} P(X \leq 1.5) &= F(1.5) \\ &= 1 - \exp\left[-\left(\frac{1.5 - 1.34}{0.04}\right)^{0.5}\right] \\ &= 1 - e^{-2} = 1 - 0.135 = 0.865 \end{aligned}$$

Therefore, the probability that an aircraft will require more than 1.5 minutes to land and clear the runway is 0.135.

7. *Triangular distribution.* A random variable X has a triangular distribution if its pdf is given by

$$f(x) = \begin{cases} \frac{2(x-a)}{(b-a)(c-a)}, & a \leq x \leq b \\ \frac{2(c-x)}{(c-b)(c-a)}, & b < x \leq c \\ 0, & \text{elsewhere} \end{cases} \tag{5.50}$$

where $a \leq b \leq c$. The mode occurs at $x = b$. A triangular pdf is shown in Figure 5.21. The parameters (a, b, c) can be related to other measures, such as the mean and the mode, as follows:

$$E(X) = \frac{a + b + c}{3} \tag{5.51}$$

From Equation (5.51) the mode can be determined as

$$\text{mode} = b = 3E(X) - (a + c) \tag{5.52}$$

Since $a \leq b \leq c$, it follows that

$$\frac{2a + c}{3} \leq E(X) \leq \frac{a + 2c}{3}$$

The mode is used more often than the mean to characterize the triangular distribution. As shown in Figure 5.21, its height is $2/(c-a)$ above the x axis. The variance, $V(X)$, of the triangular distribution is little used. Its determination is left as an exercise for the student. The cdf for the triangular distribution is given by

$$F(x) = \begin{cases} 0, & x \leq a \\ \frac{(x-a)^2}{(b-a)(c-a)}, & a < x \leq b \\ 1 - \frac{(c-x)^2}{(c-b)(c-a)}, & b < x \leq c \\ 1, & x > c \end{cases} \tag{5.53}$$

◀

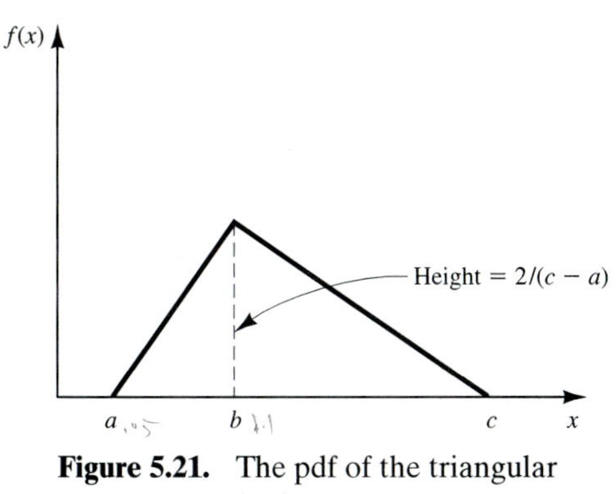

Figure 5.21. The pdf of the triangular distribution.

EXAMPLE 5.27

The central processing requirements, for programs that will execute, have a triangular distribution with $a = 0.05$ second, $b = 1.1$ seconds, and $c = 6.5$ seconds. Determine the probability that the CPU requirement for a random program is 2.5 seconds or less. The value of $F(2.5)$ is from the portion of the cdf in the interval (0.05, 1.1) plus that portion in the interval (1.1, 2.5). Using Equation (5.53), both portions can be determined at one time to yield

$$F(2.5) = 1 - \frac{(6.5 - 2.5)^2}{(6.5 - 0.05)(6.5 - 1.1)} = 0.541$$

Thus, the probability is 0.541 that the CPU requirement is 2.5 seconds or less. ◀

EXAMPLE 5.28

An electronic sensor determines the quality of memory chips, rejecting those that fail. Upon demand, the sensor will give the minimum and maximum number of rejects during each hour of production over the past 24 hours. The mean is also given. Without further information, the quality control department has assumed that the number of rejected chips can be approximated by a triangular distribution. The current dump of data indicates that the minimum number of rejected chips during any hour was zero, the maximum was 10, and the mean was 4. Knowing that $a = 0$, $c = 10$, and $E(X) = 4$, the value of b can be determined from Equation (5.52) as

$$b = 3(4) - (0 + 10) = 2$$

The height of the mode is $2/(10 - 0) = 0.2$. Thus, Figure 5.22 can be drawn.

The median is the point at which 0.5 of the area is to the left and 0.5 is to the right. The median in this example is 3.7, also shown on Figure 5.22. Determining the median of the triangular distribution requires an initial location

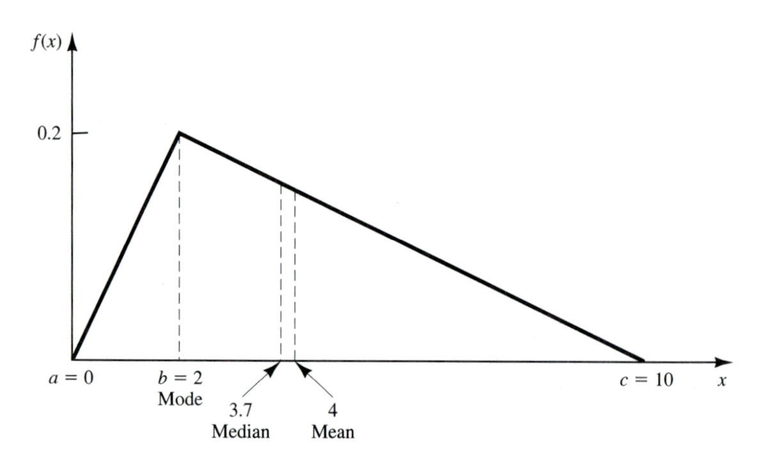

Figure 5.22. Mode, median, and mean for triangular distribution.

of the value to the left or to the right of the mode. The area to the left of the mode is determined from Equation (5.53) as

$$F(2) = \frac{2^2}{20} = 0.2$$

Thus, the median is between b and c. Setting $F(x) = 0.5$ in Equation (5.53) and solving for $x =$ median yields

$$0.5 = 1 - \frac{(10 - x)^2}{(10)(8)}$$

with

$$x = 3.7$$

This example clearly shows that the mean, mode, and median are not necessarily equal.

8. *Lognormal distribution.* A random variable X has a lognormal distribution if its pdf is given by

$$f(x) = \begin{cases} \frac{1}{\sqrt{2\pi}\sigma x} \exp\left[-\frac{(\ln x - \mu)^2}{2\sigma^2}\right], & x > 0 \\ 0, & \text{otherwise} \end{cases} \quad (5.54)$$

where $\sigma^2 > 0$. The mean and variance of a lognormal random variable are

$$E(X) = e^{\mu+\sigma^2/2} \quad (5.55)$$

$$V(X) = e^{2\mu+\sigma^2}(e^{\sigma^2} - 1) \quad (5.56)$$

Three lognormal pdf's, all with mean 1 but variances 1/2, 1, and 2, are shown in Figure 5.23.

Notice that the parameters μ and σ^2 are not the mean and variance of the lognormal. These parameters come from the fact that when Y has a $N(\mu, \sigma^2)$ distribution, then $X = e^Y$ has a lognormal distribution with parameters μ and σ^2. If the mean and variance of the lognormal are known to be μ_L and σ_L^2, respectively, then the parameters μ and σ^2 are given by

$$\mu = \ell n \left(\frac{\mu_L^2}{\sqrt{\mu_L^2 + \sigma_L^2}} \right) \quad (5.57)$$

$$\sigma^2 = \ell n \left(\frac{\mu_L^2 + \sigma_L^2}{\mu_L^2} \right) \quad (5.58)$$

◀

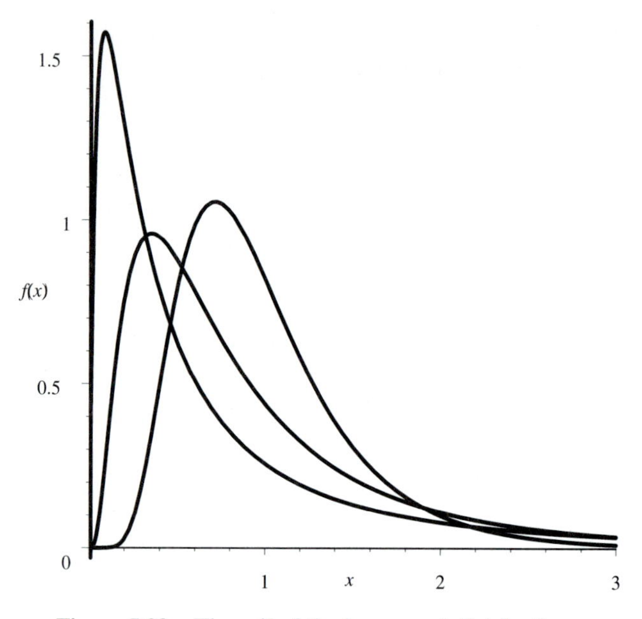

Figure 5.23. The pdf of the lognormal distribution.

EXAMPLE 5.29

The rate of return on a volatile investment is modeled as having a lognormal distribution with mean 20% and standard deviation 5%. Determine the parameters for the lognormal distribution. From the information given we have $\mu_L = 20$ and $\sigma_L^2 = 5^2$. Thus, from Equations (5.57) and (5.58)

$$\mu = \ell n \left(\frac{20^2}{\sqrt{20^2 + 5^2}} \right) \doteq 2.9654$$

$$\sigma^2 = \ell n \left(\frac{20^2 + 5^2}{20^2} \right) \doteq 0.06$$

◀

5.5 Poisson Process

Consider random events such as the arrival of jobs at a job shop, the arrival of aircraft at a runway, the arrival of ships at a port, the arrival of calls at a switchboard, the breakdown of machines in a large factory, and so on. These events may be described by a counting function $N(t)$ defined for all $t \geq 0$. This counting function will represent the number of events that occurred in $[0, t]$. Time zero is the point at which the observation began, whether or not an arrival occurred at that instant. For each interval $[0, t]$ the value $N(t)$ is an observation of a random variable, where the only possible values that can be assumed by $N(t)$ are the integers $0, 1, 2, \ldots$.

The counting process, $\{N(t), t \geq 0\}$, is said to be a Poisson process with mean rate λ if the following assumptions are fulfilled.

1. Arrivals occur one at a time.
2. $\{N(t), t \geq 0\}$ has stationary increments: The distribution of the numbers of arrivals between t and $t + s$ depends only on the length of the interval s and not on the starting point t. Thus, arrivals are completely at random without rush or slack periods.
3. $\{N(t), t \geq 0\}$ has independent increments: The numbers of arrivals during nonoverlapping time intervals are independent random variables. Thus, a large or small number of arrivals in one time interval has no effect on the number of arrivals in subsequent time intervals. Future arrivals occur completely at random, independent of the numbers of arrivals in past time intervals.

If arrivals occur according to a Poisson process, meeting the three assumptions above, it can be shown that the probability that $N(t)$ is equal to n is given by

$$P[N(t) = n] = \frac{e^{-\lambda t}(\lambda t)^n}{n!} \quad \text{for } t \geq 0 \text{ and } n = 0, 1, 2, \ldots \tag{5.59}$$

Comparing Equation (5.59) to Equation (5.18), it can be seen that $N(t)$ has the Poisson distribution with parameter $\alpha = \lambda t$. Thus, its mean and variance are given by

$$E[N(t)] = \alpha = \lambda t = V[N(t)]$$

For any times s and t such that $s < t$, the assumption of stationary increments implies that the random variable $N(t) - N(s)$, representing the number of arrivals in the interval s to t, is also Poisson distributed with mean $\lambda(t - s)$. Thus,

$$P[N(t) - N(s) = n] = \frac{e^{-\lambda(t-s)}[\lambda(t - s)]^n}{n!} \quad \text{for } n = 0, 1, 2, \ldots$$

and

$$E[N(t) - N(s)] = \lambda(t - s) = V[N(t) - N(s)].$$

Now, consider the time at which arrivals occur in a Poisson process. Let the first arrival occur at time A_1, the second at time $A_1 + A_2$, and so on, as shown in Figure 5.24. Thus, $A_1, A_2, \ldots$ are successive interarrival times. Since the first arrival occurs after time t if and only if there are no arrivals in the interval $[0, t]$, it is seen that

$$\{A_1 > t\} = \{N(t) = 0\}$$

and, therefore,

$$P(A_1 > t) = P[N(t) = 0] = e^{-\lambda t}$$

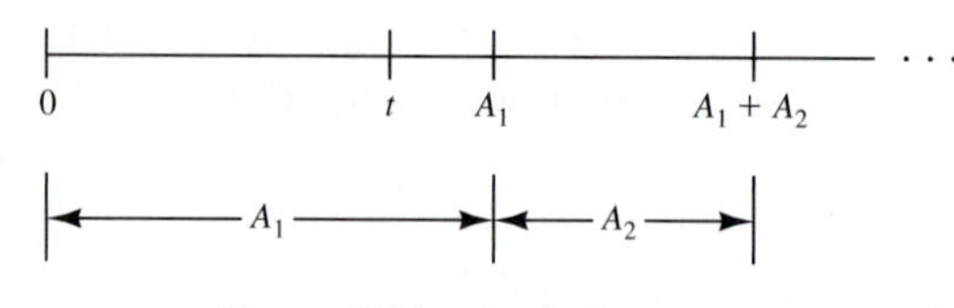

Figure 5.24. Arrival process.

the last equality following from Equation (5.59). Thus, the probability that the first arrival will occur in $[0, t]$ is given by

$$P(A_1 \le t) = 1 - e^{-\lambda t}$$

which is the cdf for an exponential distribution with parameter λ. Hence, A_1 is distributed exponentially with mean $E(A_1) = 1/\lambda$. It can also be shown that all interarrival times, $A_1, A_2, \ldots$, are exponentially distributed and independent with mean $1/\lambda$. As an alternative definition of a Poisson process, it can be shown that if interarrival times are distributed exponentially and independently, then the number of arrivals by time t, say $N(t)$, meets the three assumptions above and, therefore, is a Poisson process.

Recall that the exponential distribution is memoryless; that is, the probability of a future arrival in a time interval of length s is independent of the time of the last arrival. The probability of the arrival depends only on the length of the time interval, s. Thus, the memoryless property is related to the properties of independent and stationary increments of the Poisson process.

Additional readings concerning the Poisson process may be obtained from many sources, including Parzen [1962], Feller [1968], and Ross [1997].

Example 5.30

The jobs at a machine shop arrive according to a Poisson process with a mean of $\lambda = 2$ jobs per hour. Therefore, the interarrival times are distributed exponentially with the expected time between arrivals, $E(A) = 1/\lambda = \frac{1}{2}$ hour.

Properties of a Poisson process. Several properties of the Poisson process, discussed by Ross [1997] and others, are useful in discrete system simulation. The first of these properties concerns random splitting. Consider a Poisson process $\{N(t), t \ge 0\}$ having rate λ, as represented by the left side of Figure 5.25.

Suppose that each time an event occurs it is classified as either a type I or a type II event. Suppose further that each event is classified as a type I event with probability p and type II event with probability $1 - p$, independently of all other events.

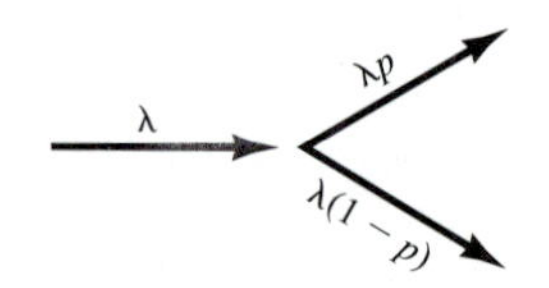

Figure 5.25. Random splitting.

Let $N_1(t)$ and $N_2(t)$ be random variables which denote respectively the number of type I and type II events occurring in $[0, t]$. Note that $N(t) = N_1(t) + N_2(t)$. It can be shown that $N_1(t)$ and $N_2(t)$ are both Poisson processes having rates λp and $\lambda(1 - p)$, as shown in Figure 5.25. Furthermore, it can be shown that the two processes are independent. ◀

Example 5.31 (Random Splitting)

Suppose that jobs arrive at a shop in accordance with a Poisson process having rate λ. Suppose further that each arrival is marked high priority with probability of 1/3 and low priority with probability of 2/3. Then a type I event would correspond to a high-priority arrival and a type II event to a low-priority arrival. If $N_1(t)$ and $N_2(t)$ are as defined above, both variables follow the Poisson process with rates $\lambda/3$ and $2\lambda/3$, respectively. ◀

Example 5.32

The rate in Example 5.31 is $\lambda = 3$ per hour. The probability that no high-priority jobs will arrive in a 2-hour period is given by the Poisson distribution with parameter $\alpha = \lambda p t = 2$. Thus,

$$P(0) = \frac{e^{-2}2^0}{0!} = 0.135$$

Now, consider the opposite situation from random splitting, namely the pooling of two arrival streams. The process of interest is shown in Figure 5.26. It can be shown that if $N_i(t)$ are random variables representing independent Poisson processes with rates λ_i, for $i = 1$ and 2, then $N(t) = N_1(t) + N_2(t)$ is a Poisson process with rate $\lambda_1 + \lambda_2$. ◀

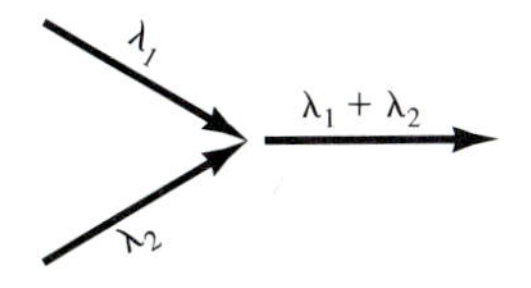

Figure 5.26. Pooled process.

Example 5.33 (Pooled Process)

A Poisson arrival stream with $\lambda_1 = 10$ arrivals per hour is combined (or pooled) with a Poisson arrival stream with $\lambda_2 = 17$ arrivals per hour. The combined process is a Poisson process with $\lambda = 27$ arrivals per hour. ◀

5.6 Empirical Distributions

An empirical distribution may be either continuous or discrete in form. It is used when it is impossible or unnecessary to establish that a random variable has any particular known distribution. One advantage of using a known distribution in simulation is the facility with which parameters can be modified to conduct a sensitivity analysis.

Table 5.3. Arrivals per Party Distribution

Arrivals per Party	*Frequency*	*Relative Frequency*	*Cumulative Relative Frequency*
1	30	0.10	0.10
2	110	0.37	0.47
3	45	0.15	0.62
4	71	0.24	0.86
5	12	0.04	0.90
6	13	0.04	0.94
7	7	0.02	0.96
8	12	0.04	1.00

Example 5.34 (Discrete)

Customers at a local restaurant arrive at lunchtime in groups ranging from one to eight persons. The number of persons per party in the last 300 groups has been observed, with the results as summarized in Table 5.3. The relative frequencies appear in Table 5.3 and again in Figure 5.27, which provides a histogram of the data that were gathered. Figure 5.28 provides a cdf of the data, which is called the empirical distribution of the given data. ◀

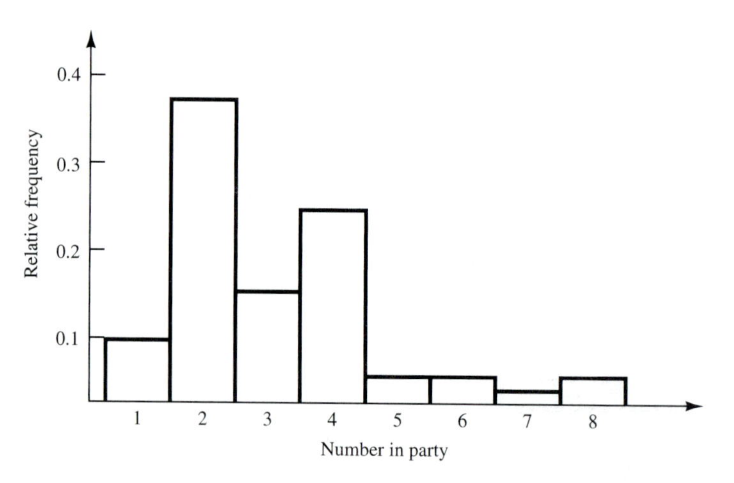

Figure 5.27. Histogram of party size.

Example 5.35 (Continuous)

The time required to repair a conveyor system which has suffered a failure has been collected for the last 100 instances with the results shown in Table 5.4. There were 21 instances in which the repair took between 0 and 0.5 hour, and so on. The empirical cdf is shown in Figure 5.29. A piecewise linear curve is formed by the connection of the points of the form $[x, F(x)]$. The points are

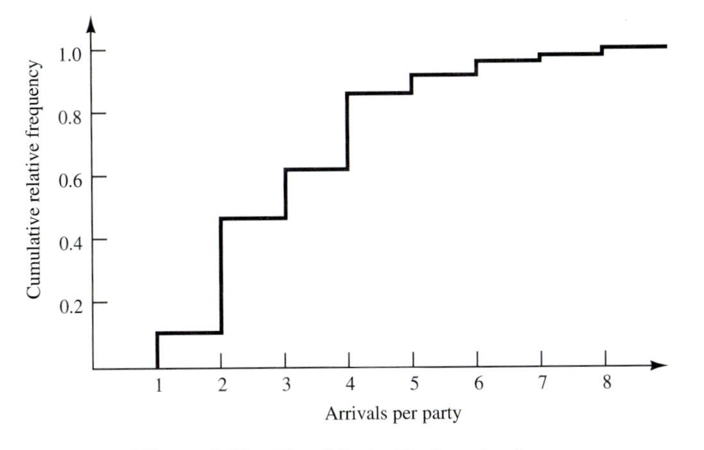

Figure 5.28. Empirical cdf of party size.

Table 5.4. Repair Times for Conveyor

Interval (Hours)	*Frequency*	*Relative Frequency*	*Cumulative Frequency*
$0 < x \leq 0.5$	21	0.21	0.21
$0.5 < x \leq 1.0$	12	0.12	0.33
$1.0 < x \leq 1.5$	29	0.29	0.62
$1.5 < x \leq 2.0$	19	0.19	0.81
$2.0 < x \leq 2.5$	8	0.08	0.89
$2.5 < x \leq 3.0$	11	0.11	1.00

connected by a straight line. The first connected pair is (0, 0) and (0.5, 0.21); then the points (0.5, 0.21) and (1.0, 0.33) are connected; and so on. ◀

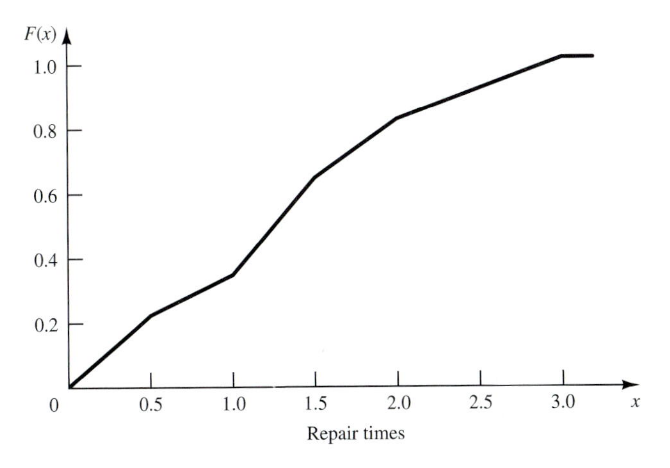

Figure 5.29. Empirical cdf for repair times.

5.7 Summary

In many instances, the world the simulation analyst sees is probabilistic rather than deterministic. The purposes of this chapter were to review several important probability distributions, to familiarize the reader with the notation used in the remainder of the text, and to show applications of the probability distributions in a simulation context.

A major task in simulation is the collection and analysis of input data. One of the first steps in this task is hypothesizing a distributional form for the input data. This is accomplished by comparing the shape of the probability density function or mass function to a histogram of the data, and by understanding that certain physical processes give rise to specific distributions. (Computer software is available to assist in this effort, as discussed in Chapter 9.) This chapter was intended to reinforce the properties of various distributions and to give insight into how these distributors arise in practice. In addition, probabilistic models of input data are used in generating random events in a simulation.

Several features which should have made a strong impression on the reader include the differences between discrete, continuous, and empirical distributions; the Poisson process and its properties; and the versatility of the gamma and the Weibull distributions.

REFERENCES

BANKS, J. AND R. G. HEIKES [1984], *Handbook of Tables and Graphs for the Industrial Engineer and Manager*, Reston Publishing, Reston, VA.

DEVORE, J. L. [1999], *Probability and Statistics for Engineers and the Sciences*, 5th. ed., Brooks/Cole, Pacific Grove, CA.

FELLER, W. [1968], *An Introduction to Probability Theory and Its Applications*, Vol. I, 3d ed., John Wiley, New York.

FISHMAN, G. S. [1973], *Concepts and Methods in Discrete Event Digital Simulation*, John Wiley, New York.

GORDON, G. [1975], *The Application of GPSS V to Discrete System Simulation*, Prentice Hall, Upper Saddle River, NJ.

HADLEY, G., AND T. M. WHITIN [1963], *Analysis of Inventory Systems*, Prentice Hall, Upper Saddle River, NJ.

HINES, W. W., AND D. C. MONTGOMERY [1990], *Probability and Statistics in Engineering and Management Science*, 3d ed., John Wiley, New York.

PAPOULIS, A. [1990], *Probability and Statistics*, Prentice Hall, Upper Saddle River, NJ.

PARZEN, E. [1962], *Stochastic Processes*, Holden-Day, San Francisco.

PEGDEN, C. D., R. E. SHANNON, and R. P. Sadowski [1995], *Introduction to Simulation Using SIMAN*, 2d ed., McGraw-Hill, New York.

ROSS, S. M. [1997], *Introduction to Probability Models*, 6th ed., Academic Press, New York.

WALPOLE, R. E. AND R. H. MYERS [1998], *Probability and Statistics for Engineers and Scientists*, 6th ed., Prentice Hall, Upper Saddle River, NJ.

EXERCISES

1. A production process manufactures alternators for outboard engines used in recreational boating. On the average, 1% of the alternators will not perform up to the required standards when tested at the engine assembly plant. When a shipment of 100 alternators is received at the plant, they are tested, and if more than two are nonconforming, the shipment is returned to the alternator manufacturer. What is the probability of returning a shipment?

2. An industrial chemical that will retard the spread of fire in paint has been developed. The local sales representative has determined, from past experience, that 48% of the sales calls will result in an order.

   (a) What is the probability that the first order will come on the fourth sales call of the day?

   (b) If eight sales calls are made in a day, what is the probability of receiving exactly six orders?

   (c) If four sales calls are made before lunch, what is the probability that one or less results in an order?

3. A recent survey indicated that 82% of single women aged 25 years old will be married in their lifetime. Using the binomial distribution, find the probability that two or three women in a sample of twenty will never be married.

4. The Hawks are currently winning 0.55 of their games. There are 5 games in the next two weeks. What is the probability that they will win more games than they lose?

5. Joe Coledge is the third-string quarterback for the University of Lower Alatoona. The probability that Joe gets into any game is 0.40.

   (a) What is the probability that the first game Joe enters is the fourth game of the season?

   (b) What is the probability that Joe plays in no more than two of the first five games?

6. For the random variables, X_1 and X_2, which are exponentially distributed with parameter $\lambda = 1$, compute $P(X_1 + X_2 > 2)$.

7. Show that the geometric distribution is memoryless.

8. The number of hurricanes hitting the coast of Florida annually has a Poisson distribution with a mean of 0.8.

   (a) What is the probability that more than two hurricanes will hit the Florida coast in a year?

   (b) What is the probability that exactly one hurricane will hit the coast of Florida in a year?

9. Arrivals at a bank teller's cage are Poisson distributed at the rate of 1.2 per minute.

   (a) What is the probability of zero arrivals in the next minute?

   (b) What is the probability of zero arrivals in the next 2 minutes?

10. Records indicate that 1.8% of the entering students at a large state university drop out of school by midterm. Using the Poisson approximation, what is the probability that three or fewer students will drop out of a random group of 200 entering students?

11. Lane Braintwain is quite a popular student. Lane receives, on the average, four phone calls a night (Poisson distributed). What is the probability that tomorrow night the number of calls received will exceed the average by more than one standard deviation?

12. Lead-time demand for condenser units is Poisson distributed with a mean of 6 units. Prepare a table for the inventory manager which will indicate the order level to achieve protection of the following levels: 50%, 80%, 90%, 95%, 97%, 97.5%, 99%, 99.5%, and 99.9%.

13. A random variable X which has pmf given by $p(x) = 1/(n+1)$ over the range $R_X = \{0, 1, 2, \ldots, n\}$ is said to have a discrete uniform distribution.

(a) Find the mean and variance of this distribution. [*Hint:*

$$\sum_{i=1}^{n} i = \frac{n(n+1)}{2} \text{ and } \sum_{i=1}^{n} i^2 = \frac{n(n+1)(2n+1)}{6}.\Bigg]$$

(b) If $R_X = \{a, a+1, a+2, \ldots, b\}$, determine the mean and variance of X.

14. The lifetime, in years, of a satellite placed in orbit is given by the following pdf:

$$f(x) = \begin{cases} 0.4e^{-0.4x}, & x \geq 0 \\ 0, & \text{otherwise} \end{cases}$$

(a) What is the probability that this satellite is still "alive" after 5 years?

(b) What is the probability that the satellite dies between 3 and 6 years from the time it is placed in orbit?

15. A mainframe computer crashes in accordance with a Poisson process with a mean rate of one crash every 36 hours. Determine the probability that the next crash will occur between 24 and 48 hours after the last crash.

16. (The Poisson distribution can be used to approximate the binomial distribution when n is large and p is small, say p less than 0.1. In utilizing the Poisson approximation, let $\lambda = np$.) In the production of ball bearings, bubbles or depressions occur, rendering the ball bearing unfit for sale. It has been determined that, on the average, one in every 800 of the ball bearings has one or more of these defects. What is the probability that a random sample of 4000 will yield fewer than three ball bearings with bubbles or depressions?

17. For an exponentially distributed random variable X, find the value of λ that satisfies the following relationship:

$$P(X \leq 3) = 0.9P(X \leq 4)$$

18. Accidents at an industrial site occur one at a time, independently, and completely at random at a mean rate of one per week. What is the probability that no accidents occur in the next three weeks?

19. A component has an exponential time-to-failure distribution with mean of 10,000 hours.

(a) The component has already been in operation for its mean life. What is the probability that it will fail by 15,000 hours?

(b) At 15,000 hours the component is still in operation. What is the probability that it will operate for another 5000 hours?

20. Suppose that a Die-Hardly Ever battery has an exponential time-to-failure distribution with a mean of 48 months. At 60 months, the battery is still operating.

(a) What is the probability that this battery is going to die in the next 12 months?

(b) What is the probability that the battery dies in an odd year of its life?

(c) If the battery is operating at 60 months, compute the expected additional months of life.

21. The time to service customers at a bank teller's cage is exponentially distributed with a mean of 50 seconds.

(a) What is the probability that the two customers in front of an arriving customer will each take less than 60 seconds to complete their transactions?

(b) What is the probability that the two customers in front will finish their transactions so that an arriving customer can reach the teller's cage within 2 minutes?

22. Determine the variance, $V(X)$, of the triangular distribution.

23. The daily use of water, in thousands of liters, at the Hardscrabble Tool and Die Works follows a gamma distribution with a shape parameter of 2 and a scale parameter of 1/4. What is the probability that the demand exceeds 4000 liters on any given day?

24. When Admiral Byrd went to the North Pole, he wore battery-powered thermal underwear. The batteries failed instantaneously rather than gradually. The batteries had a life that was exponentially distributed with a mean of 12 days. The trip took 30 days. Admiral Byrd packed three batteries. What is the probability that three batteries would be a sufficient number to keep the Admiral warm?

25. The time intervals between dial-up connections to an Internet service provider are exponentially distributed with a mean of 15 seconds. Find the probability that the third dial-up connection occurs after 30 seconds have elapsed.

26. The rail shuttle cars at Atlanta airport have a dual electrical braking system. A rail car switches to the standby system automatically if the first system fails. If both systems fail, there will be a crash! Assume that the life of a single electrical braking system is exponentially distributed with a mean of 4000 operating hours. If the systems are inspected every 5000 operating hours, what is the probability that a rail car will not crash before that time?

27. Suppose that cars arrive at a toll booth following the Poisson process with a mean interarrival time of 15 seconds. What is the probability that up to one minute will elapse until three cars have arrived?

28. Suppose that an average of 30 customers per hour arrive at the Sticky Donut Shop in accordance with a Poisson process. What is the probability that more than 5 minutes will elapse before both of the next two customers walk through the door?

29. Professor Dipsy Doodle gives six problems on each exam. Each problem requires an average of 30 minutes grading time for the entire class of 15 students. The grading time for each problem is exponentially distributed, and the problems are independent of each other.

(a) What is the probability that the Professor will finish the grading in $2\frac{1}{2}$ hours or less?

(b) What is the most likely grading time?

(c) What is the expected grading time?

30. An aircraft has dual hydraulic systems. The aircraft switches to the standby system automatically if the first system fails. If both systems have failed, the plane will crash. Assume that the life of a hydraulic system is exponentially distributed with a mean of 2000 air hours.

(a) If the hydraulic systems are inspected every 2500 hours, what is the probability that an aircraft will crash before that time?

(b) What danger would there be in moving the inspection point to 3000 hours?

31. A random variable X is beta distributed if its pdf is given by

$$f(x) = \begin{cases} \frac{(\alpha+\beta+1)!}{\alpha!\beta!}x^{\alpha}(1-x)^{\beta}, & 0 < x < 1 \\ 0, & \text{otherwise} \end{cases}$$

Show that the beta distribution becomes the uniform distribution over the unit interval when $\alpha = \beta = 0$.

32. Lead time is gamma distributed in 100s of units with a shape parameter of 3 and a scale parameter of 1. What is the probability that the lead time exceeds 2 (hundred) units during an upcoming cycle?

33. Lifetime of the video adapter card for a PC, in months, denoted by the random variable X, is gamma distributed with $\beta = 4$ and $\theta = 1/16$. What is the probability that the card will last for at least 2 years?

34. Many states have license tags which have the following format:

letter letter letter number number number

The letters indicate the weight of the automobile, but the numbers are at random, ranging from 100 to 999.

(a) What is the probability that the next two tags seen (at random) will have numbers of 500 or higher?

(b) What is the probability that the sum of the next two tags seen (at random) will have a total of 1000 or higher? [*Hint:* Approximate the discrete uniform distribution with a continuous uniform distribution. The sum of two independent uniform distributions is a triangular distribution.]

35. Let X be a random variable that is normally distributed with a mean of 10 and a variance of 4. Find the values a and b such that $P(a < X < b) = 0.90$ and $|\mu - a| = |\mu - b|$.

36. Find the probability that $6 < X < 8$ for each of the distributions:

Normal (10, 4)
Triangular (4, 10, 16)
Uniform (4, 16)

37. Lead time for an item is approximated by a normal distribution with a mean of 20 days and a variance of 4 days^2. Determine values of lead time that will be exceeded only 1%, 5%, and 10% of the time.

38. IQ scores are normally distributed throughout society with a mean of 100 and a standard deviation of 15.

(a) A person with an IQ of 140 or higher is called a "genius." What proportion of society is in the genius category?

(b) What proportion of society will miss the genius category by 5 or less points?

(c) An IQ of 110 or better is required to make it through an accredited college or university. What proportion of society could be eliminated from completing a higher education based on a low IQ score?

39. Three shafts are made and assembled in a linkage. The length of each shaft, in centimeters, is distributed as follows:

Shaft 1: $N(60, 0.09)$
Shaft 2: $N(40, 0.05)$
Shaft 3: $N(50, 0.11)$

(a) What is the distribution of the linkage?

(b) What is the probability that the linkage will be longer than 150.2 centimeters?

(c) The tolerance limits for the assembly are (149.83, 150.21). What proportion of assemblies are within the tolerance limits? [*Hint:* If $\{X_i\}$ are n independent normal random variables, and if X_i, has mean μ_i and variance σ_i^2, then the sum

$$Y = X_1 + X_2 + \cdots + X_n$$

is normal with mean $\sum_{i=1}^{n} \mu_i$ and variance $\sum_{i=1}^{n} \sigma_i^2$.]

40. The circumferences of battery posts in a nickel-cadmium battery are Weibull distributed with $\nu = 3.25$ centimeters, $\beta = 1/3$, and $\alpha = 0.005$ centimeters.

(a) Determine the probability that a battery post chosen at random will have a circumference larger than 3.40 centimeters.

(b) If battery posts are larger than 3.50 centimeters, they will not go through the hole provided; if they are smaller than 3.30 centimeters, the clamp will not tighten sufficiently. What proportion of posts will have to be scrapped for one of these reasons?

41. The time to failure of a nickel-cadmium battery is Weibull distributed with parameters $\nu = 0$, $\beta = 1/4$, and $\alpha = 1/2$ years.

(a) Find the fraction of batteries that are expected to fail prior to 1.5 years.

(b) What fraction of batteries are expected to last longer than the mean life?

(c) What fraction of batteries are expected to fail between 1.5 and 2.5 years?

42. Demand for electricity at Gipgip Pig Farm for the merry merry month of May has a triangular distribution with $a = 100$ kwh and $c = 1800$ kwh. The median kwh is 1425. Determine the modal value of kwh for the month.

43. The time to failure on an electronic subassembly can be modeled by a Weibull distribution whose location parameter is zero, $\beta = 1/2$, and $\alpha = 1000$ hours.

(a) What is the mean time to failure?

(b) What fraction of these subassemblies will fail by 3000 hours?

44. The gross weight of three-axle trucks that have been checked at the Hahira Inspection Station on Interstate Highway 85 follows a Weibull distribution with parameters $\nu = 6.8$ tons, $\beta = 1.5$, and $\alpha = 1/2$ ton. Determine the appropriate weight limit such that 0.01 of the trucks will be cited for traveling overweight.

45. The current reading on Sag Revas's gas mileage indicator is an average of 25.3 miles per gallon. Assume that gas mileage on Sag's car follows a triangular distribution with a minimum value of zero and a maximum value of 50 miles per gallon. What is the value of the median?

46. A postal letter carrier has a route consisting of five segments, with the time in minutes to complete each segment being normally distributed with mean and variance as shown:

Tennyson Place	$N(38, 16)$
Windsor Parkway	$N(99, 29)$
Knob Hill Apartments	$N(85, 25)$
Evergreen Drive	$N(73, 20)$
Chastain Shopping Center	$N(52, 12)$

In addition to the above times, the letter carrier must organize the mail at the central office, which requires a time that is distributed by $N(90, 25)$. The drive to the starting point of the route requires a time that is distributed $N(10, 4)$. The return from the route requires a time that is distributed $N(15, 4)$. The letter carrier then performs administrative tasks with a time that is distributed $N(30, 9)$.

(a) What is the expected length of the letter carrier's workday?

(b) Overtime occurs after eight hours of work on a given day. What is the probability that the letter carrier works overtime on any given day?

(c) What is the probability that the letter carrier works overtime on two or more days in a six-day week?

(d) What is the probability that the route will be completed within ± 24 minutes of eight hours on any given day? [*Hint:* See Exercise 39.]

47. The time to failure of a WD$-$1 computer chip is known to be Weibull distributed with parameters $\nu = 0$, $\beta = 1/2$, and $\alpha = 400$ days. Find the fraction expected to survive 600 days.

48. The TV sets on display at Schocker's Department Store are hooked up such that when one fails, a model exactly like the one that failed will switch on. Three such units are hooked up in this series arrangement. Their lives are independent of one another. Each TV has a life which is exponentially distributed with a mean of 10,000 hours. Find the probability that the combined life of the system is greater than 32,000 hours.

49. High temperature in Biloxi, Mississippi, on July 21, denoted by the random variable X, has the following probability density function, where x is in degrees Fahrenheit.

$$f(x) = \begin{cases} \dfrac{2(x-85)}{119}, & 85 \le x \le 92 \\ \dfrac{2(102-x)}{170}, & 92 < x \le 102 \\ 0, & \text{otherwise} \end{cases}$$

(a) What is the variance of the temperature, $V(X)$? (If you worked Exercise 22, this is quite easy.)

(b) What is the median temperature?

(c) What is the modal temperature?

50. The time to failure of Eastinghome light bulbs is Weibull distributed with $\nu = 1.8 \times 10^3$ hours, $\beta = 1/2$, and $\alpha = 1/3 \times 10^3$ hours.

(a) What fraction of bulbs are expected to last longer than the mean lifetime?

(b) What is the median lifetime of a light bulb?

51. Lead-time demand is gamma distributed in 100's of units with a shape parameter of 2 and a scale parameter of 1/4. What is the probability that the lead time exceeds 4 (hundred) units during an upcoming cycle?

6

Queueing Models

Simulation is often used in the analysis of queueing models. In a simple but typical queueing model, shown in Figure 6.1, customers arrive from time to time and join a queue, or waiting line, are eventually served, and finally leave the system. The term "customer" refers to any type of entity that can be viewed as requesting "service" from a system. Therefore, many service facilities, production systems, repair and maintenance facilities, communications and computer systems, and transport and material handling systems can be viewed as queueing systems.

Queueing models, whether solved mathematically or analyzed through simulation, provide the analyst with a powerful tool for designing and evaluating the performance of queueing systems. Typical measures of system performance include server utilization (percentage of time a server is busy), length of waiting lines, and delays of customers. Quite often, when designing or attempting to improve a queueing system, the analyst (or decision maker) is involved in trade-offs between server utilization and customer satisfaction in terms of

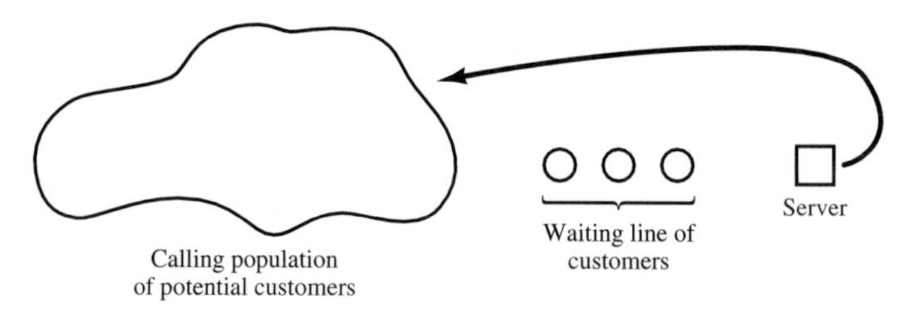

Figure 6.1. Simple queueing model.

line lengths and delays. Queueing theory and simulation analysis are used to predict these measures of system performance as a function of the input parameters. The input parameters include the arrival rate of customers, the service demands of customers, the rate at which a server works, and the number and arrangement of servers. To a certain degree, some of the input parameters are under management's direct control. Thus, the performance measures are under their indirect control, provided that the relationship between the performance measures and the input parameters is adequately understood for the given system.

For relatively simple systems, these performance measures can be computed mathematically at great savings in time and expense compared to the use of a simulation model. But for realistic models of complex systems, simulation is usually required. Nevertheless, analytically tractable models, although usually requiring many simplifying assumptions, are valuable for rough-cut estimates of system performance. These rough-cut estimates may then be refined by use of a detailed and more realistic simulation model. Simple models are also useful for developing an understanding of the dynamic behavior of queueing systems and the relationships between various performance measures. This chapter will not develop the mathematical theory of queues but instead will discuss some of the well-known models. For an elementary treatment of queueing theory, the reader is referred to the survey chapters in Hillier and Lieberman [1990], Wagner [1975], or Winston [1997]. More extensive treatments with a view toward applications are given by Cooper [1981], Gross and Harris [1997], Hall [1991] and Nelson [1995]. The latter two especially emphasize engineering and management applications.

This chapter discusses the general characteristics of queues, the meaning and relationships of the important performance measures, estimation of the mean measures of performance from a simulation, the effect of varying the input parameters, and the mathematical solution of a small number of important and basic queueing models.

6.1 Characteristics of Queueing Systems

The key elements of a queueing system are the customers and servers. The term "customer" can refer to people, machines, trucks, mechanics, patients, pallets, airplanes, e-mail, cases, orders, or dirty clothes—anything that arrives at a facility and requires service. The term "server" might refer to receptionists, repairpersons, mechanics, tool-crib clerks, medical personnel, automatic storage and retrieval machines (e.g., cranes), runways at an airport, automatic packers, order pickers, CPUs in a computer, or washing machines—any resource (person, machine, etc.) which provides the requested service. Although the terminology employed will be that of a customer arriving to a service facility, sometimes the server moves to the customer; for example, a repairperson moving to a broken machine. This in no way invalidates the models but is merely a matter of terminology. Table 6.1 lists a number of different systems

Table 6.1. Examples of Queueing Systems

System	*Customers*	*Server(s)*
Reception desk	People	Receptionist
Repair facility	Machines	Repairperson
Garage	Trucks	Mechanic
Tool crib	Mechanics	Tool-crib clerk
Hospital	Patients	Nurses
Warehouse	Pallets	Crane
Airport	Airplanes	Runway
Production line	Cases	Case packer
Warehouse	Orders	Order picker
Road network	Cars	Traffic light
Grocery	Shoppers	Checkout station
Laundry	Dirty linen	Washing machines/dryers
Job shop	Jobs	Machines/workers
Lumberyard	Trucks	Overhead crane
Saw mill	Logs	Saws
Computer	Jobs	CPU, disk, tapes
Telephone	Calls	Exchange
Ticket office	Football fans	Clerk
Mass transit	Riders	Buses, trains

together with a subsystem consisting of "arriving customers" and one or more "servers." The remainder of this section describes the elements of a queueing system in more detail.

6.1.1 The Calling Population

The population of potential customers, referred to as the *calling population*, may be assumed to be finite or infinite. For example, consider a bank of five machines that are curing tires. After an interval of time, a machine automatically opens and must be attended by a worker who removes the tire and puts an uncured tire into the machine. The machines are the "customers," who "arrive" at the instant they automatically open. The worker is the "server," who "serves" an open machine as soon as possible. The calling population is finite and consists of the five machines.

In systems with a large population of potential customers, the calling population is usually assumed to be infinite. For such systems, this assumption is usually innocuous and, furthermore, it may simplify the model. Examples of infinite populations include the potential customers of a restaurant, bank, or other similar service facility, and also very large groups of machines serviced by a technician. Even though the actual population may be finite but large, it is generally safe to use infinite population models—provided that the number of customers being served or waiting for service at any given time is a small proportion of the population of potential customers.

The main difference between finite and infinite population models is how the arrival rate is defined. In an infinite-population model, the arrival rate (i.e., the average number of arrivals per unit of time) is not affected by the number of customers who have left the calling population and joined the queueing system. When the arrival process is homogeneous over time (e.g., there are no "rush hours"), the arrival rate is usually assumed to be constant. On the other hand, for finite-calling-population models, the arrival rate to the queueing system does depend on the number of customers being served and waiting. To take an extreme case, suppose that the calling population has one member — for example, a corporate jet. When the corporate jet is being serviced by the team of mechanics who are on duty 24 hours per day, the arrival rate is zero, because there are no other potential customers (jets) who can arrive at the service facility (team of mechanics). A more typical example is that of the five tire-curing machines serviced by a single worker. When all five are closed and curing a tire, the worker is idle and the arrival rate is at a maximum, but the instant a machine opens and requires service, the arrival rate decreases. At those times when all five are open (so four machines are waiting for service while the worker is attending the other one), the arrival rate is zero; that is, no arrival is possible until the worker finishes with a machine, in which case it returns to the calling population and becomes a potential arrival. It may seem odd that the arrival rate is at its maximum when all five machines are closed. But if the arrival rate is defined as the expected number of arrivals in the next unit of time, then it becomes clear that this expectation is largest when all machines could potentially open in the next unit of time.

6.1.2 System Capacity

In many queueing systems there is a limit to the number of customers that may be in the waiting line or system. For example, an automatic car wash may have room for only 10 cars to wait in line to enter the mechanism. It may be too dangerous or illegal for cars to wait in the street. An arriving customer who finds the system full does not enter but returns immediately to the calling population. Some systems, such as concert ticket sales for students, may be considered as having unlimited capacity. There are no limits on the number of students allowed to wait to purchase tickets. As will be seen later, when a system has limited capacity, a distinction is made between the arrival rate (i.e., the number of arrivals per time unit) and the effective arrival rate (i.e., the number who arrive and enter the system per time unit).

6.1.3 The Arrival Process

The arrival process for infinite-population models is usually characterized in terms of interarrival times of successive customers. Arrivals may occur at scheduled times or at random times. When at random times, the interarrival times are usually characterized by a probability distribution. In addition, customers may arrive one at a time or in batches. The batch may be of constant size or of random size.

The most important model for random arrivals is the Poisson arrival process. If A_n represents the interarrival time between customer $n - 1$ and customer n (A_1 is the actual arrival time of the first customer), then for a Poisson arrival process, A_n is exponentially distributed with mean $1/\lambda$ time units. The arrival rate is λ customers per time unit. The number of arrivals in a time interval of length t, say $N(t)$, has the Poisson distribution with mean λt customers. For further discussion of the relationship between the Poisson distribution and the exponential distribution, the reader is referred to Section 5.5.

The Poisson arrival process has been successfully employed as a model of the arrival of people to restaurants, drive-in banks, and other service facilities; the arrival of telephone calls to a telephone exchange; the arrival of demands, or orders, for a service or product; and the arrival of failed components or machines to a repair facility.

A second important class of arrivals is the scheduled arrivals, such as patients to a physician's office or scheduled airline flight arrivals to an airport. In this case, the interarrival times $\{A_n, n = 1, 2, \ldots\}$ may be constant, or constant plus or minus a small random amount to represent early or late arrivals.

A third situation occurs when at least one customer is assumed to always be present in the queue, so that the server is never idle because of a lack of customers. For example, the "customers" may represent raw material for a product, and sufficient raw material is assumed to be always available.

For finite-population models, the arrival process is characterized in a completely different fashion. Define a customer as *pending* when that customer is outside the queueing system and a member of the potential calling population. For example, a tire-curing machine is "pending" when it is closed and curing a tire, and it becomes "not pending" the instant it opens and demands service from the worker. Define a *runtime* of a given customer as the length of time from departure from the queueing system until that customer's next arrival to the queue. Let $A_1^{(i)}, A_2^{(i)}, \ldots$ be the successive runtimes of customer i, and let $S_1^{(i)}, S_2^{(i)}, \ldots$ be the corresponding successive system times; that is, $S_n^{(i)}$ is the total time spent in the system by customer i during the nth visit. Figure 6.2 illustrates these concepts for machine 3 in the tire-curing example. The total arrival process is the superposition of the arrival times of all customers. Figure 6.2 shows the first and second arrival of machine 3, but these two times are not nec-

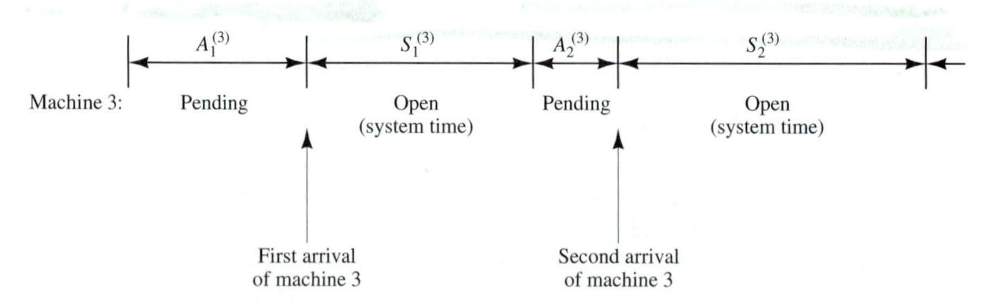

Figure 6.2. Arrival process for a finite-population model.

essarily two successive arrivals to the system. For instance, if it is assumed that all machines are pending at time 0, the first arrival to the system occurs at time $A_1 = \min\{A_1^{(1)}, A_1^{(2)}, A_1^{(3)}, A_1^{(4)}, A_1^{(5)}\}$. If $A_1 = A_1^{(2)}$, then machine 2 is the first arrival (i.e., the first to open) after time 0. As discussed earlier, the arrival rate is not constant but is a function of the number of pending customers.

One important application of finite-population models is the machine-repair problem. The machines are the customers and a runtime is also called time to failure. When a machine fails, it "arrives" at the queueing system (the repair facility) and remains there until it is "served" (repaired). Times to failure for a given class of machine have been characterized by the exponential, the Weibull, and the gamma distributions. Models with an exponential runtime are sometimes analytically tractable; an example is given in Section 6.5. Successive times to failure are usually assumed to be statistically independent, but they could depend on other factors, such as the age of a machine since its last major overhaul.

6.1.4 *Queue Behavior and Queue Discipline*

Queue behavior refers to customer actions while in a queue waiting for service to begin. In some situations, there is a possibility that incoming customers may balk (leave when they see that the line is too long), renege (leave after being in the line when they see that the line is moving too slowly), or jockey (move from one line to another if they think they have chosen a slow line).

Queue discipline refers to the logical ordering of customers in a queue and determines which customer will be chosen for service when a server becomes free. Common queue disciplines include first-in, first-out (FIFO); last-in, first-out (LIFO); service in random order (SIRO); shortest processing time first (SPT); and service according to priority (PR). In a job shop, queue disciplines are sometimes based on due dates and on expected processing time for a given type of job. Notice that a FIFO queue discipline implies that services begin in the same order as arrivals, but that customers may leave the system in a different order because of different-length service times.

6.1.5 *Service Times and the Service Mechanism*

The service times of successive arrivals are denoted by $S_1, S_2, S_3, \ldots$ They may be constant or of random duration. In the latter case, $\{S_1, S_2, \ldots\}$ is usually characterized as a sequence of independent and identically distributed random variables. The exponential, Weibull, gamma, lognormal, and truncated normal distributions have all been used successfully as models of service times in different situations. Sometimes services may be identically distributed for all customers of a given type or class or priority, while customers of different types may have completely different service-time distributions. In addition, in some systems, service times depend upon the time of day or the length of the waiting line. For example, servers may work faster than usual when the waiting line is long, thus effectively reducing the service times.

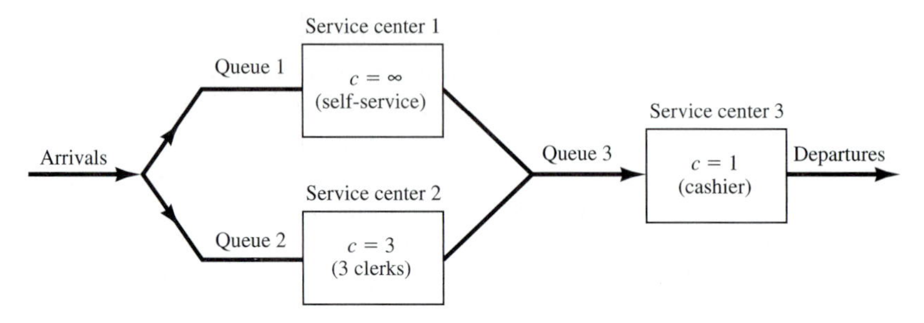

Figure 6.3. Discount warehouse with three service centers.

A queueing system consists of a number of service centers and interconnecting queues. Each service center consists of some number of servers, c, working in parallel; that is, upon getting to the head of the line, a customer takes the first available server. Parallel service mechanisms are either single server ($c = 1$), multiple server ($1 < c < \infty$), or unlimited servers ($c = \infty$). (A self-service facility is usually characterized as having an unlimited number of servers.)

Example 6.1

Consider a discount warehouse where customers may either serve themselves or wait for one of three clerks, and finally leave after paying a single cashier. The system is represented by the flow diagram in Figure 6.3. The subsystem, consisting of queue 2 and service center 2, is shown in more detail in Figure 6.4. Other variations of service mechanisms include batch service (a server serving several customers simultaneously) or a customer requiring several servers simultaneously. In the discount warehouse, a clerk may pick several small orders at the same time, but it may take two of the clerks to handle one heavy item. ◀

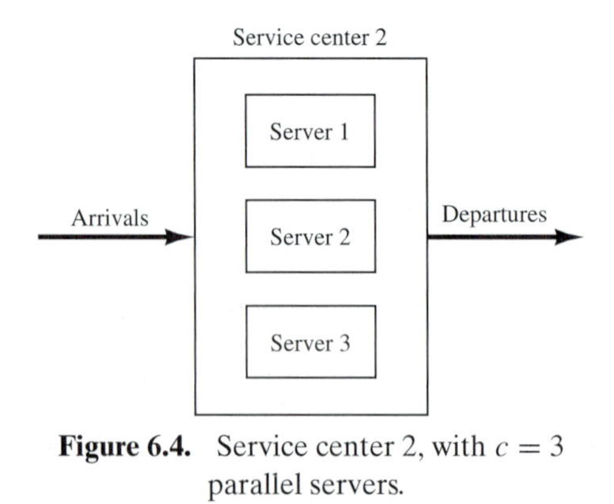

Figure 6.4. Service center 2, with $c = 3$ parallel servers.

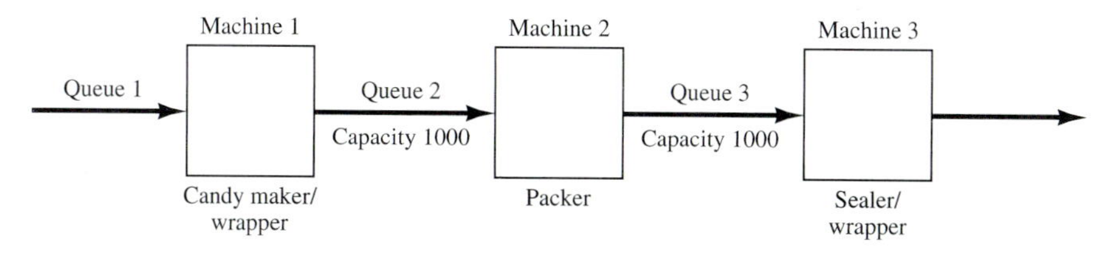

Figure 6.5. Candy production line.

Example 6.2

A candy manufacturer has a production line which consists of three machines separated by inventory-in-process buffers. The first machine makes and wraps the individual pieces of candy, the second packs 50 pieces in a box, and the third seals and wraps the box. The two inventory buffers have capacities of 1000 boxes each. As illustrated by Figure 6.5, the system is modeled as having three service centers, each center having $c = 1$ server (a machine), with queue-capacity constraints between machines. It is assumed that a sufficient supply of raw material is always available at the first queue. Because of the queue-capacity constraints, machine 1 shuts down whenever the inventory buffer (queue 2) fills to capacity, while machine 2 shuts down whenever the buffer empties. In brief, the system consists of three single-server queues in series with queue-capacity constraints and a continuous arrival stream at the first queue. ◀

6.2 Queueing Notation

Recognizing the diversity of queueing systems, Kendall [1953] proposed a notational system for parallel server systems which has been widely adopted. An abridged version of this convention is based on the format $A/B/c/N/K$. These letters represent the following system characteristics:

A represents the interarrival-time distribution.

B represents the service-time distribution.

[Common symbols for A and B include M (exponential or Markov), D (constant or deterministic), E_k (Erlang of order k), PH (phase-type), H (hyperexponential), G (arbitrary or general), and GI (general independent).]

c represents the number of parallel servers.

N represents the system capacity.

K represents the size of the calling population.

For example, $M/M/1/\infty/\infty$ indicates a single-server system that has unlimited queue capacity and an infinite population of potential arrivals. The interarrival times and service times are exponentially distributed. When N and K are infinite, they may be dropped from the notation. For example, $M/M/1/\infty/\infty$ is

Table 6.2. Queueing Notation for Parallel Server Systems

P_n	Steady-state probability of having n customers in system
$P_n(t)$	Probability of n customers in system at time t
λ	Arrival rate
λ_e	Effective arrival rate
μ	Service rate of one server
ρ	Server utilization
A_n	Interarrival time between customers $n-1$ and n
S_n	Service time of the nth arriving customer
W_n	Total time spent in system by the nth arriving customer
W_n^Q	Total time spent in the waiting line by customer n
$L(t)$	The number of customers in system at time t
$L_Q(t)$	The number of customers in queue at time t
L	Long-run time-average number of customers in system
L_Q	Long-run time-average number of customers in queue
w	Long-run average time spent in system per customer
w_Q	Long-run average time spent in queue per customer

often shortened to $M/M/1$. The tire-curing system can be initially represented by $G/G/1/5/5$.

Additional notation used throughout the remainder of this chapter for parallel server systems is listed in Table 6.2. The meanings may vary slightly from system to system. All systems will be assumed to have a FIFO queue discipline.

6.3 Long-Run Measures of Performance of Queueing Systems

The primary long-run measures of performance of queueing systems are the long-run time-average number of customers in the system (L) and in the queue (L_Q), the long-run average time spent in system (w) and in the queue (w_Q) per customer, and the server utilization, or proportion of time that a server is busy (ρ). The term "system" usually refers to the waiting line plus the service mechanism, but, in general, it can refer to any subsystem of the queueing system; whereas the term "queue" refers to the waiting line alone. Other measures of performance of interest include the long-run proportion of customers who are delayed in queue longer than t_0 time units, the long-run proportion of customers turned away because of capacity constraints, and the long-run proportion of time the waiting line contains more than k_0 customers.

This section defines the major measures of performance for a general $G/G/c/N/K$ queueing system, discusses their relationships, and shows how they can be estimated from a simulation run. There are two types of estimators: an ordinary sample average and a time-integrated (or time-weighted) sample average.

6.3.1 Time-Average Number in System L

Consider a queueing system over a period of time T, and let $L(t)$ denote the number of customers in the system at time t. A simulation of such a system is shown in Figure 6.6.

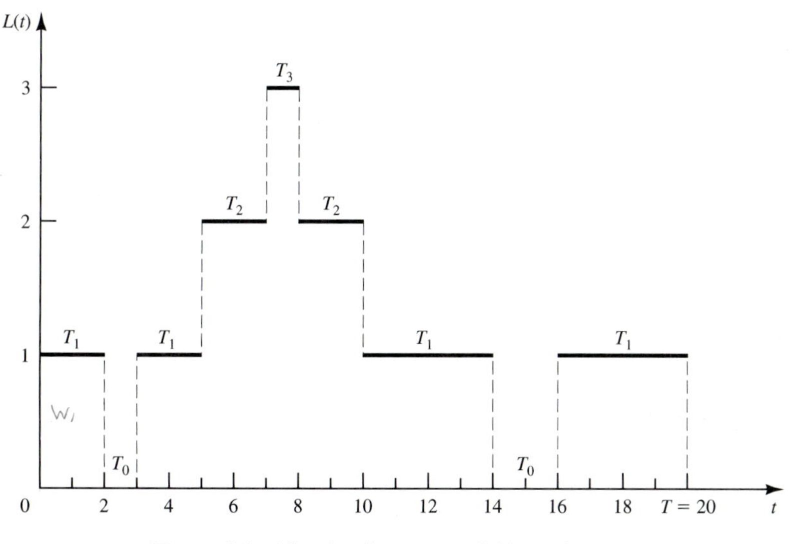

Figure 6.6. Number in system, $L(t)$, at time t.

Let T_i denote the total time during $[0, T]$ in which the system contained exactly i customers. In Figure 6.6, it is seen that $T_0 = 3$, $T_1 = 12$, $T_2 = 4$, and $T_3 = 1$. (The line segments whose lengths total $T_1 = 12$ are labeled "T_1" in Figure 6.6, etc.) In general, $\sum_{i=0}^{\infty} T_i = T$. The time-weighted-average number in system is defined by

$$\widehat{L} = \frac{1}{T}\sum_{i=0}^{\infty} i\,T_i = \sum_{i=0}^{\infty} i\left(\frac{T_i}{T}\right) \tag{6.1}$$

For Figure 6.6, $\widehat{L} = [0(3) + 1(12) + 2(4) + 3(1)]/20 = 23/20 = 1.15$ customers. Notice that T_i/T is the proportion of time the system contains exactly i customers. The estimator, $\widehat{L}$, is an example of a time-weighted average.

By considering Figure 6.6 we see that the total area under the function $L(t)$ can be decomposed into rectangles of height i and length T_i. For example, the rectangle of area $3 \times T_3$ has base running from $t = 7$ to $t = 8$ (thus $T_3 = 1$); however, most of the rectangles are broken into parts, such as the rectangle of area $2 \times T_2$, which has part of its base between $t = 5$ and $t = 7$ and the remainder from $t = 8$ to $t = 10$ (thus $T_2 = 2 + 2 = 4$). It follows that the total area is given by $\sum_{i=0}^{\infty} i\, T_i = \int_0^T L(t)\, dt$ and therefore

$$\widehat{L} = \frac{1}{T}\sum_{i=0}^{\infty} i\, T_i = \frac{1}{T}\int_0^T L(t)\, dt \tag{6.2}$$

The expressions in Equations (6.1) and (6.2) are always equal for any queueing system, regardless of the number of servers, the queue discipline, or any other special circumstances. Equation (6.2) justifies the terminology *time-integrated average.*

Many queueing systems exhibit a certain kind of long-run stability in terms of their average performance. For such systems, as time T gets large, the observed time-average number in the system $\widehat{L}$ approaches a limiting value, say L, which is called the long-run time-average number in system. That is, with probability 1

$$\widehat{L} = \frac{1}{T}\int_0^T L(t)\, dt \longrightarrow L \text{ as } T \longrightarrow \infty \tag{6.3}$$

The estimator $\widehat{L}$ is said to be strongly consistent for L. If the simulation run length T is sufficiently long, the estimator $\widehat{L}$ becomes arbitrarily close to L. Unfortunately, for $T < \infty$, $\widehat{L}$ depends on the initial conditions at time 0.

Equations (6.2) and (6.3) can be applied to any subsystem of a queueing system as well as to the whole system. If $L_Q(t)$ denotes the number of customers waiting in line, and T_i^Q denotes the total time during $[0, T]$ in which exactly i customers are waiting in line, then

$$\widehat{L}_Q = \frac{1}{T}\sum_{i=0}^{\infty} i\, T_i^Q = \frac{1}{T}\int_0^T L_Q(t)\, dt \longrightarrow L_Q \text{ as } T \longrightarrow \infty \tag{6.4}$$

where $\widehat{L}_Q$ is the observed time-average number of customers waiting in line from time 0 to time T, and L_Q is the long-run time-average number waiting in line.

EXAMPLE 6.3

Suppose that Figure 6.6 represents a single-server queue—that is, a *G/G/1/N/K* queueing system ($N \geq 3$, $K \geq 3$). Then the number of customers waiting in line is given by $L_Q(t)$, defined by

$$L_Q(t) = \begin{cases} 0 & \text{if } L(t) = 0 \\ L(t) - 1 & \text{if } L(t) \geq 1 \end{cases}$$

and shown in Figure 6.7. Thus, $T_0^Q = 5 + 10 = 15$, $T_1^Q = 2 + 2 = 4$, and $T_2^Q = 1$. Therefore,

$$\widehat{L}_Q = \frac{0(15) + 1(4) + 2(1)}{20} = 0.3 \text{ customers}$$

◀

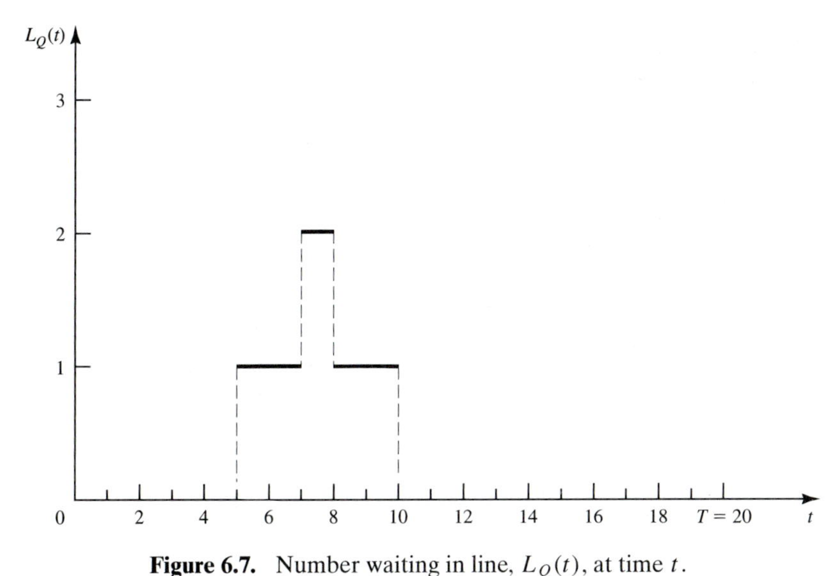

Figure 6.7. Number waiting in line, $L_Q(t)$, at time t.

6.3.2 *Average Time Spent in System per Customer, w*

If we simulate a queueing system for some period of time, say T, then we can record the time each customer spends in the system during $[0, T]$, say $W_1, W_2, \ldots, W_N$, where N is the number of arrivals during $[0, T]$. The average time spent in system per customer, called the *average system time*, is given by the ordinary sample average

$$\widehat{w} = \frac{1}{N}\sum_{i=1}^{N} W_i \tag{6.5}$$

For stable systems, as $N \longrightarrow \infty$

$$\widehat{w} \longrightarrow w \tag{6.6}$$

with probability 1, where w is called the *long-run average system time*.

If the system under consideration is the queue alone, Equations (6.5) and (6.6) are written as

$$\widehat{w}_Q = \frac{1}{N}\sum_{i=1}^{N} W_i^Q \longrightarrow w_Q \quad \text{as } N \longrightarrow \infty \tag{6.7}$$

where W_i^Q is the total time customer i spends waiting in queue, $\widehat{w}_Q$ is the observed average time spent in queue (called delay), and w_Q is the long-run average delay per customer. The estimators $\widehat{w}$ and $\widehat{w}_Q$ are influenced by initial conditions at time 0 and the run length T, analogously to $\widehat{L}$.

EXAMPLE 6.4

For the system history shown in Figure 6.6, $N = 5$ customers arrive, $W_1 = 2$, and $W_5 = 20 - 16 = 4$, but W_2, W_3, and W_4 cannot be computed unless more is known about the system. Assume that the system has a single server and a FIFO queue discipline. This implies that customers will depart from the system in the same order in which they arrived. Each jump upward of $L(t)$ in Figure 6.6 represents an arrival. Arrivals occur at times 0, 3, 5, 7, and 16. Similarly, departures occur at times 2, 8, 10, and 14. (A departure may or may not have occurred at time 20.) Under these assumptions, it is apparent that $W_2 = 8 - 3 = 5$, $W_3 = 10 - 5 = 5$, $W_4 = 14 - 7 = 7$, and therefore

$$\widehat{w} = \frac{2 + 5 + 5 + 7 + 4}{5} = \frac{23}{5} = 4.6 \text{ time units}$$

Thus, on the average, an arbitrary customer spends 4.6 time units in the system. As for time spent in the waiting line, it can be determined that $W_1^Q = 0$, $W_2^Q = 0$, $W_3^Q = 8 - 5 = 3$, $W_4^Q = 10 - 7 = 3$, and $W_5^Q = 0$; thus,

$$\widehat{w}_Q = \frac{0 + 0 + 3 + 3 + 0}{5} = 1.2 \text{ time units}$$ ◀

6.3.3 *The Conservation Equation: $L = \lambda w$*

For the system exhibited in Figure 6.6, there were $N = 5$ arrivals in $T = 20$ time units, and thus the observed arrival rate was $\widehat{\lambda} = N/T = 1/4$ customer per time unit. Recall that $\widehat{L} = 1.15$ and $\widehat{w} = 4.6$; hence, it follows that

$$\widehat{L} = \widehat{\lambda}\widehat{w} \tag{6.8}$$

This relationship between L, λ, and w is not coincidental; it holds for almost all queueing systems or subsystems, regardless of the number of servers, the queue discipline, or any other special circumstances. Allowing $T \longrightarrow \infty$ and $N \longrightarrow \infty$, Equation (6.8) becomes

$$L = \lambda w \tag{6.9}$$

where $\widehat{\lambda} \longrightarrow \lambda$, and λ is the long-run average arrival rate. Equation (6.9) is called a *conservation equation* and is usually attributed to Little [1961]. It says that the average number of customers in the system at an arbitrary point in time is equal to the average number of arrivals per time unit, times the average time spent in the system. For Figure 6.6, there is one arrival every 4 time units (on the average) and each arrival spends 4.6 time units in the system (on the average), so at an arbitrary point in time there will be (1/4)(4.6) = 1.15 customers present (on the average).

Equation (6.8) can also be derived by reconsidering Figure 6.6 in the following manner: Figure 6.8 shows system history, $L(t)$, exactly as in Figure 6.6, with each customer's time in the system, W_i, represented by a rectangle. This

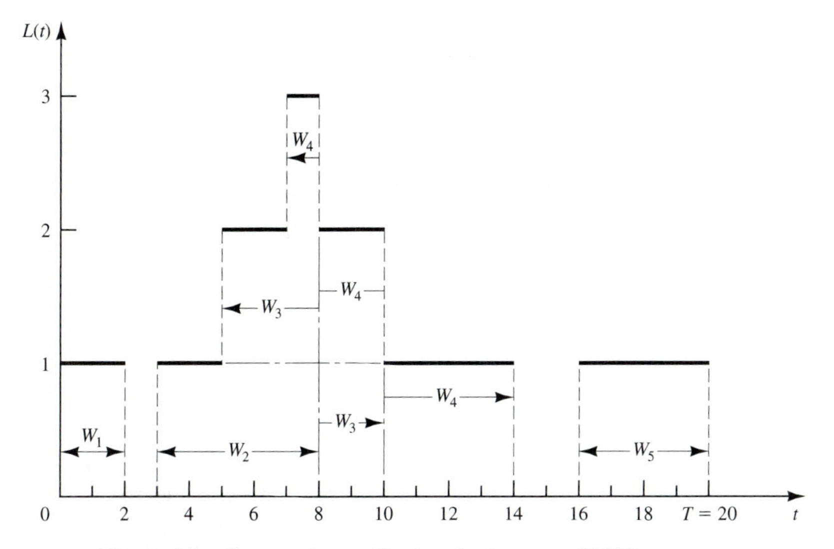

Figure 6.8. System times, W_i, for single-server FIFO system.

representation again assumes a single-server system with a FIFO queue discipline. The rectangles for the third and fourth customers are in two and three separate pieces, respectively. The ith rectangle has a height of 1 and a length of W_i for each $i = 1, 2, \ldots, N$. It follows that the total system time of all customers is given by the total area under the number-in-system function, $L(t)$; that is,

$$\sum_{i=1}^{N} W_i = \int_0^T L(t)\,dt \tag{6.10}$$

Therefore, by combining Equations (6.2) and (6.5) with $\widehat{\lambda} = N/T$, it follows that

$$\widehat{L} = \frac{1}{T}\int_0^T L(t)\,dt = \frac{N}{T}\frac{1}{N}\sum_{i=1}^{N} W_i = \widehat{\lambda}\widehat{w}$$

which is Little's Equation (6.8). The intuitive and informal derivation presented here depended on the single-server FIFO assumptions, but these assumptions are not necessary. In fact, Equation (6.10), which was the key to the derivation, holds (at least approximately) in great generality, and thus so do Equations (6.8) and (6.9). Exercises 14 and 15 ask the reader to derive Equations (6.10) and (6.8) under different assumptions.

Technical note: If, as defined in Section 6.3.2, W_i is the system time for customer i during $[0, T]$, then Equation (6.10) and hence Equation (6.8) hold exactly. Some authors choose to define W_i as total system time for customer i; this change will affect the value of W_i only for those customers i who arrive before time T but do not depart until after time T (possibly customer 5 in Figure 6.8). With this change in definition, Equations (6.10) and (6.8) only hold approximately. Nevertheless, as $T \longrightarrow \infty$ and $N \longrightarrow \infty$, the error in Equation (6.8) decreases to zero and therefore the conservation Equation (6.9), namely $L = \lambda w$, for long-run measures of performance holds exactly.

6.3.4 Server Utilization

Server utilization is defined as the proportion of time that a server is busy. Observed server utilization, denoted by $\widehat{\rho}$, is defined over a specified time interval $[0, T]$. Long-run server utilization is denoted by ρ. For systems that exhibit long-run stability,

$$\widehat{\rho} \longrightarrow \rho \text{ as } T \longrightarrow \infty$$

EXAMPLE 6.5

Referring to Figure 6.6 or 6.8, and assuming that the system has a single server, it can be seen that the server utilization is $\widehat{\rho} = (\text{total busy time})/T = (\sum_{i=1}^{\infty} T_i)/T = (T - T_0)/T = 17/20$. ◀

Server utilization in $G/G/1/\infty/\infty$ queues

Consider any single-server queueing system with average arrival rate λ customers per time unit, average service time $E(S) = 1/\mu$ time units, and infinite queue capacity and calling population. [$E(S) = 1/\mu$ implies that, when busy, the server is working at rate μ customers per time unit, on the average; μ is called the *service rate*.] The server alone is a subsystem that can be considered as a queueing system in itself, and hence the conservation Equation (6.9), $L = \lambda w$, can be applied to the server. For stable systems, the average arrival rate to the server, say λ_s, must be identical to the average arrival rate to the system, λ (certainly $\lambda_s \le \lambda$, since customers cannot be served faster than they arrive, but if $\lambda_s < \lambda$, then the waiting line would tend to grow in length at an average rate of $\lambda - \lambda_s$ customers per time unit, which is an unstable system). For the server subsystem, the average system time is $w = E(S) = \mu^{-1}$. The actual number of customers in the server subsystem is either 0 or 1, as shown in Figure 6.9 for the system represented by Figure 6.6. Hence, the average number, $\widehat{L}_s$, is given by

$$\widehat{L}_s = \frac{1}{T}\int_0^T (L(t) - L_Q(t))\,dt = \frac{T - T_0}{T}$$

In this case, $\widehat{L}_s = 17/20 = \widehat{\rho}$. In general, for a single-server queue, the average number of customers being served at an arbitrary point in time is equal to server utilization. As $T \longrightarrow \infty, \widehat{L}_s = \widehat{\rho} \longrightarrow L_s = \rho$. Combining these results into $L = \lambda w$ for the server subsystem yields

$$\rho = \lambda E(S) = \frac{\lambda}{\mu} \tag{6.11}$$

or the long-run server utilization in a single-server queue is equal to the average arrival rate divided by the average service rate. For a single-server queue to be

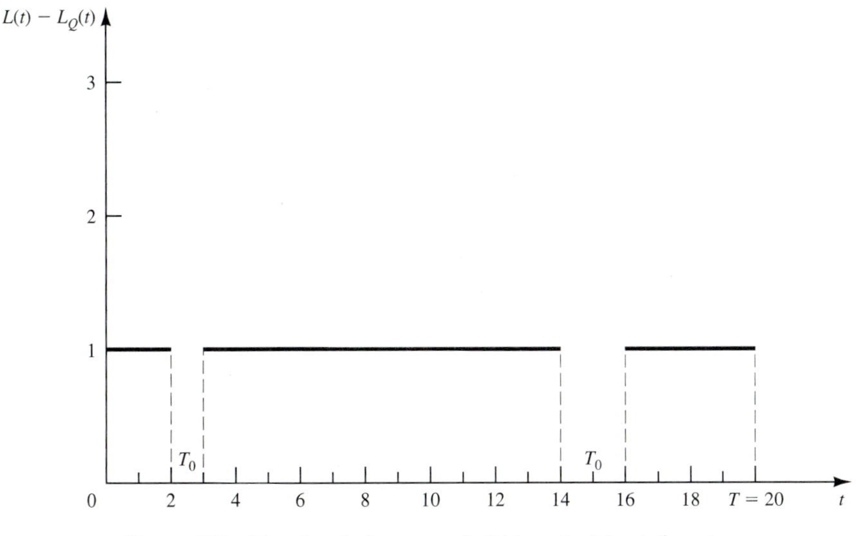

Figure 6.9. Number being served, $L(t) - L_Q(t)$, at time t.

stable, the arrival rate λ must be less than the service rate μ; that is,

$$\lambda < \mu$$

or

$$\rho = \frac{\lambda}{\mu} < 1 \tag{6.12}$$

If the arrival rate is greater than the service rate ($\lambda > \mu$), the server will eventually get further and further behind. After a time the server will always be busy and the waiting line will tend to grow in length at an average rate of $(\lambda - \mu)$ customers per time unit, because departures will be occurring at rate μ per time unit. For stable single-server systems ($\lambda < \mu$ or $\rho < 1$), long-run measures of performance such as average queue length L_Q (and also L, w, and w_Q) are well defined and have meaning. For unstable systems ($\lambda > \mu$), long-run server utilization is 1, and long-run average queue length is infinite; that is,

$$\frac{1}{T}\int_0^T L_Q(t)\,dt \longrightarrow +\infty \qquad \text{as } T \longrightarrow \infty$$

Similarly, $L = w = w_Q = \infty$, and therefore these long-run measures of performance are meaningless for unstable queues. The quantity λ/μ is also called the *offered load* and is a measure of the workload imposed on the system.

Server utilization in $G/G/c/\infty/\infty$ queues

Consider a queueing system with c identical servers in parallel. If an arriving customer finds more than one server idle, the customer chooses a server without favoring any particular server. (For example, the choice of server might be made at random.) Arrivals occur at rate λ from an infinite calling population, and each server works at rate μ customers per time unit. Using Equation (6.9), $L = \lambda w$, applied to the server subsystem alone, an argument similar to the one given for a single server leads to the result that, for systems in statistical equilibrium, the average number of busy servers, say L_s, is given by

$$L_s = \lambda E(S) = \frac{\lambda}{\mu} \tag{6.13}$$

Clearly, $0 \le L_s \le c$. The long-run average server utilization is defined by

$$\rho = \frac{L_s}{c} = \frac{\lambda}{c\mu} \tag{6.14}$$

so that $0 \le \rho \le 1$. The utilization ρ can be interpreted as the proportion of time an arbitrary server is busy in the long run.

The maximum service rate of the $G/G/c/\infty/\infty$ system is $c\mu$, which occurs when all servers are busy. For the system to be stable, the average arrival rate λ must be less than the maximum service rate $c\mu$; that is, the system is stable if and only if

$$\lambda < c\mu \tag{6.15}$$

or, equivalently, the offered load λ/μ is less than the number of servers c. If $\lambda > c\mu$, then arrivals are occurring, on the average, faster than the system can handle them, all servers will be continuously busy, and the waiting line will grow in length at an average rate of $(\lambda - c\mu)$ customers per time unit. Such a system is unstable, and the long-run performance measures (L, L_Q, w, and w_Q) are again meaningless for such systems.

Notice that condition (6.15) generalizes condition (6.12), and the equation for utilization for stable systems, Equation (6.14), generalizes Equation (6.11).

Equations (6.13) and (6.14) can also be applied when some servers work more than others — for example, when customers favor one server over others, or when certain servers serve customers only if all other servers are busy. In this case, L_s given by Equation (6.13) is still the average number of busy servers, but ρ as given by Equation (6.14) cannot be applied to an individual server. Instead, ρ must be interpreted as the average utilization of all servers.

EXAMPLE 6.6

Customers arrive at random to a license bureau at a rate of $\lambda = 50$ customers per hour. Presently there are 20 clerks, each serving $\mu = 5$ customers per hour on the average. Therefore the long-run, or steady-state, average utilization of a server, given by Equation (6.14), is

$$\rho = \frac{\lambda}{c\mu} = \frac{50}{20(5)} = 0.5$$

and the average number of busy servers is

$$L_s = \frac{\lambda}{\mu} = \frac{50}{5} = 10$$

Thus, in the long run, a typical clerk is busy serving customers only 50% of the time. The office manager asks if the number of servers can be decreased. By Equation (6.15), it follows that for the system to be stable, it is necessary for the number of servers to satisfy

$$c > \frac{\lambda}{\mu}$$

or $c > 50/5 = 10$. Thus, the possibilities for the manager to consider include $c = 11$, or $c = 12$, or $c = 13, \ldots$. Notice that $c \geq 11$ guarantees long-run stability only in the sense that all servers when busy can handle the incoming workload (i.e., $c\mu > \lambda$) on average. The office manager may well desire to have more than the minimum number of servers ($c = 11$) because of other factors, such as customer delays and the length of the waiting line. A stable queue can still have very long lines on the average. ◀

Server utilization and system performance

As will be illustrated here and in later sections, system performance can vary widely for a given value of utilization, ρ. Consider a $G/G/1/\infty/\infty$ queue — that is, a single-server queue with arrival rate λ, service rate μ, and utilization $\rho = \lambda/\mu < 1$.

At one extreme, consider the $D/D/1$ queue, which has deterministic arrival and service times. Then all interarrival times $\{A_1, A_2, \ldots\}$ are equal to $E(A) = 1/\lambda$ and all service times $\{S_1, S_2, \ldots\}$ are equal to $E(S) = 1/\mu$. Assuming that a customer arrives to an empty system at time 0, the system evolves in a completely deterministic and predictable fashion, as shown in Figure 6.10. Observe that $L = \rho = \lambda/\mu$, $w = E(S) = \mu^{-1}$, and $L_Q = w_Q = 0$. By varying λ and μ, server utilization can assume any value between 0 and 1, yet there is never any line whatsoever. What, then, causes lines to build, if not a high server utilization? In general, it is the variability of interarrival and service times that causes lines to fluctuate in length.

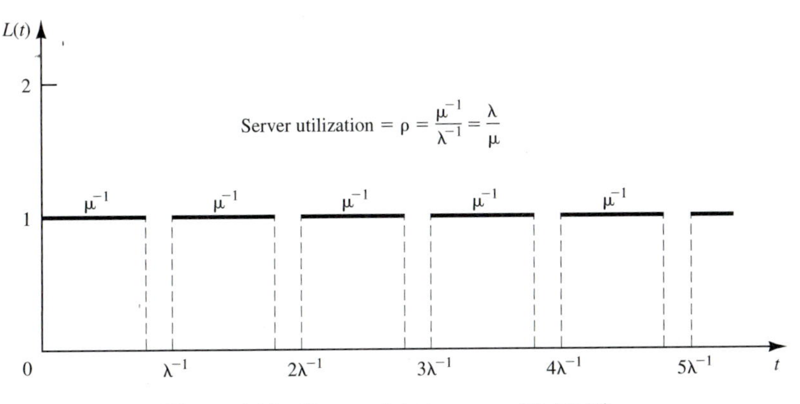

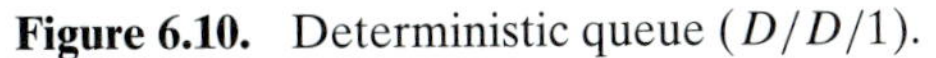
Figure 6.10. Deterministic queue ($D/D/1$).

Example 6.7

Consider a physician who schedules patients every 10 minutes and who spends S_i minutes with the ith patient, where

$$S_i = \begin{cases} 9 \text{ minutes with probability } 0.9 \\ 12 \text{ minutes with probability } 0.1 \end{cases}$$

Thus, arrivals are deterministic ($A_1 = A_2 = \cdots = \lambda^{-1} = 10$), but services are stochastic (or probabilistic), with mean and variance given by

$$E(S_i) = 9(0.9) + 12(0.1) = 9.3 \text{ minutes}$$

and

$$\begin{aligned} V(S_i) &= E(S_i^2) - [E(S_i)]^2 \\ &= 9^2(0.9) + 12^2(0.1) - (9.3)^2 \\ &= 0.81 \end{aligned}$$

Since $\rho = \lambda/\mu = E(S)/E(A) = 9.3/10 = 0.93 < 1$, the system is stable, and the physician will be busy 93% of the time in the long run. In the short run, lines will not build up as long as patients require only 9 minutes of service, but because of the variability in the service times, 10% of the patients will require 12 minutes, which in turn will cause a temporary line to form.

Suppose the system is simulated with service times $S_1 = 9$, $S_2 = 12$, $S_3 = 9$, $S_4 = 9$, $S_5 = 9, \ldots$. Assuming that at time 0 a patient arrived to find the doctor idle and subsequent patients arrived precisely at times 10, 20, 30, ..., the system evolves as in Figure 6.11. The delays in queue are $W_1^Q = W_2^Q = 0$, $W_3^Q = 22 - 20 = 2$, $W_4^Q = 31 - 30 = 1$, $W_5^Q = 0$. The occurrence of a

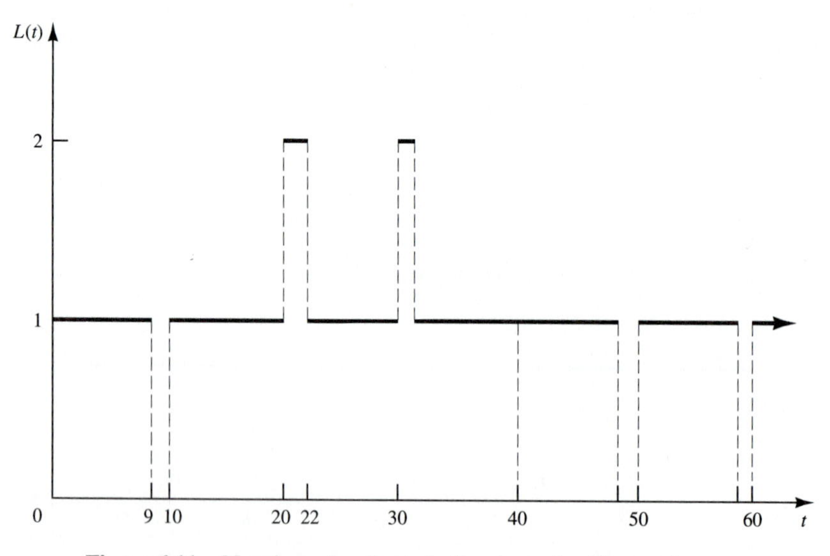

Figure 6.11. Number of patients in the doctor's office at time t.

relatively long service time (here $S_2 = 12$) caused a waiting line to temporarily form. In general, because of the variability of the interarrival and service distributions, relatively small interarrival times and relatively large service times occasionally do occur, and these in turn cause lines to lengthen. Conversely, the occurrence of a large interarrival time or a small service time will tend to shorten an existing waiting line. The relationship between utilization, service and interarrival variability, and system performance will be explored in more detail in Section 6.4. ◀

6.3.5 Costs in Queueing Problems

In many queueing situations, costs can be associated with various aspects of the waiting line or servers. Suppose that the system incurs a cost for each customer in the queue, say at a rate of \$10 per hour per customer. If customer j spends W_j^Q hours in the queue, then $\sum_{j=1}^{N}(\$10 \cdot W_j^Q)$ is the total cost of the N customers who arrive during the simulation. Thus, the average cost per customer is

$$\sum_{j=1}^{N} \frac{\$10 \cdot W_j^Q}{N} = \$10 \cdot \widehat{w}_Q$$

by Equation (6.7). If $\widehat{\lambda}$ customers per hour arrive (on the average), the average cost per hour is

$$\left(\widehat{\lambda}\frac{\text{customers}}{\text{hour}}\right)\left(\frac{\$10 \cdot \widehat{w}_Q}{\text{customer}}\right) = \$10 \cdot \widehat{\lambda}\widehat{w}_Q = \$10 \cdot \widehat{L}_Q/\text{hour}$$

the last equality following by Little's Equation (6.8). An alternative way to derive the average cost per hour is to consider Equation (6.2). If T_i^Q is the total time over the interval $[0, T]$ that the system contains exactly i customers, then $\$10 i\, T_i^Q$ is the cost incurred by the system during the time exactly i customers are present. Thus, the total cost is $\sum_{i=1}^{\infty}(\$10 \cdot i\, T_i^Q)$, and the average cost per hour is

$$\sum_{i=1}^{\infty} \frac{\$10 \cdot i\, T_i^Q}{T} = \$10 \cdot \widehat{L}_Q/\text{hour}$$

by Equation (6.4). In these cost expressions, $\widehat{L}_Q$ may be replaced by L_Q (if the long-run number in queue is known), or by L or $\widehat{L}$ (if costs are incurred while the customer is being served in addition to being delayed).

The server may also impose costs on the system. If a group of c parallel servers ($1 \le c < \infty$) have utilization ρ, and each server imposes a cost of \$5 per hour while busy, the total server cost per hour is

$$\$5 \cdot c\rho$$

since $c\rho$ is the average number of busy servers. If server cost is imposed only when the servers are idle, then the server cost per hour would be

$$\$5 \cdot c(1 - \rho)$$

since $c(1-\rho) = c - c\rho$ is the average number of idle servers. In many problems, two or more of these various costs are combined into a total cost. Such problems are illustrated by Exercises 1, 7, 8, 12, and 19. In most cases, the objective is to minimize total costs (given certain constraints) by varying those parameters under management's control, such as the number of servers, the arrival rate, the service rate, or system capacity.

6.4 Steady-State Behavior of Infinite-Population Markovian Models

This section presents the steady-state solution of a number of queueing models that can be solved mathematically. For the infinite-population models, the arrivals are assumed to follow a Poisson process with rate λ arrivals per time unit; that is, the interarrival times are assumed to be exponentially distributed with mean $1/\lambda$. Service times may be exponentially distributed (M) or arbitrary (G). The queue discipline will be FIFO. Because of the exponential distributional assumptions on the arrival process, these models are called Markovian models.

A queueing system is said to be in statistical equilibrium, or steady state, provided the probability that the system is in a given state is not time dependent; that is,

$$P(L(t) = n) = P_n(t) = P_n$$

is independent of time t. Two properties—approaching statistical equilibrium given any starting state and remaining in statistical equilibrium once it is reached—are characteristic of many stochastic models, and in particular of all the systems studied in the following subsections. On the other hand, if an analyst were interested in the transient behavior of a queue over a relatively short period of time and given some specific initial conditions (such as idle and empty), the results to be presented here are inappropriate. A transient mathematical analysis, or more likely a simulation model, would be the chosen tool of analysis.

The mathematical models whose solutions are shown in the following subsections may be used to obtain approximate results even when the assumptions of the model do not strictly hold. These results may be considered as a rough guide to the behavior of the system. A simulation may then be used for a more refined analysis. However, it should be remembered that a mathematical analysis (when it is applicable) provides the true value of the model parameter (e.g., L), whereas a simulation analysis delivers a statistical estimate (e.g., $\widehat{L}$) of the parameter. On the other hand, for complex systems, a simulation model is often a more faithful representation than a mathematical model.

For the simple models studied here, the steady-state parameter L, the time-average number of customers in the system, can be computed as

$$L = \sum_{n=0}^{\infty} n P_n \tag{6.16}$$

where $\{P_n\}$ are the steady-state probabilities (as defined in Table 6.2). As discussed in Section 6.3, and expressed in Equation (6.3), L can also be interpreted as a long-run measure of performance of the system. Once L is given, the other steady-state parameters can be readily computed by Little's Equation (6.9) applied to the whole system and the queue alone:

$$\begin{aligned} w &= \frac{L}{\lambda} \\ w_Q &= w - \frac{1}{\mu} \\ L_Q &= \lambda w_Q \end{aligned} \tag{6.17}$$

where λ is the arrival rate and μ the service rate per server.

In order for the $G/G/c/\infty/\infty$ queues considered in this section to have a statistical equilibrium, a necessary and sufficient condition is that $\lambda/(c\mu) < 1$, where λ is the arrival rate, μ is the service rate of one server, and c is the number of parallel servers. For these unlimited-capacity, infinite-calling-population models, it shall be assumed that the theoretical server utilization, $\rho = \lambda/c\mu$, satisfies $\rho < 1$. For models with finite system capacity or finite calling population, the quantity $\lambda/c\mu$ can assume any positive value.

6.4.1 *Single-Server Queues with Poisson Arrivals and Unlimited Capacity: $M/G/1$*

Suppose that service times have mean $1/\mu$ and variance σ^2, and there is one server. If $\rho = \lambda/\mu < 1$, then the $M/G/1$ queue has a steady-state probability distribution with steady-state characteristics as given in Table 6.3. In general, there is no simple expression for the steady-state probabilities $P_0, P_1, P_2, \ldots$. When $\lambda < \mu$, the quantity $\rho = \lambda/\mu$ is the server utilization, or long-run proportion of time the server is busy. As seen in Table 6.3, $1 - P_0 = \rho$ can also be interpreted as the steady-state probability that the system contains one or more customers. Notice also that $L - L_Q = \rho$ is the time-average number of customers being served.

Table 6.3. Steady-State Parameters of the $M/G/1$ Queue

ρ	$\frac{\lambda}{\mu}$
L	$\rho + \frac{\lambda^2(1/\mu^2 + \sigma^2)}{2(1-\rho)} = \rho + \frac{\rho^2(1+\sigma^2\mu^2)}{2(1-\rho)}$
w	$\frac{1}{\mu} + \frac{\lambda(1/\mu^2 + \sigma^2)}{2(1-\rho)}$
w_Q	$\frac{\lambda(1/\mu^2 + \sigma^2)}{2(1-\rho)}$
L_Q	$\frac{\lambda^2(1/\mu^2 + \sigma^2)}{2(1-\rho)} = \frac{\rho^2(1+\sigma^2\mu^2)}{2(1-\rho)}$
P_0	$1-\rho$

EXAMPLE 6.8

Widget-making machines malfunction apparently at random and require a mechanic's attention. It is assumed that malfunctions occur according to a Poisson process at rate $\lambda = 1.5$ per hour. Observation over several months has found that repairs by the single mechanic take an average time of 30 minutes with a standard deviation of 20 minutes. Thus the mean service time $1/\mu = 1/2$ hour, the service rate is $\mu = 2$ per hour, and $\sigma^2 = (20)^2$ minutes^2 = 1/9 hour^2. The "customers" are the widget makers, and the appropriate model is the $M/G/1$ queue, since only the mean and variance of service times are known, not their distribution. The proportion of time the mechanic is busy is $\rho = \lambda/\mu = 1.5/2 = 0.75$, and by Table 6.3 the steady-state time average number of broken machines is

$$L = 0.75 + \frac{(1.5)^2[(0.5)^2 + 1/9]}{2(1 - 0.75)}$$

$$= 0.75 + 1.625 = 2.375 \text{ machines}$$

Thus, an observer who noted the state of the repair system at arbitrary times would find an average of 2.375 broken machines (over the long run). ◀

A closer look at the formulas in Table 6.3 reveals the source of the waiting lines and delays in an $M/G/1$ queue. For example, L_Q may be rewritten as

$$L_Q = \frac{\rho^2}{2(1-\rho)} + \frac{\lambda^2\sigma^2}{2(1-\rho)}$$

The first term involves only the ratio of mean arrival rate λ to mean service rate μ. As shown by the second term, if λ and μ are held constant, the average length of the waiting line (L_Q) depends on the variability, σ^2, of the service times. If two systems have identical mean service times and mean interarrival times, the one with the more variable service times (larger σ^2) will tend to have longer lines on the average. Intuitively, if service times are highly variable, there is a high probability of a large service time occurring (say much larger than the mean service time), and when large service times do occur, there is a higher-than-usual tendency for lines to form and delays of customers to increase. (The reader should not confuse "steady state" with low variability or short lines; a system in steady state or statistical equilibrium can be highly variable and can have long waiting lines.)

EXAMPLE 6.9

Two workers are competing for a job. Able claims an average service time which is faster than Baker's, but Baker claims to be more consistent, if not as fast. The arrivals occur according to a Poisson process at a rate of $\lambda = 2$ per hour (1/30 per minute). Able's statistics are an average service time of 24 minutes with a standard deviation of 20 minutes. Baker's service statistics are an average service time of 25 minutes but a standard deviation of only 2 minutes. If the average length of the queue is the criterion for hiring, which worker should be hired? For Able, $\lambda = 1/30$ per minute, $1/\mu = 24$ minutes, $\sigma^2 = 20^2 = 400$ minutes^2, $\rho = \lambda/\mu = 24/30 = 4/5$, and the average queue length is computed as

$$L_Q = \frac{(1/30)^2[24^2 + 400]}{2(1 - 4/5)} = 2.711 \text{ customers}$$

For Baker, $\lambda = 1/30$ per minute, $1/\mu = 25$ minutes, $\sigma^2 = 2^2 = 4$ minutes^2, $\rho = 25/30 = 5/6$, and the average queue length is

$$L_Q = \frac{(1/30)^2[25^2 + 4]}{2(1 - 5/6)} = 2.097 \text{ customers}$$

Although working faster on the average, Able's greater service variability results in an average queue length about 30% greater than Baker's. On the basis of average queue length, L_Q, Baker wins. On the other hand, the proportion of arrivals who would find Able idle and thus experience no delay is $P_0 = 1 - \rho = 1/5 = 20\%$, while the proportion who would find Baker idle and thus experience no delay is $P_0 = 1 - \rho = 1/6 = 16.7\%$. ◀

One case of the $M/G/1$ queue that is of special note occurs when service times are exponential, as discussed next.

The $M/M/1$ queue. Suppose that service times in an $M/G/1$ queue are exponentially distributed with mean $1/\mu$. Then the variance as given by Equation (5.26) is $\sigma^2 = 1/\mu^2$. Since the mean and standard deviation of the exponential distribution are equal, the $M/M/1$ queue may often be a useful approximate model when service times have standard deviations approximately equal to their means. The steady-state parameters, given in Table 6.4, may be computed by substituting $\sigma^2 = 1/\mu^2$ into the formulas in Table 6.3. Alternatively, L may be computed by Equation (6.16) from the steady-state probabilities P_n given in Table 6.4, and then w, w_Q, and L_Q computed by Equations (6.17). The student can show that the two expressions for each parameter are equivalent by substituting $\rho = \lambda/\mu$ into the right-hand side of each equation in Table 6.4.

Table 6.4. Steady-State Parameters of the $M/M/1$ Queue

L	$\dfrac{\lambda}{\mu-\lambda} = \dfrac{\rho}{1-\rho}$
w	$\dfrac{1}{\mu-\lambda} = \dfrac{1}{\mu(1-\rho)}$
w_Q	$\dfrac{\lambda}{\mu(\mu-\lambda)} = \dfrac{\rho}{\mu(1-\rho)}$
L_Q	$\dfrac{\lambda^2}{\mu(\mu-\lambda)} = \dfrac{\rho^2}{1-\rho}$
P_n	$\left(1-\dfrac{\lambda}{\mu}\right)\left(\dfrac{\lambda}{\mu}\right)^n = (1-\rho)\rho^n$

EXAMPLE 6.10

The interarrival times as well as the service times at a single-chair unisex barbershop have been shown to be exponentially distributed. The values of λ and μ are 2 per hour and 3 per hour, respectively. That is, the time between arrivals averages 1/2 hour, exponentially distributed, and the service time averages 20 minutes, also exponentially distributed. The server utilization and the probabilities for zero, one, two, three, and four or more customers in the shop are computed as follows:

$$\rho = \frac{\lambda}{\mu} = \frac{2}{3}$$

$$P_0 = 1 - \frac{\lambda}{\mu} = \frac{1}{3}$$

$$P_1 = \left(\frac{1}{3}\right)\left(\frac{2}{3}\right) = \frac{2}{9}$$

$$P_2 = \left(\frac{1}{3}\right)\left(\frac{2}{3}\right)^2 = \frac{4}{27}$$

$$P_3 = \left(\frac{1}{3}\right)\left(\frac{2}{3}\right)^3 = \frac{8}{81}$$

$$P_{\geq 4} = 1 - \sum_{n=0}^{3} P_n = 1 - \frac{1}{3} - \frac{2}{9} - \frac{4}{27} - \frac{8}{81} = \frac{16}{81}$$

From the calculations, the probability that the barber is busy is $1 - P_0 = \rho = 0.67$, and thus the probability that the barber is idle is 0.33. The time-average number of customers in the system is given by Table 6.4 as

$$L = \frac{\lambda}{\mu-\lambda} = \frac{2}{3-2} = 2 \text{ customers}$$

The average time an arrival spends in the system can be obtained from Table 6.4 or Equation (6.17) as

$$w = \frac{L}{\lambda} = \frac{2}{2} = 1 \text{ hour}$$

The average time the customer spends in the queue can be obtained using Equation (6.17) as

$$w_Q = w - \frac{1}{\mu} = 1 - \frac{1}{3} = \frac{2}{3} \text{ hour}$$

Using Table 6.4, the time-average number in the queue is given by

$$L_Q = \frac{\lambda^2}{\mu(\mu - \lambda)} = \frac{4}{3(1)} = \frac{4}{3} \text{ customers}$$

Finally, notice that multiplying $w = w_Q + 1/\mu$ through by λ and using Little's Equation (6.9) yields

$$L = L_Q + \frac{\lambda}{\mu} = \frac{4}{3} + \frac{2}{3} = 2 \text{ customers}$$ ◄

EXAMPLE 6.11

For the $M/M/1$ queue with service rate $\mu = 10$ customers per hour, consider how L and w increase as the arrival rate, λ, increases from 5 to 8.64 by increments of 20%, and then to $\lambda = 10$.

λ	5.0	6.0	7.2	8.64	10.0
ρ	0.500	0.600	0.720	0.864	1.0
L	1.00	1.50	2.57	6.35	∞
w	0.20	0.25	0.36	0.73	∞

For any $M/G/1$ queue, if $\lambda/\mu \geq 1$, waiting lines tend to continually grow in length; the long-run measures of performance, L, w, w_Q, and L_Q, are all infinite ($L = w = w_Q = L_Q = \infty$); and a steady-state probability distribution does not exist. As shown here for $\lambda < \mu$, if ρ is close to 1, waiting lines and delays will tend to be long. Notice that the increase in average system time, w, and average number in system, L, is highly nonlinear as a function of ρ. For example, as λ increases by 20%, L increases first by 50% (from 1.00 to 1.50), then by 71% (to 2.57), and then by 147% (to 6.35). ◄

EXAMPLE 6.12

If arrivals are occurring at rate $\lambda = 10$ per hour, and management has a choice of two servers, one who works at rate $\mu_1 = 11$ customers per hour and the second at rate $\mu_2 = 12$ customers per hour, the respective utilizations are $\rho_1 = \lambda/\mu_1 = 10/11 = 0.909$ and $\rho_2 = \lambda/\mu_2 = 10/12 = 0.833$. If the $M/M/1$ queue is used as an approximate model, then with the first server the average number in the system would be, by Table 6.4,

$$L_1 = \frac{\rho_1}{1 - \rho_1} = 10$$

and with the second server the average number in the system would be

$$L_2 = \frac{\rho_2}{1 - \rho_2} = 5$$

Thus, a decrease in service rate from 12 to 11 customers per hour, a mere 8.3% decrease, would result in an increase in average number in system from 5 to 10, which is a 100% increase. ◀

The effect of utilization and service variability

For any $M/G/1$ queue, if lines are too long, they can be reduced by decreasing the server utilization, ρ, or by decreasing the service-time variability, σ^2. These remarks hold for almost all queues, not just the $M/G/1$ queue. The utilization factor ρ can be reduced by decreasing the arrival rate λ, increasing the service rate μ, or increasing the number of servers, since in general $\rho = \lambda/c\mu$, where c is the number of parallel servers. The effect of additional servers will be studied in the following subsections. Figure 6.12 illustrates the effect of service

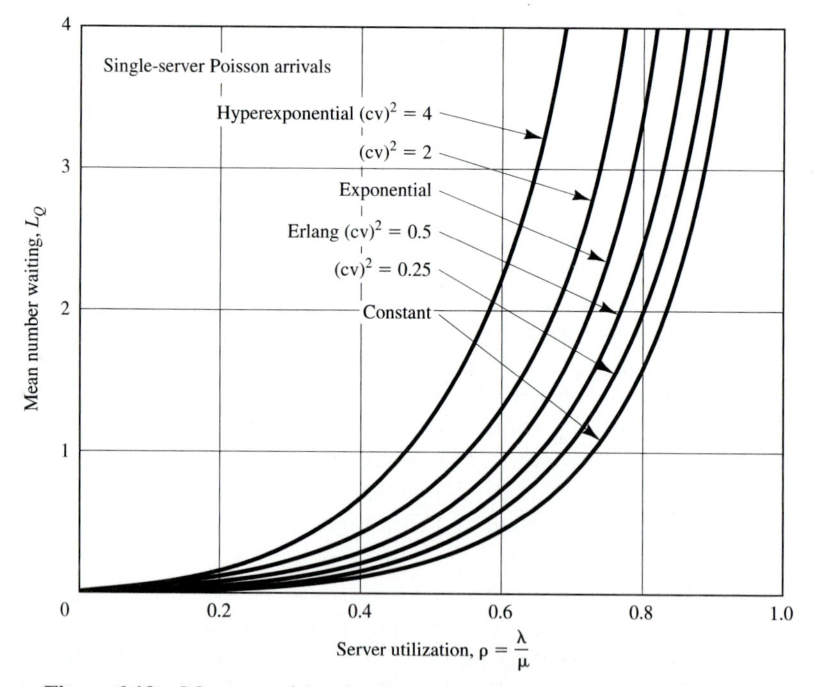

Figure 6.12. Mean number of customers waiting, L_Q, in $M/G/1$ queue having service distributions with given cv. (Adapted from Geoffrey Gordon, *System Simulation*, 2d ed., Prentice Hall, Upper Saddle River, NJ, 1978.)

variability. The mean steady-state number in the queue, L_Q, is plotted versus utilization ρ for a number of different coefficients of variation. The coefficient of variation (cv) of a positive random variable X is defined by

$$(\text{cv})^2 = \frac{V(X)}{[E(X)]^2}$$

and is a measure of the variability of a distribution. The larger its value, the more variable is the distribution relative to its expected value. For deterministic service times, $V(X) = 0$, so $\text{cv} = 0$. For Erlang service times of order k, $V(X) = 1/k\mu^2$ and $E(X) = 1/\mu$, so that $\text{cv} = 1/\sqrt{k}$. For exponential service times at service rate μ, the mean service time is $E(X) = 1/\mu$ and the variance is $V(X) = 1/\mu^2$, so that $\text{cv} = 1$. If service times have standard deviation greater than their mean (i.e., if $\text{cv} > 1$), then the hyperexponential distribution, which can achieve any desired coefficient of variation greater than 1, provides a good model. One occasion where it arises is given in Exercise 16.

The formula for L_Q for any $M/G/1$ queue can be rewritten in terms of the coefficient of variation by noticing that $(\text{cv})^2 = \sigma^2/(1/\mu)^2 = \sigma^2\mu^2$. Therefore,

$$L_Q = \frac{\rho^2(1 + \sigma^2\mu^2)}{2(1 - \rho)}$$

$$= \frac{\rho^2(1 + (\text{cv})^2)}{2(1 - \rho)} \tag{6.18}$$

$$= \left(\frac{\rho^2}{1 - \rho}\right)\left(\frac{1 + (\text{cv})^2}{2}\right) \tag{6.19}$$

The first term, $\rho^2/(1 - \rho)$, is L_Q for an $M/M/1$ queue. The second term, $(1 + (\text{cv})^2)/2$, corrects the $M/M/1$ formula to account for a nonexponential service-time distribution. The formula for w_Q can be obtained from the corresponding $M/M/1$ formula by applying the same correction factor.

6.4.2 Multiserver Queue: $M/M/c/\infty/\infty$

Suppose that there are c channels operating in parallel. Each of these channels has an independent and identical exponential service-time distribution with mean $1/\mu$. The arrival process is Poisson with rate λ. Arrivals will join a single queue and enter the first available service channel. The queueing system is shown in Figure 6.13. If the number in the system is $n < c$, an arrival will enter an available channel. However, when $n \geq c$, a queue will build if arrivals occur.

The offered load is defined by λ/μ. If $\lambda \geq c\mu$, the arrival rate is greater than or equal to the maximum service rate of the system, the service rate when all servers are busy. Thus, the system cannot handle the load put upon it, and therefore it has no statistical equilibrium. If $\lambda > c\mu$, the waiting line grows in

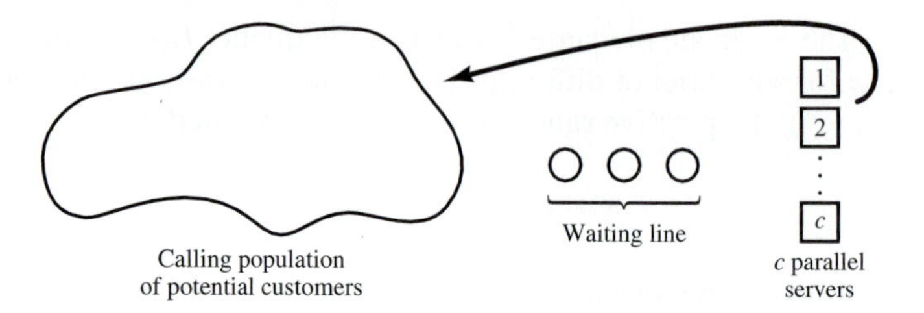

Figure 6.13. Multiserver queueing system.

length at rate $(\lambda - c\mu)$ customers per time unit, on the average. Customers are entering the system at rate λ per time unit but are leaving the system at a maximum rate of $c\mu$ per time unit.

For the $M/M/c$ queue to have statistical equilibrium, the offered load must satisfy $\lambda/\mu < c$, in which case $\lambda/(c\mu) = \rho$, the server utilization. The steady-state parameters are listed in Table 6.5. Most of the measures of performance can be expressed fairly simply in terms of P_0, the probability that the system is empty, or $\sum_{n=c}^{\infty} P_n$, the probability that all servers are busy, denoted by $P(L(\infty) \geq c)$, where $L(\infty)$ is a random variable representing the number in system in statistical equilibrium (after a very long time). Thus, $P(L(\infty) = n) = P_n, n = 0, 1, 2, \ldots$. The value of P_0 is necessary for computing all the measures of performance, and the equation for P_0 is somewhat more complex than in the previous cases. However, P_0 only depends on c and

Table 6.5. Steady-State Parameters for the $M/M/c$ Queue

ρ	$\dfrac{\lambda}{c\mu}$
P_0	$\left\{\left[\sum_{n=0}^{c-1} \dfrac{(\lambda/\mu)^n}{n!}\right] + \left[\left(\dfrac{\lambda}{\mu}\right)^c \left(\dfrac{1}{c!}\right)\left(\dfrac{c\mu}{c\mu - \lambda}\right)\right]\right\}^{-1}$ $= \left\{\left[\sum_{n=0}^{c-1} \dfrac{(c\rho)^n}{n!}\right] + \left[(c\rho)^c \left(\dfrac{1}{c!}\right)\dfrac{1}{1-\rho}\right]\right\}^{-1}$
$P(L(\infty) \geq c)$	$\dfrac{(\lambda/\mu)^c P_0}{c!(1 - \lambda/c\mu)} = \dfrac{(c\rho)^c P_0}{c!(1-\rho)}$
L	$c\rho + \dfrac{(c\rho)^{c+1} P_0}{c(c!)(1-\rho)^2} = c\rho + \dfrac{\rho P(L(\infty) \geq c)}{1-\rho}$
w	$\dfrac{L}{\lambda}$
w_Q	$w - \dfrac{1}{\mu}$
L_Q	$\lambda w_Q = \dfrac{(c\rho)^{c+1} P_0}{c(c!)(1-\rho)^2} = \dfrac{\rho P(L(\infty) \geq c)}{1-\rho}$
$L - L_Q$	$\dfrac{\lambda}{\mu} = c\rho$

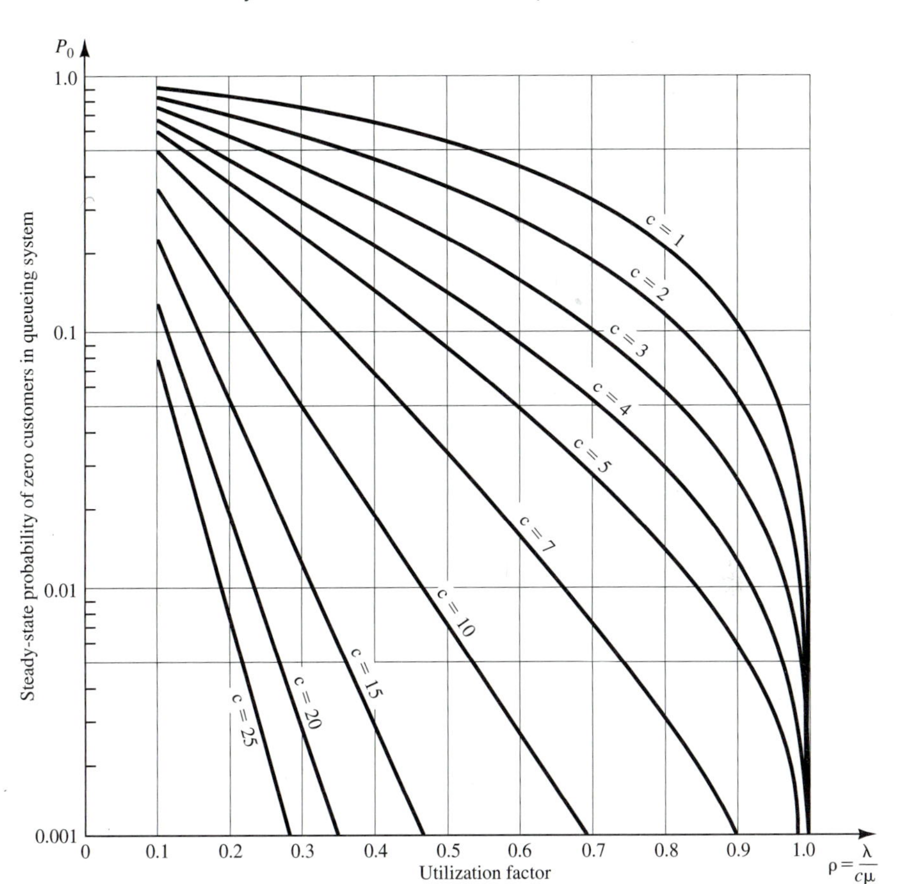

Figure 6.14. Values of P_0 for $M/M/c/\infty$ model. (From F. S. Hillier and G. J. Lieberman, *Introduction to Operations Research*, 5th ed., 1990, p. 616. Adapted with permission of McGraw-Hill, Inc., New York.)

ρ. A good approximation to P_0 can be obtained by using Figure 6.14, where P_0 is plotted versus ρ on semilog paper for various values c. Figure 6.15 is a plot of L versus ρ for different values of c.

The results in Table 6.5 simplify to those in Table 6.4 when $c = 1$, the case of a single server. Notice that the average number of busy servers, or the average number of customers being served, is given by the simple expression, $L - L_Q = \lambda/\mu = c\rho$.

Example 6.13

Many early examples of queueing theory applied to practical problems concerning tool cribs. Attendants manage the tool cribs while mechanics, assumed to be from an infinite calling population, arrive for service. Assume Poisson arrivals at rate of 2 mechanics per minute and exponentially distributed service times with mean 40 seconds.

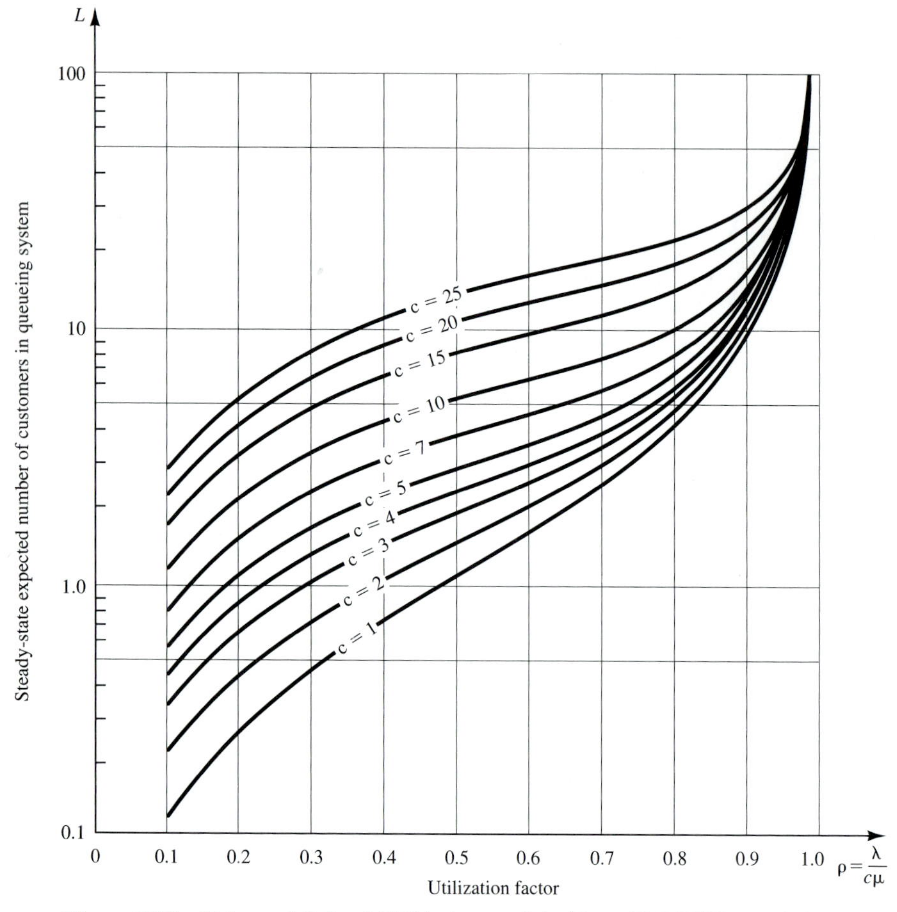

Figure 6.15. Values of L for $M/M/c/\infty$ model. (From F. S. Hillier and G. J. Lieberman, *Introduction to Operations Research*, 5th ed., 1990, p. 617. Adapted with permission of McGraw-Hill, Inc., New York.)

Now, $\lambda = 2$ per minute, and $\mu = 60/40 = 3/2$ per minute. Since the offered load is greater than 1, that is, since

$$\frac{\lambda}{\mu} = \frac{2}{3/2} = \frac{4}{3} > 1$$

more than one server is needed if the system is to have a statistical equilibrium. The requirement for steady state is that $c > \lambda/\mu = 4/3$. Thus at least $c = 2$ attendants are needed. The quantity 4/3 is the expected number of busy servers, and for $c \geq 2$, $\rho = 4/(3c)$ is the long-run proportion of time each server is busy. (What would happen if there were only $c = 1$ server?)

Let there be $c = 2$ attendants. First, P_0 is calculated as

$$P_0 = \left\{ \sum_{n=0}^{1} \frac{(4/3)^n}{n!} + \left(\frac{4}{3}\right)^2 \left(\frac{1}{2!}\right) \left[\frac{2(3/2)}{2(3/2) - 2}\right] \right\}^{-1}$$

$$= \left\{1 + \frac{4}{3} + \left(\frac{16}{9}\right)\left(\frac{1}{2}\right)(3)\right\}^{-1} = \left(\frac{15}{3}\right)^{-1} = \frac{1}{5} = 0.2$$

Proceeding, the probability that all servers are busy is given by

$$P(L(\infty) \geq 2) = \frac{(4/3)^2}{2!(1 - 2/3)}\left(\frac{1}{5}\right) = \left(\frac{8}{3}\right)\left(\frac{1}{5}\right) = \frac{8}{15} = 0.533$$

Thus, the time-average length of the waiting line of mechanics is

$$L_Q = \frac{(2/3)(8/15)}{1 - 2/3} = 1.07 \text{ mechanics}$$

and the time-average number in the system is given by

$$L = L_Q + \frac{\lambda}{\mu} = \frac{16}{15} + \frac{4}{3} = \frac{12}{5} = 2.4 \text{ mechanics}$$

Using Little's relationships, the average time a mechanic spends at the tool crib is

$$w = \frac{L}{\lambda} = \frac{2.4}{2} = 1.2 \text{ minutes}$$

while the average time spent waiting for an attendant is

$$w_Q = w - \frac{1}{\mu} = 1.2 - \frac{2}{3} = 0.533 \text{ minute}$$ ◀

Example 6.14

Using the data of Example 6.13, compute P_0 and L from Figures 6.14 and 6.15. First compute

$$\rho = \frac{\lambda}{c\mu} = \frac{2}{2(3/2)} = \frac{2}{3} = 0.667$$

Entering the utilization factor of 0.667 on the horizontal axis of Figure 6.14 gives a value for P_0 of 0.2 on the vertical axis. Similarly, a value of $L = 2.4$ is read from the vertical axis of Figure 6.15. ◀

An approximation for the $M/G/c/\infty$ queue

Recall that formulas for L_Q and w_Q for the $M/G/1$ queue can be obtained from the corresponding $M/M/1$ formulas by multiplying them by the correction factor $\left(1 + (\text{cv})^2\right)/2$, as in Equation (6.19). *Approximate* formulas for the $M/G/c$ queue can be obtained by applying the same correction factor to the $M/M/c$ formulas for L_Q and w_Q (no exact formula exists for $1 < c < \infty$). The nearer cv is to 1, the better the approximation.

EXAMPLE 6.15

Recall Example 6.13. Suppose that the service times for the mechanics at the tool crib are not exponentially distributed, but are known to have a standard deviation of 30 seconds. Then we have an $M/G/c$ model, rather than an $M/M/c$. Since the mean service time is 40 seconds, the coefficient of variation of the service time is

$$\text{cv} = \frac{30}{40} = \frac{3}{4} < 1$$

Therefore, the accuracy of L_Q and w_Q can be improved by the correction factor

$$\frac{1 + (\text{cv})^2}{2} = \frac{1 + (3/4)^2}{2} = \frac{25}{32} = 0.78$$

For example, when there are $c = 2$ attendants,

$$L_Q = (0.78)(1.07) = 0.83 \text{ mechanics}$$

Notice that since the coefficient of variation of the service time is less than 1, the congestion in the system, as measured by L_Q, is less than in the corresponding $M/M/2$ model.

The correction factor applies only to the formulas for L_Q and w_Q. Little's formula can then be used to calculate L and w. Unfortunately, there is no general method for correcting the steady-state probabilities, P_n. ◀

When the number of servers is infinite ($M/G/\infty/\infty$)

There are at least three situations in which it is appropriate to treat the number of servers as infinite:

1. When each customer is its own server — in other words, a self-service system;
2. When service capacity far exceeds service demand, a so-called ample-server system; and
3. When we want to know how many servers are required so that customers are rarely delayed.

The steady-state parameters for the $M/G/\infty$ queue are listed in Table 6.6. In the table, λ is the arrival rate of the Poisson arrival process and $1/\mu$ is the expected service time of the general service-time distribution (including exponential, constant, or any other).

EXAMPLE 6.16

Prior to introducing their new online computer information service, The Connection must plan their system capacity in terms of the number of users that can be logged on simultaneously. If the service is successful, customers are expected to log on at a rate of $\lambda = 500$ per hour, according to a Poisson process, and stay connected for an average of $1/\mu = 180$ minutes (or 3 hours). In the real system there will be an upper limit on simultaneous users, but for planning purposes The Connection can pretend that the number of simultaneous users

Table 6.6. Steady-State Parameters for the $M/G/\infty$ Queue

P_0	$e^{-\lambda/\mu}$
w	$\frac{1}{\mu}$
w_Q	0
L	$\frac{\lambda}{\mu}$
L_Q	0
P_n	$\frac{e^{-\lambda/\mu}(\lambda/\mu)^n}{n!}, \ n = 0, 1, \ldots$

is infinite. An $M/G/\infty$ model of the system implies that the expected number of simultaneous users is $L = \lambda/\mu = 500(3) = 1500$, so a capacity greater than 1500 is certainly required. To insure that they have adequate capacity 95% of the time, The Connection could allow the number of simultaneous users to be the smallest value c such that

$$P(L(\infty) \leq c) = \sum_{n=0}^{c} P_n = \sum_{n=0}^{c} \frac{e^{-1500}(1500)^n}{n!} \geq 0.95$$

A capacity of $c = 1564$ simultaneous users satisfies this requirement. ◀

6.4.3 *Multiserver Queues with Poisson Arrivals and Limited Capacity:* $M/M/c/N/\infty$

Suppose that service times are exponentially distributed at rate μ, there are c servers, and the total system capacity is $N \geq c$ customers. If an arrival occurs when the system is full, that arrival is turned away and does not enter the system. As in the preceding section, suppose that arrivals occur randomly according to a Poisson process with rate λ arrivals per time unit. For any values of λ and μ such that $\rho \neq 1$, the $M/M/c/N$ queue has a statistical equilibrium with steady-state characteristics as given in Table 6.7 (formulas for the case $\rho = 1$ can be found in Hillier and Lieberman [1990]).

The effective arrival rate, λ_e, is defined as the mean number of arrivals per time unit who enter and remain in the system. For all systems, $\lambda_e \leq \lambda$; for the unlimited-capacity systems, $\lambda_e = \lambda$; but for systems such as the present one which turn customers away when full, $\lambda_e < \lambda$. The effective arrival rate is computed by

$$\lambda_e = \lambda(1 - P_N)$$

since $1 - P_N$ is the probability that a customer, upon arrival, will find space and be able to enter the system. When using Little's Equations (6.17) to compute mean time spent in system w and in queue w_Q, λ must be replaced by λ_e.

Table 6.7. Steady-State Parameters for the $M/M/c/N$ Queue (N = System Capacity, $a = \lambda/\mu$, $\rho = \lambda/(c\mu)$)

P_0	$\left[1 + \sum_{n=1}^{c} \frac{a^n}{n!} + \frac{a^c}{c!} \sum_{n=c+1}^{N} \rho^{n-c}\right]^{-1}$
P_N	$\frac{a^N}{c!c^{N-c}} P_0$
L_Q	$\frac{P_0 a^c \rho}{c!(1-\rho)^2}\left[1 - \rho^{N-c} - (N-c)\rho^{N-c}(1-\rho)\right]$
λ_e	$\lambda(1 - P_N)$
w_Q	$\frac{L_Q}{\lambda_e}$
w	$w_Q + \frac{1}{\mu}$
L	$\lambda_e w$

Example 6.17

The unisex barbershop described in Example 6.10 can hold only three customers, one in service and two waiting. Additional customers are turned away when the system is full. The offered load is as previously determined, namely $\lambda/\mu = 2/3$.

In order to calculate the performance measures, first determine P_0 as

$$P_0 = \left[1 + \frac{2}{3} + \frac{2}{3} \sum_{n=2}^{3} \left(\frac{2}{3}\right)^{n-1}\right]^{-1} = 0.415$$

The probability that there are three customers in the system (the system is full) is computed by

$$P_N = P_3 = \frac{(2/3)^3}{1!1^2} P_0 = \frac{8}{65} = 0.123$$

Then, the average length of the queue (customers waiting for a haircut) is given by

$$L_Q = \frac{(27/65)(2/3)(2/3)}{(1 - 2/3)^2}\left[1 - (2/3)^2 - 2(2/3)^2(1 - 2/3)\right] = 0.431 \text{ customer}$$

Now, the effective arrival rate, λ_e, is given by

$$\lambda_e = 2\left(1 - \frac{8}{65}\right) = \frac{114}{65} = 1.754 \text{ customers per hour}$$

Therefore, using Little's equation, the expected time spent waiting in queue is

$$w_Q = \frac{L_Q}{\lambda_e} = \frac{28}{114} = 0.246 \text{ hour}$$

while the expected total time in the shop is

$$w = w_Q + \frac{1}{\mu} = \frac{66}{114} = 0.579 \text{ hour}$$

One last application of Little's equation gives the expected number of customers in the shop (in queue and getting a haircut) as

$$L = \lambda_e w = \frac{66}{65} = 1.015 \text{ customers}$$

Notice that $1 - P_0 = 0.585$ is the average number of customers being served, or equivalently, the probability that the single server is busy. Thus, the server utilization, or proportion of time the server is busy in the long run, is given by

$$1 - P_0 = \frac{\lambda_e}{\mu} = 0.585$$ ◀

The reader should compare these results to those of the unisex barbershop before the capacity constraint was placed on the system. Specifically, in systems with limited capacity, the offered load λ/μ can assume any positive value and no longer equals the server utilization $\rho = \lambda_e/\mu$. Notice that server utilization decreases from 67% to 58.5% when the system imposes a capacity constraint.

6.5 Steady-State Behavior of Finite-Population Models $(M/M/c/K/K)$

In many practical problems the assumption of an infinite calling population leads to invalid results, because the calling population is, in fact, small. In this situation, the presence of one or more customers in the system has a strong effect on the distribution of future arrivals, and the use of an infinite-population model can be misleading. Typical examples include a small group of machines that break down from time to time and require repair, or a small group of mechanics who line up at a counter for parts or tools. In the extreme case, if all the machines are broken, no new "arrivals" (breakdowns) of machines can occur; similarly, if all the mechanics are in line, no arrival is possible to the tool and parts counter. Contrast this to the infinite-population models in which the arrival rate, λ, of customers to the system is assumed independent of the state of the system.

Consider a finite-calling-population model with K customers. The time between the end of one service visit and the next call for service for each member of the population is assumed to be exponentially distributed with mean $1/\lambda$ time units; service times are also exponentially distributed with mean $1/\mu$ time units; there are c parallel servers, and system capacity is K, so that all arrivals remain for service. Such a system is depicted in Figure 6.16.

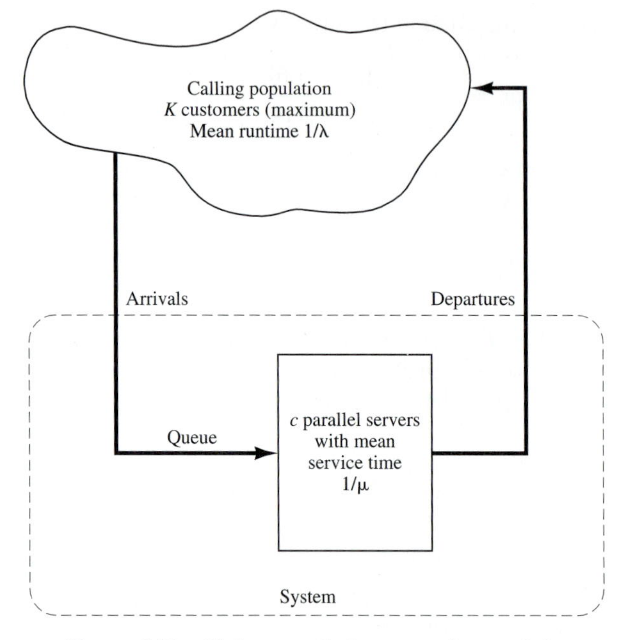

Figure 6.16. Finite-population queueing model.

The steady-state parameters for this model are listed in Table 6.8. An electronic spreadsheet or a symbolic calculation program is useful for evaluating these complex formulas. For example, Figure 6.17 is a procedure written for the symbolic calculation program Maple to calculate the steady-state probabilities for the $M/M/c/K/K$ queue. Another approach is to use precomputed queueing tables, such as those found in Banks and Heikes [1984], Hillier and Yu [1981], Peck and Hazelwood [1958] or Descloux [1962].

The effective arrival rate λ_e has several valid interpretations:

$$\begin{aligned}\lambda_e &= \text{long-run effective arrival rate of customers to the queue}\\ &= \text{long-run effective arrival rate of customers entering service}\\ &= \text{long-run rate at which customers exit service}\\ &= \text{long-run rate at which customers enter the calling population}\\ &\quad\ \text{(and begin a new runtime)}\\ &= \text{long-run rate at which customers exit the calling population}\end{aligned}$$

Example 6.18

Two workers are responsible for 10 milling machines. The machines run on the average for 20 minutes, then require an average 5-minute service period, both times exponentially distributed. Therefore, $\lambda = 1/20$ and $\mu = 1/5$. Determine the various measures of performance for this system.

Table 6.8. Steady-State Parameters for the $M/M/c/K/K$ Queue

P_0	$\left[\sum_{n=0}^{c-1}\binom{K}{n}\left(\frac{\lambda}{\mu}\right)^n + \sum_{n=c}^{K}\frac{K!}{(K-n)!c!c^{n-c}}\left(\frac{\lambda}{\mu}\right)^n\right]^{-1}$
P_n	$\begin{cases}\binom{K}{n}\left(\frac{\lambda}{\mu}\right)^n P_0, & n=0,1,\ldots,c-1\\ \frac{K!}{(K-n)!c!c^{n-c}}\left(\frac{\lambda}{\mu}\right)^n P_0, & n=c,c+1,\ldots,K\end{cases}$
L	$\sum_{n=0}^{K} nP_n$
L_Q	$\sum_{n=c+1}^{K}(n-c)P_n$
λ_e	$\sum_{n=0}^{K}(K-n)\lambda P_n$
w	L/λ_e
w_Q	L_Q/λ_e
ρ	$\frac{L-L_Q}{c} = \frac{\lambda_e}{c\mu}$

```
mmcKK := proc(lambda, mu, c, K)
         # return steady-state probabilities for M/M/c/K/K queue
         # notice that p[n+1] is P_n, n=0,..,K
         local crho, Kfac, cfac, p, n;
         p := vector(K+1,0);
         crho := lambda/mu;
         Kfac := K!;
         cfac := c!;
         p[1] := sum((Kfac/(n!*(K-n)!))*crho^n, n=0..c-1) +
                 sum((Kfac/(c^(n-c)*(K-n)!*cfac))*crho^n, n=c..K);
         p[1] := 1/p[1];
         for n from 1 to c-1
         do
            p[n+1] := p[1]*(Kfac/(n!*(K-n)!))*crho^n;
         od;
         for n from c to K
         do
            p[n+1] := p[1]*(Kfac/(c^(n-c)*(K-n)!*cfac))*crho^n;
         od;
         RETURN(evalm(p));
         end;
```

Figure 6.17 Maple procedure to calculate P_n for the $M/M/c/K/K$ queue.

All of the performance measures depend on P_0, which is

$$\left[\sum_{n=0}^{2-1} \binom{10}{n} \left(\frac{5}{20}\right)^n + \sum_{n=c}^{10} \frac{10!}{(10-n)!2!2^{n-2}} \left(\frac{5}{20}\right)^n\right]^{-1} = 0.065$$

Using P_0, we can obtain the other P_n, from which we can compute the average number of machines waiting for service:

$$L_Q = \sum_{n=3}^{10} (n-2) P_n = 1.46 \text{ machines}$$

the effective arrival rate:

$$\lambda_e = \sum_{n=0}^{10} (10-n) \left(\frac{1}{20}\right) P_n = 0.342 \text{ machines/minute}$$

and the average waiting time in the queue:

$$w_Q = L_Q/\lambda_e = 4.27 \text{ minutes}$$

Similarly, we can compute the expected number of machines being serviced or waiting to be serviced:

$$L = \sum_{n=0}^{10} n P_n = 3.17 \text{ machines}$$

The average number of machines being serviced is given by

$$L - L_Q = 3.17 - 1.46 = 1.71 \text{ machines}$$

Since the machines must be running, waiting to be serviced, or in service, the average number of running machines is given by

$$K - L = 10 - 3.17 = 6.83 \text{ machines}$$

A frequently asked question is this: What will happen if the number of servers is increased or decreased? If the number of workers in this example increases to three ($c = 3$), then the time-average number of running machines increases to

$$K - L = 7.74 \text{ machines}$$

an increase of 0.91 machine, on the average.

Conversely, what happens if the number of servers decreases to one? Then the time-average number of running machines decreases to

$$K - L = 3.98 \text{ machines}$$

The decrease from two to one server has resulted in a drop of nearly three machines running, on the average. Exercise 17 asks the reader to examine the effect on server utilization of adding or deleting one server. ◄

Example 6.18 illustrates several general relationships that have been found to hold for almost all queues. If the number of servers is decreased, delays, server utilization, and the probability of an arrival having to wait to begin service all increase.

6.6 Networks of Queues

In this chapter we have emphasized the study of single queues of the *G/G/c/N/K* type. However, many systems are naturally modeled as networks of single queues in which customers departing one queue may be routed to another. Example 6.1 (see in particular Figure 6.3) and Example 6.2 (see Figure 6.5) are illustrations.

The study of mathematical models of networks of queues is beyond the scope of this chapter; see, for instance, Gross and Harris [1997], Nelson [1995], and Kleinrock [1976]. However, a few fundamental principles are very useful for rough-cut modeling, perhaps prior to a simulation study. The following results assume a stable system with infinite calling population and no limit on system capacity:

1. Provided no customers are created or destroyed in the queue, then the departure rate out of a queue is the same as the arrival rate into the queue, over the long run.
2. If customers arrive to queue i at rate λ_i, and a fraction $0 \leq p_{ij} \leq 1$ of them are routed to queue j upon departure, then the arrival rate from queue i to queue j is $\lambda_i p_{ij}$ over the long run.
3. The overall arrival rate into queue j, λ_j, is the sum of the arrival rate from all sources. If customers arrive from outside the network at rate a_j, then

$$\lambda_j = a_j + \sum_{\text{all } i} \lambda_i p_{ij}$$

4. If queue j has $c_j < \infty$ parallel servers, each working at rate μ_j, then the long-run utilization of each server is

$$\rho_j = \frac{\lambda_j}{c_j \mu_j}$$

and $\rho_j < 1$ is required for the queue to be stable.

5. If, for each queue j, arrivals from outside the network form a Poisson process with rate a_j, and there are c_j identical servicers delivering exponentially distributed service times with mean $1/\mu_j$ (where c_j may be ∞), then in steady state queue j behaves like an $M/M/c_j$ queue with arrival rate $\lambda_j = a_j + \sum_{\text{all } i} \lambda_i p_{ij}$.

Example 6.19

Consider again the discount store described in Example 6.1 and shown in Figure 6.3. Suppose that customers arrive at a rate of 80 per hour, and of those arrivals 40% choose self-service. Then the arrival rate to service center 1 is $\lambda_1 = (80)(0.40) = 32$ per hour, and the arrival rate to service center 2 is $\lambda_2 = (80)(0.6) = 48$ per hour. Suppose that each of the $c_2 = 3$ clerks at service center 2 works at a rate of $\mu_2 = 20$ customers per hour. Then the long-run utilization of the clerks is

$$\rho_2 = \frac{48}{(3)(20)} = 0.8$$

All customers must see the cashier at service center 3. The overall arrival rate to service center 3 is $\lambda_3 = \lambda_1 + \lambda_2 = 80$ per hour, regardless of the service rate at service center 1, because over the long run the departure rate out of each service center must be equal to the arrival rate into it. If the cashier works at rate $\mu_3 = 90$ per hour, then the utilization of the cashier is

$$\rho_3 = \frac{80}{90} = 0.89$$

◀

Example 6.20

At a Driver's License branch office, drivers arrive at a rate of 50 per hour. All arrivals must first check in with one of two clerks, with the average check-in time being 2 minutes. After checking in, 15% of the drivers need to take a written test that lasts approximately 20 minutes. All arrivals must wait to have their picture taken and their license produced; this station can process about 60 drivers per hour. The branch manager wants to know whether adding a check-in clerk or a new photo station will lead to a greater reduction in customer delay.

To solve the problem, let the check-in clerks be queue 1 (with $c_1 = 2$ servers, each working at rate $\mu_1 = 30$ drivers per hour), let the testing station be queue 2 (with $c_2 = \infty$ servers, since any number of people can be taking the written test simultaneously, and service rate $\mu_2 = 3$ drivers per hour), and let the photo station be queue 3 (with $c_3 = 1$ server working at rate $\mu_3 = 60$ drivers per hour). The arrival rates to each queue are as follows:

$$\lambda_1 = a_1 + \sum_{i=1}^{3} p_{i1}\lambda_i = 50 \text{ drivers per hour}$$

$$\lambda_2 = a_2 + \sum_{i=1}^{3} p_{i2}\lambda_i = (0.15)\lambda_1 \text{ drivers per hour}$$

$$\lambda_3 = a_3 + \sum_{i=1}^{3} p_{i3}\lambda_i = (1)\lambda_2 + (0.85)\lambda_1 \text{ drivers per hour}$$

Notice that arrivals from outside the network occur only at queue 1, so $a_1 = 50$, while $a_2 = a_3 = 0$. Solving this system of equations gives $\lambda_1 = \lambda_3 = 50$ and $\lambda_2 = 7.5$.

If we approximate the arrival process as Poisson, and the service times at each queue as exponentially distributed, then the check-in clerks can be approximated as an $M/M/c_1$ queue, the testing station as an $M/M/\infty$ queue, and the photo station as an $M/M/c_3$ queue. Thus, under the current set-up the check-in station is an $M/M/2$; using the formulas in Table 6.5 gives $w_Q = 0.0758$ hours. If we add a clerk so that the model is $M/M/3$, the waiting time in queue drops to 0.0075 hours, a savings of 0.0683 hours or about 4.1 minutes.

The current photo station can be modeled as an $M/M/1$ queue, giving $w_Q = 0.0833$ hours; after adding a second photo station ($M/M/2$), the time in queue drops to 0.0035 hours, a savings of 0.0798 hours, or about 4.8 minutes. Therefore, a second photo station offers a slightly greater reduction in waiting time than adding a third clerk.

If desired, the testing station can be analyzed using the results for an $M/M/\infty$ queue in Table 6.6. For instance, the expected number of people taking the test at any time is $L = \lambda_2/\mu_2 = 7.5/3 = 2.5$. ◀

6.7 Summary

Queueing models have found widespread use in the analysis of service facilities, production and material handling systems, telephone and communications systems, and many other situations where congestion or competition for scarce resources may occur. This chapter has introduced the basic concepts of queueing models and shown how simulation, and in some cases a mathematical analysis, can be used to estimate the performance measures of a system.

A simulation can be used to generate one or more artificial histories of a complex system. This simulation-generated data in turn can be used to estimate desired performance measures of the system. Commonly used performance measures, including L, L_Q, w, w_Q, ρ, and λ_e, were introduced and formulas given for their estimation from data.

When simulating any system which evolves over time, the analyst must decide whether transient behavior or steady-state performance is to be studied. While simple formulas exist for the steady-state behavior of some queues, when estimating steady-state performance measures from simulation-generated data it is necessary to recognize and to deal with the possibly deleterious effect of the initial conditions on the estimators of steady-state performance. These estimators may be severely biased (either high or low) if the initial conditions are unrepresentative of steady state, or if simulation run length is too short. These estimation problems are discussed at greater length in Chapter 11.

Whether the analyst is interested in transient or steady-state performance of a system, it should be recognized that the estimates obtained from a simulation of a stochastic queue are exactly that—estimates. That is, an estimate contains random error, and thus a proper statistical analysis is required to assess the accuracy of the estimate. Methods for conducting such a statistical analysis are discussed in Chapters 11 and 12.

In the last three sections of this chapter it was shown that a number of simple models can be solved mathematically. Although the assumptions behind such models may not be met exactly in a practical application, these models can still be useful in providing a rough estimate of a performance measure. In many cases, models with exponentially distributed interarrival and service times will provide a conservative estimate of system behavior. For example, if the model predicts that average waiting time, w, will be 12.7 minutes, then average waiting time in the real system is likely to be less than 12.7 minutes. The conservative nature of exponential models arises because (a) performance measures, such as w and L, are generally increasing functions of the variance of interarrival times and service times (recall the $M/G/1$ queue), and (b) the exponential distribution is fairly highly variable, having its standard deviation always equal to its mean. Thus, if the arrival process or service mechanism of the real system is less variable than exhibited by exponentially distributed interarrival or service times, then it is likely that the average number in the system, L, and the average time spent in system, w, will be less than what is predicted by the exponential model. Of course, if the interarrival and service times are more variable than exponential random variables, then the M/M queueing models will underestimate congestion.

An important application of mathematical queueing models is determining the minimum number of servers needed at a workstation or service center. Quite often, if the arrival rate λ and the service rate μ are known or can be estimated, then the simple inequality $\lambda/c\mu < 1$ can be used to provide an initial estimate for the number of servers, c, at a workstation. For a large system with many workstations, it could be quite time consuming to have to simulate every possibility $(c_1, c_2, \ldots)$ for the number of servers, c_i, at work station i. Thus, a bit of mathematical analysis using rough estimates may save a great deal of computer time and analyst's time.

Finally, the qualitative behavior of the simple exponential models of queueing carries over to more complex systems. In general, it is the variability of service times and the variability of the arrival process that causes waiting lines to build up and congestion to occur. For most systems, if the arrival rate increases, or if the service rate decreases, or if the variance of service times or interarrival times increases, then the system will become more congested. Congestion can be decreased by adding more servers or by reducing the mean value and variability of service times. Simple queueing models can be a great aid in quantifying these relationships and evaluating alternative system designs.

REFERENCES

BANKS, J. AND R G. HEIKES [1984], *Handbook of Tables and Graphs for the Industrial Engineer and Manager*, Reston Publishing Company, Reston, VA.

COOPER, R. B. [1981], *Introduction to Queueing Theory*, 2d ed., North-Holland, New York.

DESCLOUX, A. [1962], *Delay Tables for Finite- and Infinite-Source Systems*, McGraw-Hill, New York.

GROSS, D. AND C. HARRIS [1997], *Fundamentals of Queueing Theory*, 3d ed., John Wiley, New York.

HALL, R. W. [1991], *Queueing Methods*, Prentice Hall, Upper Saddle River, NJ.

HILLIER, F. S. AND G. J. LIEBERMAN [1990], *Introduction to Operations Research*, 5th ed., McGraw-Hill, New York.

HILLIER, F. S., AND O. S. YU [1981], *Queueing Tables and Graphs*, Elsevier North-Holland, New York.

KENDALL, D. G. [1953], "Stochastic Processes Occurring in the Theory of Queues and Their Analysis by the Method of Imbedded Markov Chains," *Annals of Mathematical Statistics*, Vol. 24, pp. 338–54.

KLEINROCK, L. [1976], *Queueing Systems, Vol 2: Computer Applications*, John Wiley, New York.

LITTLE, J. D. C. [1961], "A Proof for the Queueing Formula $L = \lambda w$," *Operations Research*, Vol. 16, pp. 651–65.

NELSON, B. L. [1995], *Stochastic Modeling: Analysis & Simulation*, McGraw-Hill, New York.

PECK, L. G., AND R. N. HAZELWOOD [1958], *Finite Queueing Tables*, John Wiley, New York.

WAGNER, H. M. [1975], *Principles of Operations Research*, 2d ed., Prentice Hall, Upper Saddle River, NJ.

WINSTON, W. L. [1997], *Operations Research: Applications and Algorithms*, 3d ed., Wadsworth Publishing Co., Belmont, CA.

EXERCISES

1. A tool crib has exponential interarrival and service times, and it serves a very large group of mechanics. The mean time between arrivals is 4 minutes. It takes 3 minutes on the average for a tool-crib attendant to service a mechanic. The attendant is paid $10 per hour and the mechanic is paid $15 per hour. Would it be advisable to have a second tool-crib attendant?

2. A two-runway (one runway for landing, one for taking off) airport is being designed for propeller-driven aircraft. The time to land an airplane is known to be exponentially distributed with a mean of $1\frac{1}{2}$ minutes. If airplane arrivals are assumed to occur at random, what arrival rate can be tolerated if the average wait in the sky is not to exceed 3 minutes?

3. The Port of Trop can service only one ship at a time. However, there is mooring space for three more ships. Trop is a favorite port of call, but if no mooring space is available, the ships have to go to the Port of Poop. An average of seven ships arrive each week, according to a Poisson process. The Port of Trop has the capacity to handle an average of eight ships a week, with service times exponentially distributed. What is the expected number of ships waiting or in service at the Port of Trop?

4. At Metropolis City Hall, two workers "pull strings" every day. Strings arrive to be pulled on an average of one every 10 minutes throughout the day. It takes an average of 15 minutes to pull a string. Both times between arrivals and service times are exponentially distributed. What is the probability that there are no strings to be pulled in the system at a random point in time? What is the expected number of strings waiting to be pulled? What is the probability that both string pullers are busy? What is the effect on performance if a third string puller, working at the same speed as the first two, is added to the system?

5. At Tony and Cleo's bakery, one kind of birthday cake is offered. It takes 15 minutes to decorate this particular cake, and the job is performed by one particular baker. In fact, this is all this baker does. What mean time between arrivals (exponentially distributed) can be accepted if the mean length of the queue for decorating is not to exceed five cakes?

6. Patients arrive for a physical examination according to a Poisson process at the rate of one per hour. The physical examination requires three stages, each one independently and exponentially distributed with a service time of 15 minutes. A patient must go through all three stages before the next patient is admitted to the treatment facility. Determine the average number of delayed patients, L_Q, for this system. [*Hint:* The variance of the sum of independent random variables is the sum of the variance.]

7. Suppose that mechanics arrive randomly at a tool crib according to a Poisson process with rate $\lambda = 10$ per hour. It is known that the single tool clerk serves a mechanic in 4 minutes on the average, with a standard deviation of approximately 2 minutes. Suppose that mechanics make $15 per hour. Determine the steady-state average cost per hour of mechanics waiting for tools.

8. Arrivals to an airport are all directed to the same runway. At a certain time of the day, these arrivals form a Poisson process with rate 30 per hour. The time to land an aircraft is a constant 90 seconds. Determine L_Q, w_Q, L, and w for this airport. If a delayed aircraft burns $5000 worth of fuel per hour on the average, determine the average cost per aircraft of delay waiting to land.

9. A machine shop repairs small electric motors which arrive according to a Poisson process at a rate of 12 per week (5-day, 40-hour workweek). An analysis of past data indicates that engines can be repaired, on the average, in 2.5 hours, with a variance of 1 hour^2. How many working hours should a customer expect to leave a motor at the repair shop (not knowing the status of the system)? If the variance of the repair time could be controlled, what variance would reduce the expected waiting time to 6.5 hours?

10. Arrivals to a self-service gasoline pump occur in a Poisson fashion at a rate of 12 per hour. Service time has a distribution which averages 4 minutes with a standard deviation of $1\frac{1}{3}$ minutes. What is the expected number of vehicles in the system?

11. Classic Car Care has one worker who washes cars in a four-step method—soap, rinse, dry, vacuum. The time to complete each step is exponentially distributed with a mean of 9 minutes. Every car goes through every step before another car begins the process. On the average one car every 45 minutes arrives for a wash job, according to a Poisson process. What is the average time a car waits to begin the

wash job? What is the average number of cars in the car wash system? What is the average time required to wash a car?

12. A room has 10 cotton spinning looms. Once the looms are set up, they run automatically. The set-up time is exponentially distributed with a mean of 10 minutes. The machines run for an average of 40 minutes, also exponentially distributed. Loom operators are paid $10 an hour and looms not running incur a cost of $40 an hour. How many loom operators should be employed to minimize the total cost of the loom room? If the objective becomes "on the average, no loom should wait more than 1 minute for an operator," how many persons should be employed? How many operators should be employed to ensure that an average of at least 7.5 looms are running at all times?

13. Given the following information for a finite-calling-population problem with exponentially distributed runtimes and service times:

$$K = 10$$

$$\frac{1}{\mu} = 15$$

$$\frac{1}{\lambda} = 82$$

$$c = 2$$

Compute L_Q and w_Q. Determine the value of λ such that $L_Q = L/2$.

14. Suppose that Figure 6.6 represents the number in system for a last-in, first-out (LIFO) single-server system. Customers are not preempted (i.e., kicked out of service), but upon service completion the most recent arrival next begins service. For this LIFO system apportion the total area under $L(t)$ to each individual customer, as was done in Figure 6.8 for the FIFO system. Using the figure, show that Equations (6.10), (6.8), and (6.9) hold for the single-server LIFO system.

15. Repeat Exercise 14, assuming that:

(a) Figure 6.6 represents a FIFO system with $c = 2$ servers.

(b) Figure 6.6 represents a LIFO system with $c = 2$ servers.

16. Consider an $M/G/1$ queue with the following type of service distribution. Customers request one of two types of service in the proportions p and $1 - p$. Type i service is exponentially distributed at rate $\mu_i, i = 1, 2$. Let X_i denote a type i service time and X an arbitrary service time. Then $E(X_i) = 1/\mu_i$, $V(X_i) = 1/\mu_i^2$, and

$$X = \begin{cases} X_1, & \text{with probability } p \\ X_2, & \text{with probability } (1 - p) \end{cases}$$

The random variable X is said to have a hyperexponential distribution with parameters (μ_1, μ_2, p).

(a) Show that $E(X) = p/\mu_1 + (1 - p)/\mu_2$ and $E(X^2) = 2p/\mu_1^2 + 2(1 - p)/\mu_2^2$.

(b) Use $V(X) = E(X^2) - [E(X)]^2$ to show $V(X) = 2p/\mu_1^2 + 2(1 - p)/\mu_2^2 - [p/\mu_1 + (1 - p)/\mu_2]^2$.

(c) For any hyperexponential random variable, if $\mu_1 \neq \mu_2$ and $0 < p < 1$, show that its coefficient of variation is greater than 1; that is, $(\text{cv})^2 = V(X)/[E(X)]^2 > 1$. Thus, the hyperexponential distribution provides a family of statistical models for service times which are more variable than exponentially distributed service times. [*Hint*: The algebraic expression for $(\text{cv})^2$, using parts (a) and (b), can be manipulated into the form $(\text{cv})^2 = 2p(1-p)(1/\mu_1 - 1/\mu_2)^2/[E(X)]^2 + 1$.]

(d) Many choices of μ_1, μ_2, and p lead to the same overall mean $E(X)$ and $(\text{cv})^2$. If a distribution with mean $E(X) = 1$ and coefficient of variation cv $= 2$ is desired, find values of μ_1, μ_2, and p to achieve this. [*Hint*: Choose $p = 1/4$ arbitrarily; then solve the following equations for μ_1 and μ_2.]

$$\frac{1}{4\mu_1} + \frac{3}{4\mu_2} = 1$$

$$\frac{3}{8}\left(\frac{1}{\mu_1} - \frac{1}{\mu_2}\right)^2 + 1 = 4$$

17. In Example 6.18, compare the systems with $c = 1$, $c = 2$, and $c = 3$ servers on the basis of server utilization ρ (the proportion of time a typical server is busy).

18. In Example 6.18, increase the number of machines by 2 and compare the systems with $c = 1$, $c = 2$, and $c = 3$ servers on the basis of server utilization ρ (the proportion of time a typical server is busy).

19. A small lumberyard is supplied by a fleet of 10 trucks. One overhead crane is available to unload the long logs from the trucks. It takes an average of 1 hour to unload a truck. After unloading, a truck takes an average of 3 hours to get the next load of logs and return to the lumberyard.

(a) Certain distributional assumptions are needed to analyze this problem with the models of this chapter. State them and discuss their reasonableness.

(b) With one crane, what is the average number of trucks waiting to be unloaded? On the average, how many trucks arrive at the yard each hour? What percentage of trucks upon arrival find the crane busy? Is this the same as the long-run proportion of time the crane is busy?

(c) Suppose that a second crane is installed at the lumberyard. Answer the same questions as in part (b). Make a chart comparing one crane to two cranes.

(d) If the value of the logs brought to the yard is approximately \$200 per truckload and long-run crane costs are \$50 per hour per crane (whether busy or not), determine the optimal number of cranes on the basis of cost per hour.

(e) In addition to the costs in part (d), if management decides to consider the cost of idle trucks and drivers, what is the optimal number of cranes? A truck with its driver is estimated to cost approximately \$40 per hour and is considered to be idle when it is waiting for a crane.

20. A tool crib with one attendant serves a group of 10 mechanics. Mechanics work for an exponentially distributed amount of time with mean 20 minutes, then go to the crib to request a special tool. Service times by the attendant are exponentially distributed with a mean of 3 minutes. If the attendant is paid \$6 per hour and the mechanic \$10 per hour, would it be advisable to have a second attendant? (Compare to Exercise 1.)

21. This problem is based on Case 8.1 in Nelson [1995]. A large consumer shopping mall is to be constructed. During busy times the arrival rate of cars is expected to be 1000 per hour, and based on studies at other malls, it is believed that customers will spend 3 hours, on average, shopping. The mall designers would like to have sufficient parking so that there are enough spaces 99.9% of the time. How many spaces should they have? [*Hint:* Model the system as an $M/G/\infty$ queue where the spaces are servers, and find out how many spaces are adequate with probability 0.999.]

22. In Example 6.19, suppose that the overall arrival rate is expected to increase to 160 per hour. If the service rates do not change, how many clerks will be needed at service centers 2 and 3, just to keep up with the customer load?

23. A small copy shop has a self-service copier. Currently there is room for only 4 people to line up for the machine (including the person using the machine); when there are more than 4 people, then the additional people must line up outside the shop. The owners would like to avoid having people line up outside the shop as much as possible. For that reason they are thinking about adding a second self-service copier. Self-service customers have been observed to arrive at a rate of 24 per hour, and they use the machine 2 minutes, on average. Assess the impact of adding another copier. Carefully state any assumptions or approximations you make.

24. A self-service car wash has 4 washing stalls. When they are in a stall, customers may choose among three options: rinse only; wash and rinse; and wash, rinse, and wax. Each option has a fixed time to complete: rinse only (3 minutes); wash and rinse (7 minutes); and wash, rinse, and wax (12 minutes). The owners have observed that 20% of customers rinse only; 70% wash and rinse; and 10% wash, rinse, and wax. There are no scheduled appointments, and customers arrive at a rate of about 34 cars per hour. There is only room for 3 cars to wait in the parking lot, so currently many customers are lost. The owners want to know how much more business they will do if they add another stall. Adding a stall will take away one space in the parking lot.

Develop a queueing model of the system. Estimate the rate at which customers will be lost in the current and proposed system. Carefully state any assumptions or approximations you make.

25. Find examples of queueing models used in real applications. A good source is the journal *Interfaces*.

26. Study the effect of *pooling servers* (having multiple servers draw from a single queue, rather than each having their own queue) by comparing performance measures for two $M/M/1$ queues, each with arrival rate λ and service rate μ, to those for an $M/M/2$ queue with arrival rate 2λ and service rate μ for each server.

27. A repair and inspection facility consists of two stations, a repair station with two technicians, and an inspection station with 1 inspector. Each repair technician works at a rate of 3 items per hour, while the inspector can inspect 8 items per hour. Approximately 10% of all items fail inspection and are sent back to the repair station (this percentage holds even for items that have been repaired two or more times). If items arrive at the rate of 5 per hour, what is the long-run expected delay that items experience at each of the two stations, assuming a Poisson arrival process and exponentially distributed service times? What is the maximum arrival rate that the system can handle without adding personnel?

PART THREE

Random Numbers

7

Random-Number Generation

Random numbers are a necessary basic ingredient in the simulation of almost all discrete systems. Most computer languages have a subroutine, object, or function that will generate a random number. Similarly, simulation languages generate random numbers that are used to generate event times and other random variables. This chapter describes the generation of random numbers and their subsequent testing for randomness. Chapter 8 shows how random numbers are used to generate a random variable for many probability distributions.

7.1 Properties of Random Numbers

A sequence of random numbers, $R_1, R_2, \ldots$, must have two important statistical properties, uniformity and independence. Each random number R_i is an independent sample drawn from a continuous uniform distribution between zero and 1. That is, the pdf is given by

$$f(x) = \begin{cases} 1, & 0 \leq x \leq 1 \\ 0, & \text{otherwise} \end{cases}$$

This density function is shown in Figure 7.1. The expected value of each R_i is given by

$$E(R) = \int_0^1 x\,dx = \left.\frac{x^2}{2}\right|_0^1 = \frac{1}{2}$$

and the variance is given by

$$V(R) = \int_0^1 x^2 dx - [E(R)]^2 = \left.\frac{x^3}{3}\right|_0^1 - \left(\frac{1}{2}\right)^2 = \frac{1}{3} - \frac{1}{4} = \frac{1}{12}$$

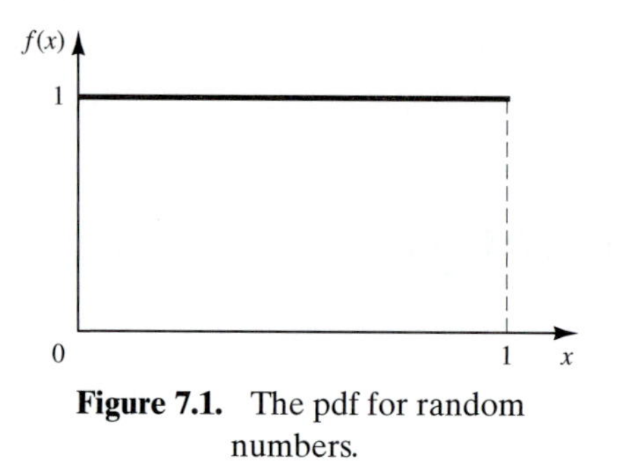

Figure 7.1. The pdf for random numbers.

Some consequences of the uniformity and independence properties are the following:

1. If the interval $(0, 1)$ is divided into n classes, or subintervals of equal length, the expected number of observations in each interval is N/n, where N is the total number of observations.
2. The probability of observing a value in a particular interval is independent of the previous values drawn.

7.2 Generation of Pseudo-Random Numbers

Notice that the title of this section has the word "pseudo" in it. "Pseudo" means false, so false random numbers are being generated! In this instance, "pseudo" is used to imply that the very act of generating random numbers by a known method removes the potential for true randomness. If the method is known, the set of random numbers can be replicated. Then an argument can be made that the numbers are not truly random. The goal of any generation scheme, however, is to produce a sequence of numbers between zero and 1 which simulates, or imitates, the ideal properties of uniform distribution and independence as closely as possible.

When generating pseudo-random numbers, certain problems or errors can occur. These errors, or departures from ideal randomness, are all related to the properties stated previously. Some examples include the following:

1. The generated numbers may not be uniformly distributed.
2. The generated numbers may be discrete-valued instead of continuous-valued.
3. The mean of the generated numbers may be too high or too low.
4. The variance of the generated numbers may be too high or too low.

5. There may be dependence. The following are examples:

(a) Autocorrelation between numbers.

(b) Numbers successively higher or lower than adjacent numbers.

(c) Several numbers above the mean followed by several numbers below the mean.

Departures from uniformity and independence for a particular generation scheme may be detected by tests such as those described in Section 7.4. If such departures are detected, the generation scheme should be dropped in favor of an acceptable generator. Generators that pass all the tests in Section 7.4, and even more stringent tests, have been developed; thus, there is no excuse for using a generator that has been found to be deficient.

Usually, random numbers are generated by a digital computer as part of the simulation. Numerous methods can be used to generate the values. In selecting among these methods, or routines, there are a number of important considerations:

1. The routine should be fast. Individual computations are inexpensive, but simulation may require many hundreds of thousands of random numbers. The total cost can be managed by selecting a computationally efficient method of random-number generation.

2. The routine should be portable to different computers, and ideally to different programming languages. This is desirable so that the simulation program produces the same results wherever it is executed.

3. The routine should have a sufficiently long cycle. The cycle length, or period, represents the length of the random-number sequence before previous numbers begin to repeat themselves in an earlier order. Thus, if 10,000 events are to be generated, the period should be many times that long.

A special case of cycling is degenerating. A routine degenerates when the same random numbers appear repeatedly. Such an occurrence is certainly unacceptable. This can happen rapidly with some methods.

4. The random numbers should be replicable. Given the starting point (or conditions), it should be possible to generate the same set of random numbers, completely independent of the system that is being simulated. This is helpful for debugging purposes and is a means of facilitating comparisons between systems (see Chapter 12). For the same reasons, it should be possible to easily specify different starting points, widely separated, within the sequence.

5. Most important, and as indicated previously, the generated random numbers should closely approximate the ideal statistical properties of uniformity and independence.

Inventing techniques that seem to generate random numbers is easy; inventing techniques that really do produce sequences that appear to be independent, uniformly distributed random numbers is incredibly difficult. There is now a vast literature and a rich theory on the topic, and many hours of testing have been devoted to establishing the properties of various generators. Even when a technique is known to be theoretically sound, it is seldom easy to implement it in a way that will be fast and portable. Therefore, this chapter aims to make the reader aware of the central issues in random-variate generation in order to enhance understanding and to show some of the techniques that are used by those working in this area.

7.3 Techniques for Generating Random Numbers

The linear congruential method of Section 7.3.1 is the most widely used technique for generating random numbers, so we describe it in detail. We also report an extension of this method that yields sequences with a longer period. Many other methods have been proposed, and they are reviewed in Bratley, Fox, and Schrage [1987], Law and Kelton [2000], and Ripley [1987].

7.3.1 Linear Congruential Method

The linear congruential method, initially proposed by Lehmer [1951], produces a sequence of integers, $X_1, X_2, \ldots$ between zero and $m-1$ according to the following recursive relationship:

$$X_{i+1} = (aX_i + c) \bmod m, \quad i = 0, 1, 2, \ldots \tag{7.1}$$

The initial value X_0 is called the seed, a is called the constant multiplier, c is the increment, and m is the modulus. If $c \neq 0$ in Equation (7.1), the form is called the *mixed congruential method*. When $c = 0$, the form is known as the *multiplicative congruential method*. The selection of the values for a, c, m, and X_0 drastically affects the statistical properties and the cycle length. Variations of Equation (7.1) are quite common in the computer generation of random numbers. An example will illustrate how this technique operates.

EXAMPLE 7.1

Use the linear congruential method to generate a sequence of random numbers with $X_0 = 27, a = 17, c = 43$, and $m = 100$. Here, the integer values generated will all be between zero and 99 because of the value of the modulus. Also, notice that random integers are being generated rather than random numbers. These random integers should appear to be uniformly distributed on the integers zero to 99. Random numbers between zero and 1 can be generated by

$$R_i = \frac{X_i}{m}, \quad i = 1, 2, \ldots \tag{7.2}$$

The sequence of X_i and subsequent R_i values is computed as follows:

$$
\begin{aligned}
X_0 &= 27 \\
X_1 &= (17 \cdot 27 + 43) \bmod 100 = 502 \bmod 100 = 2 \\
R_1 &= \frac{2}{100} = 0.02 \\
X_2 &= (17 \cdot 2 + 43) \bmod 100 = 77 \bmod 100 = 77 \\
R_2 &= \frac{77}{100} = 0.77 \\
*[3pt]X_3 &= (17 \cdot 77 + 43) \bmod 100 = 1352 \bmod 100 = 52 \\
R_3 &= \frac{52}{100} = 0.52 \\
&\vdots
\end{aligned}
$$

Recall that $a = b \bmod m$ provided that $(a - b)$ is divisible by m with no remainder. Thus, $X_1 = 502 \bmod 100$, but 502/100 equals 5 with a remainder of 2, so that $X_1 = 2$. In other words, $(502 - 2)$ is evenly divisible by $m = 100$, so $X_1 = 502$ "reduces" to $X_1 = 2 \bmod 100$. (A shortcut for the modulo, or reduction operation for the case $m = 10^b$, a power of 10, is illustrated in Example 7.3.) ◀

The ultimate test of the linear congruential method, as of any generation scheme, is how closely the generated numbers $R_1, R_2, \ldots$ approximate uniformity and independence. There are, however, several secondary properties which must be considered. These include maximum density and maximum period.

First, notice that the numbers generated from Equation (7.2) can only assume values from the set $I = \{0, 1/m, 2/m, \ldots, (m-1)/m\}$, since each X_i is an integer in the set $\{0, 1, 2, \ldots, m-1\}$. Thus, each R_i is discrete on I, instead of continuous on the interval [0, 1]. This approximation appears to be of little consequence, provided that the modulus m is a very large integer. (Values such as $m = 2^{31} - 1$ and $m = 2^{48}$ are in common use in generators appearing in many simulation languages.) By maximum density is meant that the values assumed by $R_i, i = 1, 2, \ldots$, leave no large gaps on [0, 1].

Second, to help achieve maximum density, and to avoid cycling (i.e., recurrence of the same sequence of generated numbers) in practical applications, the generator should have the largest possible period. Maximal period can be achieved by the proper choice of a, c, m, and X_0 [Fishman, 1978; Law and Kelton, 2000].

- For m a power of 2, say $m = 2^b$, and $c \neq 0$, the longest possible period is $P = m = 2^b$, which is achieved provided that c is relatively prime to m (that is, the greatest common factor of c and m is 1), and $a = 1 + 4k$, where k is an integer.

- For m a power of 2, say $m = 2^b$, and $c = 0$, the longest possible period is $P = m/4 = 2^{b-2}$, which is achieved provided that the seed X_0 is odd and the multiplier, a, is given by $a = 3 + 8k$ or $a = 5 + 8k$, for some $k = 0, 1, \ldots$.
- For m a prime number and $c = 0$, the longest possible period is $P = m-1$, which is achieved provided that the multiplier, a, has the property that the smallest integer k such that $a^k - 1$ is divisible by m is $k = m - 1$.

Example 7.2

Using the multiplicative congruential method, find the period of the generator for $a = 13$, $m = 2^6 = 64$, and $X_0 = 1, 2, 3$, and 4. The solution is given in Table 7.1. When the seed is 1 and 3, the sequence has period 16. However, a period of length eight is achieved when the seed is 2 and a period of length four occurs when the seed is 4. ◀

In Example 7.2, $m = 2^6 = 64$ and $c = 0$. The maximal period is therefore $P = m/4 = 16$. Notice that this period is achieved using odd seeds $X_0 = 1$ and $X_0 = 3$, but even seeds, $X_0 = 2$ and $X_0 = 4$, yield periods of eight and four, both less than the maximum. Notice that $a = 13$ is of the form $5 + 8k$ with $k = 1$, as required to achieve maximal period.

When $X_0 = 1$, the generated sequence assumes values from the set $\{1, 5, 9, 13, \ldots, 53, 57, 61\}$. The "gaps" in the sequence of generated random numbers, R_i, are quite large (i.e., the gap is $5/64 - 1/64$ or 0.0625). Such a gap gives rise to concern about the density of the generated sequence.

Table 7.1. Period Determination Using Various Seeds

i	X_i	X_i	X_i	X_i
0	1	2	3	4
1	13	26	39	52
2	41	18	59	36
3	21	42	63	20
4	17	34	51	4
5	29	58	23	
6	57	50	43	
7	37	10	47	
8	33	2	35	
9	45		7	
10	9		27	
11	53		31	
12	49		19	
13	61		55	
14	25		11	
15	5		15	
16	1		3	

The generator in Example 7.2 is not viable for any application—its period is too short, and its density is insufficiently low. However, the example shows the importance of properly choosing a, c, m, and X_0.

Speed and efficiency in using the generator on a digital computer are also a selection consideration. Speed and efficiency are aided by use of a modulus, m, which is either a power of 2 or close to a power of 2. Since most digital computers use a binary representation of numbers, the modulo, or remaindering, operation of Equation (7.1) can be conducted efficiently when the modulo is a power of 2 (i.e., $m = 2^b$). After ordinary arithmetic yields a value for $aX_i + c$, X_{i+1} is obtained by dropping the leftmost binary digits in $aX_i + c$ and then using only the b rightmost binary digits. The following example illustrates, by analogy, this operation using $m = 10^b$, because most human beings think in decimal representation.

EXAMPLE 7.3

Let $m = 10^2 = 100$, $a = 19$, $c = 0$, and $X_0 = 63$, and generate a sequence of random integers using Equation (7.1).

$$
\begin{aligned}
X_0 &= 63 \\
X_1 &= (19)(63) \bmod 100 = 1197 \bmod 100 = 97 \\
X_2 &= (19)(97) \bmod 100 = 1843 \bmod 100 = 43 \\
X_3 &= (19)(43) \bmod 100 = 817 \bmod 100 = 17 \\
&\vdots
\end{aligned}
$$

When m is a power of 10, say $m = 10^b$, the modulo operation is accomplished by saving the b rightmost (decimal) digits. By analogy, the modulo operation is most efficient for binary computers when $m = 2^b$ for some $b > 0$. ◀

EXAMPLE 7.4

The last example in this section is in actual use. It has been extensively tested [Learmonth and Lewis, 1973; Lewis et al., 1969]. The values for a, c, and m have been selected to ensure that the characteristics desired in a generator are most likely to be achieved. By changing X_0, the user can control the repeatability of the stream.

Let $a = 7^5 = 16{,}807$, $m = 2^{31}-1 = 2{,}147{,}483{,}647$ (a prime number), and $c = 0$. These choices satisfy the conditions that insure a period of $P = m - 1$ (well over 2 billion). Further, specify a seed, $X_0 = 123{,}457$. The first few numbers generated are as follows:

$$X_1 = 7^5(123{,}457) \bmod (2^{31} - 1) = 2{,}074{,}941{,}799 \bmod (2^{31} - 1)$$
$$X_1 = 2{,}074{,}941{,}799$$
$$R_1 = \frac{X_1}{2^{31}} = 0.9662$$
$$X_2 = 7^5(2{,}074{,}941{,}799) \bmod (2^{31} - 1) = 559{,}872{,}160$$
$$R_2 = \frac{X_2}{2^{31}} = 0.2607$$
$$X_3 = 7^5(559{,}872{,}160) \bmod (2^{31} - 1) = 1{,}645{,}535{,}613$$
$$R_3 = \frac{X_3}{2^{31}} = 0.7662$$
$$\vdots$$

Notice that this routine divides by $m + 1$ instead of m; however, for such a large value of m, the effect is negligible. ◄

7.3.2 Combined Linear Congruential Generators

As computing power has increased, the complexity of the systems that we are able to simulate has also increased. A random-number generator with period $2^{31} - 1 \approx 2 \times 10^9$, such as the popular generator described in Example 7.4, is no longer adequate for all applications. Examples include the simulation of highly reliable systems, in which hundreds of thousands of elementry events must be simulated to observe even a single failure event; and the simulation of complex computer networks, in which thousands of users are executing hundreds of programs. An area of current research is deriving generators with substantially longer periods.

One fruitful approach is to combine two or more multiplicative congruential generators in such a way that the combined generator has good statistical properties and a longer period. The following result from L'Ecuyer [1988] suggests how this can be done:

If $W_{i,1}, W_{i,2}, \ldots, W_{i,k}$ are any independent, discrete-valued random variables (not necessarily identically distributed), but one of them, say $W_{i,1}$, is uniformly distributed on the integers 0 to $m_1 - 2$, then

$$W_i = \left(\sum_{j=1}^{k} W_{i,j} \right) \bmod m_1 - 1$$

is uniformly distributed on the integers 0 to $m_1 - 2$.

To see how this result can be used to form combined generators, let $X_{i,1}, X_{i,2}, \ldots, X_{i,k}$ be the ith output from k different multiplicative congruential generators, where the jth generator has prime modulus m_j, and the multiplier a_j is chosen so that the period is $m_j - 1$. Then the jth generator is producing integers $X_{i,j}$ that are approximately uniformly distributed on 1 to $m_j - 1$, and $W_{i,j} = X_{i,j} - 1$ is approximately uniformly distributed on 0 to $m_j - 2$. L'Ecuyer [1988] therefore suggests combined generators of the form

$$X_i = \left(\sum_{j=1}^{k} (-1)^{j-1} X_{i,j} \right) \bmod m_1 - 1$$

with

$$R_i = \begin{cases} \dfrac{X_i}{m_1}, & X_i > 0 \\ \dfrac{m_1 - 1}{m_1}, & X_i = 0 \end{cases}$$

Notice that the "$(-1)^{j-1}$" coefficient implicitly performs the subtraction $X_{i,1} - 1$; for example, if $k = 2$, then $(-1)^0(X_{i,1} - 1) - (-1)^1(X_{i,2} - 1) = \sum_{j=1}^{2}(-1)^{j-1}X_{i,j}$.

The maximum possible period for such a generator is

$$P = \frac{(m_1 - 1)(m_2 - 1)\cdots(m_k - 1)}{2^{k-1}}$$

which is achieved by the following generator:

EXAMPLE 7.5

For 32-bit computers, L'Ecuyer [1988] suggests combining $k = 2$ generators with $m_1 = 2147483563$, $a_1 = 40014$, $m_2 = 2147483399$, and $a_2 = 40692$. This leads to the following algorithm:

1. Select seed $X_{1,0}$ in the range $[1, 2147483562]$ for the first generator, and seed $X_{2,0}$ in the range $[1, 2147483398]$.
   Set $j = 0$.
2. Evaluate each individual generator.

$$X_{1,j+1} = 40014 X_{1,j} \bmod 2147483563$$
$$X_{2,j+1} = 40692 X_{2,j} \bmod 2147483399$$

3. Set

$$X_{j+1} = (X_{1,j+1} - X_{2,j+1}) \bmod 2147483562$$

4. Return

$$R_{j+1} = \begin{cases} \dfrac{X_{j+1}}{2147483563}, & X_{j+1} > 0 \\ \dfrac{2147483562}{2147483563}, & X_{j+1} = 0 \end{cases}$$

5. Set $j = j + 1$ and go to step 2.

This combined generator has period $(m_1 - 1)(m_2 - 1)/2 \approx 2 \times 10^{18}$. Perhaps surprisingly, even such a long period may not be adequate for all applications. See L'Ecuyer [1996, 1999] for combined generators with periods as long as $2^{191} \approx 3 \times 10^{57}$. ◀

7.4 Tests for Random Numbers

The desirable properties of random numbers — uniformity and independence — were discussed in Section 7.1. To insure that these desirable properties are achieved, a number of tests can be performed (fortunately, the appropriate tests have already been conducted for most commercial simulation software). The tests can be placed in two categories according to the properties of interest. The first entry in the list below concerns testing for uniformity. The second through fifth entries concern testing for independence. The five types of tests discussed in this chapter are as follows:

1. *Frequency test.* Uses the Kolmogorov-Smirnov or the chi-square test to compare the distribution of the set of numbers generated to a uniform distribution.
2. *Runs test.* Tests the runs up and down or the runs above and below the mean by comparing the actual values to expected values. The statistic for comparison is the chi-square.
3. *Autocorrelation test.* Tests the correlation between numbers and compares the sample correlation to the expected correlation of zero.
4. *Gap test.* Counts the number of digits that appear between repetitions of a particular digit and then uses the Kolmogorov-Smirnov test to compare with the expected size of gaps.
5. *Poker test.* Treats numbers grouped together as a poker hand. Then the hands obtained are compared to what is expected using the chi-square test.

In testing for uniformity, the hypotheses are as follows:

$$H_0\colon R_i \sim U[0,1]$$
$$H_1\colon R_i \nsim U[0,1]$$

The null hypothesis, H_0, reads that the numbers are distributed uniformly on the interval $[0, 1]$. Failure to reject the null hypothesis means that no evidence of nonuniformity has been detected on the basis of this test. This does not imply that further testing of the generator for uniformity is unnecessary.

In testing for independence, the hypotheses are as follows:

$$H_0\colon R_i \sim \text{independently}$$
$$H_1\colon R_i \nsim \text{independently}$$

This null hypothesis, H_0, reads that the numbers are independent. Failure to reject the null hypothesis means that no evidence of dependence has been detected on the basis of this test. This does not imply that further testing of the generator for independence is unnecessary.

For each test, a level of significance α must be stated. The level α is the probability of rejecting the null hypothesis given that the null hypothesis is true, or

$$\alpha = P(\text{reject } H_0 | H_0 \text{ true})$$

The decision maker sets the value of α for any test. Frequently, α is set to 0.01 or 0.05.

If several tests are conducted on the same set of numbers, the probability of rejecting the null hypothesis on at least one test, by chance alone [i.e., making a Type I (α) error], increases. Say that $\alpha = 0.05$ and that five different tests are conducted on a sequence of numbers. The probability of rejecting the null hypothesis on at least one test, by chance alone, may be as large as 0.25.

Similarly, if one test is conducted on many sets of numbers from a generator, the probability of rejecting the null hypothesis on at least one test by chance alone [i.e., making a Type I (α) error], increases as more sets of numbers are tested. For instance, if 100 sets of numbers were subjected to the test, with $\alpha = 0.05$, it would be expected that five of those tests would be rejected by chance alone. If the number of rejections in 100 tests is close to 100α, then there is no compelling reason to discard the generator. The concept discussed in this and the preceding paragraph is discussed further at the conclusion of Example 7.12.

If one of the well-known simulation languages or random-number generators is used, it is probably unnecessary to use the tests mentioned above and described in Sections 7.4.1 through 7.4.5. (However, a generator such as RANDU, distributed by IBM in the late 1960s and still available on some computers, has been found unreliable due to autocorrelation among triplets of random numbers.) If a new method has been developed, or if the generator that is at hand is not explicitly known or documented, then the tests in this chapter should be applied to many samples of numbers from the generator. Some additional tests that are commonly used, but are not covered here, are Good's serial test for sampling numbers [1953, 1967], the median-spectrum test [Cox and Lewis, 1966; Durbin, 1967], and a variance heterogeneity test [Cox and Lewis, 1966]. Even if a set of numbers passes all the tests, it is no guarantee of randomness. It is always possible that some underlying pattern will go undetected.

In this book we emphasize empirical tests that are applied to actual sequences of numbers produced by a generator. There are also families of theoretical tests that evaluate the choices for m, a, and c without actually generating any numbers, the most common being the spectral test. Many of these tests assess how k-tuples of random numbers fill up a k-dimensional unit cube. These tests are beyond the scope of this book; see, for instance, Ripley [1987].

In the examples of tests that follow, the hypotheses are not restated. The hypotheses are as indicated in the paragraphs above.

7.4.1 Frequency Tests

A basic test that should always be performed to validate a new generator is the test of uniformity. Two different methods of testing are available. They are the Kolmogorov-Smirnov and the chi-square test. Both of these tests measure the degree of agreement between the distribution of a sample of generated random numbers and the theoretical uniform distribution. Both tests are based on the null hypothesis of no significant difference between the sample distribution and the theoretical distribution.

1. *The Kolmogorov-Smirnov test.* This test compares the continuous cdf, $F(x)$, of the uniform distribution to the empirical cdf, $S_N(x)$, of the sample of N observations. By definition,

$$F(x) = x, \quad 0 \le x \le 1$$

If the sample from the random-number generator is $R_1, R_2, \ldots, R_N$, then the empirical cdf, $S_N(x)$, is defined by

$$S_N(x) = \frac{\text{number of } R_1, R_2, \ldots, R_N \text{ which are } \le x}{N}$$

As N becomes larger, $S_N(x)$ should become a better approximation to $F(x)$, provided that the null hypothesis is true.

In Section 5.6, empirical distributions were described. The cdf of an empirical distribution is a step function with jumps at each observed value. This behavior was illustrated by Example 5.34.

The Kolmogorov-Smirnov test is based on the largest absolute deviation between $F(x)$ and $S_N(x)$ over the range of the random variable. That is, it is based on the statistic

$$D = \max |F(x) - S_N(x)| \tag{7.3}$$

The sampling distribution of D is known and is tabulated as a function of N in Table A.8. For testing against a uniform cdf, the test procedure follows these steps:

Step 1. Rank the data from smallest to largest. Let $R_{(i)}$ denote the ith smallest observation, so that

$$R_{(1)} \le R_{(2)} \le \cdots \le R_{(N)}$$

Step 2. Compute

$$D^+ = \max_{1 \le i \le N} \left\{ \frac{i}{N} - R_{(i)} \right\}$$

$$D^- = \max_{1 \le i \le N} \left\{ R_{(i)} - \frac{i-1}{N} \right\}$$

Step 3. Compute $D = \max(D^+, D^-)$.

Step 4. Determine the critical value, D_α, from Table A.8 for the specified significance level α and the given sample size N.

Step 5. If the sample statistic D is greater than the critical value, D_α, the null hypothesis that the data are a sample from a uniform distribution is rejected. If $D \le D_\alpha$, conclude that no difference has been detected between the true distribution of $\{R_1, R_2, \ldots, R_N\}$ and the uniform distribution.

EXAMPLE 7.6

Suppose that the five numbers 0.44, 0.81, 0.14, 0.05, 0.93 were generated, and it is desired to perform a test for uniformity using the Kolmogorov-Smirnov test with a level of significance α of 0.05. First, the numbers must be ranked from smallest to largest. The calculations can be facilitated by use of Table 7.2. The top row lists the numbers from smallest ($R_{(1)}$) to largest ($R_{(5)}$). The computations for D^+, namely $i/N - R_{(i)}$, and for D^-, namely $R_{(i)} - (i-1)/N$, are easily accomplished using Table 7.2. The statistics are computed as $D^+ = 0.26$ and $D^- = 0.21$. Therefore, $D = \max\{0.26, 0.21\} = 0.26$. The critical value of D, obtained from Table A.8 for $\alpha = 0.05$ and $N = 5$, is 0.565. Since the computed value, 0.26, is less than the tabulated critical value, 0.565, the hypothesis of no difference between the distribution of the generated numbers and the uniform distribution is not rejected.

Table 7.2. Calculations for Kolmogorov-Smirnov Test

$R_{(i)}$	0.05	0.14	0.44	0.81	0.93
i/N	0.20	0.40	0.60	0.80	1.00
$i/N - R_{(i)}$	0.15	0.26	0.16	—	0.07
$R_{(i)} - (i-1)/N$	0.05	—	0.04	0.21	0.13

The calculations in Table 7.2 are illustrated in Figure 7.2, where the empirical cdf, $S_N(x)$, is compared to the uniform cdf, $F(x)$. It can be seen that D^+ is the largest deviation of $S_N(x)$ above $F(x)$, and that D^- is the largest deviation of $S_N(x)$ below $F(x)$. For example, at $R_{(3)}$ the value of D^+ is given by $3/5 - R_{(3)} = 0.60 - 0.44 = 0.16$ and of D^- is given by $R_{(3)} - 2/5 = 0.44 - 0.40 = 0.04$. Although the test statistic D is defined by Equation (7.3) as the maximum deviation over all x, it can be seen from Figure 7.2 that the maximum deviation will always occur at one of the jump points $R_{(1)}, R_{(2)}, \ldots$, and thus the deviation at other values of x need not be considered. ◀

2. *The chi-square test.* The chi-square test uses the sample statistic

$$\chi_0^2 = \sum_{i=1}^{n} \frac{(O_i - E_i)^2}{E_i}$$

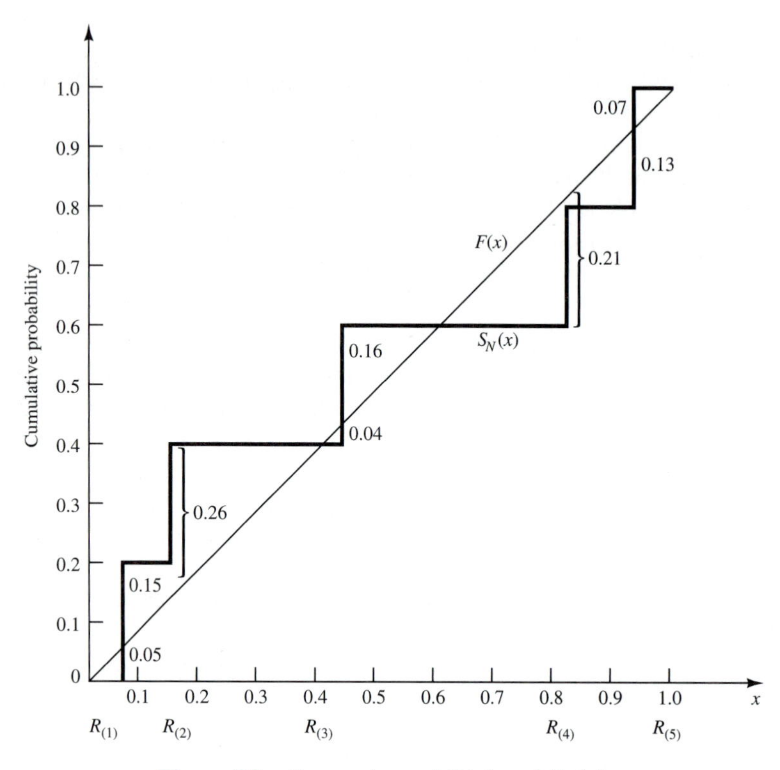

Figure 7.2. Comparison of $F(x)$ and $S_N(x)$.

where O_i is the observed number in the ith class, E_i is the expected number in the ith class, and n is the number of classes. For the uniform distribution, E_i, the expected number in each class is given by

$$E_i = \frac{N}{n}$$

for equally spaced classes, where N is the total number of observations. It can be shown that the sampling distribution of χ_0^2 is approximately the chi-square distribution with $n-1$ degrees of freedom.

Example 7.7

Use the chi-square test with $\alpha = 0.05$ to test whether the data shown below are uniformly distributed. Table 7.3 contains the essential computations. The test uses $n = 10$ intervals of equal length, namely $[0, 0.1), [0.1, 0.2), \ldots, [0.9, 1.0)$. The value of χ_0^2 is 3.4. This is compared with the critical value $\chi_{0.05,9}^2 = 16.9$. Since χ_0^2 is much smaller than the tabulated value of $\chi_{0.05,9}^2$, the null hypothesis of a uniform distribution is not rejected.

0.34	0.90	0.25	0.89	0.87	0.44	0.12	0.21	0.46	0.67
0.83	0.76	0.79	0.64	0.70	0.81	0.94	0.74	0.22	0.74
0.96	0.99	0.77	0.67	0.56	0.41	0.52	0.73	0.99	0.02
0.47	0.30	0.17	0.82	0.56	0.05	0.45	0.31	0.78	0.05
0.79	0.71	0.23	0.19	0.82	0.93	0.65	0.37	0.39	0.42
0.99	0.17	0.99	0.46	0.05	0.66	0.10	0.42	0.18	0.49
0.37	0.51	0.54	0.01	0.81	0.28	0.69	0.34	0.75	0.49
0.72	0.43	0.56	0.97	0.30	0.94	0.96	0.58	0.73	0.05
0.06	0.39	0.84	0.24	0.40	0.64	0.40	0.19	0.79	0.62
0.18	0.26	0.97	0.88	0.64	0.47	0.60	0.11	0.29	0.78

◀

Different authors have offered considerations concerning the application of the χ^2 test. In the application to a data set the size of that in Example 7.7, the considerations do not apply. That is, if 100 values are in the sample and from 5 to 10 intervals of equal length are used, the test will be acceptable. In general, it is recommended that n and N be chosen so that each $E_i \geq 5$.

Both the Kolmogorov-Smirnov and the chi-square test are acceptable for testing the uniformity of a sample of data, provided that the sample size is large. However, the Kolmogorov-Smirnov test is the more powerful of the two and is recommended. Furthermore, the Kolmogorov-Smirnov test can be applied to small sample sizes, whereas the chi-square is valid only for large samples, say $N \geq 50$.

Imagine a set of 100 numbers which are being tested for independence where the first 10 values are in the range 0.01–0.10, the second 10 values are in the range 0.11–0.20, and so on. This set of numbers would pass the frequency tests with ease, but the ordering of the numbers produced by the generator would not be random. The tests in the remainder of this chapter are concerned with the independence of random numbers which are generated. The presentation of the tests is similar to that by Schmidt and Taylor [1970].

Table 7.3. Computations for Chi-Square Test

Interval	O_i	E_i	$O_i - E_i$	$(O_i - E_i)^2$	$\frac{(O_i - E_i)^2}{E_i}$
1	8	10	−2	4	0.4
2	8	10	−2	4	0.4
3	10	10	0	0	0.0
4	9	10	−1	1	0.1
5	12	10	2	4	0.4
6	8	10	−2	4	0.4
7	10	10	0	0	0.0
8	14	10	4	16	1.6
9	10	10	0	0	0.0
10	11	10	1	1	0.1
	100	100	0		3.4

7.4.2 Runs Tests

1. *Runs up and runs down.* Consider a generator that provided a set of 40 numbers in the following sequence:

0.08 0.09 0.23 0.29 0.42 0.55 0.58 0.72 0.89 0.91
0.11 0.16 0.18 0.31 0.41 0.53 0.71 0.73 0.74 0.84
0.02 0.09 0.30 0.32 0.45 0.47 0.69 0.74 0.91 0.95
0.12 0.13 0.29 0.36 0.38 0.54 0.68 0.86 0.88 0.91

Both the Kolmogorov-Smirnov test and the chi-square test would indicate that the numbers are uniformly distributed. However, a glance at the ordering shows that the numbers are successively larger in blocks of 10 values. If these numbers are rearranged as follows, there is far less reason to doubt their independence:

0.41 0.68 0.89 0.84 0.74 0.91 0.55 0.71 0.36 0.30
0.09 0.72 0.86 0.08 0.54 0.02 0.11 0.29 0.16 0.18
0.88 0.91 0.95 0.69 0.09 0.38 0.23 0.32 0.91 0.53
0.31 0.42 0.73 0.12 0.74 0.45 0.13 0.47 0.58 0.29

The runs test examines the arrangement of numbers in a sequence to test the hypothesis of independence.

Before defining a run, a look at a sequence of coin tosses will help with some terminology. Consider the following sequence generated by tossing a coin 10 times:

$$H \; T \; T \; H \; H \; T \; T \; T \; H \; T$$

There are three mutually exclusive outcomes, or events, with respect to the sequence. Two of the possibilities are rather obvious. That is, the toss can result in a head or a tail. The third possibility is "no event." The first head is preceded by no event and the last tail is succeeded by no event. Every sequence begins and ends with no event.

A run is defined as a succession of similar events preceded and followed by a different event. The length of the run is the number of events that occur in the run. In the coin-flipping example above there are six runs. The first run is of length one, the second and third of length two, the fourth of length three, and the fifth and sixth of length one.

There are two possible concerns in a runs test for a sequence of numbers. The number of runs is the first concern and the length of runs is a second concern. The types of runs counted in the first case might be runs up and runs down. An up run is a sequence of numbers each of which is succeeded by a larger number. Similarly, a down run is a sequence of numbers each of which is succeeded by a smaller number. To illustrate the concept, consider the following sequence of 15 numbers:

$^{-}0.87$ $^{+}0.15$ $^{+}0.23$ $^{+}0.45$ $^{-}0.69$ $^{-}0.32$ $^{-}0.30$ $^{+}0.19$ $^{-}0.24$
$^{+}0.18$ $^{+}0.65$ $^{+}0.82$ $^{-}0.93$ $^{+}0.22$ 0.81

The numbers are given a "+" or a "−" depending on whether they are followed by a larger number or a smaller number. Since there are 15 numbers, and they are all different, there will be 14 +'s and −'s. The last number is followed by "no event" and hence will get neither a + nor a −. The sequence of 14 +'s and −'s is as follows:

− + + + − − − + − + + + − +

Each succession of +'s and −'s forms a run. There are eight runs. The first run is of length one, the second and third are of length three, and so on. Further, there are four runs up and four runs down.

There can be too few runs or too many runs. Consider the following sequence of numbers:

0.08 0.18 0.23 0.36 0.42 0.55 0.63 0.72 0.89 0.91

This sequence has one run, a run up. It is unlikely that a valid random-number generator would produce such a sequence. Next, consider the following sequence:

0.08 0.93 0.15 0.96 0.26 0.84 0.28 0.79 0.36 0.57

This sequence has nine runs, five up and four down. It is unlikely that a sequence of 10 numbers would have this many runs. What is more likely is that the number of runs will be somewhere between the two extremes. These two extremes can be formalized as follows: if N is the number of numbers in a sequence, the maximum number of runs is $N - 1$ and the minimum number of runs is one.

If a is the total number of runs in a truly random sequence, the mean and variance of a are given by

$$\mu_a = \frac{2N - 1}{3} \tag{7.4}$$

and

$$\sigma_a^2 = \frac{16N - 29}{90} \tag{7.5}$$

For $N > 20$, the distribution of a is reasonably approximated by a normal distribution, $N(\mu_a, \sigma_a^2)$. This approximation can be used to test the independence of numbers from a generator. In that case the standardized normal test statistic is developed by subtracting the mean from the observed number of runs, a, and dividing by the standard deviation. That is, the test statistic is

$$Z_0 = \frac{a - \mu_a}{\sigma_a}$$

Substituting Equation (7.4) for μ_a and the square root of Equation (7.5) for σ_a yields

$$Z_0 = \frac{a - [(2N - 1)/3]}{\sqrt{(16N - 29)/90}}$$

where $Z_0 \sim N(0, 1)$. Failure to reject the hypothesis of independence occurs when $-z_{\alpha/2} \le Z_0 \le z_{\alpha/2}$, where α is the level of significance. The critical values and rejection region are shown in Figure 7.3.

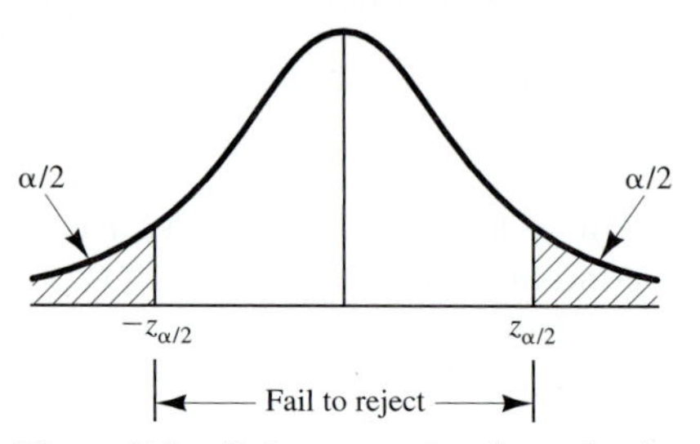

Figure 7.3. Failure to reject hypothesis.

Example 7.8

Based on runs up and runs down, determine whether the following sequence of 40 numbers is such that the hypothesis of independence can be rejected where $\alpha = 0.05$.

0.41	0.68	0.89	0.94	0.74	0.91	0.55	0.62	0.36	0.27
0.19	0.72	0.75	0.08	0.54	0.02	0.01	0.36	0.16	0.28
0.18	0.01	0.95	0.69	0.18	0.47	0.23	0.32	0.82	0.53
0.31	0.42	0.73	0.04	0.83	0.45	0.13	0.57	0.63	0.29

The sequence of runs up and down is as follows:

$$+ + + - + - + - - - + + - + - - + - +$$
$$- - + - - + - + + - - + + - + - - + + -$$

There are 26 runs in this sequence. With $N = 40$ and $a = 26$, Equations (7.4) and (7.5) yield

$$\mu_a = \frac{2(40) - 1}{3} = 26.33$$

and

$$\sigma_a^2 = \frac{16(40) - 29}{90} = 6.79$$

Then,

$$Z_0 = \frac{26 - 26.33}{\sqrt{6.79}} = -0.13$$

Now, the critical value is $z_{0.025} = 1.96$, so the independence of the numbers cannot be rejected on the basis of this test. ◀

2. *Runs above and below the mean.* The test for runs up and runs down is not completely adequate to assess the independence of a group of numbers. Consider the following 40 numbers:

0.63	0.72	0.79	0.81	0.52	0.94	0.83	0.93	0.87	0.67
0.54	0.83	0.89	0.55	0.88	0.77	0.74	0.95	0.82	0.86
0.43	0.32	0.36	0.18	0.08	0.19	0.18	0.27	0.36	0.34
0.31	0.45	0.49	0.43	0.46	0.35	0.25	0.39	0.47	0.41

The sequence of runs up and runs down is as follows:

+ + + − + − + − − − + + − + − − + − + − − + − −
+ − + + − − + + − + − − + + −

This sequence is exactly the same as that in Example 7.8. Thus, the numbers would pass the runs-up and runs-down test. However, it can be observed that the first 20 numbers are all above the mean $[(0.99 + 0.00)/2 = 0.495]$ and the last 20 numbers are all below the mean. Such an occurrence is highly unlikely. The previous runs analysis can be used to test for this condition, if the definition of a run is changed. Runs will be described as being above the mean or below the mean. A "+" sign will be used to denote an observation above the mean, and a "−" sign will denote an observation below the mean.

For example, consider the following sequence of 20 two-digit random numbers:

0.40	0.84	0.75	0.18	0.13	0.92	0.57	0.77	0.30	0.71
0.42	0.05	0.78	0.74	0.68	0.03	0.18	0.51	0.10	0.37

The pluses and minuses are as follows:

− + + − − + + + − + − − + + + − − + − −

In this case, there is a run of length one below the mean followed by a run of length two above the mean, and so on. In all, there are 11 runs, five of which are above the mean and six of which are below the mean. Let n_1 and n_2 be the number of individual observations above and below the mean and let b be the total number of runs. Notice that the maximum number of runs is $N = n_1 + n_2$, and the minimum number of runs is one. Given n_1 and n_2, the mean—with a continuity correction suggested by Swed and Eisenhart [1943] —and the variance of b for a truly independent sequence are given by

$$\mu_b = \frac{2n_1n_2}{N} + \frac{1}{2} \tag{7.6}$$

and

$$\sigma_b^2 = \frac{2n_1n_2(2n_1n_2 - N)}{N^2(N - 1)} \tag{7.7}$$

For either n_1 or n_2 greater than 20, b is approximately normally distributed. The test statistic can be formed by subtracting the mean from the number of runs and dividing by the standard deviation, or

$$Z_0 = \frac{b - (2n_1n_2/N) - 1/2}{\left[\frac{2n_1n_2(2n_1n_2 - N)}{N^2(N-1)}\right]^{1/2}}$$

Failure to reject the hypothesis of independence occurs when $-z_{\alpha/2} \leq Z_0 \leq z_{\alpha/2}$, where α is the level of significance. The rejection region is shown in Figure 7.3.

Example 7.9

Determine whether there is an excessive number of runs above or below the mean for the sequence of numbers given in Example 7.8. The assignment of +'s and −'s results in the following:

− + + + + + + + − − − + + − + − − − − −
− − + + − − − − + + − − + − + − − + + −

The values of n_1, n_2, and b are as follows:

$$n_1 = 18$$
$$n_2 = 22$$
$$N = n_1 + n_2 = 40$$
$$b = 17$$

Equations (7.6) and (7.7) are used to determine μ_b and σ_b^2 as follows:

$$\mu_b = \frac{2(18)(22)}{40} + \frac{1}{2} = 20.3$$

and

$$\sigma_b^2 = \frac{2(18)(22)[(2)(18)(22) - 40]}{(40)^2(40-1)} = 9.54$$

Since n_2 is greater than 20, the normal approximation is acceptable, resulting in a Z_0 value of

$$Z_0 = \frac{17 - 20.3}{\sqrt{9.54}} = -1.07$$

Since $z_{0.025} = 1.96$, the hypothesis of independence cannot be rejected on the basis of this test. ◀

3. *Runs test: length of runs.* Yet another concern is the length of runs. As an example of what might occur, consider the following sequence of numbers:

0.16, 0.27, 0.58, 0.63, 0.45, 0.21, 0.72, 0.87, 0.27, 0.15, 0.92, 0.85, ...

Assume that this sequence continues in a like fashion: two numbers below the mean followed by two numbers above the mean. A test of runs above and below the mean would detect no departure from independence. However, it is to be expected that runs other than of length two should occur.

Let Y_i be the number of runs of length i in a sequence of N numbers. For an independent sequence, the expected value of Y_i for runs up and down is given by

$$E(Y_i) = \frac{2}{(i+3)!}[N(i^2+3i+1) - (i^3+3i^2-i-4)], \; i \le N-2 \quad (7.8)$$

$$E(Y_i) = \frac{2}{N!}, \quad i = N-1 \quad (7.9)$$

For runs above and below the mean, the expected value of Y_i is approximately given by

$$E(Y_i) = \frac{Nw_i}{E(I)}, \quad N > 20 \quad (7.10)$$

where w_i, the approximate probability that a run has length i, is given by

$$w_i = \left(\frac{n_1}{N}\right)^i \left(\frac{n_2}{N}\right) + \left(\frac{n_1}{N}\right)\left(\frac{n_2}{N}\right)^i, \quad N > 20 \quad (7.11)$$

and where $E(I)$, the approximate expected length of a run, is given by

$$E(I) = \frac{n_1}{n_2} + \frac{n_2}{n_1}, \quad N > 20 \quad (7.12)$$

The approximate expected total number of runs (of all lengths) in a sequence of length N, $E(A)$, is given by

$$E(A) = \frac{N}{E(I)}, \quad N > 20 \quad (7.13)$$

The appropriate test is the chi-square test with O_i being the observed number of runs of length i. Then the test statistic is

$$\chi_0^2 = \sum_{i=1}^{L} \frac{[O_i - E(Y_i)]^2}{E(Y_i)}$$

where $L = N-1$ for runs up and down and $L = N$ for runs above and below the mean. If the null hypothesis of independence is true, then χ_0^2 is approximately chi-square distributed with $L-1$ degrees of freedom.

EXAMPLE 7.10

Given the following sequence of numbers, can the hypothesis that the numbers are independent be rejected on the basis of the length of runs up and down at $\alpha = 0.05$?

0.30	0.48	0.36	0.01	0.54	0.34	0.96	0.06	0.61	0.85
0.48	0.86	0.14	0.86	0.89	0.37	0.49	0.60	0.04	0.83
0.42	0.83	0.37	0.21	0.90	0.89	0.91	0.79	0.57	0.99
0.95	0.27	0.41	0.81	0.96	0.31	0.09	0.06	0.23	0.77
0.73	0.47	0.13	0.55	0.11	0.75	0.36	0.25	0.23	0.72
0.60	0.84	0.70	0.30	0.26	0.38	0.05	0.19	0.73	0.44

For this sequence the +'s and −'s are as follows:

$$
\begin{array}{l}
+ \quad - \quad - \quad + \quad - \quad + \quad - \quad + \quad + \quad - \quad + \quad - \quad + \quad + \quad - \quad + \quad + \quad - \quad + \\
- \quad + \quad - \quad - \quad + \quad - \quad + \quad - \quad - \quad + \quad - \quad - \quad + \quad + \quad + \quad - \quad - \quad - \quad + \quad + \\
- \quad - \quad - \quad + \quad - \quad + \quad - \quad - \quad - \quad + \quad - \quad + \quad - \quad - \quad - \quad + \quad - \quad + \quad + \quad -
\end{array}
$$

The length of runs in the sequence is as follows:

$$
\begin{array}{l}
1, 2, 1, 1, 1, 1, 2, 1, 1, 1, 2, 1, 2, 1, 1, 1, 1, 2, 1, 1, \\
1, 2, 1, 2, 3, 3, 2, 3, 1, 1, 1, 3, 1, 1, 1, 3, 1, 1, 2, 1
\end{array}
$$

The number of observed runs of each length is as follows:

Run Length, i	1	2	3
Observed Runs, O_i	26	9	5

The expected numbers of runs of lengths one, two, and three are computed from Equation (7.8) as

$$
\begin{aligned}
E(Y_1) &= \frac{2}{4!}[60(1 + 3 + 1) - (1 + 3 - 1 - 4)] \\
&= 25.08 \\
E(Y_2) &= \frac{2}{5!}[60(4 + 6 + 1) - (8 + 12 - 2 - 4)] \\
&= 10.77 \\
E(Y_3) &= \frac{2}{6!}[60(9 + 9 + 1) - (27 + 27 - 3 - 4)] \\
&= 3.04
\end{aligned}
$$

The mean total number of runs (up and down) is given by Equation (7.4) as

$$
\mu_a = \frac{2(60) - 1}{3} = 39.67
$$

Thus far, the $E(Y_i)$ for $i = 1, 2$, and 3 total 38.89. The expected number of runs of length 4 or more is the difference $\mu_a - \sum_{i=1}^{3} E(Y_i)$, or 0.78.

As observed by Hines and Montgomery [1990], there is no general agreement regarding the minimum value of expected frequencies in applying the chi-square test. Values of 3, 4, and 5 are widely used, and a minimum of 5 was suggested earlier in this chapter. Should an expected frequency be too small,

Table 7.4. Length of Runs Up and Down: χ^2 Test

Run Length, i	*Observed Number of Runs, O_i*	*Expected Number of Runs, $E(Y_i)$*	$\frac{[O_i - E(Y_i)]^2}{E(Y_i)}$
1	26	25.08	0.03
2	9 } 14	10.77 } 14.59	} 0.02
≥ 3	5	3.82	
	40	39.67	0.05

it can be combined with the expected frequency in an adjacent class interval. The corresponding observed frequencies would then be combined also, and L would be reduced by one. With the foregoing calculations and procedures in mind, we construct Table 7.4. The critical value $\chi^2_{0.05,1}$ is 3.84. (The degrees of freedom equals the number of class intervals minus one.) Since $\chi^2_0 = 0.05$ is less than the critical value, the hypothesis of independence cannot be rejected on the basis of this test. ◀

EXAMPLE 7.11

Given the same sequence of numbers in Example 7.10, can the hypothesis that the numbers are independent be rejected on the basis of the length of runs above and below the mean at $\alpha = 0.05$? For this sequence, the +'s and −'s are as follows:

− − − − + − + − + + − + − + + − − + − +
− + − − + + + + + + + − − + + − − − − +
+ − − + − + − − − + + + + − − − − − + −

The number of runs of each length is as follows:

Run Length, i	1	2	3	≥ 4
Observed Runs, O_i	17	9	1	5

There are 28 values above the mean ($n_1 = 28$) and 32 values below the mean ($n_2 = 32$). The probabilities of runs of various lengths, w_i, are determined from Equation (7.11) as

$$w_1 = \left(\frac{28}{60}\right)^1 \frac{32}{60} + \frac{28}{60}\left(\frac{32}{60}\right)^1 = 0.498$$

$$w_2 = \left(\frac{28}{60}\right)^2 \frac{32}{60} + \frac{28}{60}\left(\frac{32}{60}\right)^2 = 0.249$$

$$w_3 = \left(\frac{28}{60}\right)^3 \frac{32}{60} + \frac{28}{60}\left(\frac{32}{60}\right)^3 = 0.125$$

$$\vdots$$

The expected length of a run, $E(I)$, is determined from Equation (7.12) as

$$E(I) = \frac{28}{32} + \frac{32}{28} = 2.02$$

Now, Equation (7.10) can be used to determine the expected numbers of runs of various lengths as

$$E(Y_1) = \frac{60(0.498)}{2.02} = 14.79$$

$$E(Y_2) = \frac{60(0.249)}{2.02} = 7.40$$

$$E(Y_3) = \frac{60(0.125)}{2.02} = 3.71$$

The total number of runs expected is given by Equation (7.13) as $E(A) = 60/2.02 = 29.7$. This indicates that approximately 3.8 runs of length four or more can be expected. Proceeding by combining adjacent cells in which $E(Y_i) < 5$ produces Table 7.5.

Table 7.5. Length of Runs Above and Below the Mean: χ^2 Test

Run Length, i	*Observed Number of Runs, O_i*	*Expected Number of Runs, $E(Y_i)$*	$\frac{[O_i - E(Y_i)]^2}{E(Y_i)}$
1	17	14.79	0.33
2	9	7.40	0.35
3	1 } 6	3.71 } 7.51	} 0.30
≥ 4	5	3.80	
	32	29.70	0.98

The critical value $\chi^2_{0.05,2}$ is 5.99. (The degrees of freedom equals the number of class intervals minus one.) Since $\chi^2_0 = 0.98$ is less than the critical value, the hypothesis of independence cannot be rejected on the basis of this test. ◀

7.4.3 Tests for Autocorrelation

The tests for autocorrelation are concerned with the dependence between numbers in a sequence. As an example, consider the following sequence of numbers:

0.12	0.01	0.23	0.28	0.89	0.31	0.64	0.28	0.83	0.93
0.99	0.15	0.33	0.35	0.91	0.41	0.60	0.27	0.75	0.88
0.68	0.49	0.05	0.43	0.95	0.58	0.19	0.36	0.69	0.87

From a visual inspection, these numbers appear random, and they would probably pass all the tests presented to this point. However, an examination of the 5th, 10th, 15th (every five numbers beginning with the fifth), and so on, indicates a very large number in that position. Now, 30 numbers is a rather small sample size to reject a random-number generator, but the notion is that numbers in the sequence might be related. In this particular section, a method for determining whether such a relationship exists is described. The relationship would not have to be all high numbers. It is possible to have all low numbers in the locations being examined, or the numbers may alternately shift from very high to very low.

The test to be described below requires the computation of the autocorrelation between every m numbers (m is also known as the lag) starting with the ith number. Thus, the autocorrelation ρ_{im} between the following numbers would be of interest: $R_i, R_{i+m}, R_{i+2m}, \ldots, R_{i+(M+1)m}$. The value M is the largest integer such that $i + (M + 1)m \leq N$, where N is the total number of values in the sequence. (Thus, a subsequence of length $M + 2$ is being tested.)

Since a nonzero autocorrelation implies a lack of independence, the following two-tailed test is appropriate:

$$H_0: \rho_{im} = 0$$

$$H_1: \rho_{im} \neq 0$$

For large values of M, the distribution of the estimator of ρ_{im}, denoted $\widehat{\rho}_{im}$, is approximately normal if the values $R_i, R_{i+m}, R_{i+2m}, \ldots, R_{i+(M+1)m}$ are uncorrelated. Then the test statistic can be formed as follows:

$$Z_0 = \frac{\widehat{\rho}_{im}}{\sigma_{\widehat{\rho}_{im}}}$$

which is distributed normally with a mean of zero and a variance of 1, under the assumption of independence, for large M.

The formula for $\widehat{\rho}_{im}$, in a slightly different form, and the standard deviation of the estimator, $\sigma_{\widehat{\rho}_{im}}$, are given by Schmidt and Taylor [1970] as follows:

$$\widehat{\rho}_{im} = \frac{1}{M+1}\left[\sum_{k=0}^{M} R_{i+km} R_{i+(k+1)m}\right] - 0.25$$

and

$$\sigma_{\widehat{\rho}_{im}} = \frac{\sqrt{13M + 7}}{12(M + 1)}$$

After computing Z_0, do not reject the null hypothesis of independence if $-z_{\alpha/2} \leq Z_0 \leq z_{\alpha/2}$, where α is the level of significance. Figure 7.3, presented earlier, illustrates this test.

If $\rho_{im} > 0$, the subsequence is said to exhibit positive autocorrelation. In this case, successive values at lag m have a higher probability than expected of being close in value (i.e., high random numbers in the subsequence followed by high, and low followed by low). On the other hand, if $\rho_{im} < 0$, the subsequence is exhibiting negative autocorrelation, which means that low random numbers tend to be followed by high ones, and vice versa. The desired property of independence, which implies zero autocorrelation, means that there is no discernible relationship of the nature discussed here between successive random numbers at lag m.

EXAMPLE 7.12

Test whether the 3rd, 8th, 13th, and so on, numbers in the sequence at the beginning of this section are autocorrelated. (Use $\alpha = 0.05$.) Here, $i = 3$ (beginning with the third number), $m = 5$ (every five numbers), $N = 30$ (30 numbers in the sequence), and $M = 4$ (largest integer such that $3+(M+1)5 \leq 30$). Then,

$$\begin{aligned}\widehat{\rho}_{35} &= \frac{1}{4+1}[(0.23)(0.28) + (0.28)(0.33) + (0.33)(0.27) + (0.27)(0.05) \\ &\quad + (0.05)(0.36)] - 0.25 \\ &= -0.1945\end{aligned}$$

and

$$\sigma_{\widehat{\rho}_{35}} = \frac{\sqrt{13(4)+7}}{12(4+1)} = 0.1280$$

Then, the test statistic assumes the value

$$Z_0 = \frac{-0.1945}{0.1280} = -1.516$$

Now, the critical value is

$$z_{0.025} = 1.96$$

Therefore, the hypothesis of independence cannot be rejected on the basis of this test.

It can be observed that this test is not very sensitive for small values of M, particularly when the numbers being tested are on the low side. Imagine what would happen if each of the entries in the foregoing computation of $\widehat{\rho}_{im}$ were equal to zero. Then, $\widehat{\rho}_{im}$ would be equal to -0.25 and the calculated Z would have the value of -1.95, not quite enough to reject the hypothesis of independence. ◀

Many sequences can be formed in a set of data, given a large value of N. For example, beginning with the first number in the sequence, possibilities include (1) the sequence of all numbers, (2) the sequence formed from the first, third, fifth,..., numbers, (3) the sequence formed from the first, fourth, ..., numbers, and so on. If $\alpha = 0.05$, there is a probability of 0.05 of rejecting a true hypothesis. If 10 independent sequences are examined, the probability of finding no significant autocorrelation, by chance alone, is $(0.95)^{10}$ or 0.60. Thus, 40% of the time significant autocorrelation would be detected when it does not exist. If α is 0.10 and 10 tests are conducted, there is a 65% chance of finding autocorrelation by chance alone. In conclusion, when "fishing" for autocorrelation, upon performing numerous tests, autocorrelation may eventually be detected, perhaps by chance alone, even when no autocorrelation is present.

7.4.4 Gap Test

The gap test is used to determine the significance of the interval between the recurrences of the same digit. A gap of length x occurs between the recurrences of some specified digit. The following example illustrates the length of gaps associated with the digit 3:

$$
\begin{array}{l}
4,\ 1,\ \underline{3},\ 5,\ 1,\ 7,\ 2,\ 8,\ 2,\ 0,\ 7,\ 9,\ 1,\ \underline{3},\ 5,\ 2,\ 7,\ 9,\ 4,\ 1,\ 6,\ \underline{3} \\
\underline{3},\ 9,\ 6,\ \underline{3},\ 4,\ 8,\ 2,\ \underline{3},\ 1,\ 9,\ 4,\ 4,\ 6,\ 8,\ 4,\ 1,\ \underline{3},\ 8,\ 9,\ 5,\ 5,\ 7 \\
\underline{3},\ 9,\ 5,\ 9,\ 8,\ 5,\ \underline{3},\ 2,\ 2,\ \underline{3},\ 7,\ 4,\ 7,\ 0,\ \underline{3},\ 6,\ \underline{3},\ 5,\ 9,\ 9,\ 5,\ 5 \\
5,\ 0,\ 4,\ 6,\ 8,\ 0,\ 4,\ 7,\ 0,\ \underline{3},\ \underline{3},\ 0,\ 9,\ 5,\ 7,\ 9,\ 5,\ 1,\ 6,\ 6,\ \underline{3},\ 8 \\
8,\ 8,\ 9,\ 2,\ 9,\ 1,\ 8,\ 5,\ 4,\ 4,\ 5,\ 0,\ 2,\ \underline{3},\ 9,\ 7,\ 1,\ 2,\ 0,\ \underline{3},\ 6,\ 3
\end{array}
$$

To facilitate the analysis, the digit 3 has been underlined. There are eighteen 3's in the list. Thus, only 17 gaps can occur. The first gap is of length 10, the second gap is of length 7, and so on. The frequency of the gaps is of interest. The probability of the first gap is determined as follows:

$$
\begin{aligned}
P(\text{gap of } 10) &= \overbrace{P(\text{no } 3) \cdots P(\text{no } 3)}^{\text{10 of these terms}} P(3) \\
&= (0.9)^{10}(0.1)
\end{aligned}
$$

since the probability that any digit is not a 3 is 0.9, and the probability that any digit is a 3 is 0.1. In general,

$$P(t \text{ followed by exactly } x \text{ non-}t \text{ digits}) = (0.9)^x(0.1), \quad x = 0, 1, 2, \ldots$$

In the example above, only the digit 3 was examined. However, to fully analyze a set of numbers for independence using the gap test, every digit, 0, 1, 2, ..., 9, must be analyzed. The observed frequencies of the various gap sizes for all the digits are recorded and compared to the theoretical frequency using the Kolmogorov-Smirnov test for discretized data.

The theoretical frequency distribution for randomly ordered digits is given by

$$P(\text{gap} \le x) = F(x) = 0.1 \sum_{n=0}^{x} (0.9)^n = 1 - 0.9^{x+1} \qquad (7.14)$$

The procedure for the test follows the steps below. When applying the test to random numbers, class intervals such as $[0, 0.1), [0.1, 0.2), \ldots$ play the role of random digits.

Step 1. Specify the cdf for the theoretical frequency distribution given by Equation (7.14) based on the selected class interval width.

Step 2. Arrange the observed sample of gaps in a cumulative distribution with these same classes.

Step 3. Find D, the maximum deviation between $F(x)$ and $S_N(x)$ as in Equation (7.3).

Step 4. Determine the critical value, D_α, from Table A.8 for the specified value of α and the sample size N.

Step 5. If the calculated value of D is greater than the tabulated value of D_α, the null hypothesis of independence is rejected.

It should be noted that using the Kolmogorov-Smirnov test when the underlying distribution is discrete results in a reduction in the Type I error, α, and an increase in the Type II error, β. The exact value of α can be found using the methodology described by Conover [1980].

Example 7.13

Based on the frequency with which gaps occur, analyze the 110 digits above to test whether they are independent. Use $\alpha = 0.05$. The number of gaps is given by the number of data values minus the number of distinct digits, or $110 - 10 = 100$ in the example. The number of gaps associated with the various digits are as follows:

Digit	0	1	2	3	4	5	6	7	8	9
Number of Gaps	7	8	8	17	10	13	7	8	9	13

The gap test is presented in Table 7.6. The critical value of D is given by

$$D_{0.05} = \frac{1.36}{\sqrt{100}} = 0.136$$

Since $D = \max |F(x) - S_N(x)| = 0.0224$ is less than $D_{0.05}$, do not reject the hypothesis of independence on the basis of this test. ◀

Table 7.6. Gap-Test Example

Gap Length	*Frequency*	*Relative Frequency*	*Cumulative Relative Frequency*	$F(x)$	$\|F(x) - S_N(x)\|$
0–3	35	0.35	0.35	0.3439	0.0061
4–7	22	0.22	0.57	0.5695	0.0005
8–11	17	0.17	0.74	0.7176	0.0224
12–15	9	0.09	0.83	0.8147	0.0153
16–19	5	0.05	0.88	0.8784	0.0016
20–23	6	0.06	0.94	0.9202	0.0198
24–27	3	0.03	0.97	0.9497	0.0223
28–31	0	0.0	0.97	0.9657	0.0043
32–35	0	0.0	0.97	0.9775	0.0075
36–39	2	0.02	0.99	0.9852	0.0043
40–43	0	0.0	0.99	0.9903	0.0003
44–47	1	0.01	1.00	0.9936	0.0064

7.4.5 *Poker Test*

The poker test for independence is based on the frequency with which certain digits are repeated in a series of numbers. The following example shows an unusual amount of repetition:

$$0.255,\quad 0.577,\quad 0.331,\quad 0.414,\quad 0.828,\quad 0.909,\quad 0.303,\ 0.001,\quad \ldots$$

In each case, a pair of like digits appears in the number that was generated. In three-digit numbers there are only three possibilities, as follows:

1. The individual numbers can all be different.
2. The individual numbers can all be the same.
3. There can be one pair of like digits.

The probability associated with each of these possibilities is given by the following:

$$\begin{aligned}
P(\text{three different digits}) &= P(\text{second different from the first}) \\
&\quad \times P(\text{third different from the first and second}) \\
&= (0.9)(0.8) = 0.72 \\
P(\text{three like digits}) &= P(\text{second digit same as the first}) \\
&\quad \times P(\text{third digit same as the first}) \\
&= (0.1)(0.1) = 0.01 \\
P(\text{exactly one pair}) &= 1 - 0.72 - 0.01 = 0.27
\end{aligned}$$

Alternatively, the last result can be obtained as follows:

$$P(\text{exactly one pair}) = \binom{3}{2}(0.1)(0.9) = 0.27$$

The following example shows how the poker test (in conjunction with the chi-square test) is used to ascertain independence.

Example 7.14

A sequence of 1000 three-digit numbers has been generated and an analysis indicates that 680 have three different digits, 289 contain exactly one pair of like digits, and 31 contain three like digits. Based on the poker test, are these numbers independent? Let $\alpha = 0.05$. The test is summarized in Table 7.7.

The appropriate degrees of freedom are one less than the number of class intervals. Since $47.65 > \chi^2_{0.05,2} = 5.99$, the independence of the numbers is rejected on the basis of this test. ◀

Table 7.7. Poker-Test Results

Combination, i	*Observed Frequency,* O_i	*Expected Frequency,* E_i	$\frac{(O_i - E_i)^2}{E_i}$
Three different digits	680	720	2.22
Three like digits	31	10	44.10
Exactly one pair	289	270	1.33
	1000	1000	47.65

7.5 Summary

This chapter described the generation of random numbers and the subsequent testing of the generated numbers for uniformity and independence. Random numbers are used to generate random variates, the subject of Chapter 8.

Of the many types of random-number generators available, the linear congruential method is the most widely used. Of the many types of statistical tests that are used in testing random-number generators, five different types are described. Some of these tests are for uniformity, the others for testing independence.

The simulation analyst may never work directly with a random-number generator or with the testing of random numbers from a generator. Most computers and simulation languages have routines that generate a random number, or streams of random numbers, for the asking. But even generators that have been used for years, some of which are still in use, have been found to be inadequate. So this chapter calls the simulation analyst's attention to such possibilities, with a warning to investigate and confirm that the generator has been tested thoroughly. Some researchers have attained sophisticated expertise in developing methods for generating and testing random numbers and the subsequent application of these methods. However, this chapter provides only a basic introduction to the subject matter; more depth and breadth are required for the reader to become a specialist in the area. A key reference is Knuth [1981]; see also the reviews in Bratley, Fox and Schrage [1987], Law and Kelton [2000], L'Ecuyer [1998], and Ripley [1987].

One final caution is due. Even if generated numbers pass all the tests (both those covered in this chapter and those mentioned in the chapter), some underlying pattern may go undetected and the generator may not be rejected as faulty. However, the generators available in widely used simulation languages have been extensively tested and validated.

REFERENCES

BRATLEY, P., B. L. FOX, AND L. E. SCHRAGE [1987], *A Guide to Simulation,* 2d ed., Springer-Verlag, New York.

CONOVER, W. J. [1980], *Practical Nonparametric Statistics,* 2d ed., John Wiley, New York.

COX, D. R., AND P. A. W. LEWIS [1966], *The Statistical Analysis of Series of Events,* Barnes and Noble, New York.

DURBIN, J. [1967], "Tests of Serial Independence Based on the Cumulated Periodogram," *Bulletin of the International Institute of Statistics.*

FISHMAN, G. S. [1978], *Principles of Discrete Event Simulation,* John Wiley, New York.

GOOD, I. J. [1953], "The Serial Test for Sampling Numbers and Other Tests of Randomness," *Proceedings of the Cambridge Philosophical Society,* Vol. 49, pp. 276–84.

GOOD, I. J. [1967], "The Generalized Serial Test and the Binary Expansion of 4," *Journal of the Royal Statistical Society,* Ser. A, Vol. 30, No. 1, pp. 102–7.

HINES, W. W., AND D. C. MONTGOMERY [1990], *Probability and Statistics in Engineering and Management Science,* 3d ed., Prentice Hall, Upper Saddle River, NJ.

KNUTH, D. W. [1981], *The Art of Computer Programming*, Vol. 2: *Semi-numerical Algorithms,* 2d ed., Addison-Wesley, Reading, MA.

LAW, A. M., AND W. D. KELTON [2000], *Simulation Modeling & Analysis,* 3d ed., McGraw-Hill, New York.

LEARMONTH, G. P., AND P. A. W. LEWIS [1973], "Statistical Tests of Some Widely Used and Recently Proposed Uniform Random Number Generators," *Proceedings of the Conference on Computer Science and Statistics: Seventh Annual Symposium on the Interface,* Western Publishing, North Hollywood, Calif., pp. 163–71.

L'ECUYER, P. [1988], "Efficient and Portable Combined Random Number Generators," *Communications of the ACM,* Vol. 31, pp. 742–749, 774.

L'ECUYER, P. [1996], "Combined Multiple Recursive Random Number Generators," *Operations Research,* Vol. 44, pp. 816–822.

L'ECUYER, P. [1998], "Random Number Generation," Chapter 4 in *Handbook of Simulation*, John Wiley, New York.

L'ECUYER, P. [1999], "Good Parameters and Implementations for Combined Multiple Recursive Random Number Generators," *Operations Research,* Vol. 47, pp. 159–164.

LEHMER, D. H. [1951], *Proceedings of the Second Symposium on Large-Scale Digital Computing Machinery,* Harvard University Press, Cambridge, MA.

LEWIS, P. A. W., A. S. GOODMAN, AND J. M. MILLER [1969], "A Pseudo-Random Number Generator for the System/360," *IBM Systems Journal,* Vol. 8, pp. 136–45.

RIPLEY, B. D. [1987], *Stochastic Simulation,* John Wiley, New York.

SCHMIDT, J. W., AND R. E. TAYLOR [1970], *Simulation and Analysis of Industrial Systems,* Irwin, Homewood, IL.

SWED, F. S., AND C. EISENHART [1943], "Tables for Testing Randomness of Grouping in a Sequence of Alternatives," *Annals of Mathematical Statistics,* Vol. 14, pp. 66–82.

EXERCISES

1. Describe a procedure to physically generate random numbers on the interval [0, 1] with 2-digit accuracy. [*Hint:* Consider drawing something out of a hat.]
2. List applications, other than systems simulation, for pseudo-random numbers — for example, video gambling games.
3. How could random numbers that are uniform on the interval [0, 1] be transformed into random numbers that are uniform on the interval [−11, 17]? Transformations to more general distributions are described in Chapter 8.

4. Use the linear congruential method to generate a sequence of three two-digit random integers. Let $X_0 = 27$, $a = 8$, $c = 47$, and $m = 100$.

5. Do we encounter a problem in the previous exercise if $X_0 = 0$?

6. Use the multiplicative congruential method to generate a sequence of four three-digit random integers. Let $X_0 = 117$, $a = 43$, and $m = 1000$.

7. The sequence of numbers 0.54, 0.73, 0.98, 0.11, and 0.68 has been generated. Use the Kolmogorov-Smirnov test with $\alpha = 0.05$ to determine if the hypothesis that the numbers are uniformly distributed on the interval [0, 1] can be rejected.

8. Reverse the 100 two-digit random numbers in Example 7.7 to get a new set of random numbers. Thus, the first random number in the new set will be 0.43. Use the chi-square test, with $\alpha = 0.05$, to determine if the hypothesis that the numbers are uniformly distributed on the interval [0, 1] can be rejected.

9. Consider the first 50 two-digit values in Example 7.7. Based on runs up and runs down, determine whether the hypothesis of independence can be rejected, where $\alpha = 0.05$.

10. Consider the last 50 two-digit values in Example 7.7. Determine whether there is an excessive number of runs above or below the mean. Use $\alpha = 0.05$.

11. Consider the first 50 two-digit values in Example 7.7. Can the hypothesis that the numbers are independent be rejected on the basis of the length of runs up and down when $\alpha = 0.05$?

12. Consider the last 50 two-digit values in Example 7.7. Can the hypothesis that the numbers are independent be rejected on the basis of the length of runs above and below the mean, where $\alpha = 0.05$?

13. Consider the 60 values in Example 7.10. Test whether the 2nd, 9th, 16th, ... numbers in the sequence are autocorrelated, where $\alpha = 0.05$.

14. Consider the following sequence of 120 digits:

1 3 7 4 8 6 2 5 1 6 4 4 3 3 4 2 1 5 8 7
0 7 6 2 6 0 5 7 8 0 1 1 2 6 7 6 3 7 5 9
0 8 8 2 6 7 8 1 3 5 3 8 4 0 9 0 3 0 9 2
2 3 6 5 6 0 0 1 3 4 4 6 9 9 8 5 6 0 1 7
5 6 7 9 4 9 3 1 8 3 3 6 6 7 8 2 3 5 9 6
6 7 0 3 1 0 2 4 2 0 6 4 0 3 9 3 6 8 1 5

Test whether these digits can be assumed to be independent based on the frequency with which gaps occur. Use $\alpha = 0.05$.

15. Develop the poker test for:

(a) Four-digit numbers

(b) Five-digit numbers

16. A sequence of 1000 four-digit numbers has been generated and an analysis indicates the following combinations and frequencies.

Combination i	*Observed Frequency, O_i*
Four different digits	565
One pair	392
Two pairs	17
Three like digits	24
Four like digits	2
	1000

Based on the poker test, test whether these numbers are independent. Use $\alpha = 0.05$.

17. Determine whether the linear congruential generators shown below can achieve a maximum period. Also, state restrictions on X_0 to obtain this period.

(a) The mixed congruential method with

$$a = 2{,}814{,}749{,}767{,}109$$

$$c = 59{,}482{,}661{,}568{,}307$$

$$m = 2^{48}$$

(b) The multiplicative congruential generator with

$$a = 69{,}069$$

$$c = 0$$

$$m = 2^{32}$$

(c) The mixed congruential generator with

$$a = 4951$$

$$c = 247$$

$$m = 256$$

(d) The multiplicative congruential generator with

$$a = 6507$$

$$c = 0$$

$$m = 1024$$

18. Use the mixed congruential method to generate a sequence of three two-digit random numbers with $X_0 = 37$, $a = 7$, $c = 29$, and $m = 100$.

19. Use the mixed congruential method to generate a sequence of three two-digit random integers between 0 and 24 with $X_0 = 13$, $a = 9$, and $c = 35$.

20. Write a computer program that will generate four-digit random numbers using the multiplicative congruential method. Allow the user to input values of X_0, a, c and m.

21. If $X_0 = 3579$ in Exercise 17(c), generate the first random number in the sequence. Compute the random number to four-place accuracy.

22. Investigate the random-number generator in a spreadsheet program on a computer to which you have access. In many spreadsheets, random numbers are generated by a function called RAND or @RAND.

(a) Check the user's manual to see if it describes how the random numbers are generated.

(b) Write macros to conduct each of the tests described in this chapter. Generate 100 sets of random numbers, each set containing 100 random numbers. Perform each test on each set of random numbers. Draw conclusions.

23. Consider the multiplicative congruential generator under the following circumstances:

(a) $a = 11, m = 16, X_0 = 7$.

(b) $a = 11, m = 16, X_0 = 8$.

(c) $a = 7, m = 16, X_0 = 7$.

(d) $a = 7, m = 16, X_0 = 8$.

Generate enough values in each case to complete a cycle. What inferences can be drawn? Is maximum period achieved?

24. For 16-bit computers, L'Ecuyer [1988] recommends combining three multiplicative generators with $m_1 = 32363$, $a_1 = 157$, $m_2 = 31727$, $a_2 = 146$, $m_3 = 31657$, and $a_3 = 142$. The period of this generator is approximately 8×10^{12}. Generate 5 random numbers with the combined generator using initial seeds $X_{i,0} = 100, 300, 500$ for the individual generators $i = 1, 2, 3$.

25. Apply the tests described in this chapter to the generator given in the previous exercise.

26. Use the principles described in this chapter to develop your own linear congruential random-number generator.

27. Use the principles described in this chapter to develop your own combined linear congruential random-number generator.

28. Test the following sequence of numbers for uniformity and independence using procedures you learned in this chapter: 0.594, 0.928, 0.515, 0.055, 0.507, 0.351, 0.262, 0.797, 0.788, 0.442, 0.097, 0.798, 0.227, 0.127, 0.474, 0.825, 0.007, 0.182, 0.929, 0.852.

29. In some applications it is useful to be able to quickly skip ahead in a pseudo-random number sequence without actually generating all of the intermediate values. (a) For a linear congruential generator with $c = 0$, show that $X_{i+n} = (a^n X_i) \bmod m$. (b) Next show that $(a^n X_i) \bmod m = (a^n \bmod m)X_i \bmod m$ (this result is useful because $a^n \bmod m$ can be precomputed, making it easy to skip ahead n random numbers from any point in the sequence). (c) In Example 7.3, use this result to compute X_5 starting with $X_0 = 63$. Check your answer by computing X_5 in the usual way.

8

Random-Variate Generation

This chapter deals with procedures for sampling from a variety of widely used continuous and discrete distributions. Previous discussions and examples indicated the usefulness of statistical distributions to model activities that are generally unpredictable or uncertain. For example, interarrival times and service times at queues, and demands for a product, are quite often unpredictable in nature, at least to a certain extent. Usually, such variables are modeled as random variables with some specified statistical distribution, and standard statistical procedures exist for estimating the parameters of the hypothesized distribution and for testing the validity of the assumed statistical model. Such procedures are discussed in Chapter 9.

In this chapter it is assumed that a distribution has been completely specified, and ways are sought to generate samples from this distribution to be used as input to a simulation model. The purpose of the chapter is to explain and illustrate some widely used techniques for generating random variates, not to give a state-of-the-art survey of the most efficient techniques. In practice, most simulation modelers will use existing routines available in programming libraries, or the routines built into the simulation language being used. However, some programming languages do not have built-in routines for all of the regularly used distributions, and some computer installations do not have random-variate-generation libraries, in which case the modeler must construct an acceptable routine. Even though the chance of this happening is small, it is nevertheless worthwhile to understand how random-variate generation occurs.

This chapter discusses the inverse transform technique, the convolution method, and, more briefly, the acceptance-rejection technique. Another technique, the composition method, is discussed by Fishman [1978] and Law and

Kelton [2000]. All the techniques in this chapter assume that a source of uniform (0,1) random numbers, $R_1, R_2, \ldots$ is readily available, where each R_i has pdf

$$f_R(x) = \begin{cases} 1, & 0 \le x \le 1 \\ 0, & \text{otherwise} \end{cases}$$

and cdf

$$F_R(x) = \begin{cases} 0, & x < 0 \\ x, & 0 \le x \le 1 \\ 1, & x > 1 \end{cases}$$

Throughout this chapter R and $R_1, R_2, \ldots$ represent random numbers uniformly distributed on (0,1) and generated by one of the techniques in Chapter 7 or taken from a random-number table such as Table A.1 described in Chapter 2.

8.1 Inverse Transform Technique

The inverse transform technique can be used to sample from the exponential, the uniform, the Weibull, and the triangular distributions and empirical distributions. Additionally, it is the underlying principle for sampling from a wide variety of discrete distributions. The technique will be explained in detail for the exponential distribution and then applied to other distributions. It is the most straightforward, but not always the most efficient, technique computationally.

8.1.1 Exponential Distribution

The exponential distribution, discussed in Section 5.4, has probability density function (pdf) given by

$$f(x) = \begin{cases} \lambda e^{-\lambda x}, & x \ge 0 \\ 0, & x < 0 \end{cases}$$

and cumulative distribution function (cdf) given by

$$F(x) = \int_{-\infty}^{x} f(t)dt = \begin{cases} 1 - e^{-\lambda x}, & x \ge 0 \\ 0, & x < 0 \end{cases}$$

The parameter λ can be interpreted as the mean number of occurrences per time unit. For example, if interarrival times $X_1, X_2, X_3, \ldots$ had an exponential distribution with rate λ, then λ could be interpreted as the mean number of arrivals per time unit, or the arrival rate. Notice that for any i

$$E(X_i) = \frac{1}{\lambda}$$

so that $1/\lambda$ is the mean interarrival time. The goal here is to develop a procedure for generating values $X_1, X_2, X_3, \ldots$ which have an exponential distribution.

The inverse transform technique can be utilized, at least in principle, for any distribution, but it is most useful when the cdf, $F(x)$, is of such simple form that its inverse, F^{-1}, can be easily computed.[1] A step-by-step procedure for the inverse transform technique, illustrated by the exponential distribution, is as follows:

Step 1. Compute the cdf of the desired random variable X. For the exponential distribution, the cdf is $F(x) = 1 - e^{-\lambda x}$, $x \geq 0$.

Step 2. Set $F(X) = R$ on the range of X. For the exponential distribution, it becomes $1 - e^{-\lambda X} = R$ on the range $x \geq 0$. Since X is a random variable (with the exponential distribution in this case), it follows that $1 - e^{-\lambda X}$ is also a random variable, here called R. As will be shown later, R has a uniform distribution over the interval (0, 1).

Step 3. Solve the equation $F(X) = R$ for X in terms of R. For the exponential distribution, the solution proceeds as follows:

$$\begin{aligned} 1 - e^{-\lambda X} &= R \\ e^{-\lambda X} &= 1 - R \\ -\lambda X &= \ell n(1 - R) \\ X &= -\frac{1}{\lambda}\ell n(1 - R) \end{aligned} \tag{8.1}$$

Equation (8.1) is called a random-variate generator for the exponential distribution. In general, Equation (8.1) is written as $X = F^{-1}(R)$. Generating a sequence of values is accomplished through step 4.

Step 4. Generate (as needed) uniform random numbers $R_1, R_2, R_3, \ldots$ and compute the desired random variates by

$$X_i = F^{-1}(R_i)$$

For the exponential case, $F^{-1}(R) = (-1/\lambda)\ell n(1 - R)$ by Equation (8.1), so that

$$X_i = -\frac{1}{\lambda}\ell n(1 - R_i) \tag{8.2}$$

for $i = 1, 2, 3, \ldots$. One simplification that is usually employed in Equation (8.2) is to replace $1 - R_i$ by R_i to yield

$$X_i = -\frac{1}{\lambda}\ell n R_i \tag{8.3}$$

which is justified since both R_i and $1 - R_i$ are uniformly distributed on (0, 1).

[1] The notation F^{-1} denotes the solution of the equation $r = F(x)$ in terms of r, not $1/F$.

Table 8.1. Generation of Exponential Variates X_i with Mean 1, Given Random Numbers R_i

i	1	2	3	4	5
R_i	0.1306	0.0422	0.6597	0.7965	0.7696
X_i	0.1400	0.0431	1.078	1.592	1.468

EXAMPLE 8.1

Table 8.1 gives a sequence of random numbers from Table A.1 and the computed exponential variates, X_i, given by Equation (8.2) with a value of $\lambda = 1$. Figure 8.1(a) is a histogram of 200 values, $R_1, R_2, \ldots, R_{200}$ from the uniform dis-

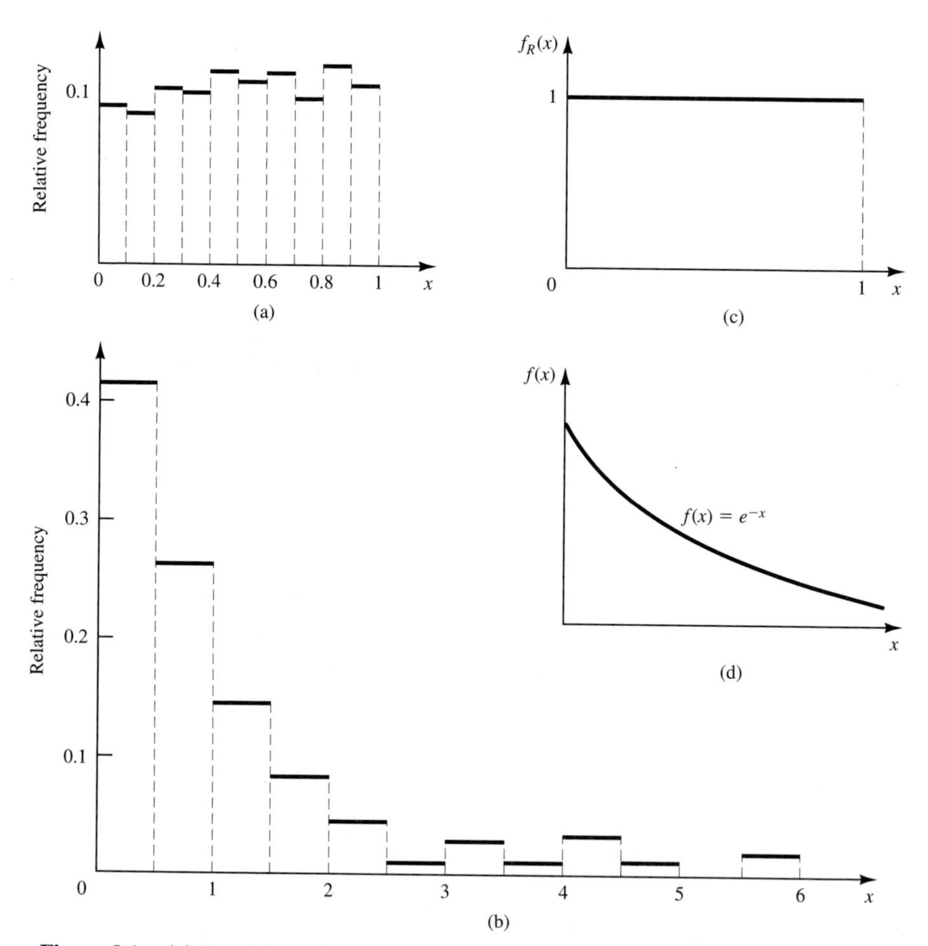

Figure 8.1. (a) Empirical histogram of 200 uniform random numbers; (b) empirical histogram of 200 exponential variates; (c) theoretical uniform density on (0, 1); (d) theoretical exponential density with mean 1.

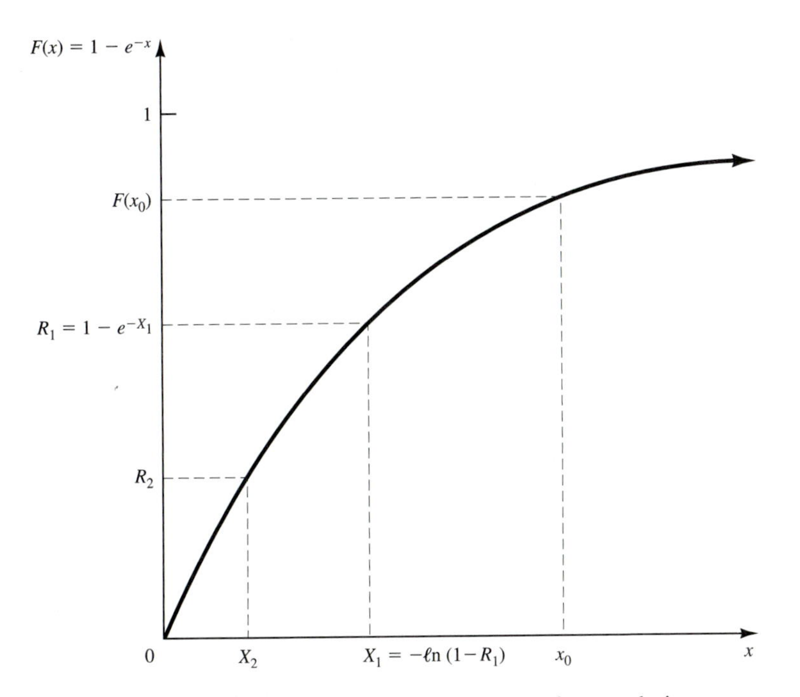

Figure 8.2. Graphical view of the inverse transform technique.

tribution, and Figure 8.1(b) is a histogram of the 200 values, $X_1, X_2, \ldots, X_{200}$, computed by Equation (8.2). Compare these empirical histograms with the theoretical density functions in Figure 8.1(c) and (d). As illustrated here, a histogram is an estimate of the underlying density function. (This fact is used in Chapter 9 as a way to identify distributions.) ◀

Figure 8.2 gives a graphical interpretation of the inverse transform technique. The cdf shown is $F(x) = 1 - e^{-x}$, an exponential distribution with rate $\lambda = 1$. To generate a value X_1 with cdf $F(x)$, first a random number R_1 between 0 and 1 is generated, a horizontal line is drawn from R_1 to the graph of the cdf, then a vertical line is dropped to the x-axis to obtain X_1, the desired result. Notice the inverse relation between R_1 and X_1, namely

$$R_1 = 1 - e^{-X_1}$$

and

$$X_1 = -\ell n(1 - R_1)$$

In general, the relation is written as

$$R_1 = F(X_1)$$

and

$$X_1 = F^{-1}(R_1)$$

Why does the random variable X_1 generated by this procedure have the desired distribution? Pick a value x_0 and compute the cumulative probability

$$P(X_1 \le x_0) = P(R_1 \le F(x_0)) = F(x_0) \tag{8.4}$$

To see the first equality in Equation (8.4), refer to Figure 8.2, where the fixed numbers x_0 and $F(x_0)$ are drawn on their respective axes. It can be seen that $X_1 \le x_0$ when and only when $R_1 \le F(x_0)$. Since $0 \le F(x_0) \le 1$, the second equality in Equation (8.4) follows immediately from the fact that R_1 is uniformly distributed on (0, 1). Equation (8.4) shows that the cdf of X_1 is F; hence, X_1 has the desired distribution.

8.1.2 Uniform Distribution

Consider a random variable X that is uniformly distributed on the interval $[a, b]$. A reasonable guess for generating X is given by

$$X = a + (b - a)R \tag{8.5}$$

[Recall that R is always a random number on (0, 1).] The pdf of X is given by

$$f(x) = \begin{cases} \dfrac{1}{b-a}, & a \le x \le b \\ 0, & \text{otherwise} \end{cases}$$

The derivation of Equation (8.5) follows steps 1 through 3 of Section 8.1.1:

Step 1. The cdf is given by

$$F(x) = \begin{cases} 0, & x < a \\ \dfrac{x-a}{b-a}, & a \le x \le b \\ 1, & x > b \end{cases}$$

Step 2. Set $F(X) = (X - a)/(b - a) = R$.

Step 3. Solving for X in terms of R yields $X = a + (b - a)R$, which agrees with Equation (8.5).

8.1.3 Weibull Distribution

The Weibull distribution was introduced in Section 5.4 as a model for time to failure for machines or electronic components. When the location parameter ν is set to 0, its pdf is given by Equation (5.46) as

$$f(x) = \begin{cases} \dfrac{\beta}{\alpha^\beta} x^{\beta-1} e^{-(x/\alpha)^\beta}, & x \ge 0 \\ 0, & \text{otherwise} \end{cases}$$

where $\alpha > 0$ and $\beta > 0$ are the scale and shape parameters of the distribution. To generate a Weibull variate, follow steps 1 through 3 of Section 8.1.1:

Step 1. The cdf is given by $F(X) = 1 - e^{-(x/\alpha)^\beta}, x \ge 0$.

Step 2. Let $F(X) = 1 - e^{-(X/\alpha)^\beta} = R$.
Step 3. Solving for X in terms of R yields

$$X = \alpha[-\ell n(1 - R)]^{1/\beta} \tag{8.6}$$

The derivation of Equation (8.6) is left as Exercise 10 for the reader. By comparing Equations (8.6) and (8.1), it can be seen that if X is a Weibull variate, then X^β is an exponential variate with mean α^β. Conversely, if Y is an exponential variate with mean μ, then $Y^{1/\beta}$ is a Weibull variate with shape parameter β and scale parameter $\alpha = \mu^{1/\beta}$.

8.1.4 Triangular Distribution

Consider a random variable X which has pdf

$$f(x) = \begin{cases} x, & 0 \le x \le 1 \\ 2 - x, & 1 < x \le 2 \\ 0, & \text{otherwise} \end{cases}$$

as shown in Figure 8.3. This distribution is called a triangular distribution with endpoints (0, 2) and mode at 1. Its cdf is given by

$$F(x) = \begin{cases} 0, & x \le 0 \\ \dfrac{x^2}{2}, & 0 < x \le 1 \\ 1 - \dfrac{(2-x)^2}{2}, & 1 < x \le 2 \\ 1, & x > 2 \end{cases}$$

For $0 \le X \le 1$,

$$R = \frac{X^2}{2} \tag{8.7}$$

and for $1 \le X \le 2$,

$$R = 1 - \frac{(2 - X)^2}{2} \tag{8.8}$$

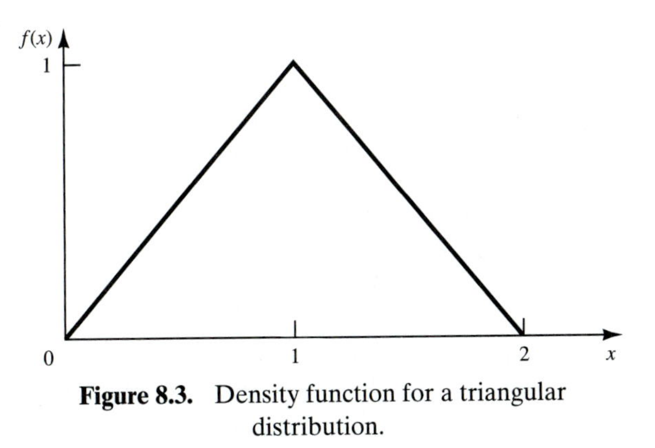

Figure 8.3. Density function for a triangular distribution.

By Equation (8.7), $0 \le X \le 1$ implies that $0 \le R \le \frac{1}{2}$, in which case $X = \sqrt{2R}$. By Equation (8.8), $1 \le X \le 2$ implies that $\frac{1}{2} \le R \le 1$, in which case $X = 2 - \sqrt{2(1-R)}$. Thus, X is generated by

$$X = \begin{cases} \sqrt{2R}, & 0 \le R \le \frac{1}{2} \\ 2 - \sqrt{2(1-R)}, & \frac{1}{2} < R \le 1 \end{cases} \tag{8.9}$$

Exercises 2, 3, and 4 give the student practice in dealing with other triangular distributions. Notice that if the pdf and cdf of the random variable X come in parts (i.e., require different formulas over different parts of the range of X), then the application of the inverse transform technique for generating X will result in separate formulas over different parts of the range of R, as in Equation (8.9). A general form of the triangular distribution was discussed in Section 5.4.

8.1.5 Empirical Continuous Distributions

If the modeler has been unable to find a theoretical distribution that provides a good model for the input data, then it may be necessary to use the empirical distribution of the data. One possibility is to simply resample the observed data itself. This is known as using the *empirical distribution*, and it makes particularly good sense when the input process is known to take on a finite number of values. See Section 8.1.7 for an example of this type of situation and a method for generating random inputs.

On the other hand, if the data are drawn from what is believed to be a continuous-valued input process, then it makes sense to interpolate between the observed data points to fill in the gaps. This section describes a method for defining and generating data from a continuous empirical distribution.

EXAMPLE 8.2

Five observations of fire crew response times (in minutes) to incoming alarms have been collected to be used in a simulation investigating possible alternative staffing and crew scheduling policies. The data are

2.76 1.83 0.80 1.45 1.24

Before collecting more data, it is desired to develop a preliminary simulation model which uses a response-time distribution based on these five observations. Thus, a method for generating random variates from the response-time distribution is needed. Initially, it will be assumed that response times X have a range $0 \le X \le c$, where c is unknown, but will be estimated by $\widehat{c} = \max\{X_i: i = 1, \ldots, n\} = 2.76$, where $\{X_i, i = 1, \ldots, n\}$ are the raw data and $n = 5$ is the number of observations.

Arrange the data from smallest to largest and let $x_{(1)} \le x_{(2)} \le \cdots \le x_{(n)}$ denote these sorted values. Since the smallest possible value is believed to be 0, define $x_{(0)} = 0$. Assign a probability of $1/n = 1/5$ to each interval

Table 8.2. Summary of Fire Crew Response-Time Data

i	*Interval,* $x_{(i-1)} < x \leq x_{(i)}$	*Probability,* $1/n$	*Cumulative Probability,* i/n	*Slope,* a_i
1	$0.0 < x \leq 0.80$	0.2	0.2	4.00
2	$0.80 < x \leq 1.24$	0.2	0.4	2.20
3	$1.24 < x \leq 1.45$	0.2	0.6	1.05
4	$1.45 < x \leq 1.83$	0.2	0.8	1.90
5	$1.83 < x \leq 2.76$	0.2	1.0	4.65

$x_{(i-1)} < x \leq x_{(i)}$, as shown in Table 8.2. The resulting empirical cdf, $\widehat{F}(x)$, is illustrated in Figure 8.4. The slope of the ith line segment is given by

$$a_i = \frac{x_{(i)} - x_{(i-1)}}{i/n - (i-1)/n} = \frac{x_{(i)} - x_{(i-1)}}{1/n}$$

The inverse cdf is calculated by

$$X = \widehat{F}^{-1}(R) = x_{(i-1)} + a_i\left(R - \frac{(i-1)}{n}\right) \tag{8.10}$$

when $(i-1)/n < R \leq i/n$.

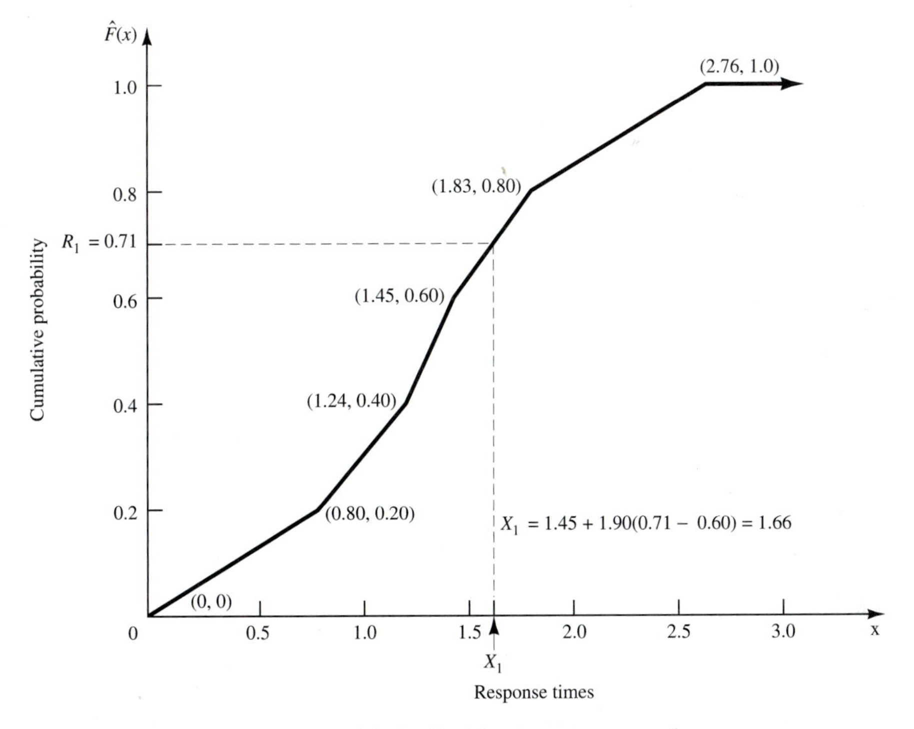

Figure 8.4. Empirical cdf of fire crew response times.

For example, if a random number $R_1 = 0.71$ is generated, then R_1 is seen to lie in the fourth interval (between $3/5 = 0.60$ and $4/5 = 0.80$), so that by Equation (8.10),

$$\begin{aligned} X_1 &= x_{(4-1)} + a_4(R_1 - (4-1)/n) \\ &= 1.45 + 1.90(0.71 - 0.60) \\ &= 1.66 \end{aligned}$$

The reader is referred to Figure 8.4 for a graphical view of the generation procedure. ◀

In Example 8.2 each data point was represented in the empirical cdf. If a large sample of data is available (and sample sizes from several hundred to tens of thousands are possible with modern, automated data collection), then it may be more convenient and computationally efficient to first summarize the data into a frequency distribution with a much smaller number of intervals and then fit a continuous empirical cdf to the frequency distribution. Only a slight generalization of Equation (8.10) is required to accomplish this. Now the slope of the ith line segment is given by

$$a_i = \frac{x_{(i)} - x_{(i-1)}}{c_i - c_{i-1}}$$

where c_i is the cumulative probability of the first i intervals of the frequency distribution and $x_{(i-1)} < x \leq x_{(i)}$ is the ith interval. The inverse cdf is calculated by

$$X = \widehat{F}^{-1}(R) = x_{(i-1)} + a_i\,(R - c_{i-1}) \tag{8.11}$$

when $c_{i-1} < R \leq c_i$.

EXAMPLE 8.3

Suppose that 100 broken-widget repair times have been collected. The data are summarized in Table 8.3 in terms of the number of observations in various intervals. For example, there were 31 observations between 0 and 0.5 hour, 10 between 0.5 and 1 hour, and so on.

Suppose it is known that all repairs take at least 15 minutes, so that $X \geq 0.25$ hour always. Then we set $x_{(0)} = 0.25$, as shown in Table 8.3 and Figure 8.5.

Table 8.3. Summary of Repair-Time Data

i	*Interval (Hours)*	*Frequency*	*Relative Frequency*	*Cumulative Frequency, c_i*	*Slope, a_i*
1	$0.25 \leq x \leq 0.5$	31	0.31	0.31	0.81
2	$0.5 < x \leq 1.0$	10	0.10	0.41	5.0
3	$1.0 < x \leq 1.5$	25	0.25	0.66	2.0
4	$1.5 < x \leq 2.0$	34	0.34	1.00	1.47

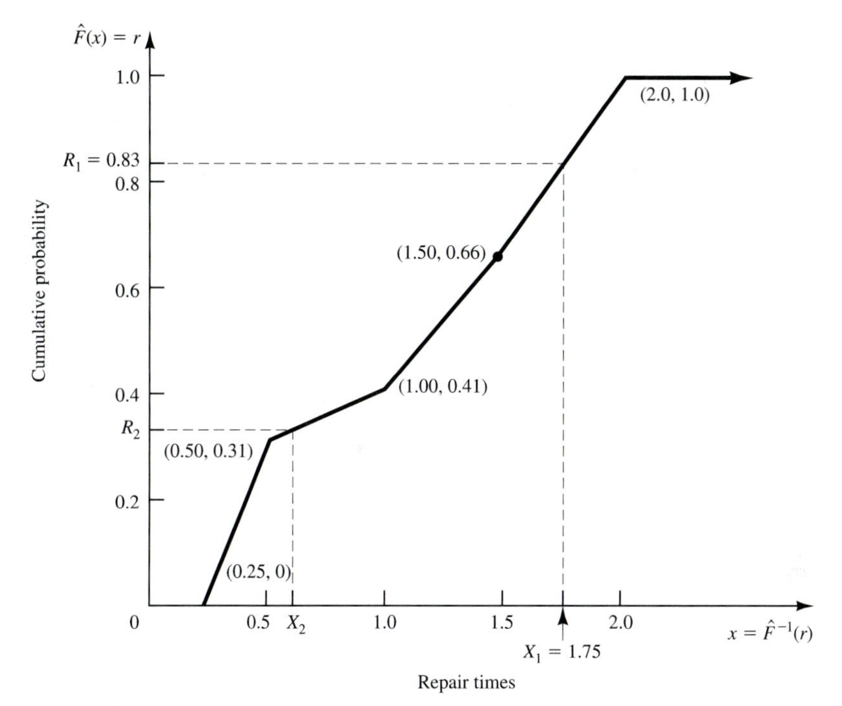

Figure 8.5. Generating variates from the empirical distribution function for repair-time data ($X \geq 0.25$).

For example, suppose the first random number generated is $R_1 = 0.83$. Then since R_1 is between $c_3 = 0.66$ and $c_4 = 1.00$, X_1 is

$$X_1 = x_{(4-1)} + a_4(R_1 - c_{4-1}) = 1.5 + 1.47(0.83 - 0.66) = 1.75 \quad (8.12)$$

As another illustration, suppose that $R_2 = 0.33$. Since $c_1 = 0.31 < R_2 \leq 0.41 = c_2$,

$$\begin{aligned} X_2 &= x_{(1)} + a_2(R_2 - c_1) \\ &= 0.5 + 5.0(0.33 - 0.31) \\ &= 0.6 \end{aligned}$$

The point ($R_2 = 0.33$, $X_2 = 0.6$) is also shown in Figure 8.5. ◀

Now reconsider the data of Table 8.3. The data are restricted in the range $0.25 \leq X \leq 2.0$, but the underlying distribution may have a wider range. This provides one important reason for attempting to find a theoretical statistical distribution (such as the gamma or Weibull) for the data, since these distributions allow a wider range, namely $0 \leq X < \infty$. On the other hand, an empirical distribution adheres closely to what is present in the data itself, and the data are often the best source of information available.

When data are summarized in terms of frequency intervals, it is recommended that relatively short intervals be used, as this results in a more accurate portrayal of the underlying cdf. For example, for the repair-time data of Table 8.3, for which there were $n = 100$ observations, a much more accurate estimate could have been obtained by using 10 to 20 intervals, certainly not an excessive number, rather than the four fairly wide intervals actually used here for purposes of illustration.

Several comments are in order:

1. A computerized version of the procedure will become more inefficient as the number of intervals, n, increases. A systematic computerized version is often called a table-lookup generation scheme, because given a value of R, the computer program must search an array of c_i values to find the interval i in which R lies, namely the interval i satisfying

$$c_{i-1} < R \leq c_i$$

The more intervals there are, the longer on the average the search will take if it is implemented in the crude way described here. The analyst should consider this trade-off between accuracy of the estimating cdf and computational efficiency when programming the procedure. If a large number of observations are available, the analyst may well decide to group the observations from 20 to 50 intervals (say) and then use the procedure of Example 8.3. Or a more efficient table-lookup procedure may be used, such as the one described in Law and Kelton [2000].

2. In Example 8.2 it was assumed that response times X satisfied $0 \leq X \leq 2.76$. This assumption led to the inclusion of the points $x_{(0)} = 0$ and $x_{(5)} = 2.76$ in Figure 8.4 and Table 8.2. If it is known a priori that X falls in some other range, for example, if it is known that response times are always between 15 seconds and 3 minutes, that is,

$$0.25 \leq X \leq 3.0$$

then the points $x_{(0)} = 0.25$ and $x_{(6)} = 3.0$ would be used to estimate the empirical cdf of response times. Notice that because of inclusion of the new point $x_{(6)}$ there are now six intervals instead of five and each interval is assigned probability $1/6 = 0.167$. Exercise 12 illustrates the use of these additional assumptions.

8.1.6 Continuous Distributions without a Closed-Form Inverse

A number of useful continuous distributions do not have a closed form expression for their cdf or its inverse; examples include the normal, gamma, and beta distributions. For this reason, it is often stated that the inverse transform technique for random-variate generation cannot be used for these distributions. This is not true, provided we are willing to *approximate* the inverse cdf, or numerically integrate and search the cdf. Although this may sound imprecise, notice that even a closed-form inverse requires approximation in order to evaluate it on a computer. For example, generating exponentially distributed

Table 8.4. Comparison of Approximate Inverse with Exact Values (to Four Decimal Places) for the Standard Normal Distribution

R	*Approximate Inverse*	*Exact Inverse*
0.01	−2.3263	−2.3373
0.10	−1.2816	−1.2813
0.25	−0.6745	−0.6713
0.50	0.0000	0.0000
0.75	0.6745	0.6713
0.90	1.2816	1.2813
0.99	2.3263	2.3373

random variates via the inverse cdf $X = F^{-1}(R) = -\ell n(1 - R)/\lambda$ requires a numerical approximation for the logarithm function. Thus, there is no essential difference between using an approximate inverse cdf and approximately evaluating a closed-form inverse. The problem with using an approximate inverse cdf is that some of them are computationally slow to evaluate.

To illustrate the idea, consider the simple approximation to the inverse cdf of the standard normal distribution proposed by Schmeiser [1979]:

$$X = F^{-1}(R) \approx \frac{R^{0.135} - (1 - R)^{0.135}}{0.1975}$$

This approximation gives at least one-decimal-place accuracy for $0.0013499 \leq R \leq 0.9986501$. Table 8.4 compares the approximation with exact values (to four decimal places) obtained by numerical integration for several values of R. Much more accurate approximations exist that are only slightly more complicated. A good source of these approximations for a number of distributions is Bratley, Fox, and Schrage [1987].

8.1.7 Discrete Distributions

All discrete distributions can be generated using the inverse transform technique, either numerically through a table-lookup procedure, or in some cases algebraically with the final generation scheme in terms of a formula. Other techniques are sometimes used for certain distributions, such as the convolution technique for the binomial distribution. Some of these methods are discussed in later sections. This subsection gives examples covering both empirical distributions and two of the standard discrete distributions, the (discrete) uniform and the geometric. Highly efficient table-lookup procedures for these and other distributions are found in Bratley, Fox, and Schrage [1987] and Ripley [1987].

Table 8.5. Distribution of Number of Shipments, X

x	$p(x)$	$F(x)$
0	0.50	0.50
1	0.30	0.80
2	0.20	1.00

Example 8.4 (An Empirical Discrete Distribution)

At the end of the day, the number of shipments on the loading dock of the IHW Company (whose main product is the famous, incredibly huge widget) is either 0, 1, or 2, with observed relative frequency of occurrence of 0.50, 0.30, and 0.20, respectively. Internal consultants have been asked to develop a model to improve the efficiency of the loading and hauling operations, and as part of this model they will need to be able to generate values, X, to represent the number of shipments on the loading dock at the end of each day. The consultants decide to model X as a discrete random variable with distribution as given in Table 8.5 and shown in Figure 8.6.

The probability mass function (pmf), $p(x)$, is given by

$$p(0) = P(X = 0) = 0.50$$
$$p(1) = P(X = 1) = 0.30$$
$$p(2) = P(X = 2) = 0.20$$

and the cdf, $F(x) = P(X \leq x)$, is given by

$$F(x) = \begin{cases} 0, & x < 0 \\ 0.5, & 0 \leq x < 1 \\ 0.8, & 1 \leq x < 2 \\ 1.0, & 2 \leq x \end{cases}$$

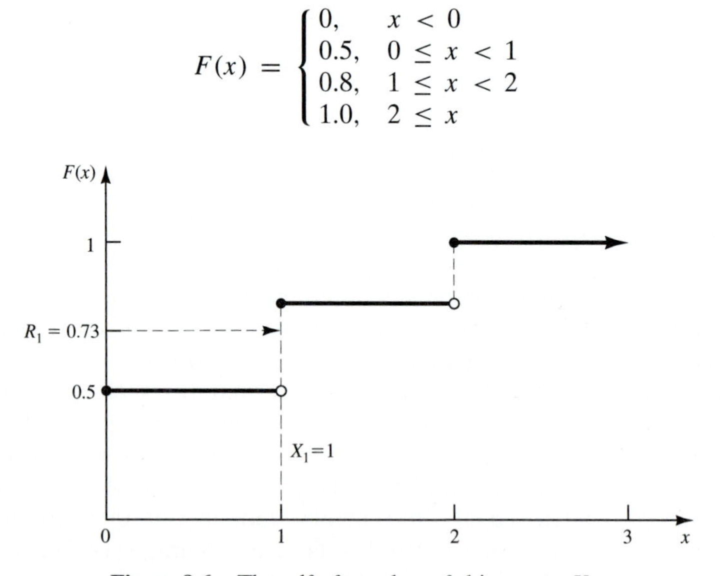

Figure 8.6. The cdf of number of shipments, X.

Table 8.6. Table for Generating the Discrete Variate X

i	*Input,* r_i	*Output,* x_i
1	0.50	0
2	0.80	1
3	1.00	2

Recall that the cdf of a discrete random variable always consists of horizontal line segments with jumps of size $p(x)$ at those points, x, which the random variable can assume. For example, in Figure 8.6 there is a jump of size $p(0) = 0.5$ at $x = 0$, of size $p(1) = 0.3$ at $x = 1$, and of size $p(2) = 0.2$ at $x = 2$.

For generating discrete random variables, the inverse transform technique becomes a table-lookup procedure, but unlike the case of continuous variables, interpolation is not required. To illustrate the procedure, suppose that $R_1 = 0.73$ is generated. Graphically, as illustrated in Figure 8.6, first locate $R_1 = 0.73$ on the vertical axis, next draw a horizontal line segment until it hits a "jump" in the cdf, and then drop a perpendicular to the horizontal axis to get the generated variate. Here $R_1 = 0.73$ is transformed to $X_1 = 1$. This procedure is analogous to the procedure used for empirical continuous distributions in Section 8.1.5 and illustrated in Figure 8.5, except that the final step of linear interpolation is eliminated.

The table-lookup procedure is facilitated by construction of a table such as Table 8.6. When $R_1 = 0.73$ is generated, first find the interval in which R_1 lies. In general, for $R = R_1$, if

$$F(x_{i-1}) = r_{i-1} < R \leq r_i = F(x_i) \tag{8.13}$$

then set $X_1 = x_i$. Here $r_0 = 0$, $x_0 = -\infty$, while $x_1, x_2, \ldots, x_n$ are the possible values of the random variable, and $r_k = p(x_1) + \cdots + p(x_k)$, $k = 1, 2, \ldots, n$. For this example, $n = 3$, $x_1 = 0$, $x_2 = 1$, $x_3 = 2$, and hence $r_1 = 0.5$, $r_2 = 0.8$, and $r_3 = 1.0$. (Notice that $r_n = 1.0$ in all cases.)

Since $r_1 = 0.5 < R_1 = 0.73 \leq r_2 = 0.8$, set $X_1 = x_2 = 1$. The generation scheme is summarized as follows:

$$X = \begin{cases} 0, & R \leq 0.5 \\ 1, & 0.5 < R \leq 0.8 \\ 2, & 0.8 < R \leq 1.0 \end{cases}$$ ◀

Example 8.4 illustrates the table-lookup procedure, while the next example illustrates an algebraic approach that can be used for certain distributions.

EXAMPLE 8.5 (A Discrete Uniform Distribution)

Consider the discrete uniform distribution on $\{1, 2, \ldots, k\}$ with pmf and cdf given by

$$p(x) = \frac{1}{k}, \quad x = 1, 2, \ldots, k$$

and

$$F(x) = \begin{cases} 0, & x < 1 \\ \frac{1}{k}, & 1 \le x < 2 \\ \frac{2}{k}, & 2 \le x < 3 \\ \vdots & \vdots \\ \frac{k-1}{k}, & k-1 \le x < k \\ 1, & k \le x \end{cases}$$

Let $x_i = i$ and $r_i = p(1) + \cdots + p(x_i) = F(x_i) = i/k$ for $i = 1, 2, \ldots, k$. Then by using Inequality (8.13) it can be seen that if the generated random number R satisfies

$$r_{i-1} = \frac{i-1}{k} < R \le r_i = \frac{i}{k} \tag{8.14}$$

then X is generated by sctting $X = i$. Now, Inequality (8.14) can be solved for i:

$$i - 1 < Rk \le i$$

$$Rk \le i < Rk + 1 \tag{8.15}$$

Let $\lceil y \rceil$ denote the smallest integer $\ge y$. For example, $\lceil 7.82 \rceil = 8$, $\lceil 5.13 \rceil = 6$, and $\lceil -1.32 \rceil = -1$. For $y \ge 0$, $\lceil y \rceil$ is a function that rounds up. This notation and Inequality (8.15) yield a formula for generating X, namely

$$X = \lceil Rk \rceil \tag{8.16}$$

For example, consider generating a random variate X, uniformly distributed on $\{1, 2, \ldots, 10\}$. The variate, X, might represent the number of pallets to be loaded onto a truck. Using Table A.1 as a source of random numbers, R, and Equation (8.16) with $k = 10$ yields

$$R_1 = 0.78, \quad X_1 = \lceil 7.8 \rceil = 8$$

$$R_2 = 0.03, \quad X_2 = \lceil 0.3 \rceil = 1$$

$$R_3 = 0.23, \quad X_3 = \lceil 2.3 \rceil = 3$$

$$R_4 = 0.97, \quad X_4 = \lceil 9.7 \rceil = 10$$

The procedure discussed here can be modified to generate a discrete uniform random variate with any range consisting of consecutive integers. Exercise 13 asks the student to devise a procedure for one such case. ◀

Example 8.6

Consider the discrete distribution with pmf given by

$$p(x) = \frac{2x}{k(k+1)}, \quad x = 1, 2, \ldots, k$$

(This example is taken from Schmidt and Taylor [1970].) For integer values of x in the range $\{1, 2, \ldots, k\}$, the cdf is given by

$$\begin{aligned} F(x) &= \sum_{i=1}^{x} \frac{2i}{k(k+1)} \\ &= \frac{2}{k(k+1)} \sum_{i=1}^{x} i \\ &= \frac{2}{k(k+1)} \frac{x(x+1)}{2} \\ &= \frac{x(x+1)}{k(k+1)} \end{aligned}$$

Generate R and use Inequality (8.13) to conclude that $X = x$ whenever

$$F(x-1) = \frac{(x-1)x}{k(k+1)} < R \leq \frac{x(x+1)}{k(k+1)} = F(x)$$

or whenever

$$(x-1)x < k(k+1)R \leq x(x+1)$$

To solve this inequality for x in terms of R, first find a value of x that satisfies

$$(x-1)x = k(k+1)R$$

or

$$x^2 - x - k(k+1)R = 0$$

Then by rounding up, the solution is $X = \lceil x - 1 \rceil$. By the quadratic formula, namely

$$x = \frac{-b \pm \sqrt{b^2 - 4ac}}{2a}$$

with $a = 1, b = -1$, and $c = -k(k+1)R$, the solution to the quadratic equation is

$$x = \frac{1 \pm \sqrt{1 + 4k(k+1)R}}{2} \tag{8.17}$$

The positive root in Equation (8.17) is the correct one to use (why?), so X is generated by

$$X = \left\lceil \frac{1 + \sqrt{1 + 4k(k+1)R}}{2} - 1 \right\rceil \tag{8.18}$$

Exercise 14 asks the student to generate a few values from this distribution. ◀

Example 8.7 (The Geometric Distribution)

Consider the geometric distribution with pmf

$$p(x) = p(1-p)^x, \quad x = 0, 1, 2, \ldots$$

where $0 < p < 1$. Its cdf is given by

$$\begin{aligned} F(x) &= \sum_{j=0}^{x} p(1-p)^j \\ &= \frac{p\{1 - (1-p)^{x+1}\}}{1 - (1-p)} \\ &= 1 - (1-p)^{x+1} \end{aligned}$$

for $x = 0, 1, 2, \ldots$ Using the inverse transform technique [i.e., Inequality (8.13)], recall that a geometric random variable X will assume the value x whenever

$$F(x-1) = 1 - (1-p)^x < R \le 1 - (1-p)^{x+1} = F(x) \tag{8.19}$$

where R is a generated random number assumed $0 < R < 1$. Solving Inequality (8.19) for x proceeds as follows:

$$\begin{gathered} (1-p)^{x+1} \le 1 - R < (1-p)^x \\ (x+1)\ell n(1-p) \le \ell n(1-R) < x \ell n(1-p) \end{gathered}$$

But $1 - p < 1$ implies that $\ell n(1-p) < 0$, so that

$$\frac{\ell n(1-R)}{\ell n(1-p)} - 1 \le x < \frac{\ell n(1-R)}{\ell n(1-p)} \tag{8.20}$$

Thus, $X = x$ for that integer value of x satisfying Inequality (8.20), or, in brief, using the round-up function $\lceil \cdot \rceil$

$$X = \left\lceil \frac{\ell n(1-R)}{\ell n(1-p)} - 1 \right\rceil \tag{8.21}$$

Since p is a fixed parameter, let $\beta = -1/\ell n(1-p)$. Then $\beta > 0$ and, by Equation (8.21), $X = \lceil -\beta \ell n(1-R) - 1 \rceil$. By Equation (8.1), $-\beta \ell n(1-R)$ is an exponentially distributed random variable with mean β, so that one way of generating a geometric variate with parameter p is to generate (by any method) an exponential variate with parameter $\beta^{-1} = -\ell n(1-p)$, subtract one, and round up.

Occasionally, a geometric variate X is needed which can assume values $\{q, q+1, q+2, \ldots\}$ with pmf $p(x) = p(1-p)^{x-q}$ $(x = q, q+1, \ldots)$. Such a variate, X can be generated, using Equation (8.21), by

$$X = q + \left\lceil \frac{\ell n(1-R)}{\ell n(1-p)} - 1 \right\rceil \tag{8.22}$$

One of the most common cases is $q = 1$. ◀

EXAMPLE 8.8

Generate three values from a geometric distribution on the range $\{X \geq 1\}$ with mean 2. Such a geometric distribution has pmf $p(x) = p(1-p)^{x-1} (x = 1, 2, \ldots)$ with mean $1/p = 2$, or $p = 1/2$. Thus, X can be generated by Equation (8.22) with $q = 1$, $p = 1/2$, and $1/\ell n(1-p) = -1.443$. Using Table A.1, $R_1 = 0.932$, $R_2 = 0.105$, and $R_3 = 0.687$, which yields

$$\begin{aligned} X_1 &= 1 + \lceil -1.443 \ell n(1 - 0.932) - 1 \rceil \\ &= 1 + \lceil 3.878 - 1 \rceil = 4 \\ X_2 &= 1 + \lceil -1.443 \ell n(1 - 0.105) - 1 \rceil = 1 \\ X_3 &= 1 + \lceil -1.443 \ell n(1 - 0.687) - 1 \rceil = 2 \end{aligned}$$

Exercise 15 deals with an application of the geometric distribution. ◀

8.2 Direct Transformation for the Normal and Lognormal Distributions

Many methods have been developed for generating normally distributed random variates. The inverse transform technique cannot be applied easily, however, because the inverse cdf cannot be written in closed form. The standard normal cdf is given by

$$\Phi(x) = \int_{-\infty}^{x} \frac{1}{\sqrt{2\pi}} e^{-t^2/2} dt, \quad -\infty < x < \infty$$

This section describes an intuitively appealing direct transformation that produces an independent pair of standard normal variates with mean zero and variance 1. The method is due to Box and Muller [1958]. Although not as efficient as many more modern techniques, it is easy to program in a scientific language such as FORTRAN, C, or Pascal. We then show how to transform a standard normal variate into a normal variate with mean μ and variance σ^2. Once we have a method (this or any other) for generating X from a $N(\mu, \sigma^2)$ distribution, then we can generate a lognormal random variate Y with parameters μ and σ^2 using the direct transformation $Y = e^X$ [recall that μ and σ^2 are *not* the mean and variance of the lognormal; see Equations (5.57) and (5.58)].

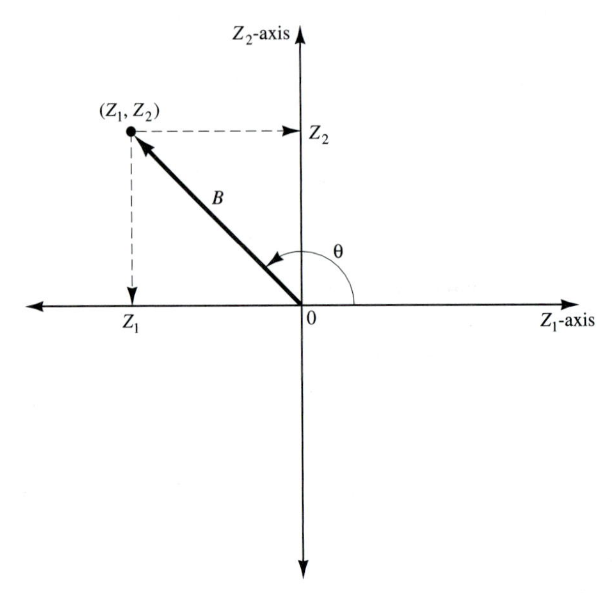

Figure 8.7. Polar representation of a pair of standard normal variables.

Consider two standard normal random variables, Z_1 and Z_2, plotted as a point in the plane as shown in Figure 8.7 and represented in polar coordinates as

$$\begin{aligned} Z_1 &= B \cos \theta \\ Z_2 &= B \sin \theta \end{aligned} \tag{8.23}$$

It is known that $B^2 = Z_1^2 + Z_2^2$ has the chi-square distribution with 2 degrees of freedom, which is equivalent to an exponential distribution with mean 2. Thus, the radius, B, can be generated by use of Equation (8.3):

$$B = (-2\ell n R)^{1/2} \tag{8.24}$$

By the symmetry of the normal distribution, it seems reasonable to suppose, and indeed it is the case, that the angle is uniformly distributed between 0 and 2π radians. In addition, the radius, B, and the angle, θ, are mutually independent. Combining Equations (8.23) and (8.24) gives a direct method for generating two independent standard normal variates, Z_1 and Z_2, from two independent random numbers R_1 and R_2:

$$\begin{aligned} Z_1 &= (-2\ell n R_1)^{1/2} \cos (2\pi R_2) \\ Z_2 &= (-2\ell n R_1)^{1/2} \sin (2\pi R_2) \end{aligned} \tag{8.25}$$

To illustrate the generation scheme, consider Equation (8.25) with $R_1 = 0.1758$ and $R_2 = 0.1489$. Two standard normal random variates are generated as follows:

$$Z_1 = [-2\ell n(0.1758)]^{1/2} \cos(2\pi 0.1489) = 1.11$$

$$Z_2 = [-2\ell n(0.1758)]^{1/2} \sin(2\pi 0.1489) = 1.50$$

To obtain normal variates X_i with mean μ and variance σ^2, we then apply the transformation

$$X_i = \mu + \sigma Z_i \tag{8.26}$$

to the standard normal variates. For example, to transform the two standard normal variates into normal variates with mean $\mu = 10$ and variance $\sigma^2 = 4$ we compute

$$X_1 = 10 + 2(1.11) = 12.22$$

$$X_2 = 10 + 2(1.50) = 13.00$$

8.3 Convolution Method

The probability distribution of a sum of two or more independent random variables is called a convolution of the distributions of the original variables. The convolution method thus refers to adding together two or more random variables to obtain a new random variable with the desired distribution. This technique can be applied to obtain Erlang variates and binomial variates. What is important is not the cdf of the desired random variable, but rather its relation to other more easily generated variates.

8.3.1 Erlang Distribution

As discussed in Section 5.4, an Erlang random variable X with parameters (K, θ) can be shown to be the sum of K independent exponential random variables, $X_i (i = 1, \ldots, K)$, each having mean $1/K\theta$; that is,

$$X = \sum_{i=1}^{K} X_i$$

Since each X_i can be generated by Equation (8.3) with $1/\lambda = 1/K\theta$, an Erlang variate can be generated by

$$X = \sum_{i=1}^{K} -\frac{1}{K\theta} \ell n R_i$$

$$= -\frac{1}{K\theta} \ell n \left(\prod_{i=1}^{K} R_i \right) \tag{8.27}$$

In Equation (8.27), $\prod$ stands for product. It is more efficient computationally to multiply all the random numbers first and then to compute only one logarithm.

EXAMPLE 8.9

Trucks arrive at a large warehouse in a completely random fashion which is modeled as a Poisson process with arrival rate $\lambda = 10$ trucks per hour. The guard at the entrance sends trucks alternately to the north and south docks. An analyst has developed a model to study the loading/unloading process at the south docks and needs a model of the arrival process at the south docks alone. An interarrival time X between successive truck arrivals at the south docks is equal to the sum of two interarrival times at the entrance and thus it is the sum of two exponential random variables, each with mean 0.1 hour, or 6 minutes. Thus, X has the Erlang distribution with $K = 2$ and mean $1/\theta = 2/\lambda$ $= 0.2$ hour. To generate the variate X, first obtain $K = 2$ random numbers from Table A.1, say $R_1 = 0.937$ and $R_2 = 0.217$. Then by Equation (8.27),

$$\begin{aligned} X &= -0.1\ell n[0.937(0.217)] \\ &= 0.159 \text{ hour } = 9.56 \text{ minutes} \end{aligned}$$

◀

In general, Equation (8.27) implies that K uniform random numbers are needed for each Erlang variate generated. If K is large, it is more efficient to generate Erlang variates by other techniques, such as one of the many acceptance-rejection techniques for the gamma distribution given by Bratley, Fox, and Schrage [1987], Fishman [1978], and Law and Kelton [2000].

8.4 Acceptance-Rejection Technique

Suppose that an analyst needed to devise a method for generating random variates, X, uniformly distributed between 1/4 and 1. One way to proceed would be to follow these steps:

Step 1. Generate a random number R.
Step 2a. If $R \geq 1/4$, accept $X = R$, then go to step 3.
Step 2b. If $R < 1/4$, reject R, and return to step 1.
Step 3. If another uniform random variate on [1/4, 1] is needed, repeat the procedure beginning at step 1. If not, stop.

Each time step 1 is executed, a new random number R must be generated. Step 2a is an "acceptance" and step 2b is a "rejection" in this acceptance-rejection technique. To summarize the technique, random variates (R) with some distribution (here uniform on [0, 1]) are generated until some condition ($R > 1/4$) is satisfied. When the condition is finally satisfied, the desired random variate, X (here uniform on [1/4, 1]), can be computed ($X = R$). This procedure can be shown to be correct by recognizing that the accepted values of R are conditioned values; that is, R itself does not have the desired distribution, but R conditioned on the event $\{R \geq 1/4\}$ does have the desired distribution. To show this, take $1/4 \leq a < b \leq 1$; then

$$P(a < R \leq b | 1/4 \leq R \leq 1) = \frac{P(a < R \leq b)}{P(1/4 \leq R \leq 1)} = \frac{b - a}{3/4} \tag{8.28}$$

which is the correct probability for a uniform distribution on [1/4, 1]. Equation (8.28) says that the probability distribution of R, given that R is between 1/4 and 1 (all other values of R are thrown out), is the desired distribution. Therefore, if $1/4 \leq R \leq 1$, set $X = R$.

The efficiency of an acceptance-rejection technique depends heavily on being able to minimize the number of rejections. In this example, the probability of a rejection is $P(R < 1/4) = 1/4$, so that the number of rejections is a geometrically distributed random variable with probability of "success" being $p = 3/4$ and mean number of rejections $(1/p - 1) = 4/3 - 1 = 1/3$. (Example 8.7 discussed the geometric distribution.) The mean number of random numbers R required to generate one variate X is one more than the number of rejections; hence, it is $4/3 = 1.33$. In other words, to generate 1000 values of X would require approximately 1333 random numbers R.

In the present situation an alternative procedure exists for generating a uniform variate on [1/4, 1], namely Equation (8.5), which reduces to $X = 1/4 + (3/4)R$. Whether the acceptance-rejection technique or an alternative procedure such as the inverse-transform technique [Equation (8.5)] is the more efficient depends on several considerations. The computer being used, the skills of the programmer, and the relative efficiency of generating the additional (rejected) random numbers needed by acceptance-rejection should be compared to the computations required by the alternative procedure. In practice, concern with generation efficiency is left to specialists who conduct extensive tests comparing alternative methods.

For the uniform distribution on [1/4, 1], the inverse transform technique of Equation (8.5) is undoubtedly much easier to apply and more efficient than the acceptance-rejection technique. The main purpose of this example was to explain and motivate the basic concept of the acceptance-rejection technique. However, for some important distributions such as the normal, gamma, and beta, the inverse cdf does not exist in closed form and therefore the inverse transform technique is difficult. These more advanced techniques are summarized by Bratley, Fox, and Schrage [1987], Fishman [1978], and Law and Kelton [2000].

In the following subsections, the acceptance-rejection technique is illustrated for the generation of random variates for the Poisson and gamma distributions.

8.4.1 Poisson Distribution

A Poisson random variable, N, with mean $\alpha > 0$ has pmf

$$p(n) = P(N = n) = \frac{e^{-\alpha}\alpha^n}{n!}, \qquad n = 0, 1, 2, \ldots$$

but more important, N can be interpreted as the number of arrivals from a Poisson arrival process in one unit of time. Recall from Section 5.5 that the interarrival times, $A_1, A_2, \ldots$ of successive customers are exponentially distributed with rate α (i.e., α is the mean number of arrivals per unit time); in addition, an exponential variate can be generated by Equation (8.3). Thus there is a relationship between the (discrete) Poisson distribution and the (continuous) exponential distribution, namely:

$$N = n \tag{8.29}$$

if and only if

$$A_1 + A_2 + \cdots + A_n \leq 1 < A_1 + \cdots + A_n + A_{n+1} \tag{8.30}$$

Equation (8.29), $N = n$, says there were exactly n arrivals during one unit of time; but relation (8.30) says that the nth arrival occurred before time 1 while the $(n+1)$st arrival occurred after time 1. Clearly, these two statements are equivalent. Proceed now by generating exponential interarrival times until some arrival, say $n+1$, occurs after time 1; then set $N = n$.

For efficient generation purposes, relation (8.30) is usually simplified by first using Equation (8.3), $A_i = (-1/\alpha)\ell n R_i$, to obtain

$$\sum_{i=1}^{n} -\frac{1}{\alpha}\ell n R_i \leq 1 < \sum_{i=1}^{n+1} -\frac{1}{\alpha}\ell n R_i$$

Next multiply through by $-\alpha$, which reverses the sign of the inequality, and use the fact that a sum of logarithms is the logarithm of a product, to get

$$\ell n \prod_{i=1}^{n} R_i = \sum_{i=1}^{n} \ell n R_i \geq -\alpha > \sum_{i=1}^{n+1} \ell n R_i = \ell n \prod_{i=1}^{n+1} R_i$$

Finally, use the relation $e^{\ell n x} = x$ for any number x to obtain

$$\prod_{i=1}^{n} R_i \geq e^{-\alpha} > \prod_{i=1}^{n+1} R_i \tag{8.31}$$

which is equivalent to relation (8.30). The procedure for generating a Poisson random variate, N, is given by the following steps:

Step 1. Set $n = 0$, $P = 1$.

Step 2. Generate a random number R_{n+1} and replace P by $P \cdot R_{n+1}$.

Step 3. If $P < e^{-\alpha}$, then accept $N = n$. Otherwise, reject the current n, increase n by one, and return to step 2.

Notice that upon completion of step 2, P is equal to the rightmost expression in relation (8.31). The basic idea of a rejection technique is again exhibited; if $P \geq e^{-\alpha}$ in step 3, then n is rejected and the generation process must proceed through at least one more trial.

How many random numbers will be required, on the average, to generate one Poisson variate, N? If $N = n$, then $n + 1$ random numbers are required, so the average number is given by

$$E(N + 1) = \alpha + 1$$

which is quite large if the mean, α, of the Poisson distribution is large.

EXAMPLE 8.10

Generate three Poisson variates with mean $\alpha = 0.2$. First compute $e^{-\alpha} = e^{-0.2} = 0.8187$. Next get a sequence of random numbers R from Table A.1 and follow steps 1 to 3 above:

Step 1. Set $n = 0$, $P = 1$.
Step 2. $R_1 = 0.4357$, $P = 1 \cdot R_1 = 0.4357$.
Step 3. Since $P = 0.4357 < e^{-\alpha} = 0.8187$, accept $N = 0$.
Step 1–3. ($R_1 = 0.4146$ leads to $N = 0$.)
Step 1. Set $n = 0$, $P = 1$.
Step 2. $R_1 = 0.8353$, $P = 1 \cdot R_1 = 0.8353$.
Step 3. Since $P \geq e^{-\alpha}$, reject $n = 0$ and return to step 2 with $n = 1$.
Step 2. $R_2 = 0.9952$, $P = R_1 R_2 = 0.8313$.
Step 3. Since $P \geq e^{-\alpha}$, reject $n = 1$ and return to step 2 with $n = 2$.
Step 2. $R_3 = 0.8004$, $P = R_1 R_2 R_3 = 0.6654$.
Step 3. Since $P < e^{-\alpha}$, accept $N = 2$.

The calculations required for the generation of these three Poisson random variates are summarized as follows:

n	R_{n+1}	P	*Accept/Reject*	*Result*
0	0.4357	0.4357	$P < e^{-\alpha}$ (accept)	$N = 0$
0	0.4146	0.4146	$P < e^{-\alpha}$ (accept)	$N = 0$
0	0.8353	0.8353	$P \geq e^{-\alpha}$ (reject)	
1	0.9952	0.8313	$P \geq e^{-\alpha}$ (reject)	
2	0.8004	0.6654	$P < e^{-\alpha}$ (accept)	$N = 2$

It took five random numbers, R, to generate three Poisson variates here ($N = 0$, $N = 0$, and $N = 2$), but in the long run to generate, say, 1000 Poisson variates with mean $\alpha = 0.2$ it would require approximately $1000(\alpha + 1)$ or 1200 random numbers. ◀

EXAMPLE 8.11

Buses arrive at the bus stop at Peachtree and North Avenue according to a Poisson process with a mean of one bus per 15 minutes. Generate a random variate, N, which represents the number of arriving buses during a 1-hour time slot. Now, N is Poisson distributed with a mean of four buses per hour. First compute $e^{-\alpha} = e^{-4} = 0.0183$. Using a sequence of 12 random numbers from Table A.1 yields the following summarized results:

n	R_{n+1}	P	*Accept/Reject*	*Result*
0	0.4357	0.4357	$P \geq e^{-\alpha}$ (reject)	
1	0.4146	0.1806	$P \geq e^{-\alpha}$ (reject)	
2	0.8353	0.1508	$P \geq e^{-\alpha}$ (reject)	
3	0.9952	0.1502	$P \geq e^{-\alpha}$ (reject)	
4	0.8004	0.1202	$P \geq e^{-\alpha}$ (reject)	
5	0.7945	0.0955	$P \geq e^{-\alpha}$ (reject)	
6	0.1530	0.0146	$P < e^{-\alpha}$ (accept)	$N = 6$

It is immediately seen that a larger value of α (here $\alpha = 4$) usually requires more random numbers; if 1000 Poisson variates were desired, approximately $1000(\alpha + 1) = 5000$ random numbers would be required. ◀

When α is large, say $\alpha \geq 15$, the rejection technique outlined here becomes quite expensive, but fortunately an approximate technique based on the normal distribution works quite well. When the mean, α, is large,

$$Z = \frac{N - \alpha}{\sqrt{\alpha}}$$

is approximately normally distributed with mean zero and variance 1, which suggests an approximate technique. First generate a standard normal variate Z, by Equation (8.25), then generate the desired Poisson variate, N, by

$$N = \lceil \alpha + \sqrt{\alpha} Z - 0.5 \rceil \tag{8.32}$$

where $\lceil \cdot \rceil$ is the round-up function described in Section 8.1.7. (If $\alpha + \sqrt{\alpha} Z - 0.5 < 0$, then set $N = 0$.) The "0.5" used in the formula makes the round-up function become a "round to the nearest integer" function. Equation (8.32) is not an acceptance rejection technique, but, used as an alternative to the acceptance rejection method, it provides a fairly efficient and accurate method for generating Poisson variates with a large mean.

8.4.2 Gamma Distribution

Several acceptance-rejection techniques for generating gamma random variates have been developed (Bratley, Fox, and Schrage [1987]; Fishman [1978]; Law and Kelton [2000]). One of the most efficient is by Cheng [1977]; the mean number of trials is between 1.13 and 1.47 for any value of the shape parameter $\beta \geq 1$.

If the shape parameter β is an integer, say $\beta = k$, one possibility is to use the convolution technique in Section 8.3.1, since the Erlang distribution is a special case of the more general gamma distribution. On the other hand, the acceptance-rejection technique described here would be a highly efficient method for the Erlang distribution, especially if $\beta = k$ were large. The routine generates gamma random variates with scale parameter θ and shape parameter β — that is, with mean $1/\theta$ and variance $1/\beta\theta^2$. The steps are as follows:

Step 1. Compute $a = (2\beta - 1)^{1/2}$, $b = 2\beta - \ell n4 + 1/a$.
Step 2. Generate R_1 and R_2.
Step 3. Compute $X = \beta[R_1/(1 - R_1)]^a$.
Step 4a. If $X > b - \ell n(R_1^2 R_2)$, reject X and return to step 2.
Step 4b. If $X \le b - \ell n(R_1^2 R_2)$, use X as the desired variate. The generated variates from step 4b will have mean and variance both equal to β. If it is desired to have mean $1/\theta$ and variance $1/\beta\theta^2$ as in Section 5.4, then include
Step 5. Replace X by $X/\beta\theta$.

The basic idea of all acceptance-rejection methods is again illustrated here, but the proof of this example is beyond the scope of this book. In step 3, $X = \beta[R_1/(1 - R_1)]^a$ is not gamma distributed, but rejection of certain values of X in step 4a guarantees that the accepted values in step 4b do have the gamma distribution.

EXAMPLE 8.12

Downtimes for a high-production candy-making machine have been found to be gamma distributed with mean 2.2 minutes and variance 2.10 minutes^2. Thus, $1/\theta = 2.2$ and $1/\beta\theta^2 = 2.10$, which implies that $\beta = 2.30$ and $\theta = 0.4545$.

Step 1. $a = 1.90$, $b = 3.74$.
Step 2. Generate $R_1 = 0.832$, $R_2 = 0.021$.
Step 3. Compute $X = 2.3(0.832/0.168)^{1.9} = 48.1$.
Step 4. $X = 48.1 > 3.74 - \ell n[(0.832)^2 0.021] = 7.97$, so reject X and return to step 2.
Step 2. Generate $R_1 = 0.434$, $R_2 = 0.716$.
Step 3. Compute $X = 2.3(0.434/0.566)^{1.9} = 1.389$.
Step 4. Since $X = 1.389 \le 3.74 - \ell n[(0.434)^2 0.716] = 5.74$, accept X.
Step 5. Divide X by $\beta\theta = 1.045$ to get $X = 1.329$.

This example took two trials (i.e., one rejection) to generate an acceptable gamma-distributed random variate, but on the average to generate, say, 1000 gamma variates, the method will require between 1130 and 1470 trials, or equivalently, between 2260 and 2940 random numbers. The method is somewhat cumbersome for hand calculations, but is easy to program on the computer and is one of the most efficient gamma generators known. ◀

8.5 Summary

The basic principles of random-variate generation using the inverse transform technique, the convolution method, and acceptance-rejection techniques have been introduced and illustrated by examples. Methods for generating many of the important continuous and discrete distributions, as well as empirical distributions, have been given. See Schmeiser [1980] for an excellent survey; for a state-of-the-art treatment, the reader is referred to Devroye [1986] or Dagpunar [1988].

REFERENCES

BOX, G. E. P., AND M. F. MULLER [1958], "A Note on the Generation of Random Normal Deviates," *Annals of Mathematical Statistics,* Vol. 29, pp. 610–11.

BRATLEY, P., B. L. FOX, AND L. E. SCHRAGE [1987], *A Guide to Simulation*, 2d ed., Springer-Verlag, New York.

CHENG, R. C. H. [1977], "The Generation of Gamma Variables," *Applied Statistician,* Vol. 26, No. 1, pp. 71–75.

DAGPUNAR, J. [1988], *Principles of Random Variate Generation*, Clarendon Press, Oxford.

DEVROYE, L. [1986], *Non-Uniform Random Variate Generation*, Springer-Verlag, New York.

FISHMAN, G. S. [1978], *Principles of Discrete Event Simulation,* John Wiley, New York.

LAW, A. M., AND W. D. KELTON [2000], *Simulation Modeling & Analysis*, 3d ed., McGraw-Hill, New York.

RIPLEY, B. D. [1987], *Stochastic Simulation*, John Wiley, New York.

SCHMEISER, B. W. [1979], "Approximations to the Inverse Cumulative Normal Function for Use on Hand Calculators," *Applied Statistics*, Vol. 28, pp. 175–176.

SCHMEISER, B. W. [1980], "Random Variate Generation: A Survey," in *Simulation with Discrete Models*: *A State of the Art View*, eds. T. I. Ören, C. M. Shub, and P. F. Roth, IEEE.

SCHMIDT, J. W., AND R. E. TAYLOR [1970], *Simulation and Analysis of Industrial Systems*, Irwin, Homewood, lll.

EXERCISES

1. Develop a random-variate generator for a random variable X with the pdf

$$f(x) = \begin{cases} e^{2x}, & -\infty < x \le 0 \\ e^{-2x}, & 0 < x < \infty \end{cases}$$

2. Develop a generation scheme for the triangular distribution with pdf

$$f(x) = \begin{cases} \frac{1}{2}(x-2), & 2 \le x \le 3 \\ \frac{1}{2}\left(2 - \frac{x}{3}\right), & 3 < x \le 6 \\ 0, & \text{otherwise} \end{cases}$$

Generate 10 values of the random variate, compute the sample mean, and compare it to the true mean of the distribution.

3. Develop a generator for a triangular distribution with range (1, 10) and mode at $x = 4$.

4. Develop a generator for a triangular distribution with range (1, 10) and a mean of 4.

5. Given the following cdf for a continuous variable with range −3 to 4, develop a generator for the variable.

$$F(x) = \begin{cases} 0, & x \le -3 \\ \frac{1}{2} + \frac{x}{6}, & -3 < x \le 0 \\ \frac{1}{2} + \frac{x^2}{32}, & 0 < x \le 4 \\ 1, & x > 4 \end{cases}$$

6. Given the cdf $F(x) = x^4/16$ on $0 \le x \le 2$, develop a generator for this distribution.

7. Given the pdf $f(x) = x^2/9$ on $0 \le x \le 3$, develop a generator for this distribution.

8. Develop a generator for a random variable whose pdf is

$$f(x) = \begin{cases} \frac{1}{3}, & 0 \le x \le 2 \\ \frac{1}{24}, & 2 < x \le 10 \\ 0, & \text{otherwise} \end{cases}$$

9. The cdf of a discrete random variable X is given by

$$F(x) = \frac{x(x+1)(2x+1)}{n(n+1)(2n+1)}, \quad x = 1, 2, \ldots, n$$

When $n = 4$, generate three values of X using $R_1 = 0.83$, $R_2 = 0.24$, and $R_3 = 0.57$.

10. Times to failure for an automated production process have been found to be randomly distributed with a Weibull distribution with parameters $\beta = 2$ and $\alpha = 10$. Derive Equation (8.6) and then use it to generate five values from this Weibull distribution, using five random numbers taken from Table A.1.

11. Data have been collected on service times at a drive-in bank window at the Shady Lane National Bank. This data are summarized into intervals as follows:

Interval (Seconds)	*Frequency*
15–30	10
30–45	20
45–60	25
60–90	35
90–120	30
120–180	20
180–300	10

Set up a table like Table 8.2 for generating service times by the table-lookup method and generate five values of service time using four-digit random numbers.

12. In Example 8.2, assume that fire crew response times satisfy $0.25 \le x \le 3$. Modify Table 8.2 to accommodate this assumption. Then generate five values of response time using four-digit uniform random numbers from Table A.1.

13. For a preliminary version of a simulation model, the number of pallets, X, to be loaded onto a truck at a loading dock was assumed to be uniformly distributed between 8 and 24. Devise a method for generating X, assuming that the loads on successive trucks are independent. Use the technique of Example 8.5 for discrete uniform distributions. Finally, generate loads for 10 successive trucks by using four-digit random numbers.

14. After collecting more data, it was found that the distribution of Example 8.6 was a better approximation to the number of pallets loaded than was the uniform distribution, as was assumed in Exercise 13. Using Equation (8.18) generate loads for 10 successive trucks using the same random numbers as were used in Exercise 13. Compare the results to the results of Exercise 13.

15. The weekly demand, X, for a slow-moving item has been found to be well approximated by a geometric distribution on the range $\{0, 1, 2, \ldots\}$ with mean weekly demand of 2.5 items. Generate 10 values of X, demand per week, using random numbers from Table A.1. [*Hint*: For a geometric distribution on the range $\{q, q+1, \ldots\}$ with parameter p, the mean is $1/p + q - 1$.]

16. In Exercise 15, suppose that the demand has been found to have a Poisson distribution with mean 2.5 items per week. Generate 10 values of X, demand per week, using random numbers from Table A.1. Discuss the differences between the geometric and the Poisson distributions.

17. Lead times have been found to be exponentially distributed with mean 3.7 days. Generate five random lead times from this distribution.

18. Regular maintenance of a production routine has been found to vary and has been modeled as a normally distributed random variable with mean 33 minutes and variance 4 minutes^2. Generate five random maintenance times, with the given distribution.

19. A machine is taken out of production if it fails, or after 5 hours, whichever comes first. By running similar machines until failure, it has been found that time to failure, X, has the Weibull distribution with $\alpha = 8$, $\beta = 0.75$, and $\nu = 0$ (refer to Sections 5.4 and 8.1.3). Thus, the time until the machine is taken out of production can be represented as $Y = \min(X, 5)$. Develop a step-by-step procedure for generating Y.

20. The time until a component is taken out of service is uniformly distributed on 0 to 8 hours. Two such independent components are put in series, and the whole system goes down when one of the components goes down. If X_i $(i = 1, 2)$ represents the component runtimes, then $Y = \min(X_1, X_2)$ represents the system lifetime. Devise two distinct ways to generate Y. [*Hint*: One way is relatively straightforward. For a second method, first compute the cdf of Y: $F_Y(y) = P(Y \le y) = 1 - P(Y > y)$, for $0 \le y \le 8$. Use the equivalence $\{Y > y\} = \{X_1 > y \text{ and } X_2 > y\}$ and the independence of X_1, and X_2. After finding $F_Y(y)$, proceed with the inverse transform technique.]

21. In Exercise 20, component lifetimes are exponentially distributed, one with mean 2 hours and the other with mean 6 hours. Rework Exercise 20 under this new assumption. Discuss the relative efflciency of the two generation schemes devised.

22. Develop a technique for generating a binomial random variable, X, using the convolution technique. [*Hint*: X can be represented as the number of successes in n independent Bernoulli trials, each success having probability p. Thus, $X = \sum_{i=1}^{n} X_i$, where $P(X_i = 1) = p$ and $P(X_i = 0) = 1 - p$.]

23. Develop an acceptance-rejection technique for generating a geometric random variable, X, with parameter p on the range $\{0, 1, 2, \ldots\}$. (*Hint*: X can be thought of as the number of trials before the first success occurs in a sequence of independent Bernoulli trials.)

24. Write a computer routine to generate standard normal variates by the exact method discussed in this chapter. Use it to generate 1000 values. Compare the true probability, $\Phi(z)$, that a value lies in $(-\infty, z)$ to the actual observed relative frequency that values were $\leq z$, for $z = -4, -3, -2, -1, 0, 1, 2, 3$ and 4.

25. Write a computer routine to generate gamma variates with shape parameter β and scale parameter θ. Generate 1000 values with $\beta = 2.5$ and $\theta = 0.2$ and compare the true mean, $1/\theta = 5$, to the sample mean.

26. Write a computer routine to generate 200 values from one of the variates in Exercises 1 to 23. Make a histogram of the 200 values and compare it to the theoretical density function (or probability mass function for discrete random variables).

27. Many spreadsheet, symbolic calculation, and statistical analysis programs have built-in routines for generating random variates from standard distributions. Try to find out what variate-generation methods are used in one of these packages by looking at the documentation. Should you trust a variate generator if the method is not documented?

28. Suppose that somehow we have available a source of exponentially distributed random variates with mean 1. Write an algorithm to generate random variates with a triangular distribution by transforming the exponentially distributed random variates. [*Hint*: First transform to obtain uniformly distributed random variates.]

PART FOUR

Analysis of Simulation Data

9

Input Modeling

Input data provide the driving force for a simulation model. In the simulation of a queueing system, typical input data are the distributions of time between arrivals and service times. For an inventory system simulation, input data include the distributions of demand and lead time. For the simulation of a reliability system, the distribution of time-to-failure of a component is an example of input data.

In the examples and exercises in Chapters 2 and 3, the appropriate distributions were specified for you. In real-world simulation applications, however, determining appropriate distributions for input data is a major task from the standpoint of time and resource requirements. Regardless of the sophistication of the analyst, faulty models of the inputs will lead to outputs whose interpretation may give rise to misleading recommendations.

There are four steps in the development of a useful model of input data:

1. Collect data from the real system of interest. This often requires a substantial time and resource commitment. Unfortunately, in some situations it is not possible to collect data (for example, when time is extremely limited, when the input process does not yet exist, or when laws or rules prohibit the collection of data). When data are not available, expert opinion and knowledge of the process must be used to make educated guesses.
2. Identify a probability distribution to represent the input process. When data are available, this step typically begins by developing a frequency distribution, or histogram, of the data. Based on the frequency distribution and structural knowledge of the process, a family of distributions is chosen. Fortunately, as described in Chapter 5, several well-known distributions often provide good approximations in practice.

3. Choose parameters that determine a specific instance of the distribution family. When data are available, these parameters may be estimated from the data.
4. Evaluate the chosen distribution and the associated parameters for goodness-of-fit. Goodness-of-fit may be evaluated informally via graphical methods, or formally via statistical tests. The chi-square and the Kolmogorov-Smirnov tests are standard goodness-of-fit tests. If not satisfied that the chosen distribution is a good approximation of the data, then the analyst returns to the second step, chooses a different family of distributions, and repeats the procedure. If several iterations of this procedure fail to yield a fit between an assumed distributional form and the collected data, the empirical form of the distribution may be used as described in Section 8.1.5 of the previous chapter.

Each of these steps is discussed in this chapter. Although software is now widely available to accomplish steps 2, 3 and 4 — including standalone programs such as ExpertFit® and Stat::Fit™, and integrated programs such as Arena's Input Processor and @Risk's BestFit — it is still important to understand what the software does so that it can be used appropriately. Unfortunately, software is not as readily available for input modeling when there is a relationship between two or more variables of interest, or when no data are available. These two topics are discussed toward the end of the chapter.

9.1 Data Collection

Problems are found at the end of each chapter, as exercises for the reader, in mathematics, physics, chemistry, and other technical subject texts. Years and years of working these problems may give the reader the impression that data are readily available. Nothing could be further from the truth. Data collection is one of the biggest tasks in solving a real problem. It is one of the most important and difficult problems in simulation. And even when data are available, they have rarely been recorded in a form that is directly useful for simulation input modeling.

"GIGO," or "garbage-in, garbage-out," is a basic concept in computer science and it applies equally in the area of discrete system simulation. Many are fooled by a pile of computer output or a sophisticated animation, as if these were the absolute truth. Even if the model structure is valid, if the input data are inaccurately collected, inappropriately analyzed, or not representative of the environment, the simulation output data will be misleading and possibly damaging or costly when used for policy or decision making.

EXAMPLE 9.1 (The Laundromat)

As budding simulation students, the first two authors had assignments to simulate the operation of an ongoing system. One of these systems, which seemed to be a rather simple operation, was a self-service laundromat with 10 washing machines and six dryers.

However, the data-collection aspect of the problem rapidly became rather enormous. The interarrival-time distribution was not homogeneous; that is, the distribution changed by time of day and by day of week. The laundromat was open 7 days a week for 16 hours per day, or 112 hours per week. It would have been impossible to cover the operation of the laundromat with the limited resources available (two students who were also taking four other courses) and with a tight time constraint (the simulation was to be completed in a 4-week period). Additionally, the distribution of time between arrivals during one week may not have been followed during the next week. As a compromise, a sample of times was selected, and the interarrival-time distributions were determined and classified according to arrival rate (perhaps inappropriately) as "high," "medium," and "low."

Service-time distributions also presented a difficult problem from many perspectives. The proportion of customers demanding the various service combinations had to be observed and recorded. The simplest case was the customer desiring one washer followed by one dryer. However, a customer might choose two washing machines followed by one dryer, one dryer only, and so on. Since the customers used numbered machines, it was possible to follow them using that reference, rather than remembering them by personal characteristics. Because of the dependence between washer demand and dryer demand for an individual customer, it would have been inappropriate to treat the service times for washers and dryers separately as independent variables.

Some customers waited patiently for their clothes to complete the washing or drying cycle, and then they removed their clothes promptly. Others left the premises and returned after their clothes had finished their cycle on the machine being used. In a very busy period, the manager would remove a customer's clothes after the cycle and set them aside in a basket. It was decided that service termination would be measured as the point in time when the machine was emptied of its contents.

Also, machines would break down from time to time. The length of the breakdown varied from a few moments, when the manager repaired the machine, to several days (a breakdown on Friday night, requiring a part not in the laundromat storeroom, would not be fixed until the following Monday). The short-term repair times were recorded by the student team. The long-term repair completion times were estimated by the manager. Breakdowns then became part of the simulation. ◀

Many lessons can be learned from an actual experience in data collection. The first five exercises at the end of this chapter suggest some situations in which the student can gain such experience.

The following suggestions may enhance and facilitate data collection, although they are not all-inclusive.

1. A useful expenditure of time is in planning. This could begin by a practice or preobserving session. Try to collect data while preobserving. Devise forms for this purpose. It is very likely that these forms will have to be

modified several times before the actual data collection begins. Watch for unusual circumstances and consider how they will be handled. When possible, videotape the system and extract the data later by viewing the tape. Planning is important even if data will be collected automatically (e.g., via computer data collection) to insure that the appropriate data are available. When data have already been collected by someone else, be sure to allow plenty of time for converting the data into a usable format.

2. Try to analyze the data as they are being collected. Determine if the data being collected are adequate to provide the distributions needed as input to the simulation. Determine if any data being collected are useless to the simulation. There is no need to collect superfluous data.
3. Try to combine homogeneous data sets. Check data for homogeneity in successive time periods and during the same time period on successive days. For example, check for homogeneity of data from 2:00 P.M. to 3:00 P.M. and 3:00 P.M. to 4:00 P.M., and check to see if the data are homogeneous for 2:00 P.M. to 3:00 P.M. on Thursday and Friday. When checking for homogeneity, an initial test is to see if the means of the distributions (the average interarrival times, for example) are the same. The two-sample t test can be used for this purpose. A more thorough analysis would require a determination of the equivalence of the distributions using, perhaps, a quantile-quantile plot (described later).
4. Be aware of the possibility of data censoring, in which a quantity of interest is not observed in its entirety. This problem most often occurs when the analyst is interested in the time required to complete some process (for example, produce a part, treat a patient, or have a component fail), but the process begins prior to, or finishes after the completion of, the observation period. Censoring can result in especially long process times being left out of the data sample.
5. To determine whether there is a relationship between two variables, build a scatter diagram. Sometimes an eyeball scan of the scatter diagram will indicate if there is a relationship between two variables of interest. Section 9.6 describes models for statistically dependent input data.
6. Consider the possibility that a sequence of observations which appear to be independent may possess autocorrelation. Autocorrelation may exist in successive time periods or for successive customers. For example, the service time for the ith customer may be related to the service time for the $(i+n)$th customer. A brief introduction to autocorrelation was provided in Section 7.4.3, and some input models that account for autocorrelation are presented in Section 9.6.
7. Keep in mind the difference between input data and output or performance data, and be sure to collect input data. Input data typically represent the uncertain quantities that are largely beyond the control of the system and will not be altered by changes made to improve the system. Output data, on the other hand, represent the performance of the system when subjected to the inputs, performance that we may be trying to

improve. In a queueing simulation, the customer arrival times are usually inputs, while the customer delay is an output. Performance data are useful for model validation, however (see Chapter 10).

Again, these are just a few suggestions. As a rule, data collection and analysis must be approached with great care.

9.2 Identifying the Distribution with Data

In this section we discuss methods for selecting families of input distributions when data are available. The specific distribution within a family is specified by estimating its parameters, as described in Section 9.3. Section 9.5 takes up the case when no data are available.

9.2.1 Histograms

A frequency distribution or histogram is useful in identifying the shape of a distribution. A histogram is constructed as follows:

1. Divide the range of the data into intervals (intervals are usually of equal width; however, unequal widths may be used if the heights of the frequencies are adjusted).
2. Label the horizontal axis to conform to the intervals selected.
3. Determine the frequency of occurrences within each interval.
4. Label the vertical axis so that the total occurrences can be plotted for each interval.
5. Plot the frequencies on the vertical axis.

The number of class intervals depends on the number of observations and the amount of scatter or dispersion in the data. Hines and Montgomery [1990] state that choosing the number of class intervals approximately equal to the square root of the sample size often works well in practice. If the intervals are too wide, the histogram will be coarse, or blocky, and its shape and other details will not show well. If the intervals are too narrow, the histogram will be ragged and will not smooth the data. Examples of a ragged, coarse, and appropriate histogram using the same data are shown in Figure 9.1. Modern data-analysis software often allows the interval sizes to be changed easily and interactively until a good choice is found.

The histogram for continuous data corresponds to the probability density function of a theoretical distribution. If continuous, a line drawn through the center point of each class interval frequency should result in a shape like that of a pdf.

Histograms for discrete data, where there are a large number of data points, should have a cell for each value in the range of the data. However, if there are few data points, it may be necessary to combine adjacent cells to eliminate the ragged appearance of the histogram. If the histogram is associated with discrete data, it should look like a probability mass function.

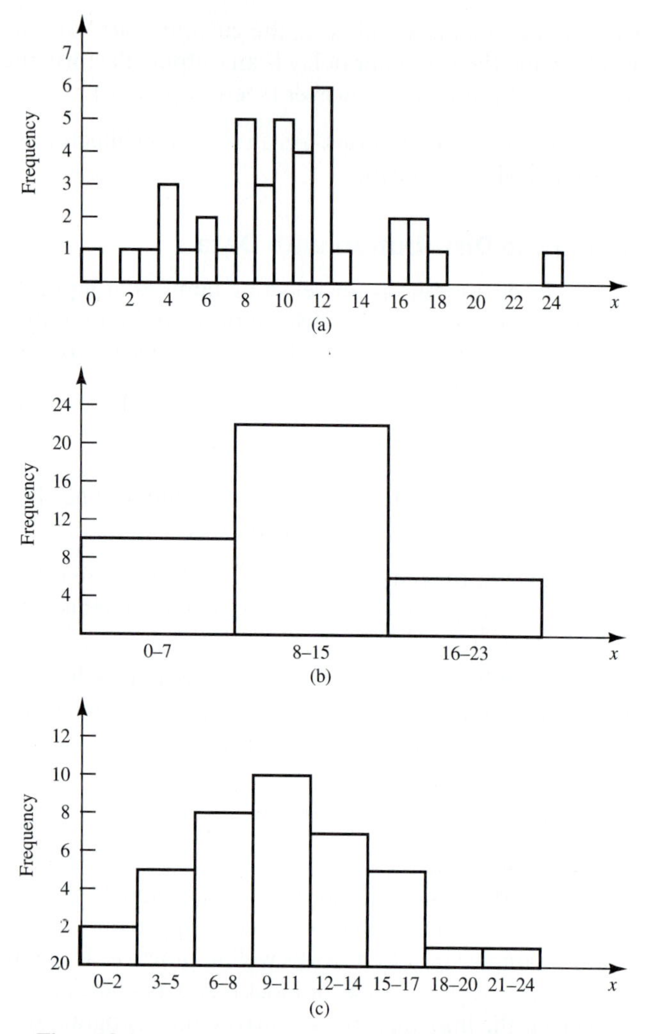

Figure 9.1. Ragged, coarse, and appropriate histograms: (a) original data — too ragged; (b) combining adjacent cells — too coarse; (c) combining adjacent cells — appropriate.

EXAMPLE 9.2 (Discrete Data)

The number of vehicles arriving at the northwest corner of an intersection in a 5-minute period between 7:00 A.M. and 7:05 A.M. was monitored for five workdays over a 20-week period. Table 9.1 shows the resulting data. The first entry in the table indicates that there were 12 5-minute periods during which zero vehicles arrived, 10 periods during which one vehicle arrived, and so on.

Since the number of automobiles is a discrete variable, and since there are ample data, the histogram can have a cell for each possible value in the range of the data. The resulting histogram is shown in Figure 9.2. ◀

Table 9.1. Number of Arrivals in a 5-Minute Period

Arrivals per Period	*Frequency*	*Arrivals per Period*	*Frequency*
0	12	6	7
1	10	7	5
2	19	8	5
3	17	9	3
4	10	10	3
5	8	11	1

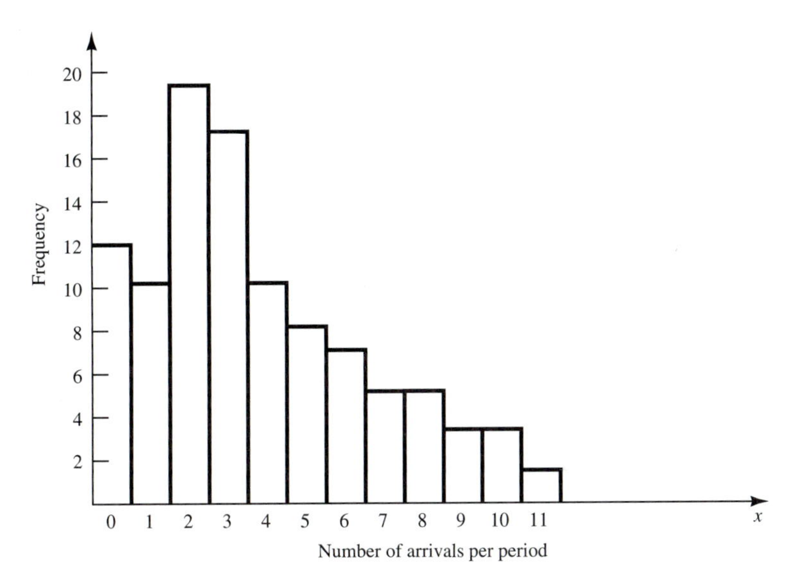

Figure 9.2. Histogram of number of arrivals per period.

EXAMPLE 9.3 (Continuous Data)

Life tests were performed on a random sample of electronic chips at 1.5 times the nominal voltage, and their lifetime (or time to failure) in days was recorded:

79.919	3.081	0.062	1.961	5.845
3.027	6.505	0.021	0.013	0.123
6.769	59.899	1.192	34.760	5.009
18.387	0.141	43.565	24.420	0.433
144.695	2.663	17.967	0.091	9.003
0.941	0.878	3.371	2.157	7.579
0.624	5.380	3.148	7.078	23.960
0.590	1.928	0.300	0.002	0.543
7.004	31.764	1.005	1.147	0.219
3.217	14.382	1.008	2.336	4.562

Lifetime, usually considered a continuous variable, is recorded here to three-decimal-place accuracy. The histogram is prepared by placing the data in class intervals. The range of the data is rather large, from 0.002 day to 144.695 days. However, most of the values (30 of 50) are in the zero-to–5-day range. Using intervals of width three results in Table 9.2. The data of Table 9.2 are then used to prepare the histogram shown in Figure 9.3. ◀

Table 9.2. Electronic Chip Data

Chip Life (Days)	*Frequency*
$0 \leq x_j < 3$	23
$3 \leq x_j < 6$	10
$6 \leq x_j < 9$	5
$9 \leq x_j < 12$	1
$12 \leq x_j < 15$	1
$15 \leq x_j < 18$	2
$18 \leq x_j < 21$	0
$21 \leq x_j < 24$	1
$24 \leq x_j < 27$	1
$27 \leq x_j < 30$	0
$30 \leq x_j < 33$	1
$33 \leq x_j < 36$	1
.	.
.	.
.	.
$42 \leq x_j < 45$	1
.	.
.	.
.	.
$57 \leq x_j < 60$	1
.	.
.	.
.	.
$78 \leq x_j < 81$	1
.	.
.	.
.	.
$144 \leq x_j < 147$	1

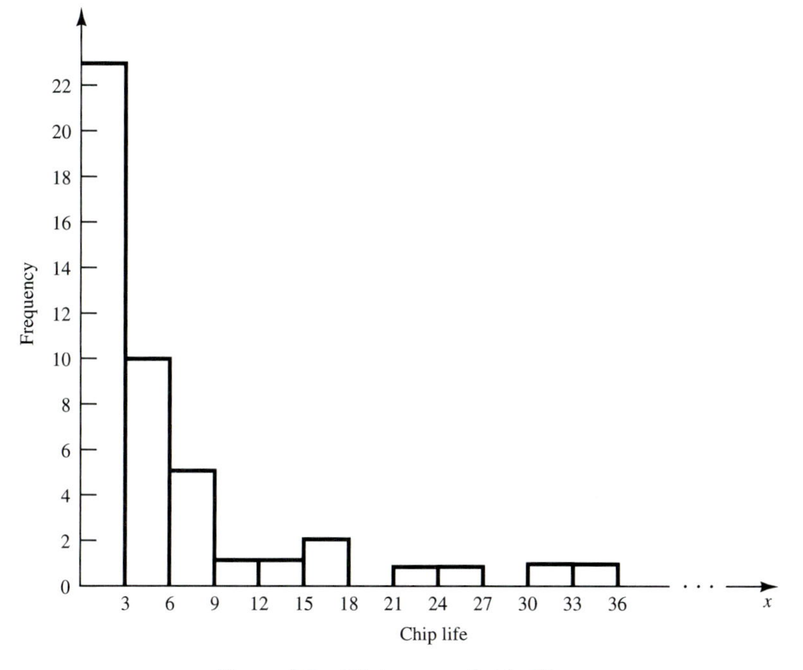

Figure 9.3. Histogram of chip life.

9.2.2 Selecting the Family of Distributions

In Chapter 5 some distributions that often arise in simulation were described. Additionally, the shapes of these distributions were displayed. The purpose of preparing a histogram is to infer a known pdf or pmf. A family of distributions is selected on the basis of what might arise in the context being investigated along with the shape of the histogram. Thus, if interarrival-time data have been collected, and the histogram has a shape similar to the pdf in Figure 5.9, the assumption of an exponential distribution would be warranted. Similarly, if measurements of the weights of pallets of freight are being made, and the histogram appears symmetric about the mean with a shape like that shown in Figure 5.12, the assumption of a normal distribution would be warranted.

The exponential, normal, and Poisson distributions are frequently encountered and are not difficult to analyze from a computational standpoint. Although more difficult to analyze, the gamma and Weibull distributions provide a wide array of shapes and should not be overlooked when modeling an underlying probabilistic process. Perhaps an exponential distribution was assumed, but it was found not to fit the data. The next step would be to examine where the lack of fit occurred. If the lack of fit was in one of the tails of the distribution, perhaps a gamma or Weibull distribution would more adequately fit the data.

Literally hundreds of probability distributions have been created, many with some specific physical process in mind. One aid to selecting distributions is to use the physical basis of the distributions as a guide. Here are some examples:

Binomial Models the number of successes in n trials, when the trials are independent with common success probability, p; for example, the number of defective computer chips found in a lot of n chips.

Negative Binomial (includes the geometric distribution) Models the number of trials required to achieve k successes; for example, the number of computer chips that we must inspect to find 4 defective chips.

Poisson Models the number of independent events that occur in a fixed amount of time or space; for example, the number of customers that arrive to a store during 1 hour, or the number of defects found in 30 square meters of sheet metal.

Normal Models the distribution of a process that can be thought of as the sum of a number of component processes; for example, the time to assemble a product which is the sum of the times required for each assembly operation. Notice that the normal distribution admits negative values, which may be impossible for process times.

Lognormal Models the distribution of a process that can be thought of as the product of (meaning to multiply together) a number of component processes; for example, the rate of return on an investment, when interest is compounded, is the product of the returns for a number of periods.

Exponential Models the time between independent events, or a process time which is memoryless (knowing how much time has passed gives no information about how much additional time will pass before the process is complete); for example, the times between the arrivals of a large number of customers who act independently of each other. The exponential is a highly variable distribution and is sometimes overused because it often leads to mathematically tractable models. Recall that, if the time between events is exponentially distributed, then the number of events in a fixed period of time is Poisson.

Gamma An extremely flexible distribution used to model nonnegative random variables. The gamma can be shifted away from 0 by adding a constant.

Beta An extremely flexible distribution used to model bounded (fixed upper and lower limits) random variables. The beta can be shifted away from 0 by adding a constant and can have a larger range than [0, 1] by multiplying by a constant.

Erlang Models processes that can be viewed as the sum of several exponentially distributed processes; for example, a computer network fails when a computer and two backup computers fail, and each has a time to failure that is exponentially distributed. The Erlang is a special case of the gamma.

Weibull Models the time to failure for components; for example, the time to failure for a disk drive. The exponential is a special case of the Weibull.

Discrete or Continuous Uniform Models complete uncertainty, since all outcomes are equally likely. This distribution is often overused when there are no data.

Triangular Models a process when only the minimum, most-likely, and maximum values of the distribution are known; for example, the minimum, most-likely, and maximum time required to test a product.

Empirical Resamples from the actual data collected; often used when no theoretical distribution seems appropriate.

Do not ignore physical characteristics of the process when selecting distributions. Is the process naturally discrete or continuous valued? Is it bounded or is there no natural bound? This knowledge, which does not depend on data, can help narrow the family of distributions from which to choose. And keep in mind that there is no "true" distribution for any stochastic input process. An input model is an approximation of reality, so the goal is to obtain an approximation that yields useful results from the simulation experiment.

The reader is encouraged to complete Exercises 6 through 11 to learn more about the shapes of the distributions mentioned in this section. Examining the variations in shape, as the parameters change will be very instructive.

9.2.3 Quantile-Quantile Plots

The construction of histograms, as discussed in Section 9.2.1, and the recognition of a distributional shape, as discussed in Section 9.2.2, are necessary ingredients for selecting a family of distributions to represent a sample of data. However, a histogram is not as useful for evaluating the *fit* of the chosen distribution. When there are a small number of data points, say 30 or fewer, a histogram can be rather ragged. Further, our perception of the fit depends on the widths of the histogram intervals. But even if the intervals are well chosen, grouping data into cells makes it difficult to compare a histogram to a continuous probabiliy density function. A quantile-quantile (q-q) plot is a useful tool for evaluating distribution fit that does not suffer from these problems.

If X is a random variable with cdf F, then the q-quantile of X is that value γ such that $F(\gamma) = P(X \leq \gamma) = q$, for $0 < q < 1$. When F has an inverse, we write $\gamma = F^{-1}(q)$.

Now let $\{x_i, i = 1, 2, \ldots, n\}$ be a sample of data from X. Order the observations from the smallest to the largest, and denote these as $\{y_j, j = 1, 2, \ldots, n\}$, where $y_1 \leq y_2 \leq \cdots \leq y_n$. Let j denote the ranking or order number. Therefore, $j = 1$ for the smallest and $j = n$ for the largest. The q-q plot is based on the fact that y_j is an estimate of the $(j - 1/2)/n$ quantile of X. In other words,

$$y_j \text{ is approximately } F^{-1}\left(\frac{j - \frac{1}{2}}{n}\right)$$

Now suppose that we have chosen a distribution with cdf F as a possible representation of the distribution of X. If F is a member of an appropriate family of distributions, then a plot of y_j versus $F^{-1}((j-1/2)/n)$ will be *approximately a straight line*. If F is from an appropriate family of distributions and also has appropriate parameter values, then the line will have slope 1. On the other hand, if the assumed distribution is inappropriate, the points will deviate from a straight line, usually in a systematic manner. The decision of whether or not to reject some hypothesized model is subjective.

EXAMPLE 9.4 (Normal Q-Q Plot)

A robot is used to install the doors on automobiles along an assembly line. It was thought that the installation times followed a normal distribution. The robot is capable of accurately measuring installation times. A sample of 20 installation times was automatically taken by the robot with the following results, where the values are in seconds:

99.79	99.56	100.17	100.33
100.26	100.41	99.98	99.83
100.23	100.27	100.02	100.47
99.55	99.62	99.65	99.82
99.96	99.90	100.06	99.85

The sample mean is 99.99 seconds, and the sample variance is $(0.2832)^2$ seconds^2. These values can serve as the parameter estimates for the mean and variance of the normal distribution. The observations are now ordered from smallest to largest as follows:

j	*Value*	j	*Value*	j	*Value*	j	*Value*
1	99.55	6	99.82	11	99.98	16	100.26
2	99.56	7	99.83	12	100.02	17	100.27
3	99.62	8	99.85	13	100.06	18	100.33
4	99.65	9	99.90	14	100.17	19	100.41
5	99.79	10	99.96	15	100.23	20	100.47

The ordered observations are then plotted versus $F^{-1}((j-1/2)/20)$, for $j = 1, 2, \ldots, 20$, where F is the cdf of the normal distribution with mean 99.99 and variance $(0.2832)^2$ to obtain a q-q plot. The plotted values are shown in Figure 9.4, along with a histogram of the data that has the density function of the normal distribution superimposed. Notice that it is difficult to tell if the data are well represented by a normal distribution by looking at the histogram, but the general perception of a straight line is quite clear in the q-q plot, supporting the hypothesis of a normal distribution. ◀

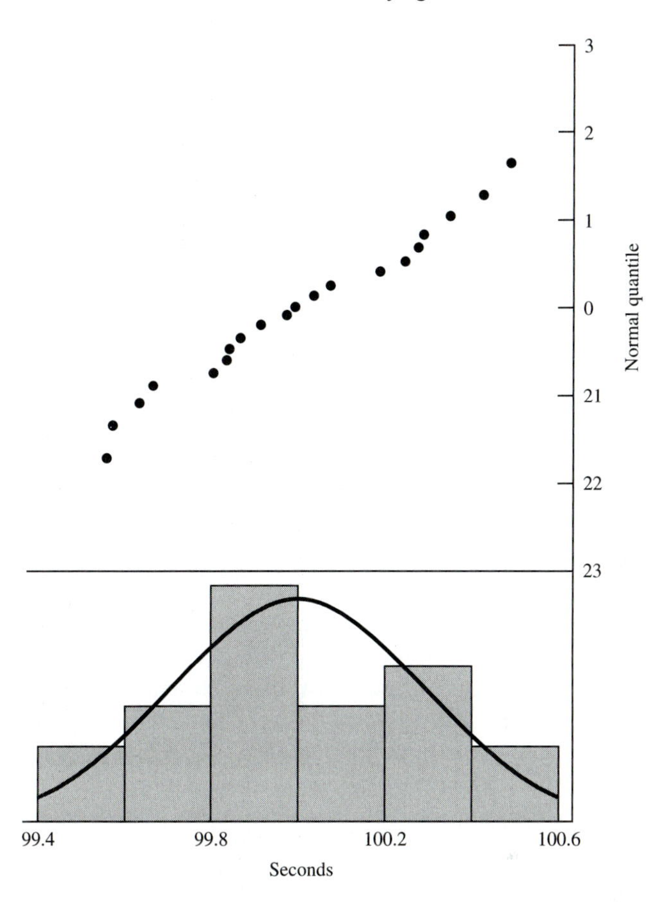

Figure 9.4. Histogram and q-q plot of the installation times.

In the evaluation of the linearity of a q-q plot, the following should be considered:

1. The observed values will never fall exactly on a straight line.
2. The ordered values are not independent, since they have been ranked. Hence, if one point is above a straight line, it is likely that the next point will also lie above the line. And it is unlikely that the points will be scattered about the line.
3. The variances of the extremes (largest and smallest values) are much higher than the variances in the middle of the plot. Greater discrepancies can be accepted at the extremes. The linearity of the points in the middle of the plot is more important than the linearity at the extremes.

Modern data-analysis software often includes tools for generating q-q plots, especially for the normal distribution. The q-q plot can also be used to compare two samples of data to see if they can be represented by the same distribution (that is, they are homogeneous). If $x_1, x_2, \ldots, x_n$ are a sample of the random variable X, and $z_1, z_2, \ldots, z_n$ are a sample of the random variable Z, then plotting the ordered values of X versus the ordered values of Z will reveal approximately a straight line if both samples are well represented by the same distribution (Chambers, Cleveland, Kleiner, and Tukey [1983]).

9.3 Parameter Estimation

After a family of distributions has been selected, the next step is to estimate the parameters of the distribution. Estimators for many useful distributions are described in this section. In addition, many software packages—some of them integrated into simulation languages—are now available to compute these estimates.

9.3.1 Preliminary Statistics: Sample Mean and Sample Variance

In a number of instances the sample mean, or the sample mean and sample variance, are used to estimate the parameters of a hypothesized distribution; see Example 9.4. In the following paragraphs, three sets of equations are given for computing the sample mean and sample variance. Equations (9.1) and (9.2) can be used when discrete or continuous raw data are available. Equations (9.3) and (9.4) are used when the data are discrete and have been grouped in a frequency distribution. Equations (9.5) and (9.6) are used when the data are discrete or continuous and have been placed in class intervals. Equations (9.5) and (9.6) are approximations and should be used only when the raw data are unavailable.

If the observations in a sample of size n are $X_1, X_2, \ldots, X_n$, the sample mean ($\bar{X}$) is defined by

$$\bar{X} = \frac{\sum_{i=1}^{n} X_i}{n} \tag{9.1}$$

and the sample variance, S^2, is defined by

$$S^2 = \frac{\sum_{i=1}^{n} X_i^2 - n\bar{X}^2}{n-1} \tag{9.2}$$

If the data are discrete and grouped in a frequency distribution, Equations (9.1) and (9.2) can be modified to provide for much greater computational efficiency. The sample mean can be computed by

$$\bar{X} = \frac{\sum_{j=1}^{k} f_j X_j}{n} \tag{9.3}$$

and the sample variance by

$$S^2 = \frac{\sum_{j=1}^{k} f_j X_j^2 - n\bar{X}^2}{n-1} \tag{9.4}$$

where k is the number of distinct values of X and f_j is the observed frequency of the value X_j of X.

EXAMPLE 9.5 (Grouped Data)

The data in Table 9.1 can be analyzed to obtain $n = 100$, $f_1 = 12$, $X_1 = 0$, $f_2 = 10$, $X_2 = 1, \ldots, \sum_{j=1}^{k} f_j X_j = 364$, and $\sum_{j=1}^{k} = f_j X_j^2 = 2080$. From Equation (9.3),

$$\bar{X} = \frac{364}{100} = 3.64$$

and from Equation (9.4),

$$S^2 = \frac{2080 - 100(3.64)^2}{99} = 7.63$$

The sample standard deviation, S, is just the square root of the sample variance. In this case $S = \sqrt{7.63} = 2.76$. Equations (9.1) and (9.2) would have yielded exactly the same results for $\bar{X}$ and S^2. ◀

It is preferable to use the raw data, if possible, when the values are continuous. However, the data may have been received after they have been placed in class intervals. Then it is no longer possible to obtain the exact sample mean and variance. In such cases, the sample mean and sample variance are approximated from the following equations:

$$\bar{X} \doteq \frac{\sum_{j=1}^{c} f_j m_j}{n} \tag{9.5}$$

and

$$S^2 \doteq \frac{\sum_{j=1}^{c} f_j m_j^2 - n\bar{X}^2}{n - 1} \tag{9.6}$$

where f_j is the observed frequency in the jth class interval, m_j is the midpoint of the jth interval, and c is the number of class intervals.

EXAMPLE 9.6 (Continuous Data in Class Intervals)

Assume that the raw data on chip life shown in Example 9.3 were either discarded or lost. However, the data shown in Table 9.2 are still available. To determine approximate values of $\bar{X}$ and S^2 Equations (9.5) and (9.6) are used. The following values are determined: $f_1 = 23$, $m_1 = 1.5$, $f_2 = 10$, $m_2 = 4.5, \ldots,$ $\sum_{j=1}^{49} f_j m_j = 614$, and $\sum_{j=1}^{49} f_j m_j^2 = 37{,}226.5$. With $n = 50$, $\bar{X}$ is approximated from Equation (9.5) as

$$\bar{X} \doteq \frac{614}{50} = 12.28$$

Then, S^2 is approximated from Equation (9.6) as

$$S^2 \doteq \frac{37{,}226.5 - 50(12.28)^2}{49} = 605.849$$

and

$$S \doteq 24.614$$

Applying Equations (9.1) and (9.2) to the original data in Example 9.3 results in $\bar{X} = 11.894$ and $S = 24.953$. Thus, when the raw data are either discarded or lost, some inaccuracies may result. ◀

9.3.2 Suggested Estimators

Numerical estimates of the distribution parameters are needed to reduce the family of distributions to a specific distribution and to test the resulting hypothesis. Table 9.3 contains suggested estimators for distributions often used in simulation, all of which were described in Chapter 5. Except for an adjustment to remove bias in the estimate of σ^2 for the normal distribution, these estimators are the maximum-likelihood estimators based on the raw data. (If the data are in class intervals, these estimators must be modified.) The reader is referred to Fishman [1973] and Law and Kelton [2000] for parameter estimates for the beta, uniform, binomial, and negative binomial distributions. The triangular distribution is usually employed when no data are available, with the parameters obtained from educated guesses for the minimum, most likely, and maximum possible values; the uniform distribution may also be used in this way if only minimum and maximum values are available.

Examples of the use of the estimators are given in the following paragraphs. The reader should keep in mind that a parameter is an unknown constant, but the estimator is a statistic or random variable because it depends on the sample values. To distinguish the two clearly, if, say, a parameter is denoted by α, the estimator will be denoted by $\widehat{\alpha}$.

EXAMPLE 9.7 (Poisson Distribution)

Assume that the arrival data in Table 9.1 require analysis. By comparison to Figure 5.7, an examination of Figure 9.2 suggests a Poisson distributional assumption with unknown parameter α. From Table 9.3, the estimator of α is $\bar{X}$, which was determined in Example 9.5. Thus, $\widehat{\alpha} = 3.64$. Recall that the true mean and variance are equal for the Poisson distribution. In Example 9.5, the sample variance was estimated by $S^2 = 7.63$. However, it should never be expected that the sample mean and the sample variance will be precisely equal, since both are random variables. ◀

EXAMPLE 9.8 (Lognormal Distribution)

The percentage rates of return on 10 investments in a portfolio are 18.8, 27.9, 21.0, 6.1, 37.4, 5.0, 22.9, 1.0, 3.1, and 8.3. To estimate the parameters of a lognormal model of this data, we first take the natural log of the data to obtain 2.9, 3.3, 3.0, 1.8, 3.6, 1.6, 3.1, 0, 1.1, and 2.1. Then set $\widehat{\mu} = \bar{X} = 2.3$ and $\widehat{\sigma}^2 = S^2 = 1.3$. ◀

EXAMPLE 9.9 (Normal Distribution)

The parameters of the normal distribution, μ and σ^2, are estimated by $\bar{X}$ and S^2, as shown in Table 9.3. The q-q plot in Example 9.4 leads to a distributional assumption that the installation times are normal. Using Equations (9.1) and (9.2), the data in Example 9.4 yield $\widehat{\mu} = \bar{X} = 99.9865$ and $\widehat{\sigma}^2 = S^2 = (0.2832)^2$ second^2. ◀

EXAMPLE 9.10 (Gamma Distribution)

The estimator, $\widehat{\beta}$, for the gamma distribution is determined by the use of Table A.9 from Choi and Wette [1969]. Table A.9 requires the computation of the quantity $1/M$, where

$$M = \ell n \bar{X} - \frac{1}{n}\sum_{i=1}^{n} \ell n X_i \tag{9.7}$$

Also, it can be seen in Table 9.3 that $\widehat{\theta}$ is given by

$$\widehat{\theta} = \frac{1}{\bar{X}} \tag{9.8}$$

In Chapter 5 it was stated that lead time is often gamma distributed. Suppose that the lead times (in days) associated with 20 orders have been

Table 9.3. Suggested Estimators for Distributions Often Used in Simulation

Distribution	*Parameter(s)*	*Suggested Estimator(s)*
Poisson	α	$\hat{\alpha} = \bar{X}$
Exponential	λ	$\hat{\lambda} = \dfrac{1}{\bar{X}}$
Gamma	β, θ	$\hat{\beta}$ (see Table A.9) $\hat{\theta} = \dfrac{1}{\bar{X}}$
Normal	μ, σ^2	$\hat{\mu} = \bar{X}$ $\hat{\sigma}^2 = S^2$ (unbiased)
Lognormal	μ, σ^2	$\hat{\mu} = \bar{X}$ (after taking ℓn of the data) $\hat{\sigma}^2 = S^2$ (after taking ℓn of the data)
Weibull with $\nu = 0$	α, β	$\hat{\beta}_0 = \dfrac{\bar{X}}{S}$ $\hat{\beta}_j = \hat{\beta}_{j-1} - \dfrac{f(\hat{\beta}_{j-1})}{f'(\hat{\beta}_{j-1})}$ See Equations (9.12) and (9.15) for $f(\hat{\beta})$ and $f'(\hat{\beta})$. Iterate until convergence: $\hat{\alpha} = \left(\dfrac{1}{n}\sum_{i=1}^{n} X_i^{\widehat{\beta}}\right)^{1/\widehat{\beta}}$

accurately measured as follows:

Order	*Lead Time (Days)*	Order	*Lead Time (Days)*
1	70.292	11	30.215
2	10.107	12	17.137
3	48.386	13	44.024
4	20.480	14	10.552
5	13.053	15	37.298
6	25.292	16	16.314
7	14.713	17	28.073
8	39.166	18	39.019
9	17.421	19	32.330
10	13.905	20	36.547

To determine $\widehat{\beta}$ and $\widehat{\theta}$, it is first necessary to determine M using Equation (9.7). Here, $\bar{X}$ is determined from Equation (9.1) to be

$$\bar{X} = \frac{564.32}{20} = 28.22$$

Then,

$$\ell n \bar{X} = 3.34$$

Next,

$$\sum_{i=1}^{20} \ell n X_i = 63.99$$

Then,

$$M = 3.34 - \frac{63.99}{20} = 0.14$$

and

$$1/M = 7.14$$

By interpolation in Table A.9, $\widehat{\beta} = 3.728$. Finally, Equation (9.8) results in

$$\widehat{\theta} = \frac{1}{28.22} = 0.035$$

◀

EXAMPLE 9.11 (Exponential Distribution)

Assuming that the data in Example 9.3 come from an exponential distribution, the parameter estimate, $\widehat{\lambda}$, can be determined. In Table 9.3, $\widehat{\lambda}$ is obtained using $\bar{X}$ as follows:

$$\widehat{\lambda} = \frac{1}{\bar{X}} = \frac{1}{11.894} = 0.084 \text{ per day}$$ ◀

EXAMPLE 9.12 (Weibull Distribution)

Suppose that a random sample of size n, $X_1, X_2, \ldots, X_n$, has been taken, and the observations are assumed to come from a Weibull distribution. The likelihood function derived using the pdf given by Equation (5.46) can be shown to be

$$L(\alpha, \beta) = \frac{\beta^n}{\alpha^{\beta n}} \left[\prod_{i=1}^{n} X_i^{(\beta-1)} \right] \exp\left[-\sum_{i=1}^{n} \left(\frac{X_i}{\alpha} \right)^{\beta} \right] \tag{9.9}$$

The maximum-likelihood estimates are those values of $\widehat{\alpha}$ and $\widehat{\beta}$ that maximize $L(\alpha, \beta)$, or equivalently maximize $\ell n L(\alpha, \beta)$, denoted by $l(\alpha, \beta)$. The maximum value of $l(\alpha, \beta)$ is obtained by taking the partial derivatives $\partial l(\alpha, \beta)/\partial \alpha$ and $\partial l(\alpha, \beta)/\partial \beta$, setting each to zero, and solving the resulting equations, which after substitution become

$$f(\beta) = 0 \tag{9.10}$$

and

$$\alpha = \left(\frac{1}{n} \sum_{i=1}^{n} X_i^{\beta} \right)^{1/\beta} \tag{9.11}$$

where

$$f(\beta) = \frac{n}{\beta} + \sum_{i=1}^{n} \ell n X_i - \frac{n \sum_{i=1}^{n} X_i^{\beta} \ell n X_i}{\sum_{i=1}^{n} X_i^{\beta}} \tag{9.12}$$

The maximum-likelihood estimates, $\widehat{\alpha}$ and $\widehat{\beta}$, are the solutions of Equations (9.10) and (9.11). First $\widehat{\beta}$ is determined through the iterative procedure explained below. Then $\widehat{\alpha}$ is determined using Equation (9.11) with $\beta = \widehat{\beta}$.

Since Equation (9.10) is nonlinear, it is necessary to use a numerical analysis technique to solve it. In Table 9.3 an iterative method for determining $\widehat{\beta}$ is given as

$$\widehat{\beta}_j = \widehat{\beta}_{j-1} - \frac{f(\widehat{\beta}_{j-1})}{f'(\widehat{\beta}_{j-1})} \tag{9.13}$$

Equation (9.13) employs Newton's method in reaching $\widehat{\beta}$, where $\widehat{\beta}_j$ is the jth iteration beginning with an initial estimate for $\widehat{\beta}_0$, given in Table 9.3, as follows:

$$\widehat{\beta}_0 = \frac{\bar{X}}{S} \tag{9.14}$$

If the initial estimate, $\widehat{\beta}_0$, is sufficiently close to the solution $\widehat{\beta}$, then $\widehat{\beta}_j$ approaches $\widehat{\beta}$ as $j \longrightarrow \infty$. When using Newton's method, $\widehat{\beta}$ is approached through increments of size $f(\widehat{\beta}_{j-1})/f'(\widehat{\beta}_{j-1})$. Equation (9.12) is used to compute $f(\widehat{\beta}_{j-1})$ and Equation (9.15) is used to compute $f'(\widehat{\beta}_{j-1})$ as follows:

$$f'(\beta) = -\frac{n}{\beta^2} - \frac{n \sum_{i=1}^{n} X_i^{\beta} (\ell n X_i)^2}{\sum_{i=1}^{n} X_i^{\beta}} + \frac{n \left(\sum_{i=1}^{n} X_i^{\beta} \ell n X_i \right)^2}{\left(\sum_{i=1}^{n} X_i^{\beta} \right)^2} \tag{9.15}$$

Equation (9.15) can be derived from Equation (9.12) by differentiating $f(\beta)$ with respect to β. The iterative process continues until $f(\widehat{\beta}_j) \doteq 0$, for example, until $|f(\widehat{\beta}_j)| \leq 0.001$.

Consider the data given in Example 9.3. These data concern the failure of electronic components and may come from an exponential distribution. In Example 9.11, the parameter $\widehat{\lambda}$ was estimated on the hypothesis that the data were from an exponential distribution. If the hypothesis that the data came from an exponential distribution is rejected, an alternative hypothesis is that the data come from a Weibull distribution. The Weibull distribution is suspected, since the data pertain to electronic component failures which occur suddenly.

Equation (9.14) is used to determine $\widehat{\beta}_0$. For the data in Example 9.3, $n = 50$, $\bar{X} = 11.894$, $\bar{X}^2 = 141.467$, and $\sum_{i=1}^{50} X_1^2 = 37{,}575.850$, so that S^2 is found by Equation (9.2) to be

$$S^2 = \frac{37{,}578.850 - 50(141.467)}{49} = 622.650$$

and $S = 24.953$. Thus,

$$\widehat{\beta}_0 = \frac{11.894}{24.953} = 0.477$$

To compute $\widehat{\beta}_1$ using Equation (9.13) requires the determination of $f(\widehat{\beta}_0)$ and $f'(\widehat{\beta}_0)$ using Equations (9.12) and (9.15). The following additional values are needed: $\sum_{i=1}^{50} X_i^{\widehat{\beta}_0} = 115.125$, $\sum_{i=1}^{50} \ell n X_i = 38.294$, $\sum_{i=1}^{50} X_i^{\widehat{\beta}_0} \ell n X_i = 292.629$, and $\sum_{i=1}^{50} X_i^{\widehat{\beta}_0} (\ell n X_i)^2 = 1057.781$. Thus,

$$f(\widehat{\beta}_0) = \frac{50}{0.477} + 38.294 - \frac{50(292.629)}{115.125} = 16.024$$

and

$$f'(\widehat{\beta}_0) = \frac{-50}{(0.477)^2} - \frac{50(1057.781)}{115.125} + \frac{50(292.629)^2}{(115.125)^2} = -356.110$$

Then, by Equation (9.13),

$$\widehat{\beta}_1 = 0.477 - \frac{16.024}{-356.110} = 0.522$$

After four iterations, $|f(\widehat{\beta}_3)| \leq 0.001$, at which point $\widehat{\beta} \doteq \widehat{\beta}_4 = 0.525$ is the approximate solution to Equation (9.10). Table 9.4 contains the values needed to complete each iteration.

Now, $\widehat{\alpha}$ can be determined using Equation (9.11) with $\beta = \widehat{\beta} = 0.525$ as follows:

$$\widehat{\alpha} = \left[\frac{130.608}{50}\right]^{1/0.525} = 6.227$$

If $\widehat{\beta}_0$ is sufficiently close to $\widehat{\beta}$, the procedure converges quickly, usually in four to five iterations. However, if the procedure appears to be diverging,

Table 9.4. Iterative Estimation of Parameters of the Weibull Distribution

j	$\hat{\beta}_j$	$\sum_{i=1}^{50} X_i^{\hat{\beta}_j}$	$\sum_{i=1}^{50} X_i^{\hat{\beta}_j} \ell n X_i$	$\sum_{i=1}^{50} X_i^{\hat{\beta}_j} (\ell n X_i)^2$	$f(\hat{\beta}_j)$	$f'(\hat{\beta}_j)$	$\hat{\beta}_{j+1}$
0	0.477	115.125	292.629	1057.781	16.024	−356.110	0.522
1	0.522	129.489	344.713	1254.111	1.008	−313.540	0.525
2	0.525	130.603	348.769	1269.547	0.004	−310.853	0.525
3	0.525	130.608	348.786	1269.614	0.000	−310.841	0.525

try other initial guesses for $\widehat{\beta}_0$ — for example, one-half the initial estimate or twice the initial estimate.

The difficult task of determining parameters for the Weibull distribution by hand emphasizes the value of having software support for input modeling. ◀

9.4 Goodness-of-Fit Tests

Hypothesis testing was discussed in Section 7.4 with respect to testing random numbers. In Section 7.4.1 the Kolmogorov-Smirnov test and the chi-square test were introduced. These two tests are applied in this section to hypotheses about distributional forms of input data.

Goodness-of-fit tests provide helpful guidance for evaluating the suitability of a potential input model. However, since there is no single correct distribution in a real application, you should not be a slave to the verdict of such tests. It is especially important to understand the effect of sample size. If very little data are available, then a goodness-of-fit test is unlikely to reject *any* candidate distribution; but if a lot of data are available, then a goodness-of-fit test will likely reject *all* candidate distributions. Therefore, failing to reject a candidate distribution should be taken as one piece of evidence in favor of that choice, while rejecting an input model is only one piece of evidence against the choice.

9.4.1 Chi-Square Test

One procedure for testing the hypothesis that a random sample of size n of the random variable X follows a specific distributional form is the chi-square goodness-of-fit test. This test formalizes the intuitive idea of comparing the histogram of the data to the shape of the candidate density or mass function. The test is valid for large sample sizes, for both discrete and continuous distributional assumptions, when parameters are estimated by maximum likelihood. The test procedure begins by arranging the n observations into a set of k class intervals or cells. The test statistic is given by

$$\chi_0^2 = \sum_{i=1}^{k} \frac{(O_i - E_i)^2}{E_i} \tag{9.16}$$

where O_i is the observed frequency in the ith class interval and E_i is the expected frequency in that class interval. The expected frequency for each class interval is computed as $E_i = np_i$, where p_i is the theoretical, hypothesized probability associated with the ith class interval.

It can be shown that χ_0^2 approximately follows the chi-square distribution with $k - s - 1$ degrees of freedom, where s represents the number of parameters of the hypothesized distribution estimated by the sample statistics. The hypotheses are:

H_0: the random variable, X, conforms to the distributional assumption with the parameter(s) given by the parameter estimate(s)

H_1: the random variable X does not conform

The critical value $\chi^2_{\alpha,k-s-1}$ is found in Table A.6. The null hypothesis, H_0, is rejected if $\chi_0^2 > \chi^2_{\alpha,k-s-1}$.

In applying the test, if expected frequencies are too small, χ_0^2 will reflect not only the departure of the observed from the expected frequency but the smallness of the expected frequency as well. Although there is no general agreement regarding the minimum size of E_i, values of 3, 4, and 5 have been widely used. In Section 7.4.1, when the chi-square test was discussed, the minimum expected frequency of five was suggested. If an E_i value is too small, it can be combined with expected frequencies in adjacent class intervals. The corresponding O_i values should also be combined, and k should be reduced by one for each cell that is combined.

If the distribution being tested is discrete, each value of the random variable should be a class interval, unless it is necessary to combine adjacent class intervals to meet the minimum expected cell-frequency requirement. For the discrete case, if combining adjacent cells is not required,

$$p_i = p(x_i) = P(X = x_i)$$

Otherwise, p_i is determined by summing the probabilities of appropriate adjacent cells.

If the distribution being tested is continuous, the class intervals are given by $[a_{i-1}, a_i)$, where a_{i-1} and a_i are the endpoints of the ith class interval. For the continuous case with assumed pdf $f(x)$, or assumed cdf $F(x)$, p_i can be computed by:

$$p_i = \int_{a_{i-1}}^{a_i} f(x)\,dx = F(a_i) - F(a_{i-1})$$

For the discrete case, the number of class intervals is determined by the number of cells resulting after combining adjacent cells as necessary. However, for the continuous case the number of class intervals must be specified. Although there are no general rules to be followed, the recommendations in Table 9.5 are made to aid in determining the number of class intervals for continuous data.

Table 9.5. Recommendations for Number of Class Intervals for Continuous Data

Sample Size, n	*Number of Class Intervals,* k
20	Do not use the chi-square test
50	5 to 10
100	10 to 20
>100	$\sqrt{n}$ to $n/5$

EXAMPLE 9.13 (Chi-Square Test Applied to Poisson Assumption)

In Example 9.7, the vehicle-arrival data presented in Example 9.2 were analyzed. Since the histogram of the data, shown in Figure 9.2, appeared to follow a Poisson distribution, the parameter, $\widehat{\alpha} = 3.64$, was determined. Thus, the following hypotheses are formed:

H_0: the random variable is Poisson distributed
H_1: the random variable is not Poisson distributed

The pmf for the Poisson distribution was given in Equation (5.18) as follows:

$$p(x) = \begin{cases} \dfrac{e^{-\alpha}\alpha^x}{x!}, & x = 0, 1, 2, \ldots \\ 0, & \text{otherwise} \end{cases} \tag{9.17}$$

For $\alpha = 3.64$, the probabilities associated with various values of x are obtained using Equation (9.17) with the following results:

$$\begin{aligned} p(0) &= 0.026, & p(6) &= 0.085 \\ p(1) &= 0.096, & p(7) &= 0.044 \\ p(2) &= 0.174, & p(8) &= 0.020 \\ p(3) &= 0.211, & p(9) &= 0.008 \\ p(4) &= 0.192, & p(10) &= 0.003 \\ p(5) &= 0.140, & p(11) &= 0.001 \end{aligned}$$

With this information, Table 9.6 is constructed. The value of E_1 is given by $np_0 = 100(0.026) = 2.6$. In a similar manner, the remaining E_i values are determined. Since $E_1 = 2.6 < 5$, E_1 and E_2 are combined. In that case O_1 and O_2 are also combined and k is reduced by one. The last five class intervals are also combined for the same reason, and k is further reduced by four.

The calculated χ_0^2 is 27.68. The degrees of freedom for the tabulated value of χ^2 is $k - s - 1 = 7 - 1 - 1 = 5$. Here, $s = 1$, since one parameter, $\widehat{\alpha}$, was estimated from the data. At the 0.05 level of significance, the critical value $\chi^2_{0.05,5}$ is 11.1. Thus, H_0 would be rejected at level of significance 0.05. The analyst may therefore want to search for a better-fitting model or use the empirical distribution of the data. ◀

Table 9.6. Chi-Square Goodness-of-Fit Test for Example 9.2

x_i	Observed Frequency, O_i		Expected Frequency, E_i		$\frac{(O_i - E_i)^2}{E_i}$
0	12	22	2.6	12.2	7.87
1	10		9.6		
2	19		17.4		0.15
3	17		21.1		0.80
4	10		19.2		4.41
5	8		14.0		2.57
6	7		8.5		0.26
7	5	17	4.4	7.6	11.62
8	5		2.0		
9	3		0.8		
10	3		0.3		
11	1		0.1		
	100		100.0		27.68

9.4.2 *Chi-Square Test with Equal Probabilities*

If a continuous distributional assumption is being tested, class intervals that are equal in probability rather than equal in width of interval should be used. This has been recommended by a number of authors (Mann and Wald [1942]; Gumbel [1943]; Law and Kelton [2000]; Stuart, Ord, and Arnold [1998]). It should be noted that the procedure is not applicable to data collected in class intervals, where the raw data have been discarded or lost.

Unfortunately, there is as yet no method for determining the probability associated with each interval that maximizes the power for a test of a given size. (The power of a test is defined as the probability of rejecting a false hypothesis.) However, if using equal probabilities, then $p_i = 1/k$. Since we recommend

$$E_i = np_i \geq 5$$

substituting for p_i yields

$$\frac{n}{k} \geq 5$$

and solving for k yields

$$k \leq \frac{n}{5} \tag{9.18}$$

Equation (9.18) was used in determining the recommended maximum number of class intervals in Table 9.5.

If the assumed distribution is normal, exponential, or Weibull, the method described in this section is straightforward. Example 9.14 indicates how the procedure is accomplished for the exponential distribution. If the assumed distribution is gamma (but not Erlang), or certain other distributions, then the computation of endpoints for class intervals is complex and may require numerical integration of the density function. Statistical-analysis software is very helpful in such cases.

EXAMPLE 9.14 (Chi-Square Test for Exponential Distribution)

In Example 9.11, the failure data presented in Example 9.3 were analyzed. Since the histogram of the data, shown in Figure 9.3, appeared to follow an exponential distribution, the parameter $\widehat{\lambda} = 1/\bar{X} = 0.084$ was determined. Thus, the following hypotheses are formed:

H_0: the random variable is exponentially distributed

H_1: the random variable is not exponentially distributed

In order to perform the chi-square test with intervals of equal probability, the endpoints of the class intervals must be determined. Equation (9.18) indicates that the number of intervals should be less than or equal to $n/5$. Here, $n = 50$, so that $k \leq 10$. In Table 9.5, it is recommended that 7 to 10 class intervals be used. Let $k = 8$, then each interval will have probability $p = 0.125$. The endpoints for each interval are computed from the cdf for the exponential distribution, given in Equation (5.27), as follows:

$$F(a_i) = 1 - e^{-\lambda a_i} \tag{9.19}$$

where a_i represents the endpoint of the ith interval, $i = 1, 2, \ldots, k$. Since $F(a_i)$ is the cumulative area from zero to a_i, $F(a_i) = ip$, so Equation (9.19) can be written as

$$ip = 1 - e^{-\lambda a_i}$$

or

$$e^{-\lambda a_i} = 1 - ip$$

Taking the logarithm of both sides and solving for a_i gives a general result for the endpoints of k equiprobable intervals for the exponential distribution, namely

$$a_i = -\frac{1}{\lambda}\ell n(1 - ip), \quad i = 0, 1, \ldots, k \tag{9.20}$$

Regardless of the value of λ, Equation (9.20) will always result in $a_0 = 0$ and $a_k = \infty$. With $\widehat{\lambda} = 0.084$ and $k = 8$, a_1 is determined from Equation (9.20) as

$$a_1 = -\frac{1}{0.084}\ell n(1 - 0.125) = 1.590$$

Continued application of Equation (9.20) for $i = 2, 3, \ldots, 7$ results in $a_2, \ldots, a_7$ as 3.425, 5.595, 8.252, 11.677, 16.503, and 24.755. Since $k = 8$, $a_8 = \infty$. The first interval is [0, 1.590), the second interval is [1.590, 3.425), and so on. The expectation is that 0.125 of the observations will fall in each interval. The observations, expectations, and the contributions to the calculated value of χ_0^2 are shown in Table 9.7. The calculated value of χ_0^2 is 39.6. The degrees of

Table 9.7. Chi-Square Goodness-of-Fit Test for Example 9.14

Class Interval	*Observed Frequency, O_i*	*Expected Frequency, E_i*	$\frac{(O_i - E_i)^2}{E_i}$
[0, 1.590)	19	6.25	26.01
[1.590, 3.425)	10	6.25	2.25
[3.425, 5.595)	3	6.25	0.81
[5.595, 8.252)	6	6.25	0.01
[8.252, 11.677)	1	6.25	4.41
[11.677, 16.503)	1	6.25	4.41
[16.503, 24.755)	4	6.25	0.81
[24.755, ∞)	6	6.25	0.01
	50	50	39.6

freedom are given by $k - s - 1 = 8 - 1 - 1 = 6$. At $\alpha = 0.05$, the tabulated value of $\chi^2_{0.05,6}$ is 12.6. Since $\chi^2_0 > \chi^2_{0.05,6}$, the null hypothesis is rejected. (The value of $\chi^2_{0.01,6}$ is 16.8, so the null hypothesis would also be rejected at level of significance $\alpha = 0.01$.) ◀

9.4.3 Kolmogorov-Smirnov Goodness-of-Fit Test

The chi-square goodness-of-fit test can accommodate the estimation of parameters from the data with a resultant decrease in the degrees of freedom (one for each parameter estimated). The chi-square test requires that the data be placed in class intervals, and in the case of a continuous distributional assumption, this grouping is arbitrary. Changing the number of classes and the interval width affects the value of the calculated and tabulated chi-square. A hypothesis may be accepted when the data are grouped one way but rejected when grouped another way. Also, the distribution of the chi-square test statistic is known only approximately, and the power of the test is sometimes rather low. As a result of these considerations, goodness-of-fit tests, other than the chi-square, are desired. The Kolmogorov-Smirnov test formalizes the idea behind examining a q-q plot.

The Kolmogorov-Smirnov test was presented in Section 7.4.1 to test for the uniformity of numbers and again in Section 7.4.4 to perform the gap test. Both of these uses fall into the category of testing for goodness-of-fit. Any continuous distributional assumption can be tested for goodness-of-fit using the method of Section 7.4.1, while discrete distributional assumptions can be tested using the method of Section 7.4.4.

The Kolmogorov-Smirnov test is particularly useful when sample sizes are small and when no parameters have been estimated from the data. When parameter estimates have been made, the critical values in Table A.8 are biased; in particular, they are too conservative. In this context "conservative" means that the critical values will be too large, resulting in smaller Type I (α) errors than those specified. The exact value of α can be determined in some instances as discussed at the end of this section.

The Kolmogorov-Smirnov test does not take any special tables when an exponential distribution is assumed. The following example indicates how the test is applied in this instance. (Notice that it is not necessary to estimate the parameter of the distribution in this example, permitting the use of Table A.8.)

EXAMPLE 9.15 (Kolmogorov-Smirnov Test for Exponential Distribution)

Suppose that 50 interarrival times (in minutes) are collected over the following 100-minute interval (arranged in order of occurrence):

0.44	0.53	2.04	2.74	2.00	0.30	2.54	0.52	2.02	1.89	1.53	0.21
2.80	0.04	1.35	8.32	2.34	1.95	0.10	1.42	0.46	0.07	1.09	0.76
5.55	3.93	1.07	2.26	2.88	0.67	1.12	0.26	4.57	5.37	0.12	3.19
1.63	1.46	1.08	2.06	0.85	0.83	2.44	2.11	3.15	2.90	6.58	0.64

The null hypothesis and its alternate are formed as follows:

H_0: the interarrival times are exponentially distributed
H_1: the interarrival times are not exponentially distributed

The data were collected over the interval 0 to $T = 100$ minutes. It can be shown that if the underlying distribution of interarrival times $\{T_1, T_2, \ldots\}$ is exponential, the arrival times are uniformly distributed on the interval $(0, T)$. The arrival times T_1, $T_1 + T_2$, $T_1 + T_2 + T_3, \ldots,$ $T_1 + \cdots + T_{50}$ are obtained by adding interarrival times. The arrival times are then normalized to a (0, 1) interval so that the Kolmogorov-Smirnov test, as presented in Section 7.4.1, can be applied. On a (0, 1) interval, the points will be $[T_1/T, (T_1+T_2)/T, \ldots, (T_1+\cdots+T_{50})/T]$. The resulting 50 data points are as follows:

0.0044	0.0097	0.0301	0.0575	0.0775	0.0805	0.1059	0.1111	0.1313	0.1502
0.1655	0.1676	0.1956	0.1960	0.2095	0.2927	0.3161	0.3356	0.3366	0.3508
0.3553	0.3561	0.3670	0.3746	0.4300	0.4694	0.4796	0.5027	0.5315	0.5382
0.5494	0.5520	0.5977	0.6514	0.6526	0.6845	0.7008	0.7154	0.7262	0.7468
0.7553	0.7636	0.7880	0.7982	0.8206	0.8417	0.8732	0.9022	0.9680	0.9744

Following the procedure in Example 7.6 yields a D^+ of 0.1054 and a D^- of 0.0080. Therefore, the Kolmogorov-Smirnov statistic is $D = \max(0.1054, 0.0080) = 0.1054$. The critical value of D obtained from Table A.8 for a level of significance of $\alpha = 0.05$ and $n = 50$ is $D_{0.05} = 1.36/\sqrt{n} = 0.1923$. Since $D = 0.1054$, the hypothesis that the interarrival times are exponentially distributed cannot be rejected. ◀

The Kolmogorov-Smirnov test has been modified so that it can be used in several situations where the parameters are estimated from the data. The computation of the test statistic is the same, but different tables of critical values are used. Different tables of critical values are required for different distributional assumptions. Lilliefors [1967] developed a test for normality. The null hypothesis states that the population is one of the family of normal distributions without specifying the parameters of the distribution. The interested reader may wish to study Lilliefors' original work, as he describes how simulation was used to develop the critical values.

Lilliefors [1969] also modified the critical values of the Kolmogorov-Smirnov test for the exponential distribution. Lilliefors again used random sampling to obtain approximate critical values, but Durbin [1975] subsequently obtained the exact distribution. Connover [1980] gives examples of Kolmogorov-Smirnov tests for the normal and exponential distributions. He also refers to several other Kolmogorov-Smirnov-type tests which may be of interest to the reader.

A test that is similar in spirit to the Kolmogorov-Smirnov test is the *Anderson-Darling test.* Like the Kolmogorov-Smirnov test, the Anderson-Darling test is based on the difference between the empirical cdf and the fitted cdf; unlike the Kolmogorov-Smirnov test, the Anderson-Darling test is based on a more comprehensive measure of difference (not just the maximum difference) and is more sensitive to discrepancies in the tails of the distributions. The critical values for the Anderson-Darling test also depend on the candidate distribution and on whether or not parameters have been estimated. Fortunately, this test and the Kolmogorov-Smirnov test have been implemented in a number of software packages that support simulation input modeling.

9.4.4 *p-Values and "Best Fits"*

To apply a goodness-of-fit test a significance level must be chosen. Recall that the significance level is the probability of falsely rejecting H_0: the random variable conforms to the distributional assumption. The traditional significance levels are 0.1, 0.05, and 0.01. Prior to the availability of high-speed computing, having a small set of standard values made it possible to produce tables of useful critical values. Now most statistical software computes critical values as needed, rather than storing them in tables. Thus, if the analyst prefers a level of significance of, say, 0.07, then he or she can choose it.

However, rather than require a prespecified significance level, many software packages compute a *p-value* for the test statistic. The p-value is the significance level at which one would *just reject* H_0 for the given value of the test statistic. Therefore, a large p-value tends to indicate a good fit (we would have to accept a large chance of error in order to reject), while a small p-value suggests a poor fit (to accept we would have to insist on almost no risk).

Recall Example 9.13, in which a chi-square test was used to check the Poisson assumption for the vehicle arrival data. The value of the test statistic was $\chi_0^2 = 27.68$ with 5 degrees of freedom. The p-value for this test statistic is 0.00004, meaning that we would reject the hypothesis that the data are Poisson at the 0.00004 significance level (recall that we rejected the hypothesis at the 0.05 level; now we know that we would also reject it at even lower levels).

The p-value can be viewed as a measure of fit, with larger values being better. This suggests that we could fit every distribution at our disposal, compute a test statistic for each fit, and then choose the distribution that yields the largest p-value. While we know of no input modeling software that implements this specific algorithm, many such packages do include a "best-fit" option in which the software recommends an input model to the user based on evaluating all feasible models. While the software may also take into account other factors — such as whether the data are discrete or continuous, bounded or unbounded — in the end some summary measure of fit, like the p-value, is used to rank the distributions. There is nothing wrong with this, but there are several things to keep in mind:

1. The software may know nothing about the physical basis of the data, and that information can suggest distribution families that are appropriate (see the list in Section 9.2.2). Remember that the goal of input modeling is often to fill in gaps or smooth the data, rather than find an input model that conforms as closely as possible to the given sample.
2. Recall that both the Erlang and the exponential distributions are special cases of the gamma, while the exponential is also a special case of the more flexible Weibull. Automated best-fit procedures tend to choose the more flexible distributions (gamma and Weibull over Erlang and exponential) because the extra flexibility allows closer conformance to the data and a better summary measure of fit. But again, close conformance to the data may not always lead to the most appropriate input model.
3. A summary statistic, like the p-value, is just that, a summary measure. It says little or nothing about where the lack of fit occurs (in the body of the distribution, in the right tail or in the left tail). A human, using graphical tools, can see where the lack of fit occurs and decide whether or not it is important for the application at hand.

Our recommendation is that automated distribution selection be used as one of several ways to suggest candidate distributions. Always inspect the automatic selection using graphical methods, and remember that the final choice is yours.

9.5 Selecting Input Models without Data

Unfortunately, it is often necessary in practice to develop a simulation model —perhaps for demonstration purposes or a preliminary study—before any process data are available. In this case the modeler must be resourceful in choosing input models and must carefully check the sensitivity of results to the chosen models.

There are a number of ways to obtain information about a process even if data are not available:

Engineering data Often a product or process has performance ratings provided by the manufacturer (for example, the mean time to failure of a disk drive is 5000 hours; a laser printer can produce 4 pages/minute; the cutting speed of a tool is 1 inch/second, etc.). Company rules may specify time or production standards. These values provide a starting point for input modeling by fixing a central value.

Expert option Talk to people who are experienced with the process or similar processes. Often they can provide optimistic, pessimistic, and most likely times. They may also be able to say if the process is nearly constant or highly variable, and they may be able to define the source of variability.

Physical or conventional limitations Most real processes have physical limits on performance (for example, computer data entry cannot be faster than a person can type). Because of company policies, there may be upper limits on how long a process may take. Do not ignore obvious limits or bounds that narrow the range of the input process.

The nature of the process The description of the distributions in Section 9.2.2 can be used to justify a particular choice even when no data are available.

When data are not available, the uniform, triangular, and beta distributions are often used as input models. The uniform can be a poor choice, because the upper and lower bounds are rarely just as likely as the central values in real processes. If, in addition to upper and lower bounds, a most-likely value can be given, then the triangular distribution can be used. The triangular distribution places much of its probability near the most-likely value and much less near the extremes (see Section 5.4). If a beta distribution is used, then be sure to plot the density function of the selected distribution, since the beta can take unusual shapes.

A useful refinement is obtained when a minimum, maximum, and one or more "breakpoints" can be given. A breakpoint is an intermediate value and a probability of being less than or equal to that value. The following example illustrates how breakpoints are used.

Example 9.16

For a production planning simulation, the sales volume of various products is required. The salesperson responsible for product XYZ–123 says that no fewer than 1000 units will be sold because of existing contracts, and no more than 5000 units will be sold because that is the entire market for the product. Based on her experience, she believes that there is a 90% chance of selling more than 2000 units, a 25% chance of selling more than 3500 units, and only a 1% chance of selling more than 4500 units.

Table 9.8 summarizes this information. Notice that the chances of exceeding certain sales goals have been translated into the cumulative probability of being less than or equal to those goals. With the information in this form the method of Section 8.1.5 can be employed to generate simulation input data. ◀

Table 9.8. Summary of Sales Information

i	*Interval (Hours)*	*Cumulative Frequency,* c_i
1	$1000 \leq x \leq 2000$	0.10
2	$2000 < x \leq 3500$	0.75
3	$3500 < x \leq 4500$	0.99
4	$4500 < x \leq 5000$	1.00

When input models have been selected without data, it is especially important to test the sensitivity of simulation results to the distribution chosen. Check sensitivity not only to the center of the distribution but also to the variability or limits. Extreme sensitivity of output results to the input model provides a convincing argument against making critical decisions based on the results, and in favor of undertaking data collection.

For additional discussion of input modeling in the absence of data, see Pegden, Shannon, and Sadowski [1995].

9.6 Multivariate and Time-Series Input Models

In Sections 9.1–9.4, the random variables presented were considered to be independent of any other variables within the context of the problem. However, variables may be related, and if the variables appear in a simulation model as inputs, the relationship should be determined and taken into consideration.

EXAMPLE 9.17

An inventory simulation includes the lead time and annual demand for industrial robots. An increase in demand results in an increase in lead time, since the final assembly of the robots must be made according to the specifications of the purchaser. Therefore, rather than treat lead time and demand as independent random variables, a multivariate input model should be developed. ◀

EXAMPLE 9.18

A simulation of the web-based trading site of a stock broker includes the time between arrivals of orders to buy and sell. Since investors tend to react to what other investors are doing, these buy and sell orders arrive in bursts. Therefore, rather than treat the time between arrivals as independent random variables, a time-series model should be developed. ◀

We distinguish between *multivariate input models* of a fixed, finite number of random variables (such as the two random variables lead time and annual demand in Example 9.17), and *time-series input models* of a (conceptually infinite) sequence of related random variables (such as the successive times between orders in Example 9.18). We describe input models appropriate for these examples after reviewing two measures of dependence, the covariance and correlation.

9.6.1 *Covariance and Correlation*

Let X_1 and X_2 be two random variables, and let $\mu_i = \mathrm{E}(X_i)$ and $\sigma_i^2 = \mathrm{Var}(X_i)$ be the mean and variance of X_i, respectively. The *covariance* and *correlation* are measures of the linear dependence between X_1 and X_2. In other words, the covariance and correlation indicate how well the relationship between X_1 and X_2 is described by the model

$$(X_1 - \mu_1) = \beta(X_2 - \mu_2) + \epsilon$$

where ϵ is a random variable with mean 0 that is independent of X_2. If, in fact, $(X_1 - \mu_1) = \beta(X_2 - \mu_2)$, then this model is perfect. On the other hand, if X_1 and X_2 are statistically independent, then $\beta = 0$ and the model is of no value. In general, a positive value of β indicates that X_1 and X_2 tend to be above or below their means together, while a negative value of β indicates that they tend to be on opposite sides of their means.

The covariance between X_1 and X_2 is defined to be

$$\mathrm{cov}(X_1, X_2) = \mathrm{E}[(X_1 - \mu_1)(X_2 - \mu_2)] = \mathrm{E}(X_1 X_2) - \mu_1\mu_2 \qquad (9.21)$$

A value of $\mathrm{cov}(X_1, X_2) = 0$ implies $\beta = 0$ in our model of dependence, while $\mathrm{cov}(X_1, X_2) < 0$ (> 0) implies $\beta < 0$ (> 0).

The covariance can take any value between $-\infty$ and ∞. The correlation standardizes the covariance to be between -1 and 1:

$$\rho = \mathrm{corr}(X_1, X_2) = \frac{\mathrm{cov}(X_1, X_2)}{\sigma_1\sigma_2} \qquad (9.22)$$

Again, a value of $\mathrm{corr}(X_1, X_2) = 0$ implies $\beta = 0$ in our model, while $\mathrm{corr}(X_1, X_2) < 0$ (> 0) implies $\beta < 0$ (> 0). The closer ρ is to -1 or 1, the stronger the linear relationship is between X_1 and X_2.

Now suppose that we have a sequence of random variables $X_1, X_2, X_3, \ldots$ that are identically distributed (implying that they all have the same mean and variance) but may be dependent. We refer to such a sequence as a *time series*, and to $\mathrm{cov}(X_t, X_{t+h})$ and $\mathrm{corr}(X_t, X_{t+h})$ as the *lag-h autocovariance* and *lag-h autocorrelation*, respectively. If the value of the autocovariance depends only on h and not on t, then we say that the time series is *covariance stationary*; this concept is discussed further in Chapter 11. For a covariance-stationary time series we use the shorthand notation

$$\rho_h = \mathrm{corr}(X_t, X_{t+h})$$

for the lag-h autocorrelation. Notice that autocorrelation measures the dependence between random variables that are separated by $h - 1$ others in the time series.

9.6.2 *Multivariate Input Models*

If X_1 and X_2 each are normally distributed, then dependence between them can be modeled by the bivariate normal distribution with parameters $\mu_1, \mu_2, \sigma_1^2, \sigma_2^2,$

and $\rho = \text{corr}(X_1, X_2)$. Estimation of μ_1, μ_2, σ_1^2, and σ_2^2 was described in Section 9.3.2. To estimate ρ, suppose that we have n independent and identically distributed pairs $(X_{11}, X_{21}), (X_{12}, X_{22}), \ldots, (X_{1n}, X_{2n})$. Then the sample covariance is

$$\widehat{\text{cov}}(X_1, X_2) = \frac{1}{n-1} \sum_{j=1}^{n} (X_{1j} - \bar{X}_1)(X_{2j} - \bar{X}_2)$$

$$= \frac{1}{n-1} \left(\sum_{j=1}^{n} X_{1j} X_{2j} - n\bar{X}_1\bar{X}_2 \right) \tag{9.23}$$

where $\bar{X}_1$ and $\bar{X}_2$ are the sample means. The correlation is estimated by

$$\widehat{\rho} = \frac{\widehat{\text{cov}}(X_1, X_2)}{\widehat{\sigma}_1 \widehat{\sigma}_2} \tag{9.24}$$

where $\widehat{\sigma}_1$ and $\widehat{\sigma}_2$ are the sample standard deviations.

EXAMPLE 9.19 (Example 9.17 Continued)

Let X_1 represent the average lead time to deliver (in months), and X_2 the annual demand, for industrial robots. The following data were available on demand and lead time for the last ten years:

Lead time	*Demand*
6.5	103
4.3	83
6.9	116
6.0	97
6.9	112
6.9	104
5.8	106
7.3	109
4.5	92
6.3	96

Standard calculations give $\bar{X}_1 = 6.14$, $\widehat{\sigma}_1 = 1.02$, $\bar{X}_2 = 101.80$, and $\widehat{\sigma}_2 = 9.93$ as estimates of μ_1, σ_1, μ_2, and σ_2, respectively. To estimate the correlation we need

$$\sum_{j=1}^{10} X_{1j} X_{2j} = 6328.5$$

Therefore, $\widehat{\text{cov}} = [6328.5 - (10)(6.14)(101.80)]/(10 - 1) = 8.66$ and

$$\widehat{\rho} = \frac{8.66}{(1.02)(9.93)} = 0.86$$

Clearly, lead time and demand are strongly dependent. Before accepting this model, however, lead time and demand should be checked individually to see if they are well represented by normal distributions. In particular, since demand is a discrete-valued quantity, the continuous normal distribution is certainly an approximation. ◀

The following algorithm can be used to generate bivariate normal random variables:

1. Generate Z_1 and Z_2, independent standard normal random variables (see Section 8.2).
2. Set $X_1 = \mu_1 + \sigma_1 Z_1$.
3. Set $X_2 = \mu_2 + \sigma_2 \left(\rho Z_1 + \sqrt{1-\rho^2}\, Z_2\right)$.

Obviously the bivariate normal distribution will not be appropriate for all multivariate-input modeling problems. It can be generalized to the k-variate normal distribution to model the dependence among more than two random variables, but in many instances a normal distribution is not appropriate in any form. Good references for other models are Johnson [1987] and Nelson and Yamnitsky [1998].

9.6.3 Time-Series Input Models

If $X_1, X_2, X_3, \ldots$ is a sequence of identically distributed, but dependent and covariance-stationary random variables, then there are a number of time series models that can be used to represent the process. We will describe two models that have the characteristic that the autocorrelations take the form

$$\rho_h = \text{corr}(X_t, X_{t+h}) = \rho^h$$

for $h = 1, 2, \ldots$. Notice that the lag-h autocorrelation decreases geometrically as the lag increases, so that observations far apart in time are nearly independent. For one model below each X_t is normally distributed; for the other model each X_t is exponentially distributed. More general time-series input models are described in Nelson and Yamnitsky [1998].

AR(1) MODEL

Consider the time-series model

$$X_t = \mu + \phi(X_{t-1} - \mu) + \varepsilon_t \tag{9.25}$$

for $t = 2, 3, \ldots$, where $\varepsilon_2, \varepsilon_3, \ldots$ are independent and identically normally distributed with mean 0 and variance σ_ε^2, and $-1 < \phi < 1$. If the initial value X_1 is chosen appropriately (see below), then $X_1, X_2, \ldots$ are all normally distributed with mean μ, variance $\sigma_\varepsilon^2/(1-\phi^2)$, and

$$\rho_h = \phi^h$$

for $h = 1, 2, \ldots$. This time-series model is called the autoregressive order–1 model, or AR(1) for short.

Estimation of the parameter ϕ can be obtained from the fact that

$$\phi = \rho^1 = \text{corr}(X_t, X_{t+1})$$

the lag–1 autocorrelation. Therefore to estimate ϕ, we first estimate the lag–1 autocovariance by

$$\begin{aligned} \widehat{\text{cov}}(X_t, X_{t+1}) &= \frac{1}{n-1}\sum_{t=1}^{n-1}(X_t - \bar{X})(X_{t+1} - \bar{X}) \\ &\doteq \frac{1}{n-1}\left(\sum_{t=1}^{n-1} X_t X_{t+1} - (n-1)\bar{X}^2\right) \end{aligned} \tag{9.26}$$

and the variance $\sigma^2 = \text{var}(X)$ by the usual estimator $\widehat{\sigma}^2$. Then

$$\widehat{\phi} = \frac{\widehat{\text{cov}}(X_t, X_{t+1})}{\widehat{\sigma}^2}$$

Finally, estimate μ and σ_ε^2 by $\widehat{\mu} = \bar{X}$ and

$$\widehat{\sigma}_\varepsilon^2 = \widehat{\sigma}^2(1 - \widehat{\phi}^2)$$

respectively.

The following algorithm generates a stationary AR(1) time series, given values of the parameters ϕ, μ and σ_ε^2:

1. Generate X_1 from the normal distribution with mean μ and variance $\sigma_\varepsilon^2/(1-\phi^2)$. Set $t = 2$.
2. Generate ε_t from the normal distribution with mean 0 and variance σ_ε^2.
3. Set $X_t = \mu + \phi(X_{t-1} - \mu) + \varepsilon_t$.
4. Set $t = t + 1$ and go to 2.

EAR(1) MODEL

Consider the time-series model

$$X_t = \begin{cases} \phi X_{t-1}, & \text{with probability } \phi \\ \phi X_{t-1} + \varepsilon_t, & \text{with probability } 1 - \phi \end{cases} \tag{9.27}$$

for $t = 2, 3, \ldots$, where $\varepsilon_2, \varepsilon_3, \ldots$ are independent and identically exponentially distributed with mean $1/\lambda$ and $0 \le \phi < 1$. If the initial value X_1 is chosen appropriately (see below), then $X_1, X_2, \ldots$ are all exponentially distributed with mean $1/\lambda$ and

$$\rho_h = \phi^h$$

for $h = 1, 2, \ldots$. This time-series model is called the exponential autoregressive order–1 model, or EAR(1) for short. Only autocorrelations greater than 0 can be represented by this model. Estimation of the parameters proceeds as for the AR(1) by setting $\widehat{\phi} = \widehat{\rho}$, the estimated lag–1 autocorrelation, and setting $\widehat{\lambda} = 1/\bar{X}$.

The following algorithm generates a stationary EAR(1) time series, given values of the parameters ϕ and λ:

1. Generate X_1 from the exponential distribution with mean $1/\lambda$. Set $t = 2$.
2. Generate U from the uniform distribution on $[0, 1]$. If $U \leq \phi$, then set

$$X_t = \phi X_{t-1}$$

Otherwise, generate ε_t from the exponential distribution with mean $1/\lambda$ and set

$$X_t = \phi X_{t-1} + \varepsilon_t$$

3. Set $t = t + 1$ and go to 2.

EXAMPLE 9.20 (Example 9.18 Continued)

The stock brokerage would typically have a large sample of data, but for the sake of illustration suppose that the following twenty time gaps between customer buy and sell orders had been recorded (in seconds): 1.95, 1.75, 1.58, 1.42, 1.28, 1.15, 1.04, 0.93, 0.84, 0.75, 0.68, 0.61, 11.98, 10.79, 9.71, 14.02, 12.62, 11.36, 10.22, 9.20. Standard calculations give $\bar{X} = 5.2$ and $\widehat{\sigma}^2 = 26.7$. To estimate the lag–1 autocorrelation we need

$$\sum_{j=1}^{19} X_t X_{t+1} = 924.1$$

Thus, $\widehat{\text{cov}} = [924.1 - (20 - 1)(5.2)^2]/(20 - 1) = 21.6$ and

$$\widehat{\rho} = \frac{21.6}{26.7} = 0.8$$

Therefore, we could model the interarrival times as an EAR(1) process with $\widehat{\lambda} = 1/5.2 = 0.192$ and $\widehat{\phi} = 0.8$, provided an exponential distribution is a good model for the individual gaps. ◀

9.7 Summary

Input data collection and analysis require major time and resource commitments in a discrete-event simulation project. However, regardless of the validity or sophistication of the simulation model, unreliable inputs may lead to outputs whose subsequent interpretation could result in faulty recommendations.

This chapter discussed four steps in the development of models of input data: collecting the raw data, identifying the underlying statistical distribution, estimating the parameters, and testing for goodness-of-fit.

Some suggestions were given for facilitating the data-collection step. However, experience, such as that obtained by completing any of Exercises 1 through 5, will increase awareness of the difficulty of problems that may arise in data collection and the need for planning.

Once the data have been collected, a statistical model should be hypothesized. Constructing a histogram is very useful at this point if sufficient data are available. Based on the underlying process and the shape of the histogram, a distribution can usually be selected for further investigation.

The investigation proceeds with the estimation of parameters for the hypothesized distribution. Suggested estimators were given for distributions often used in simulation. In a number of instances, these are functions of the sample mean and sample variance.

The last step in the process is the testing of the distributional hypothesis. The q-q plot is a useful graphical method for assessing fit. The Kolmogorov-Smirnov, chi-square, and Anderson-Darling goodness-of-fit tests can be applied to many distributional assumptions. When a distributional assumption is rejected, another distribution is tried. When all else fails, the empirical distribution may be used in the model.

Unfortunately, in some situations a simulation study must be undertaken when we do not have the time or resources to collect data on which to base input models. When this happens the analyst must use any available information —such as manufacturer specifications and expert opinion—to construct the input models. When input models are derived without the benefit of data, it is particularly important to examine the sensitivity of the results to the models chosen.

Many, but not all, input processes can be represented as sequences of independent and identically distributed random variables. When inputs should exhibit dependence, then multivariate input models are required. The bivariate normal distribution (and more generally the multivariate normal distribution) is often used to represent a finite number of dependent random variables. Time-series models are useful for representing a (conceptually infinite) sequence of dependent inputs.

For a recent reference on models for particularly complex input modeling problems, see Nelson and Yamnitsky [1998].

REFERENCES

CHAMBERS, J. M., W. S. CLEVELAND, B. KLEINER AND P. A. TUKEY [1983], *Graphical Methods for Data Analysis*, Duxbury Press, Boston.

CHOI, S. C., AND R. WETTE [1969], "Maximum Likelihood Estimation of the Parameters of the Gamma Distribution and Their Bias," *Technometrics*, Vol. 11, No. 4, pp. 683–890.

CONNOVER, W. J. [1980], *Practical Nonparametric Statistics*, 2d ed., John Wiley, New York.

DURBIN, J. [1975], "Kolmogorov-Smirnov Tests When Parameters Are Estimated with Applications to Tests of Exponentiality and Tests on Spacings," *Biometrika*, Vol. 65, pp. 5–22.

FISHMAN, G. S. [1973], *Concepts and Methods in Discrete Event Digital Simulation*, John Wiley, New York.

GUMBEL, E. J. [1943], "On the Reliability of the Classical Chi-squared Test," *Annals of Mathematical Statistics*, Vol. 14, pp. 253*ff.*

HINES, W. W., AND D. C. MONTGOMERY [1990], *Probability and Statistics in Engineering and Management Science*, 3d ed., John Wiley, New York.

JOHNSON, M. E. [1987], *Multivariate Statistical Simulation*, John Wiley, New York.

LAW, A. M., AND W. D. KELTON [2000], *Simulation Modeling & Analysis*, 3d ed., McGraw-Hill, New York.

LILLIEFORS, H. W. [1967], "On the Kolmogorov-Smirnov Test for Normality with Mean and Variance Unknown," *Journal of the American Statistical Association*, Vol. 62, pp. 339–402.

LILLIEFORS, H. W. [1969], "On the Kolmogorov-Smirnov Test for the Exponential Distribution with Mean Unknown," *Journal of the American Statistical Association*, Vol. 64, pp. 387–389.

MANN, H. B., AND A. WALD [1942], "On the Choice of the Number of Intervals in the Application of the Chi-squared Test," *Annals of Mathematical Statistics*, Vol. 18, p. 50*ff.*

NELSON, B. L., AND M. YAMNITSKY [1998], "Input Modeling Tools for Complex Problems," in *Proceedings of the 1998 Winter Simulation Conference*, eds. D. Medeiros, E. Watson, J. Carson and M. Manivannan, pp. 105–112, The Institute for Electrical and Electronics Engineers, Piscataway, NJ.

PEGDEN, C. D., R. E. SHANNON, AND R. P. SADOWSKI [1995], *Introduction to Simulation using SIMAN*, 2d ed. McGraw-Hill, New York.

STUART, A., J. K. ORD AND E. ARNOLD [1998], *Kendall's Advanced Theory of Statistics*, 6th ed., Vol. 2, Oxford University Press, Oxford.

EXERCISES

1. Go to a small appliance store and determine the interarrival and service-time distributions. If there are several workers, how do the service-time distributions compare to each other? Do service-time distributions need to be constructed for each type of appliance? (Make sure that the management gives permission to perform this study.)

2. Go to a cafeteria and collect data on the distributions of interarrival and service times. The distribution of interarrival times is probably different for each of the three daily meals and may also vary during the meal; that is, the interarrival time distribution for 11:00 A.M. to 12:00 noon may be different than from 12:00 noon to 1:00 P.M. Define service time as the time from when the customer reaches the point at which the first selection could be made until he or she exits from the cafeteria line. (Any reasonable modification of this definition is acceptable.) The service-time distribution probably changes for each meal. Can times of the day or days

of the week for either distribution be grouped due to homogeniety of the data? (Make sure that the management gives permission to perform this study.)

3. Go to a major traffic intersection and determine the interarrival-time distributions from each direction. Some arrivals want to go straight, some turn left, some turn right. The interarrival-time distribution varies during the day and by day of the week. Every now and then an accident occurs.
4. Go to a grocery store and determine the interarrival and service distributions at the checkout counters. These distributions may vary by time of day and by day of week. Record, also, the number of service channels available at all times. (Make sure that the management gives permission to perform this study.)
5. Go to a laundromat and "relive" the authors' data-collection experience as discussed in Example 9.1. (Make sure that the management gives permission to perform this study.)
6. Prepare four theoretical normal density functions, all on the same figure, each distribution having mean zero, but let the standard deviations be 1/4, 1/2, 1, and 2.
7. On one figure, draw the pdfs of the Erlang distribution where $\theta = 1/2$ and $k = 1, 2, 4,$ and 8.
8. On one figure, draw the pdfs of the Erlang distribution where $\theta = 2$ and $k = 1, 2, 4,$ and 8.
9. Draw the pmf of the Poisson distribution that results when the parameter α is equal to the following:
   (a) $\alpha = 1/2$
   (b) $\alpha = 1$
   (c) $\alpha = 2$
   (d) $\alpha = 4$
10. On one figure draw the two exponential pdf's that result when the parameter, λ, equals 0.6 and 1.2.
11. On one figure draw the three Weibull pdf's which result when $\nu = 0, \alpha = 1/2$, and $\beta = 1, 2,$ and 4
12. The following data are randomly generated from a gamma distribution:

1.691	1.437	8.221	5.976
1.116	4.435	2.345	1.782
3.810	4.589	5.313	10.90
2.649	2.432	1.581	2.432
1.843	2.466	2.833	2.361

    Determine the maximum-likelihood estimators $\widehat{\beta}$ and $\widehat{\theta}$.

13. The following data are randomly generated from a Weibull distribution where $\nu = 0$:

7.936	5.224	3.937	6.513
4.599	7.563	7.172	5.132
5.259	2.759	4.278	2.696
6.212	2.407	1.857	5.002
4.612	2.003	6.908	3.326

Determine the maximum-likelihood estimators $\widehat{\alpha}$ and $\widehat{\beta}$. (This exercise requires a programmable calculator, a computer, or a lot of patience.)

14. The highway between Atlanta, Georgia, and Athens, Georgia, has a high incidence of accidents along its 100 kilometers. Public safety officers say that the occurrence of accidents along the highway is randomly (uniformly) distributed, but the news media say otherwise. The Georgia Department of Public Safety published records for the month of September. These records indicated the point at which 30 accidents involving an injury or death occurred, as follows (the data points represent the distance from the city limits of Atlanta):

88.3	40.7	36.3	27.3	36.8
91.7	67.3	7.0	45.2	23.3
98.8	90.1	17.2	23.7	97.4
32.4	87.8	69.8	62.6	99.7
20.6	73.1	21.6	6.0	45.3
76.6	73.2	27.3	87.6	87.2

Use the Kolmogorov-Smirnov test to determine whether the distribution of location of accidents is uniformly distributed for the month of September.

15. Show that the Kolmogorov-Smirnov test statistic for Example 9.15 is $D = 0.1054$.

16. Records pertaining to the monthly number of job-related injuries at an underground coal mine were being studied by a federal agency. The values for the past 100 months were as follows:

Injuries per Month	*Frequency of Occurrence*
0	35
1	40
2	13
3	6
4	4
5	1
6	1

(a) Apply the chi-square test to these data to test the hypothesis that the underlying distribution is Poisson. Use a level of significance of $\alpha = 0.05$.

(b) Apply the chi-square test to these data to test the hypothesis that the distribution is Poisson with mean 1.0. Again let $\alpha = 0.05$.

(c) What are the differences in parts (a) and (b), and when might each case arise?

17. The time required for 50 different employees to compute and record the number of hours worked during the week was measured with the following results in minutes:

Employee	*Time (Minutes)*	Employee	*Time (Minutes)*
1	1.88	26	0.04
2	0.54	27	1.49
3	1.90	28	0.66
4	0.15	29	2.03
5	0.02	30	1.00
6	2.81	31	0.39
7	1.50	32	0.34
8	0.53	33	0.01
9	2.62	34	0.10
10	2.67	35	1.10
11	3.53	36	0.24
12	0.53	37	0.26
13	1.80	38	0.45
14	0.79	39	0.17
15	0.21	40	4.29
16	0.80	41	0.80
17	0.26	42	5.50
18	0.63	43	4.91
19	0.36	44	0.35
20	2.03	45	0.36
21	1.42	46	0.90
22	1.28	47	1.03
23	0.82	48	1.73
24	2.16	49	0.38
25	0.05	50	0.48

Use the chi-square test (as in Example 9.14) to test the hypothesis that these service times are exponentially distributed. Let the number of class intervals be $k = 6$. Use a level of significance of $\alpha = 0.05$.

18. Studentwiser Beer Company is trying to determine the distribution of the breaking strength of their glass bottles. Fifty bottles are selected at random and tested for breaking strength, with the following results (in pounds per square inch):

218.95	232.75	212.80	231.10	215.95
237.55	235.45	228.25	218.65	212.80
230.35	228.55	216.10	229.75	229.00
199.75	225.10	208.15	213.85	205.45
219.40	208.15	198.40	238.60	219.55
243.10	198.85	224.95	212.20	222.90
218.80	203.35	223.45	213.40	206.05
229.30	239.20	201.25	216.85	207.25
204.85	219.85	226.15	230.35	211.45
227.95	229.30	225.25	201.25	216.10

Using input modeling sofware, apply as many tests for normality as are available in the software. If the chi-square test is available, apply it with at least two different choices for the number of intervals. Do all of the tests reach the same conclusion?

19. The Crosstowner is a bus that cuts a diagonal path from northeast Atlanta to southwest Atlanta. The time required to complete the route is maintained by the bus operator. The bus runs Monday through Friday. The times of the last fifty 8:00 A.M. runs, in minutes, are as follows:

92.3	92.8	106.8	108.9	106.6
115.2	94.8	106.4	110.0	90.9
104.6	72.0	86.0	102.4	99.8
87.5	111.4	105.9	90.7	99.2
97.8	88.3	97.5	97.4	93.7
99.7	122.7	100.2	106.5	105.5
80.7	107.9	103.2	116.4	101.7
84.8	101.9	99.1	102.2	102.5
111.7	101.5	95.1	92.8	88.5
74.4	98.9	111.9	96.5	95.9

How are these run times distributed? Develop and test a suitable model.

20. The time required for the transmission of a message (in minutes) is sampled electronically at a communications center. The last 50 values in the sample are as follows:

7.936	4.612	2.407	4.278	5.132
4.599	5.224	2.003	1.857	2.696
5.259	7.563	3.937	6.908	5.002
6.212	2.759	7.172	6.513	3.326
8.761	4.502	6.188	2.566	5.515
3.785	3.742	4.682	4.346	5.359
3.535	5.061	4.629	5.298	6.492
3.502	4.266	3.129	1.298	3.454
5.289	6.805	3.827	3.912	2.969
4.646	5.963	3.829	4.404	4.924

How are the transmission times distributed? Develop and test an appropriate model.

21. The time (in minutes) between requests for the hookup of electric service was accurately maintained at the Gotwatts Flash and Flicker Company with the following results for the last 50 requests:

0.661	4.910	8.989	12.801	20.249
5.124	15.033	58.091	1.543	3.624
13.509	5.745	0.651	0.965	62.146
15.512	2.758	17.602	6.675	11.209
2.731	6.892	16.713	5.692	6.636
2.420	2.984	10.613	3.827	10.244
6.255	27.969	12.107	4.636	7.093
6.892	13.243	12.711	3.411	7.897
12.413	2.169	0.921	1.900	0.315
4.370	0.377	9.063	1.875	0.790

How are the times between requests for service distributed? Develop and test a suitable model.

22. Daily demands for transmission overhaul kits for the D–3 dragline were maintained by Earth Moving Tractor Company with the following results:

0	2	0	0	0
1	0	1	1	1
0	1	0	0	0
2	0	1	0	1
0	1	0	0	2
1	0	1	0	0
0	0	0	0	0
1	0	1	0	1
0	0	3	0	1
1	0	0	0	0

How are the daily demands distributed? Develop and test an appropriate model.

23. A simulation is to be conducted of a job shop that performs two operations, milling and planing, in that order. It would be possible to collect data about processing times for each operation, then generate random occurrences from each distribution. However, the shop manager says that the times might be related; large milling jobs take lots of planing. Data are collected for the next 25 orders with the following results in minutes:

Order	*Milling Time (Minutes)*	*Planing Time (Minutes)*	*Order*	*Milling Time (Minutes)*	*Planing Time (Minutes)*
1	12.3	10.6	14	24.6	16.6
2	20.4	13.9	15	28.5	21.2
3	18.9	14.1	16	11.3	9.9
4	16.5	10.1	17	13.3	10.7
5	8.3	8.4	18	21.0	14.0
6	6.5	8.1	19	19.5	13.0
7	25.2	16.9	20	15.0	11.5
8	17.7	13.7	21	12.6	9.9
9	10.6	10.2	22	14.3	13.2
10	13.7	12.1	23	17.0	12.5
11	26.2	16.0	24	21.2	14.2
12	30.4	18.9	25	28.4	19.1
13	9.9	7.7			

(a) Plot milling time on the horizontal axis and planing time on the vertical axis. Do these data seem dependent?

(b) Compute the sample correlation between milling time and planing time.

(c) Fit a bivariate normal distribution to this data.

24. Write a computer program to compute the maximum-likelihood estimates $(\widehat{\alpha}, \widehat{\beta})$ of the Weibull distribution. Inputs to the program should include the sample size, n; the observations, $x_1, x_2, \ldots, x_n$; a stopping criterion, ϵ (stop when $|f(\widehat{\beta}_j)| \leq \epsilon$); and a print option, OPT (usually set = 0). Output would be the estimates $\widehat{\alpha}$ and $\widehat{\beta}$. If OPT = 1, additional output would be printed as in Table 9.4 showing convergence. Make the program as "user friendly" as possible.

25. Examine a computer software library or simulation support environment to which you have access. Obtain documentation on data-analysis software that would be useful in solving Exercises 7 through 24. Use the software as an aid in solving selected problems.

26. The numbers of patrons staying at a small hotel on 20 successive nights were observed to be 20, 14, 21, 19, 14, 18, 21, 25, 27, 26, 22, 18, 13, 18, 18, 18, 25, 23, 20, 21. Fit both an AR(1) and an EAR(1) model to this data. Decide which model provides a better fit by looking at a histogram of the data.

27. The following data represent the time to perform transactions in a bank, measured in minutes: 0.740, 1.28, 1.46, 2.36, 0.354, 0.750, 0.912, 4.44, 0.114, 3.08, 3.24, 1.10, 1.59, 1.47, 1.17, 1.27, 9.12, 11.5, 2.42, 1.77. Develop an input model for this data.

28. Two types of jobs (A and B) are released to the input buffer of a job shop as orders arrive, and the arrival of orders is uncertain. The following data are available from the last week of production:

Day	*Number of Jobs*	*Number of A's*
1	83	53
2	93	62
3	112	66
4	65	41
5	78	55

Develop an input model for the number of new arrivals of each type each day.

29. The following data are available on the processing time at a machine (in minutes): 0.64, 0.59, 1.1, 3.3, 0.54, 0.04, 0.45, 0.25, 4.4, 2.7, 2.4, 1.1, 3.6, 0.61, 0.20, 1.0, 0.27, 1.7, 0.04, 0.34. Develop an input model for the processing time.

10

Verification and Validation of Simulation Models

One of the most important and difficult tasks facing a model developer is the verification and validation of the simulation model. The engineers and analysts who use the model outputs to aid in making design recommendations and the managers who make decisions based on these recommendations justifiably look upon a model with some degree of skepticism about its validity. It is the job of the model developer to work closely with the end users throughout the period of development and validation to reduce this skepticism and to increase the model's credibility. The goal of the validation process is twofold: (1) to produce a model that represents true system behavior closely enough for the model to be used as a substitute for the actual system for the purpose of experimenting with the system; and (2) to increase to an acceptable level the credibility of the model, so that the model will be used by managers and other decision makers.

Validation should not be seen as an isolated set of procedures that follows model development, but rather as an integral part of model development. Conceptually, however, the verification and validation process consists of the following components:

1. Verification is concerned with building the model right. It is utilized in the comparison of the conceptual model to the computer representation that implements that conception. It asks the questions: Is the model implemented correctly in the computer? Are the input parameters and logical structure of the model correctly represented?
2. Validation is concerned with building the right model. It is utilized to determine that a model is an accurate representation of the real system. Validation is usually achieved through the calibration of the model, an

iterative process of comparing the model to actual system behavior and using the discrepancies between the two, and the insights gained, to improve the model. This process is repeated until model accuracy is judged to be acceptable.

This chapter describes methods that have been recommended and used in the verification and validation process. Most of the methods are informal subjective comparisons, while a few are formal statistical procedures. The use of the latter procedures involves issues related to output analysis, the subject of Chapters 11 and 12. Output analysis refers to analyzing the data produced by a simulation and drawing inferences from these data about the behavior of the real system. To summarize their relationship, validation is the process by which model users gain confidence that output analysis is making valid inferences about the real system under study.

Many articles and chapters in texts have been written on verification and validation. For discussion of the main issues, the reader is referred to Balci [1994, 1998], Carson [1986], Gass [1983], Kleijnen [1995], Law and Kelton [2000], Naylor and Finger [1967], Oren [1981], Sargent [1994], Shannon [1975], and van Horn [1969, 1971]. For statistical techniques relevant to different aspects of validation, the reader can obtain the foregoing references plus those by Balci and Sargent [1982a, b; 1984a], Kleijnen [1987], and Schruben [1980]. For case studies in which validation is emphasized, the reader is referred to Carson et al. [1981a,b], Gafarian and Walsh [1970], Kleijnen [1993], and Shechter and Lucas [1980]. Bibliographies on validation have been published by Balci and Sargent [1984b] and Youngblood [1993].

10.1 Model Building, Verification, and Validation

The first step in model building consists of observing the real system and the interactions among its various components and collecting data on its behavior. But observation alone seldom yields sufficient understanding of system behavior. Persons familiar with the system, or any subsystem, should be questioned to take advantage of their special knowledge. Operators, technicians, repair and maintenance personnel, engineers, supervisors, and managers understand certain aspects of the system which may be unfamiliar to others. As model development proceeds, new questions may arise, and the model developers will return to this step of learning true system structure and behavior.

The second step in model building is the construction of a conceptual model—a collection of assumptions on the components and the structure of the system, plus hypotheses on the values of model input parameters. As illustrated by Figure 10.1, conceptual validation is the comparison of the real system to the conceptual model.

The third step is the translation of the operational model into a computer-recognizable form — the computerized model. In actuality, model building is not a linear process with three steps. Instead, the model builder will return

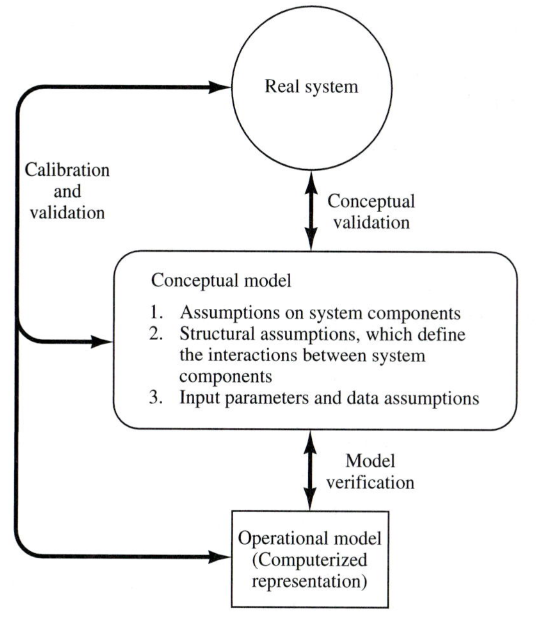

Figure 10.1. Model building, verification, and validation.

to each of these steps many times while building, verifying, and validating the model. Figure 10.1 depicts the ongoing model-building process in which the need for verification and validation causes continual comparison of the real system to the conceptual model and to the operational model, and repeated modification of the model to improve its accuracy.

10.2 Verification of Simulation Models

The purpose of model verification is to assure that the conceptual model is reflected accurately in the computerized representation. The conceptual model quite often involves some degree of abstraction about system operations, or some amount of simplification of actual operations. Verification asks the question: Is the conceptual model (assumptions on system components and system structure, parameter values, abstractions and simplifications) accurately represented by the operational model (i.e., by the computerized representation)?

Many common-sense suggestions can be given for use in the verification process.

1. Have the computerized representation checked by someone other than its developer.
2. Make a flow diagram which includes each logically possible action a system can take when an event occurs, and follow the model logic for each action for each event type. (An example of a logic flow diagram is given in Figures 2.2 and 2.3 for the model of a single-server queue.)

3. Closely examine the model output for reasonableness under a variety of settings of the input parameters. Have the computerized representation print out a wide variety of output statistics.
4. Have the computerized representation print the input parameters at the end of the simulation, to be sure that these parameter values have not been changed inadvertently.
5. Make the computerized representation as self-documenting as possible. Give a precise definition of every variable used and a general description of the purpose of each major section of code.
6. If the computerized representation is animated, verify that what is seen in the animation imitates the actual system. Examples of errors that can be observed through animation are automated guided vehicles (AGVs) that pass through one another at an intersection and entities that disappear (unintentionally) during a simulation.
7. The interactive run controller (IRC) or debugger is an essential component of successful simulation model building. Even the best of simulation analysts makes mistakes or commits logical errors when building a model. The IRC assists in finding and correcting those errors in the following ways:
   - **(a)** The simulation can be monitored as it progresses. This can be accomplished by advancing the simulation until a desired time has elapsed, then displaying model information at that time. Another possibility is to advance the simulation until a particular condition is in effect and then display information.
   - **(b)** Attention can be focused on a particular line of logic or multiple lines of logic that constitute a procedure or a particular entity. For instance, every time that an entity enters a specified block, the simulation will pause so that information can be gathered. As another example, every time that a specified entity becomes active, the simulation will pause.
   - **(c)** Values of selected model components can be observed. When the simulation has paused, the current value or status of variables, attributes, queues, resources, counters, etc., can be observed.
   - **(d)** The simulation can be temporarily suspended, or paused, not only to view information but also to reassign values or redirect entities.
8. Graphical interfaces are recommended for accomplishing verification and validation (Bortscheller and Saulnier [1992]). The graphical representation of the model is essentially a form of self-documentation. It simplifies the task of model understanding.

These suggestions are basically the same ones any software engineer would follow.

Among these common-sense suggestions, one that is most easily implemented, but quite often overlooked, especially by students who are learning simulation, is a close and thorough examination of model output for reasonableness (suggestion 3). For example, consider a model of a complex network of queues consisting of many service centers in series and parallel configurations. Suppose that the modeler is interested mainly in the response time, defined as the time required for a customer to pass through a designated part of the network. During the verification (and calibration) phase of model development, it is recommended that the program collect and print out many statistics in addition to response times, such as utilizations of servers and time-average number of customers in various subsystems. Examination of the utilization of a server, for example, may reveal that it is unreasonably low (or high), a possible error that may be caused by wrong specification of mean service time, or a mistake in model logic that sends too few (or too many) customers to this particular server, or any number of other possible parameter misspecifications or errors in logic.

In a simulation language which automatically collects many standard statistics (average queue lengths, average waiting times, etc.), it takes little or no extra programming effort to display almost all statistics of interest. The effort required can be considerably greater in a general-purpose language such as FORTRAN, C, or C++, which does not have statistics-gathering capabilities to aid the programmer.

Two sets of statistics that can give a quick indication of model reasonableness are *current contents* and *total count*. These statistics apply to any system having items of some kind flowing through it, whether these items are called customers, transactions, inventory, or vehicles. Current contents refers to the number of items in each component of the system at a given time. Total count refers to the total number of items that have entered each component of the system by a given time. In some simulation software, such as GPSS/H and AutoMod, these statistics are automatically kept and can be displayed at any point in simulation time. In other simulation software, simple counters may have to be added to the computerized model and displayed at appropriate times. If the current contents in some portion of the system is high, this indicates that a large number of entities are delayed. If the output is displayed for successively longer simulation run times and the current contents tends to grow in a more or less linear fashion, it is highly likely that a queue is unstable and the server(s) will fall further behind as time continues. This indicates that possibly the number of servers is too small or a service time is misspecified. (Unstable queues were discussed in Chapter 6.) On the other hand, if the total count for some subsystem is zero, this indicates that no items entered that subsystem — again a highly suspect occurrence. Another possibility is that the current count and total count are equal to one. This may indicate that an entity has captured a resource but never freed it. Careful evaluation of these statistics for various run lengths can aid in the detection of mistakes in model logic and data misspecifications. Checking for output reasonableness will usually fail to detect the more subtle errors, but it is one of the quickest ways to discover gross errors. To aid in error detection, it is best if the model developer forecasts a reasonable range for the value of selected output statistics before making a run of the model. Such a forecast reduces the possibility of rationalizing a discrepancy and failing to investigate the cause of unusual output.

For certain models, it is possible to consider more than whether a particular statistic is reasonable. It is possible to compute certain long-run measures of performance. For example, as seen in Chapter 6, the analyst can compute the long-run server utilization for a large number of queueing systems without any special assumptions regarding interarrival or service-time distributions. Typically, the only information needed is the network configuration plus arrival and service rates. Any measure of performance that can be computed analytically and then compared to its simulated counterpart provides another valuable tool for verification. Presumably, the objective of the simulation is to estimate some measure of performance, such as mean response time, which cannot be computed analytically. But as illustrated by the formulas in Chapter 6 for a number of special queues ($M/M/1$, $M/G/1$, etc.), all the measures of performance in a queueing system are interrelated. Thus, if a simulation model is predicting one measure (such as utilization) correctly, then confidence in the model's predictive ability for other related measures (such as response time) is increased (even though the exact relation between the two measures is, of course, unknown in general and varies from model to model). Conversely, if a model incorrectly predicts utilization, its prediction of other quantities, such as mean response time, is highly suspect.

Another important way to aid the verification process is the oft-neglected documentation phase. If a model builder writes brief comments in the computerized model, plus definitions of all variables and parameters, and descriptions of each major section of the computerized model, it becomes much simpler for someone else, or the model builder at a later date, to verify the model logic. Documentation is also important as a means of clarifying the logic of a model and verifying its completeness.

A more sophisticated technique is the use of a trace. In general, a trace is a detailed computer printout which gives the value of every variable (in a specified set of variables) in a computer program, every time that one of these variables changes in value. A trace designed specifically for use in a simulation program would give the value of selected variables each time the simulation clock was incremented (i.e., each time an event occurred). Thus, a simulation trace is nothing more than a detailed printout of the state of the simulation model as it changes over time.

Example 10.1

When verifying the computer implementation (in a general-purpose language such as FORTRAN, Pascal, C, or C++ or in most simulation languages) of the single-server queue model of Example 2.1, an analyst made a run over 16 units of time and observed that the time-average length of the waiting line was

Definition of Variables:

CLOCK = Simulation clock

EVTYP = Event type (start, arrival, departure, or stop)

NCUST = Number of customers in system at time 'CLOCK'

STATUS = Status of server (1–busy, 0–idle)

State of System Just after the Named Event Occurs:

CLOCK = 0	EVTYP = 'Start'	NCUST = 0	STATUS = 0
CLOCK = 3	EVTYP = 'Arrival'	NCUST = 1	STATUS = 0
CLOCK = 5	EVTYP = 'Depart'	NCUST = 0	STATUS = 0
CLOCK = 11	EVTYP = 'Arrival'	NCUST = 1	STATUS = 0
CLOCK = 12	EVTYP = 'Arrival'	NCUST = 2	STATUS = 1
CLOCK = 16	EVTYP = 'Depart'	NCUST = 1	STATUS = 1

.
.
.

Figure 10.2 Simulation of trace of Example 2.1.

$\hat{L}_Q = 0.4375$ customer, which is certainly reasonable for a short run of only 16 time units. Nevertheless, the analyst decided that a more detailed verification would be of value.

The trace in Figure 10.2 gives the hypothetical printout from simulation time CLOCK = 0 to CLOCK = 16 for the simple single-server queue of Example 2.1. This example illustrates how an error can be found with a trace, when no error was apparent from the examination of the summary output statistics (such as $\hat{L}_Q$). Note that at simulation time CLOCK = 3, the number of customers in the system is NCUST = 1, but the server is idle (STATUS = 0). The source of this error could be incorrect logic, or simply not setting the attribute STATUS to a value of 1 (when coding in a general-purpose language or in most simulation languages).

In any case the error must be found and corrected. Note that the less sophisticated practice of examining the summary measures, or output, did not detect the error. By using Equation (6.1), the reader can verify that $\hat{L}_Q$ was computed correctly from the data ($\hat{L}_Q$ is the time-average value of NCUST minus STATUS):

$$\hat{L}_Q = \frac{(0-0)3+(1-0)2+(0-0)6+(1-0)1+(2-1)4}{3+2+6+1+4}$$

$$= \frac{7}{16} = 0.4375$$

as previously mentioned. Thus, the output measure, $\hat{L}_Q$, had a reasonable value and was computed correctly from the data, but its value was indeed wrong because the attribute STATUS was not assuming correct values. As seen from Figure 10.2, a trace yields information on the actual history of the model which is more detailed and informative than the summary measures alone.

Most simulation software has a built-in capability to conduct a trace without the programmer having to do any extensive programming. In addition, a 'print' or 'write' statement can be used to implement a tracing capability in a general-purpose language.

As can be easily imagined, a trace over a large span of simulation time can quickly produce an extremely large amount of computer printout, which would be extremely cumbersome to check in detail for correctness. The purpose of the trace is to verify the correctness of the computer program by making detailed paper-and-pencil calculations. To make this practical, a simulation with a trace is usually restricted to a very short period of time. It is desirable, of course, to ensure that each type of event (such as ARRIVAL) occurs at least once, so that its consequences and effect on the model can be checked for accuracy. If an event is especially rare in occurrence, it may be necessary to use artificial data to force it to occur during a simulation of short duration. This is legitimate, as the purpose is to verify that the effect on the system of the rare event is as anticipated.

Some software allows a selective trace. For example, a trace could be set for specific locations in the computerized model. Any time an entity goes through that location or those locations, a message is written. Another example of a selective trace is to set it on a particular entity. Any time that entity becomes active, the trace is on and a message is written. This trace is very useful in following one entity through the entire computerized model. Another example of a selective trace is to set it for the existence of a particular condition. For example, whenever the queue before a certain resource reaches five or more, turn on the trace. This allows running the simulation until something unusual occurs, then examining the behavior from that point forward in time.

Of the three classes of techniques — the common-sense techniques, thorough documentation, and traces — it is recommended that the first two always be carried out. Close examination of model output for reasonableness is especially valuable and informative. A trace can also provide information if it is selective. The generalized trace can be extremely time consuming. ◄

10.3 Calibration and Validation of Models

Verification and validation, although conceptually distinct, usually are conducted simultaneously by the modeler. Validation is the overall process of comparing the model and its behavior to the real system and its behavior. Calibration is the iterative process of comparing the model to the real system, making adjustments (or even major changes) to the model, comparing the revised model to reality, making additional adjustments, comparing again, and so on. Figure 10.3 shows the relationship of model calibration to the overall validation process. The comparison of the model to reality is carried out by a variety of tests—some subjective and others objective. Subjective tests usually

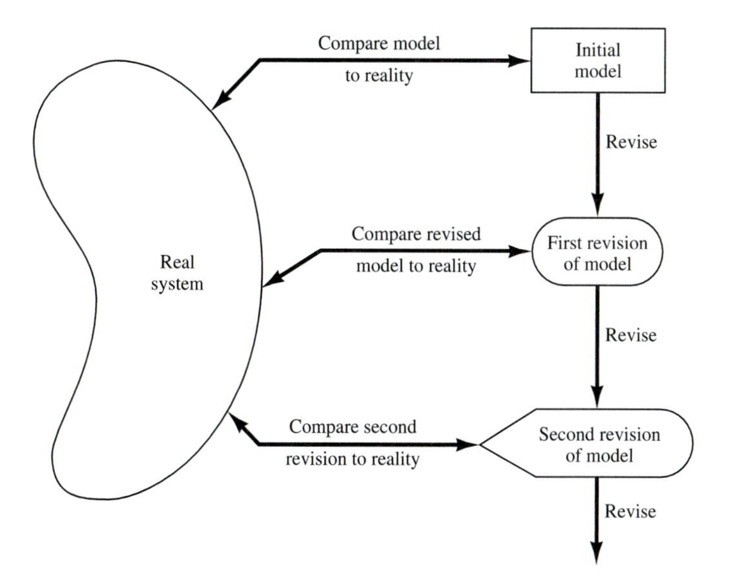

Figure 10.3. Iterative process of calibrating a model.

involve people, who are knowledgeable about one or more aspects of the system, making judgments about the model and its output. Objective tests always require data on the system's behavior plus the corresponding data produced by the model. Then one or more statistical tests are performed to compare some aspect of the system data set to the same aspect of the model data set. This iterative process of comparing model and system, and revising both the conceptual and operational models to accommodate any perceived model deficiencies, is continued until the model is judged to be sufficiently accurate.

A possible criticism of the calibration phase, were it to stop at this point, is that the model has been validated only for the one data set used; that is, the model has been "fit" to one data set. One way to alleviate this criticism is to collect a new set of system data (or to reserve a portion of the original system data) to be used at this final stage of validation. That is, after the model has been calibrated using the original system data set, a "final" validation is conducted using the second system data set. If unacceptable discrepancies between the model and the real system are discovered in the "final" validation effort, the modeler must return to the calibration phase and modify the model until it becomes acceptable.

Validation is not an either/or proposition—no model is ever totally representative of the system under study. In addition, each revision of the model, as pictured in Figure 10.3, involves some cost, time, and effort. The modeler must weigh the possible, but not guaranteed, increase in model accuracy versus the cost of increased validation effort. Usually, the modeler (and model users) have some maximum discrepancy between model predictions and system behavior that would be acceptable. If this level of accuracy cannot be obtained within the budget constraints, either expectations of model accuracy must be lowered, or the model must be abandoned.

Yücesan and Jacobson [1992] indicate that verifying simulation models is so difficult as to be intractable. They offer theorems to confirm this intractibility.

As an aid in the validation process, Naylor and Finger [1967] formulated a three-step approach which has been widely followed:

1. Build a model that has high face validity.
2. Validate model assumptions.
3. Compare the model input-output transformations to corresponding input-output transformations for the real system.

The next five subsections investigate these three steps in detail.

10.3.1 Face Validity

The first goal of the simulation modeler is to construct a model that appears reasonable on its face to model users and others who are knowledgeable about the real system being simulated. The potential users of a model should be involved in model construction from its conceptualization to its implementation to ensure that a high degree of realism is built into the model through reasonable assumptions regarding system structure, and reliable data. Potential users and knowledgeable persons can also evaluate model output for reasonableness and can aid in identifying model deficiencies. Thus, the users can be involved in the calibration process as the model is iteratively improved, based on the insights gained from the initial model deficiencies. Another advantage of user involvement is the increase in the model's perceived validity, or credibility, without which a manager would not be willing to trust simulation results as a basis for decision making.

Sensitivity analysis can also be used to check a model's face validity. The model user is asked if the model behaves in the expected way when one or more input variables is changed. For example, in most queueing systems, if the arrival rate of customers (or demands for service) were to increase, it would be expected that utilizations of servers, lengths of lines, and delays would tend to increase (although by how much might well be unknown). Based on experience and observations on the real system (or similar related systems), the model user and model builder would probably have some notion at least of the direction of change in model output when an input variable is increased or decreased. For most large-scale simulation models, there are many input variables and thus many possible sensitivity tests. The model builder must attempt to choose the most critical input variables for testing if it is too expensive or time consuming to vary all input variables. If real system data are available for at least two settings of the input parameters, objective scientific sensitivity tests can be conducted using appropriate statistical techniques.

10.3.2 Validation of Model Assumptions

Model assumptions fall into two general classes: structural assumptions and data assumptions. Structural assumptions involve questions of how the system operates and usually involve simplifications and abstractions of reality. For example, consider the customer queueing and service facility in a bank. Customers may form one line, or there may be an individual line for each teller. If there are many lines, customers may be served strictly on a first-come, first-served basis, or some customers may change lines if one is moving faster. The number of tellers may be fixed or variable. These structural assumptions should be verified by actual observation during appropriate time periods together with discussions with managers and tellers regarding bank policies and actual implementation of these policies.

Data assumptions should be based on the collection of reliable data and correct statistical analysis of the data. (Example 9.1 discussed similar issues for a model of a laundromat.) For example, in the bank study previously mentioned, data were collected on:

1. Interarrival times of customers during several 2-hour periods of peak loading ("rush-hour" traffic)
2. Interarrival times during a slack period
3. Service times for commercial accounts
4. Service times for personal accounts

The reliability of the data was verified by consultation with bank managers, who identified typical rush hours and typical slack times. When combining two or more data sets collected at different times, data reliability can be further enhanced by objective statistical tests for homogeneity of data. (Do two data sets $\{X_i\}$ and $\{Y_i\}$ on service times for personal accounts, collected at two different times, come from the same parent population? If so, the two sets can be combined.) Additional tests may be required to test for correlation in the data. As soon as the analyst is assured of dealing with a random sample (i.e., correlation is not present), the statistical analysis can begin.

The procedures for analyzing input data from a random sample were discussed in detail in Chapter 9. Whether by hand, or using computer software for the purpose, the analysis consists of three steps:

1. Identifying the appropriate probability distribution
2. Estimating the parameters of the hypothesized distribution
3. Validating the assumed statistical model by a goodness-of-fit test, such as the chi-square or Kolmogorov-Smirnov test, and by graphical methods

The use of goodness-of-fit tests is an important part of the validation of the model assumptions.

10.3.3 Validating Input-Output Transformations

The ultimate test of a model, and in fact the only objective test of the model as a whole, is its ability to predict the future behavior of the real system when the model input data match the real inputs and when a policy implemented in the model is implemented at some point in the system. Furthermore, if the level of some input variables (e.g., the arrival rate of customers to a service facility) were to increase or decrease, the model should accurately predict what would happen in the real system under similar circumstances. In other words, the structure of the model should be accurate enough for the model to make good predictions, not just for one input data set, but for the range of input data sets which are of interest.

In this phase of the validation process, the model is viewed as an input–output transformation. That is, the model accepts values of the input parameters and transforms these inputs into output measures of performance. It is this correspondence that is being validated.

Instead of validating the model input-output transformations by predicting the future, the modeler may use past historical data which have been reserved for validation purposes only; that is, if one data set has been used to develop and calibrate the model, it is recommended that a separate data set be used as the final validation test. Thus, accurate "prediction of the past" may replace prediction of the future for the purpose of validating the model.

A model is usually developed with primary interest in a specific set of system responses to be measured under some range of input conditions. For example, in a queueing system, the responses may be server utilization and customer delay, and the range of input conditions (or input variables) may include two or three servers at some station and a choice of scheduling rules. In a production system, the response may be throughput (i.e., production per hour) and the input conditions may be a choice of several machines that run at different speeds, with each machine having its own breakdown and maintenance characteristics. In any case, the modeler should use the main responses of interest as the criteria for validating a model. If the model is used later for a purpose different from its original purpose, the model should be revalidated in terms of the new responses of interest and under the possibly new input conditions.

A necessary condition for the validation of input-output transformations is that some version of the system under study exists, so that system data under at least one set of input conditions can be collected to compare to model predictions. If the system is in the planning stages and no system operating data can be collected, complete input-output validation is not possible. Other types of validation should be conducted, to the extent possible. In some cases, subsystems of the planned system may exist and a partial input-output validation can be conducted.

Presumably, the model will be used to compare alternative system designs, or to investigate system behavior under a range of new input conditions. Assume for now that some version of the system is operating, and that the model of the existing system has been validated. What, then, can be said about the validity of the model of a nonexistent proposed system, or the model of the existing system under new input conditions?

First, the responses of the two models under similar input conditions will be used as the criteria for comparison of the existing system to the proposed system. Validation increases the modeler's confidence that the model of the existing system is accurate. Second, in many cases, the proposed system is a modification of the existing system, and the modeler hopes that confidence in the model of the existing system can be transferred to the model of the new system. This transfer of confidence usually can be justified if the new model is a relatively minor modification of the old model in terms of changes to the computerized representation of the system (it may be a major change for the actual system). Changes in the computerized representation of the system, ranging from relatively minor to relatively major, include:

1. Minor changes of single numerical parameters, such as the speed of a machine, the arrival rate of customers (with no change in distributional form of interarrival times), or the number of servers in a parallel service center
2. Minor changes of the form of a statistical distribution, such as the distribution of a service time or a time to failure of a machine
3. Major changes in the logical structure of a subsystem, such as a change in queue discipline for a waiting-line model, or a change in the scheduling rule for a job shop model
4. Major changes involving a different design for the new system, such as a computerized inventory control system replacing an older noncomputerized system, or an automatic computerized storage and retrieval system replacing a warehouse system in which workers pick items manually

If the change to the computerized representation of the system is minor, such as in items 1 or 2, these changes can be carefully verified and output from the new model accepted with considerable confidence. If a sufficiently similar subsystem exists elsewhere, it may be possible to validate the submodel that represents the subsystem and then to integrate this submodel with other validated submodels to build a complete model. In this way, partial validation of the substantial model changes in items 3 and 4 may be possible. Unfortunately, there is no way to completely validate the input-output transformations of a model of a nonexisting system. In any case, within time and budget constraints the modeler should use as many validation techniques as possible, including input-output validation of subsystem models if operating data can be collected on such subsystems.

Example 10.2 will illustrate some of the techniques that are possible for input-output validation, and will discuss the concepts of an input variable, uncontrollable variable, decision variable, output or response variable, and input-output transformation in more detail.

EXAMPLE 10.2 (The Fifth National Bank of Jaspar)

The Fifth National Bank of Jaspar, as shown in Figure 10.4, is planning to expand its drive-in service at the corner of Main Street. Currently, there is one drive-in window serviced by one teller. Only one or two transactions are allowed at the drive-in window, so it was assumed that each service time was a random sample from some underlying population. Service times $\{S_i, i = 1, 2, \ldots, 90\}$ and interarrival times $\{A_i, i = 1, 2, \ldots, 90\}$ were collected for the 90 customers who arrived between 11:00 A.M. and 1:00 P.M. on a Friday. This time slot was selected for data collection after consultation with management and the teller because it was felt to be representative of a typical rush hour.

Data analysis (as outlined in Chapter 9) led to the conclusion that arrivals could be modeled as a Poisson process at a rate of 45 customers per hour, and that service times were approximately normally distributed with a mean of 1.1 minutes and a standard deviation of 0.2 minute. Thus, the model has two input variables:

1. Interarrival times, exponentially distributed (i.e., a Poisson arrival process) at rate $\lambda = 45$ per hour
2. Service times, assumed to be $N(1.1, (0.2)^2)$

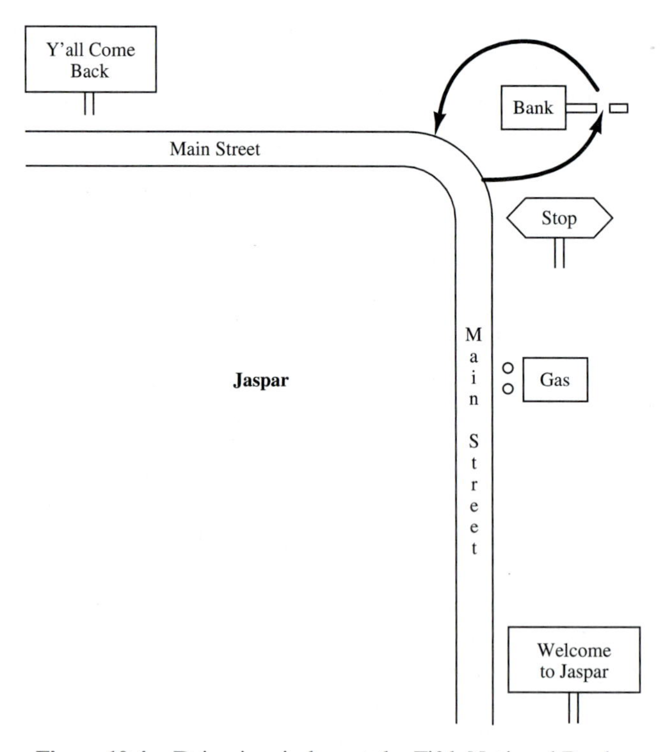

Figure 10.4. Drive-in window at the Fifth National Bank.

Each input variable has a level: the rate ($\lambda = 45$ per hour) for the interarrival times, and the mean 1.1 minutes and standard deviation 0.2 minute for the service times. The interarrival times are examples of uncontrollable variables (i.e., uncontrollable by management in the real system). The service times are also uncontrollable variables, although the level of the service times may be partially controllable. If the mean service time could be decreased to 0.9 minute by increasing the technology, the level of the service-time variable becomes a decision variable or controllable parameter. Setting all decision variables at some level constitutes a policy. For example, the current bank policy is one teller ($D_1 = 1$), mean service time $D_2 = 1.1$ minutes, and one line for waiting cars ($D_3 = 1$). ($D_1, D_2, \ldots$ are used to denote decision variables.) Decision variables are under management's control; the uncontrollable variables, such as arrival rate and actual arrival times, are not under management's control. The arrival rate may change from time to time, but such change is due to external factors not under management's control.

A model of current bank operations was developed and verified in close consultation with bank management and employees. Model assumptions were validated, as discussed in Section 10.3.2. The resulting model is now viewed as a "black box" which takes all input variable specifications and transforms them into a set of output or response variables. The output variables consist of all statistics of interest generated by the simulation about the model's behavior. For example, management is interested in the teller's utilization at the drive-in window (percent of time the teller is busy at the window), average delay in minutes of a customer from arrival to beginning of service, and the maximum length of the line during the rush hour. These input and output variables are shown in Figure 10.5, and are listed in Table 10.1 together with some additional output variables. The uncontrollable input variables are denoted by X, the decision variables by D, and the output variables by Y. From the "black-box" point of view, the model takes the inputs X and D and produces the outputs Y, namely

$$(X, D) \xrightarrow{f} Y$$

or

$$f(X, D) = Y$$

Here f denotes the transformation that is due to the structure of the model. For the Fifth National Bank study, the exponentially distributed interarrival time generated in the model (by the methods of Chapter 8) between customer $n-1$ and customer n is denoted by X_{1n}. (Do not confuse X_{1n} with A_n; the latter was an observation made on the real system.) The normally distributed service time generated in the model for customer n is denoted by X_{2n}. The set of decision variables, or policy, is $D = (D_1, D_2, D_3) = (1, 1.1, 1)$ for current operations. The output, or response, variables are denoted by $Y = (Y_1, Y_2, \ldots, Y_7)$ and are defined in Table 10.1.

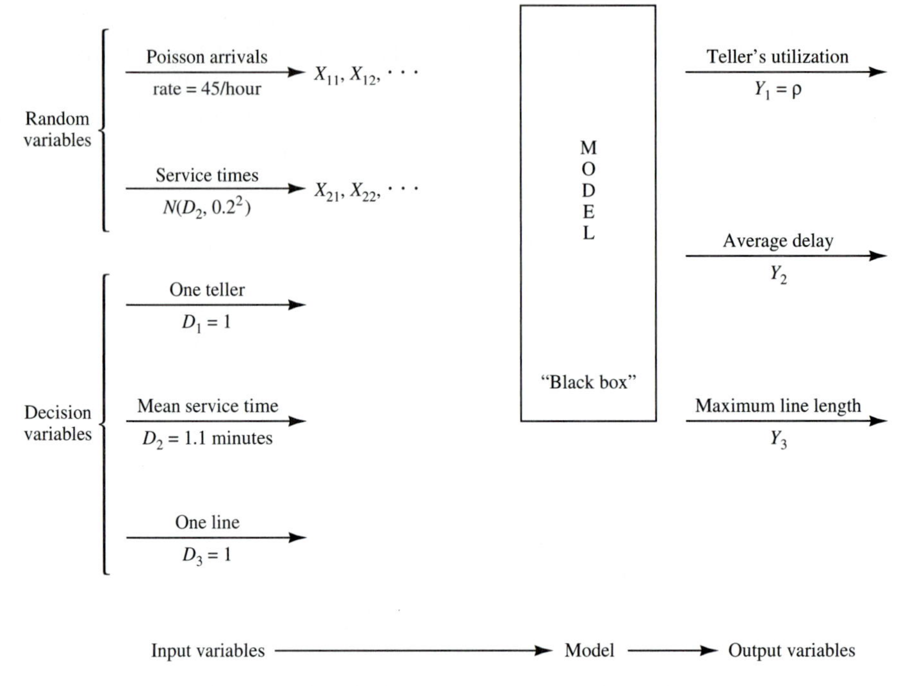

Figure 10.5. Model input-output transformation.

For validation of the input-output transformations of the bank model to be possible, real system data must be available, comparable to at least some of the model output Y of Table 10.1. The system responses should have been collected during the same time period (from 11:00 A.M. to 1:00 P.M. on the

Table 10.1. Input and Output Variables for Model of Current Bank Operations

Input Variables	*Model Output Variables, Y*
D = decision variables	Variables of primary interest
X = other variables	to management (Y_1, Y_2, Y_3):
	Y_1 = teller's utilization
Poisson arrivals at rate = 45/hour	Y_2 = average delay
$X_{11}, X_{12}, \ldots$	Y_3 = maximum line length
[3pt] Service times, $N(D_2, 0.2^2)$	Other output variables of
$X_{21}, X_{22}, \ldots$	secondary interest:
	Y_4 = observed arrival rate
$D_1 = 1$ (one teller)	Y_5 = average service time
$D_2 = 1.1$ minutes (mean service time)	Y_6 = sample standard deviation of service times
$D_3 = 1$ (one line)	
	Y_7 = average length of time

Table 10.2. Results of Six Replications of the First Bank Model

Replication	Y_4 *(Arrivals/Hour)*	Y_5 *(Minutes)*	Y_2 = *Average Delay (Minutes)*
1	51	1.07	2.79
2	40	1.12	1.12
3	45.5	1.06	2.24
4	50.5	1.10	3.45
5	53	1.09	3.13
6	49	1.07	2.38
Sample mean			2.51
Standard deviation			0.82

same Friday) in which the input data $\{A_i, S_i\}$ were collected. This is important because if system response data were collected on a slower day (say, an arrival rate of 40 per hour), the system responses such as teller utilization (Z_1), average delay (Z_2), and maximum line length (Z_3) would be expected to be lower than the same variables during a time slot when the arrival rate was 45 per hour, as observed. Suppose that the delay of successive customers was measured on the same Friday between 11:00 A.M. and 1:00 P.M., and that the average delay was found to be $Z_2 = 4.3$ minutes.

When the model is run using generated random variates X_{1n} and X_{2n}, it is expected that observed values of average delay, Y_2, should be close to $Z_2 = 4.3$ minutes. However, the generated input values (X_{1n} and X_{2n}) cannot be expected to replicate exactly the actual input values (A_n and S_n) of the real system, but they are expected to replicate the statistical pattern of the actual inputs. Hence, simulation-generated values of Y_2 are expected to be consistent with the observed system variable $Z_2 = 4.3$ minutes. Now consider how the modeler might test this consistency.

The modeler makes a small number of statistically independent replications of the model. Statistical independence is guaranteed by using nonoverlapping sets of random numbers produced by the random-number generator, or by choosing seeds for each replication independently (from a random-number table). The results of six independent replications, each of 2 hours duration, are given in Table 10.2.

Observed arrival rate Y_4 and sample average service time Y_5 for each replication of the model are also noted, to be compared to the specified values of 45/hour and 1.1 minutes, respectively. The validation test consists of comparing the system response, namely average delay $Z_2 = 4.3$ minutes, to the model responses, Y_2. Formally, a statistical test of the null hypothesis

$$H_0\colon\ E(Y_2) = 4.3 \text{ minutes}$$

versus (10.1)

$$H_1\colon\ E(Y_2) \neq 4.3 \text{ minutes}$$

is conducted. If H_0 is not rejected, then on the basis of this test there is no reason to consider the model invalid. If H_0 is rejected, the current version of the model is rejected and the modeler is forced to seek ways to improve the model, as illustrated by Figure 10.3. As formulated here, the appropriate statistical test is the t-test, which is conducted in the following manner:

Choose a level of significance α and a sample size n. For the bank model, choose

$$\alpha = 0.05, \quad n = 6$$

Compute the sample mean $\bar{Y}_2$ and the sample standard deviation S over the n replications by Equations (9.1) and (9.2):

$$\bar{Y}_2 = \frac{1}{n}\sum_{i=1}^{n} Y_{2i} = 2.51 \text{ minutes}$$

and

$$S = \left[\frac{\sum_{i=1}^{n}(Y_{2i} - \bar{Y}_2)^2}{n - 1}\right]^{1/2} = 0.82 \text{ minute}$$

where $Y_{2i}, i = 1, \ldots, 6$, are as shown in Table 10.2.

Get the critical value of t from Table A.5. For a two-sided test such as that in Equation (10.1), use $t_{\alpha/2,n-1}$; for a one-sided test, use $t_{\alpha,n-1}$ or $-t_{\alpha,n-1}$ as appropriate ($n - 1$ is the degrees of freedom). From Table A.5, $t_{0.025,5} = 2.571$ for a two-sided test.

Compute the test statistic

$$t_0 = \frac{\bar{Y}_2 - \mu_0}{S/\sqrt{n}} \tag{10.2}$$

where μ_0 is the specified value in the null hypothesis, H_0. Here $\mu_0 = 4.3$ minutes, so that

$$t_0 = \frac{2.51 - 4.3}{0.82/\sqrt{6}} = -5.34$$

For the two-sided test, if $|t_0| > t_{\alpha/2,n-1}$, reject H_0. Otherwise, do not reject H_0. [For the one-sided test with H_1: $E(Y_2) > \mu_0$, reject H_0 if $t > t_{\alpha,n-1}$; with H_1: $E(Y_2) < \mu_0$, reject H_0 if $t < -t_{\alpha,n-1}$.]

Since $|t| = 5.34 > t_{0.025,5} = 2.571$, reject H_0 and conclude that the model is inadequate in its prediction of average customer delay.

Recall that when testing hypotheses, rejection of the null hypothesis H_0 is a strong conclusion, because

$$P(H_0 \text{ rejected} \,|\, H_0 \text{ is true}) = \alpha \tag{10.3}$$

and the level of significance α is chosen small, say $\alpha = 0.05$, as was done here. Equation (10.3) says that the probability of making the error of rejecting H_0 when H_0 is in fact true is low ($\alpha = 0.05$); that is, the probability is small of declaring the model invalid when it is valid (with respect to the variable being tested). The assumptions justifying a t-test are that the observations (Y_{2i}) are normally and independently distributed. Are these assumptions met in the present case?

1. The ith observation Y_{2i} is the average delay of all drive-in customers who began service during the ith simulation run of 2 hours, and thus by a central-limit-theorem effect, it is reasonable to assume that each observation Y_{2i} is approximately normally distributed, provided that the number of customers it is based on is not too small.
2. The observations $Y_{2i}, i = 1, \ldots, 6$, are statistically independent by design, that is, by choice of the random-number seeds independently for each replication, or by use of nonoverlapping streams.
3. The t-statistic computed by Equation (10.2) is a robust statistic; that is, it is approximately distributed as the t-distribution with $n - 1$ degrees of freedom, even when $Y_{21}, Y_{22}, \ldots$ are not exactly normally distributed, and thus the critical values in Table A.5 can reliably be used.

Now that the model of the Fifth National Bank of Jaspar has been found lacking, what should the modeler do? Upon further investigation, the modeler realized that the model contained two unstated assumptions:

1. When a car arrived to find the window immediately available, the teller began service immediately.
2. There is no delay between one service ending and the next beginning, when a car is waiting.

Assumption 2 was found to be approximately correct because a service time was considered to begin when the teller actually began service but was not considered to have ended until the car had exited the drive-in window and the next car, if any, had begun service, or the teller saw that the line was empty. On the other hand, assumption 1 was found to be incorrect because the teller had other duties—mainly serving walk-in customers if no cars were present—and tellers always finished with a previous customer before beginning service on a car. It was found that walk-in customers were always present during rush hour; that the transactions were mostly commercial in nature, taking a considerably longer time than the time required to service drive-up customers; and that when an arriving car found no other cars at the window, it had to wait until the teller finished with the present walk-in customer. To correct this model inadequacy, the structure of the model was changed to include the additional demand on the teller's time, and data were collected on service times of walk-in customers. Analysis of these data found that they were approximately exponentially distributed with a mean of 3 minutes.

The revised model was run, yielding the results in Table 10.3. A test of the null hypothesis H_0: $E(Y_2) = 4.3$ minutes [as in Equation (10.1)] was again conducted, according to the procedure previously outlined.

Table 10.3. Results of Six Replications of the Revised Bank Model

Replication	Y_4 *(Arrivals/Hour)*	Y_5 *(Minutes)*	Y_2 = *Average Delay (Minutes)*
1	51	1.07	5.37
2	40	1.11	1.98
3	45.5	1.06	5.29
4	50.5	1.09	3.82
5	53	1.08	6.74
6	49	1.08	5.49
Sample mean			4.78
Standard deviation			1.66

Choose $\alpha = 0.05$ and $n = 6$ (sample size).

Compute $\bar{Y}_2 = 4.78$ minutes, $S = 1.66$ minutes.

From Table A.5, the critical value is $t_{0.25,5} = 2.571$.

Compute the test statistic $t_0 = (\bar{Y}_2 - \mu_0)/(S/\sqrt{n}) = 0.710$.

Since $|t_0| < t_{0.025,5} = 2.571$, do not reject H_0, and thus tentatively accept the model as valid.

Failure to reject H_0 must be considered as a weak conclusion unless the power of the test has been estimated and found to be high (close to 1). That is, it can only be concluded that the data at hand ($Y_{21}, \ldots, Y_{26}$) were not sufficient to reject the hypothesis H_0: $\mu_0 = 4.3$ minutes. In other words, this test detects no inconsistency between the sample data ($Y_{21}, \ldots, Y_{26}$) and the specified mean μ_0.

The power of a test is the probability of detecting a departure from H_0: $\mu = \mu_0$ when in fact such a departure exists. In the validation context, the power of the test is the probability of detecting an invalid model. The power may also be expressed as 1 minus the probability of a Type II, or β, error, where $\beta = P(\text{Type II error}) = P(\text{failing to reject } H_0 | \, H_1 \text{ is true})$ is the probability of accepting the model as valid when it is not valid.

To consider failure to reject H_0 as a strong conclusion, the modeler would want β to be small. Now, β depends on the sample size n and on the true difference between $E(Y_2)$ and $\mu_0 = 4.3$ minutes — that is, on

$$\delta = \frac{|E(Y_2) - \mu_0|}{\sigma}$$

where σ, the population standard deviation of an individual Y_{2i}, is estimated by S. Tables A.10 and A.11 are typical operating-characteristic (OC) curves, which are graphs of the probability of a Type II error $\beta(\delta)$ versus δ for given sample size n. Table A.10 is for a two-sided t-test while Table A.11 is for a one-sided t-test. Suppose that the modeler would like to reject H_0 (model validity) with probability at least 0.90 if the true mean delay of the model, $E(Y_2)$, differed from the average delay in the system, $\mu_0 = 4.3$ minutes, by 1 minute. Then δ is estimated by

$$\hat{\delta} = \frac{|E(Y_2) - \mu_0|}{S} = \frac{1}{1.66} = 0.60$$

For the two-sided test with $\alpha = 0.05$, use of Table A.10 results in

$$\beta(\hat{\delta}) = \beta(0.6) = 0.75 \text{ for } n = 6$$

To guarantee that $\beta(\hat{\delta}) \leq 0.10$, as was desired by the modeler, Table A.10 reveals that a sample size of approximately $n = 30$ independent replications would be required. That is, for a sample size $n = 6$ and assuming that the population standard deviation is 1.66, the probability of accepting H_0 (model validity), when in fact the model is invalid ($|E(Y_2) - \mu_0| = 1$ minute), is $\beta = 0.75$, which is quite high. If a 1-minute difference is critical, and if the modeler wants to control the risk of declaring the model valid when model predictions are as much as 1 minute off, a sample size of $n = 30$ replications is required to achieve a power of 0.9. If this sample size is too high, either a higher β risk (lower power) or a larger difference δ must be considered. ◀

In general, it is always best to control the Type II error, or β error, by specifying a critical difference δ and choosing a sample size by making use of an appropriate OC curve. (Computation of power and use of OC curves for a wide range of tests is discussed in Hines and Montgomery [1990].) In summary, in the context of model validation, the Type I error is the rejection of a valid model and is easily controlled by specifying a small level of significance α (say $\alpha = 0.2, 0.1, 0.05$, or 0.01). The Type II error is the acceptance of a model as valid when it is invalid. For a fixed sample size n, increasing α will decrease β, the probability of a Type II error. Once α is set, and the critical difference to be detected is selected, the only way to decrease β is to increase the sample size. A Type II error is the more serious of the two types of errors, and thus it is important to design the simulation experiments to control the risk of accepting an invalid model. The two types of error are summarized in Table 10.4, which compares statistical terminology to modeling terminology.

Note that validation is not to be viewed as an either/or proposition, but rather should be viewed in the context of calibrating a model, as conceptually exhibited in Figure 10.3. If the current version of the bank model produces estimates of average delay (Y_2) that are not close enough to real system behavior

Table 10.4. Types of Error in Model Validation

Statistical Terminology	*Modeling Terminology*	*Associated Risk*
Type I: rejecting H_0 when H_0 is true	Rejecting a valid model	α
Type II: failure to reject H_0 when H_1 is true	Failure to reject an invalid model	β

($\mu_0 = 4.3$ minutes), the source of the discrepancy is sought, and the model is revised in light of this new knowledge. This iterative process is repeated until model accuracy is judged adequate.

10.3.4 Input-Output Validation: Using Historical Input Data

When using artificially generated data as input data, as was done to test the validity of the bank models in Section 10.3.3, the modeler expects the model to produce event patterns that are compatible with, but not identical to, the event patterns that occurred in the real system during the period of data collection. Thus, in the bank model, artificial input data $\{X_{1n}, X_{2n}, n = 1, 2, \ldots\}$ for interarrival and service times were generated and replicates of the output data Y_2 were compared to what was observed in the real system by means of the hypothesis test stated in Equation (10.1). An alternative to generating input data is to use the actual historical record, $\{A_n, S_n, n = 1, 2, \ldots\}$, to drive the simulation model and then to compare model output to system data.

To implement this technique for the bank model, the data $A_1, A_2, \ldots$ and $S_1, S_2, \ldots$ would have to be entered into the model into arrays, or stored on a file to be read as the need arose. Just after customer n arrived at time $t_n = \sum_{i=1}^{n} A_i$, customer $n + 1$ would be scheduled on the future event list to arrive at future time $t_n + A_{n+1}$ (without any random numbers being generated). If customer n were to begin service at time t'_n, a service completion would be scheduled to occur at time $t'_n + S_n$. This event scheduling without random-number generation could be implemented quite easily in most simulation languages by using arrays to store the data.

When using this technique, the modeler hopes that the simulation will duplicate as closely as possible the important events that occurred in the real system. In the model of the Fifth National Bank of Jaspar, the arrival times and service durations will exactly duplicate what happened in the real system on that Friday between 11:00 A.M. and 1:00 P.M. If the model is sufficiently accurate, then the delays of customers, lengths of lines, utilizations of servers, and departure times of customers predicted by the model will be close to what actually happened in the real system. It is, of course, the model builder's and model user's judgment that determines the level of accuracy required.

To conduct a validation test using historical input data, it is important that all the input data ($A_n, S_n, \ldots$) and all the system response data, such as average delay (Z_2), be collected during the same time period. Otherwise, the comparison of model responses to system responses, such as the comparison of average delay in the model (Y_2) to that in the system (Z_2), could be misleading. The responses (Y_2 and Z_2) depend on the inputs (A_n and S_n) as well as on the structure of the system, or model. Implementation of this technique could be difficult for a large system because of the need for simultaneous data collection of all input variables and those response variables of primary interest. In some systems, electronic counters and devices are used to ease the data-collection task by automatically recording certain types of data. The following example was based on two simulation models reported in Carson et al. [1981a, b], in which simultaneous data collection and the subsequent validation were both completed suceessfully.

EXAMPLE 10.3 (The Candy Factory)

The production line at the Sweet Li'l Things Candy Factory in Decatur consists of three machines which make, package, and box their famous candy. One machine (the candy maker) makes and wraps individual pieces of candy and sends them by conveyor to the packer. The second machine (the packer) packs the individual pieces into a box. A third machine (the box maker) forms the boxes and supplies them by conveyor to the packer. The system is illustrated in Figure 10.6.

Each machine is subject to random breakdowns due to jams and other causes. These breakdowns cause the conveyor to begin to empty or fill. The conveyors between the two makers and the packer are used as a temporary storage buffer for in-process inventory. In addition to the randomly occurring breakdowns, if the candy conveyor empties, a packer runtime is interrupted and the packer remains idle until more candy is produced. If the box conveyor empties because of a long random breakdown of the box machine, an operator manually places racks of boxes onto the packing machine. If the conveyor fills, the corresponding maker becomes idle. The purpose of the model is to investigate the frequency of these operator interventions which require manual loading of racks of boxes, as a function of various combinations of individual machines and lengths of conveyor. Different machines have different production speeds and breakdown characteristics, and longer conveyors can hold more in-process inventory. The goal is to hold operator interventions to an acceptable level while maximizing production. As machine stoppages (due to a full or empty conveyor) cause increased damage to the product, this is also a factor in production.

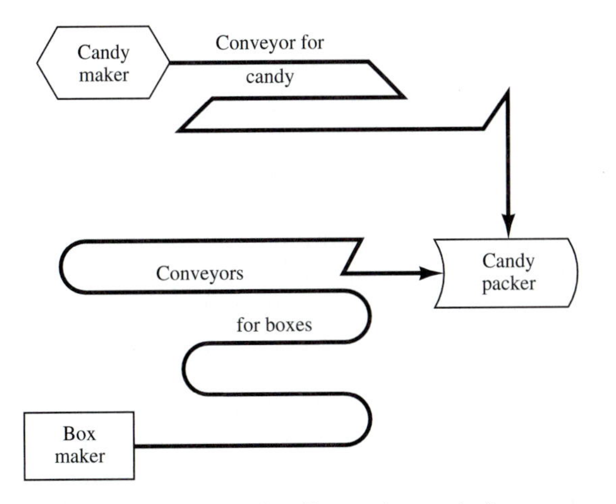

Figure 10.6. Production line at the candy factory.

Table 10.5. Validation of the Candy Factory Model

Response, i	*System, Z_i*	*Model, Y_i*
1. Production level	897, 208	883, 150
2. Number of operator interventions	3	3
3. Time of occurrence	7:22, 8:41, 10:10	7:24, 8:42, 10:14

A simulation model of the Candy Factory was developed and a validation effort using historical inputs was conducted. Engineers in the Candy Factory set aside a 4-hour time slot from 7:00 A.M. to 11:00 A.M. to collect data on an existing production line. For each machine, say machine i, time to failure and random downtime data

$$T_{i1}, D_{i1}, T_{i2}, D_{i2}, \ldots$$

were collected. For machine $i(i = 1, 2, 3)$, T_{ij} is the jth runtime (or time to failure), and D_{ij} is the successive random downtime. A runtime, T_{ij}, can be interrupted due to a full or empty conveyor (as appropriate) but resumes when conditions are right. Initial system conditions at 7:00 A.M. were recorded so that they could be duplicated in the model as initial conditions at time 0. Additionally, system responses of primary interest—the production level (Z_1) and the number (Z_2) and time of occurrence (Z_3) of operator interventions—were recorded for comparison with model predictions.

The system input data, T_{ij} and D_{ij}, were fed into the model and used as runtimes and random downtimes. The structure of the model determined the occurrence of shutdowns due to a full or empty conveyor, and the occurrence of operator interventions. Model response variables ($Y_i, i = 1, 2, 3$) were collected for comparison to the corresponding system response variables ($Z_i, i = 1, 2, 3$).

The closeness of model predictions to system performance aided the engineering staff considerably in convincing management of the validity of the model. These results are shown in Table 10.5. A simple display such as Table 10.5 can be quite effective in convincing skeptical engineers and managers of a model's validity—perhaps more effective than the most sophisticated statistical methods! ◀

With only one set of historical input and output data, only one set of simulated output data can be obtained, and thus no simple statistical tests are possible based on summary measures. But if K historical input data sets are collected, and K observations $Z_{i1}, Z_{i2}, \ldots, Z_{iK}$ of some system response variable, Z_i, are collected, such that the output measure Z_{ij} corresponds to the jth input set, an objective statistical test becomes possible. For example, Z_{ij} could be the average delay of all customers who were served during the time the jth input data set was collected. With the K input data sets in hand, the modeler now runs the model K times, once for each input set, and observes the simulated results $W_{i1}, W_{i2}, \ldots, W_{iK}$ corresponding to $Z_{ij}, j = 1, \ldots, K$.

Table 10.6. Comparison of System and Model Output Measures When Using Identical Historical Inputs

Input Data Set	System Output, Z_{ij}	Model Output, W_{ij}	Observed Difference, d_j	Squared Deviation from Mean, $(d_j - \bar{d})^2$
1	Z_{i1}	W_{i1}	$d_1 = Z_{i1} - W_{i1}$	$(d_1 - \bar{d})^2$
2	Z_{i2}	W_{i2}	$d_2 = Z_{i2} - W_{i2}$	$(d_2 - \bar{d})^2$
3	Z_{i3}	W_{i3}	$d_3 = Z_{i3} - W_{i3}$	$(d_3 - \bar{d})^2$
.	.	.	.	.
.	.	.	.	.
.	.	.	.	.
K	Z_{iK}	W_{iK}	$d_K = Z_{iK} - W_{iK}$	$(d_K - \bar{d})^2$
			$\bar{d} = \frac{1}{K}\sum_{j=1}^{K} d_j$	$S_d^2 = \frac{1}{K-1}\sum_{j=1}^{K}(d_j - \bar{d})^2$

Continuing the same example, W_{ij} would be the average delay predicted by the model when using the jth input set. The available data for comparison appears as in Table 10.6.

If the K input data sets are fairly homogeneous, it is reasonable to assume that the K observed differences $d_j = Z_{ij} - W_{ij}$, $j = 1, \ldots, K$, are identically distributed. Furthermore, if the collection of the K sets of input data was separated in time, say on different days, it is reasonable to assume that the K differences $d_1, \ldots, d_K$ are statistically independent, and hence the differences $d_1, \ldots, d_K$ constitute a random sample. In many cases, each Z_i and W_i is a sample average over customers, so that (by the central limit theorem) the differences $d_j = Z_{ij} - W_{ij}$ are approximately normally distributed with some mean μ_d and variance σ_d^2. The appropriate statistical test is then a t-test of the null hypothesis of no mean difference:

$$H_0\colon \mu_d = 0$$

versus the alternative of significant difference:

$$H_1\colon \mu_d \neq 0$$

The proper test is a paired t-test (Z_{i1} is paired with W_{i1}, since each was produced by the first input data set, and so on). First, compute the sample mean difference $\bar{d}$, and the sample variance S_d^2 by the formulas given in Table 10.6. Then compute the t-statistic by

$$t_0 = \frac{\bar{d} - \mu_d}{S_d/\sqrt{K}} \tag{10.4}$$

(with $\mu_d = 0$), and get the critical value $t_{\alpha/2,K-1}$ from Table A.5, where α is the prespecified significance level and $K - 1$ is the number of degrees of freedom. If $|t_0| > t_{\alpha/2,K-1}$, reject the hypothesis H_0 of no mean difference and conclude that the model is inadequate. If $|t_0| \leq t_{\alpha/2,K-1}$, do not reject H_0 and hence conclude that this test provides no evidence of model inadequacy.

Example 10.4 (The Candy Factory, Continued)

Engineers at the Sweet Li'l Things Candy Factory decided to expand the initial validation effort reported in Example 10.3. Electronic devices were installed which could automatically monitor one of the production lines, and the validation effort of Example 10.3 was repeated with $K = 5$ sets of input data. The system and the model were compared on the basis of production level. The results are shown in Table 10.7.

Table 10.7. Validation of the Candy Factory Model (Continued)

Input Data Set, j	*System Production,* Z_{1j}	*Model Production,* W_{1j}	*Observed Difference,* d_j	*Squared Deviation from Mean,* $(d_j - \bar{d})^2$
1	897,208	883,150	14,058	7.594×10^7
2	629,126	630,550	−1,424	4.580×10^7
3	735,229	741,420	−6,191	1.330×10^8
4	797,263	788,230	9,033	1.362×10^7
5	825,430	814,190	11,240	3.4772×10^7
			$\bar{d} = 5{,}343.2$	$S_d^2 = 7.580 \times 10^7$

A paired t-test was conducted to test H_0: $\mu_d = 0$, or equivalently, H_0: $E(Z_1) = E(W_1)$, where Z_1 is the system production level and W_1 is the production level predicted by the simulated model. Let the level of significance be $\alpha = 0.05$. Using the results in Table 10.7, the test statistic, as given by Equation (10.4), is

$$t_0 = \frac{\bar{d}}{S_d/\sqrt{K}} = \frac{5343.2}{8705.85/\sqrt{5}} = 1.37$$

From Table A.5, the critical value is $t_{\alpha/2,K-1} = t_{0.025,4} = 2.78$. Since $|t_0| = 1.37 < t_{0.025,4} = 2.78$, the null hypothesis cannot be rejected on the basis of this test; that is, no inconsistency is detected between system response and model predictions in terms of mean production level. If H_0 had been rejected, the modeler would have searched for the cause of the discrepancy and revised the model, in the spirit of Figure 10.3. ◀

10.3.5 Input-Output Validation: Using a Turing Test

In addition to statistical tests, or when no statistical test is readily applicable, persons knowledgeable about system behavior can be used to compare model output to system output. For example, suppose that five reports of system performance over five different days are prepared, and simulation output data are used to produce five "fake" reports. The 10 reports should all be in exactly the same format and should contain information of the type that managers and engineers have previously seen on the system. The 10 reports are randomly shuffled and given to the engineer, who is asked to decide which reports are fake and which are real. If the engineer identifies a substantial number of the fake reports, the model builder questions the engineer and uses the information gained to improve the model. If the engineer cannot distinguish between fake and real reports with any consistency, the modeler will conclude that this test provides no evidence of model inadequacy. For further discussion and an application to a real simulation, the reader is referred to Schruben [1980]. This type of validation test is commonly called a Turing test. Its use as model development proceeds can be a valuable tool in detecting model inadequacies, and eventually in increasing model credibility as the model is improved and refined.

10.4 Summary

Validation of simulation models is of great importance. Decisions are made on the basis of simulation results; thus, the accuracy of these results should be subject to question and investigation.

Quite often simulations appear realistic on the surface because simulation models, unlike analytic models, can incorporate any level of detail about the real system. To avoid being "fooled" by this apparent realism, it is best to compare system data to model data, and to make the comparison using a wide variety of techniques, including an objective statistical test, if at all possible.

As discussed by Van Horn [1969, 1971], some of the possible validation techniques, in order of increasing cost-to-value ratios, include:

1. Develop models with high face validity by consulting persons knowledgeable about system behavior on model structure, model input, and model output. Use any existing knowledge in the form of previous research and studies, observation, and experience.
2. Conduct simple statistical tests of input data for homogeneity, randomness, and goodness-of-fit to assumed distributional forms.
3. Conduct a Turing test. Have knowledgeable people (engineers, managers) compare model output to system output and attempt to detect the difference.
4. Compare model output to system output by means of statistical tests.
5. After model development, collect new system data and repeat techniques 2 to 4.
6. Build the new system (or redesign the old one) based on simulation results, collect data on the new system, and use this data to validate the model. (Not recommended if this is the only technique used.)
7. Do little or no validation. Implement simulation results without validating. (Not recommended.)

It is usually too difficult, too expensive, or too time consuming to use all possible validation techniques for every model that is developed. It is an important part of the model builder's task to choose those validation techniques that are most appropriate, both to assure model accuracy and to assure model credibility.

REFERENCES

BALCI, O. [1994], "Validation, Verification and Testing Techniques throughout the Life Cycle of a Simulation Study," *Annals of Operations Research*, Vol. 53, pp. 121–174.

BALCI, O. [1998] "Verification, Validation, and Testing," in *Handbook of Simulation*, ed. J. Banks, John Wiley, New York.

BALCI, O., AND R. G. SARGENT [1982a], "Some Examples of Simulation Model Validation Using Hypothesis Testing," *Proceedings of the Winter Simulation Conference*, ed. H. J. Highland, Y. W. Chao, and O. S. Madrigal, pp. 620–629, Association for Computing Machinery, New York.

BALCI, O., AND R. G. SARGENT [1982b], "Validation of Multivariate Response Models Using Hotelling's Two-Sample T^2 Test," *Simulation*, Vol. 39, No. 6 (Dec.), pp. 185–192.

BALCI, O., AND R. G. SARGENT [1984a], "Validation of Simulation Models via Simultaneous Confidence Intervals," *American Journal of Mathematical Management Sciences*, Vol. 4, Nos. 3 and 4, pp. 375–406.

BALCI, O., AND R. G. SARGENT [1984b], "A Bibliography on the Credibility Assessment and Validation of Simulation and Mathematical Models," *Simuletter*, Vol. 15, No. 3, pp. 15–27.

BORTSCHELLER, B. J., AND E. T. SAULNIER [1992], "Model Reusability in a Graphical Simulation Package," *Proceedings of the Winter Simulation Conference*, ed. J. J. Swain, D. Goldsman, R. C. Crain, J. R. Wilson, pp. 764–772, Association for Computing Machinery, New York.

CARSON, J. S., N. WILSON, D. CARROLL, AND C. H. WYSOWSKI [1981a], "A Discrete Simulation Model of a Cigarette Fabrication Process," *Proceedings of the Twelfth Modeling and Simulation Conference*, University of Pittsburgh.

CARSON, J. S., N. WILSON, D. CARROLL, AND C. H. WYSOWSKI [1981b], "Simulation of a Filter Rod Manufacturing Process," *Proceedings of the 1981 Winter Simulation Conference*, ed. T. I. Oren, C. M. Delfosse, and C. M. Shub, pp. 535–541, Association for Computing Machinery, New York.

CARSON, J. S., [1986], "Convincing Users of Model's Validity Is Challenging Aspect of Modeler's Job," *Industrial Engineering*, June, pp. 76–85.

GAFARIAN, A. V., AND J. E. WALSH [1970], "Methods for Statistical Validation of a Simulation Model for Freeway Traffic near an On-Ramp," *Transportation Research*, Vol. 4, p. 379–384.

GASS, S. I. [1983], "Decision-Aiding Models: Validation, Assessment, and Related Issues for Policy Analysis," *Operations Research*, Vol. 31, No. 4, pp. 601–663.

HINES, W. W., AND D. C. MONTGOMERY [1990], *Probability and Statistics in Engineering and Management Science*, 3d ed., John Wiley, New York.

KLEIJNEN, J. P. C. [1987], *Statistical Tools for Simulation Practitioners*, Marcel Dekker, New York.

KLEIJNEN, J. P. C. [1993] "Simulation and Optimization in Production Planning: A Case Study," *Decision Support Systems*, Vol. 9, pp. 269–280.

KLEIJNEN, J. P. C. [1995], "Theory and Methodology: Verification and Validation of Simulation Models," *European Journal of Operational Research*, Vol. 82, No. 1, pp. 145–162.

LAW, A. M. AND W. D. KELTON [2000], *Simulation Modeling and Analysis*, 3d ed., McGraw-Hill, New York.

NAYLOR, T. H., AND J. M. FINGER [1967], "Verification of Computer Simulation Models," *Management Science*, Vol. 2, pp. B92–B101.

OREN, T. [1981], "Concepts and Criteria to Assess Acceptability of Simulation Studies: A Frame of Reference," *Communications of the Association for Computing Machinery*, Vol. 24, No. 4, pp. 180–89.

SARGENT, R. G. [1994], "Verification and Validation of Simulation Models," *Proceedings of the Winter Simulation Conference*, ed. A. Seila, S. Manivannan, J. Tew, D. Sadowski, pp. 77–87, Association for Computing Machinery, New York.

SCHECTER, M., AND R. C. LUCAS [1980], "Validating a Large Scale Simulation Model of Wilderness Recreation Travel," *Interfaces*, Vol. 10, pp. 11–18.

SCHRUBEN, L. W. [1980], "Establishing the Credibility of Simulations," *Simulation*, Vol. 34, pp. 101–105.

SHANNON, R. E. [1975], *Systems Simulation: The Art and Science*. Prentice Hall, Upper Saddle River, NJ.

VAN HORN, R. L. [1969], "Validation," in *The Design of Computer Simulation Experiments*, ed. T. H. Naylor, Duke University Press, Durham, NC.

VAN HORN, R. L. [1971], "Validation of Simulation Results," *Management Science*, Vol. 17, pp. 247–258.

YOUNGBLOOD, S. M. [1993] "Literature Review and Commentary on the Verification, Validation and Accreditation of Models," Johns Hopkins University, Laurel, MD.

YÜCESAN, E., AND S. H. JACOBSON [1992], "Building Correct Simulation Models is Difficult," *Proceedings of the Winter Simulation Conference*, ed. J. J. Swain, D. Goldsman, R. C. Crain, and J. R. Wilson, pp. 783–790, Association for Computing Machinery, New York.

EXERCISES

1. A simulation model of a job shop was developed to investigate different scheduling rules. To validate the model, the currently used scheduling rule was incorporated into the model and the resulting output compared to observed system behavior. By searching the previous year's computerized records it was estimated that the average number of jobs in the shop was 22.5 on a given day. Seven independent replications of the model were run, each of 30 days duration, with the following results for average number of jobs in the shop:

18.9 22.0 19.4 22.1 19.8 21.9 20.2

(a) Develop and conduct a statistical test to determine if model output is consistent with system behavior. Use a level of significance of $\alpha = 0.05$.

(b) What is the power of this test if a difference of two jobs is viewed as critical? What sample size is needed to guarantee a power of 0.8 or higher? (Use $\alpha = 0.05$.)

2. System data for the job shop of Exercise 1 revealed that the average time spent by a job in the shop was approximately 4 working days. The model made the following predictions on seven independent replications, for average time spent in the shop:

3.70 4.21 4.35 4.13 3.83 4.32 4.05

(a) Is model output consistent with system behavior? Conduct a statistical test using a level of significance $\alpha = 0.01$.

(b) If it is important to detect a difference of 0.5 day, what sample size is needed to have a power of 0.90? Interpret your results in terms of model validity or invalidity. (Use $\alpha = 0.01$.)

3. For the job shop of Exercise 1, four sets of input data were collected over four different 10-day periods, together with the average number of jobs in the shop (Z_i) for each period. The input data were used to drive the simulation model for four runs of 10 days each, and model predictions of average number of jobs in the shop (Y_i) were collected, with these results:

i	1	2	3	4
Z_i	21.7	19.2	22.8	19.4
Y_i	24.6	21.1	19.7	24.9

(a) Conduct a statistical test to check the consistency of system output and model output. Use a level of significance of $\alpha = 0.05$.

(b) If a difference of two jobs is viewed as important to detect, what sample size is required to guarantee a probability of at least 0.80 of detecting this difference if it indeed exists? (Use $\alpha = 0.05$.)

4. Obtain at least two of the papers or reports listed in the References dealing with validation and verification. Write a short essay comparing and contrasting the various philosophies and approaches to the topic of verification and validation.

5. Find several examples of actual simulations reported in the literature in which the authors discuss validation of their model. Is enough detail given to judge the adequacy of the validation effort? If so, compare the reported validation to the criteria set forth in this chapter. Did the authors use any validation technique not discussed in this chapter? [Several potential sources of articles on simulation applications include the two journals *Interfaces* and *Simulation*, and the *Winter Simulation Conference Proceedings*.]

6. Compare and contrast the various simulation languages in their capability to aid the modeler in the often arduous task of debugging and verification (articles discussing the nature of simulation languages may be found in the *Winter Simulation Conference Proceedings*).

7. **(a)** Compare validation in simulation to the validation of theories in the physical sciences.

(b) Compare the issues involved and the techniques available for validation of models of physical systems versus models of social systems.

(c) Contrast the difficulties, and compare the techniques, in validating a model of a manually operated warehouse versus a model of an automated storage and retrieval system.

(d) Repeat (c) for a model of a production system involving considerable manual labor and human decision making, versus a model of the same production system after it has been automated.

11

Output Analysis for a Single Model

Output analysis is the examination of data generated by a simulation. Its purpose is to predict the performance of a system or to compare the performance of two or more alternative system designs. This chapter deals with the analysis of a single system, while Chapter 12 deals with the comparison of two or more systems. The need for statistical output analysis is based on the observation that the output data from a simulation exhibits random variability when random-number generators are used to produce the values of the input variables: that is, two different streams or sequences of random numbers will produce two sets of outputs which (probably) will differ. If the performance of the system is measured by a parameter θ, the result of a set of simulation experiments will be an estimator $\widehat{\theta}$ of θ. The precision of the estimator $\widehat{\theta}$ can be measured by the variance (or standard error) of $\widehat{\theta}$. The purpose of the statistical analysis is to estimate this variance, or to determine the number of observations required to achieve a desired precision.

Consider a typical output variable, Y, the total cost per week of an inventory system; Y should be treated as a random variable with an unknown distribution. A simulation run of length 1 week provides a single sample observation from the population of all possible observations on Y. By increasing the run length, the sample size can be increased to n observations, $Y_1, Y_2, \ldots, Y_n$, based on a run length of n weeks. However, these observations do not constitute a random sample, in the classical sense, because they are not statistically independent. In this case, the inventory on hand at the end of one week is the beginning inventory on hand for the next week, and thus the value of Y_i has some influence on the value of Y_{i+1}. Thus, the sequence of random variables $Y_1, Y_2, \ldots, Y_n$ may be autocorrelated (i.e., correlated with itself). This autocorrelation, which is a measure of a lack of statistical independence, means that classical methods of statistics which assume independence are not directly applicable to the analysis of this output data. The methods must be properly modified and the simulation experiments properly designed for valid inferences to be made.

In addition to the autocorrelation present in most simulation output data, the specification of the initial conditions of the system at time 0 may pose a problem for the simulation analyst, and may influence the output data. For example, the inventory on hand and the number of backorders at time 0 would most likely influence the value of Y_1, the total cost for week 1. Because of the autocorrelation, these initial conditions would also influence the costs $(Y_2, \ldots, Y_n)$ for subsequent weeks. The specified initial conditions, if not chosen well, may have an especially deleterious effect when attempting to estimate the steady-state (long-run) performance of a simulation model. For purposes of statistical analysis, the effect of the initial conditions is that the output observations may not be identically distributed, and the initial observations may not be representative of steady-state behavior of the system.

Section 11.1 distinguishes between two types of simulation—transient versus steady state—and defines commonly used measures of system performance for each type of simulation. Section 11.2 illustrates by example the inherent variability in a stochastic (i.e., probabilistic) discrete-event simulation, and thus demonstrates the need for a statistical analysis of the output; it also discusses the effect of autocorrelation on variance estimation. Section 11.3 discusses the statistical estimation of performance measures. Section 11.4 discusses the analysis of transient simulations, and Section 11.5 the analysis of steady-state simulations.

11.1 Types of Simulations with Respect to Output Analysis

When analyzing simulation output data, a distinction is made between terminating or transient simulations and steady-state simulations. A *terminating* simulation is one that runs for some duration of time T_E, where E is a specified event (or set of events) which stops the simulation. Such a simulated system "opens" at time 0 under well-specified *initial conditions* and "closes" at the *stopping time* T_E. The next four examples are terminating simulations.

Example 11.1

The Shady Grove Bank opens at 8:30 A.M. (time 0) with no customers present and 8 of the 11 tellers working (initial conditions), and closes at 4:30 P.M. (time $T_E = 480$ minutes). Here, the event E is merely the fact that the bank has been open for 480 minutes. The simulation analyst is interested in modeling the interaction between customers and tellers over the entire day, including the effect of starting up and of closing down at the end of the day. ◀

EXAMPLE 11.2

Consider the Shady Grove Bank of Example 11.1 restricted to the period from 11:30 A.M. (time 0) to 1:30 P.M., when it is especially busy. The simulation run length is $T_E = 120$ minutes. The initial conditions at time 0 (11:30 A.M.) could be specified in essentially two ways: (1) the real system could be observed at 11:30 on a number of different days and a distribution of number of customers in system (at 11:30 A.M.) could be estimated; then these data could be used to load the simulation model with customers at time 0; or (2) the model could be simulated from 8:30 A.M. to 11:30 A.M. without collecting output statistics, and the ending conditions at 11:30 A.M. used as initial conditions for the 11:30 A.M. to 1:30 P.M. simulation. ◀

EXAMPLE 11.3

A communications system consists of several components plus several backup components. It is represented schematically in Figure 11.1. Consider the system over a period of time, T_E, until the system fails. The stopping event E is defined by $E = \{A \text{ fails, or } D \text{ fails, or } (B \text{ and } C \text{ both fail})\}$. Initial conditions are that all components are new at time 0. ◀

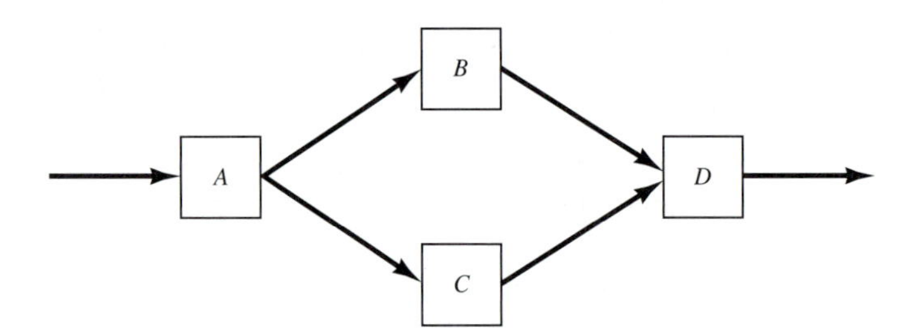

Figure 11.1. Example of a communications system.

Notice that in the bank model of Example 11.1, the stopping time $T_E = 480$ minutes is known, but in Example 11.3, the stopping time T_E is generally unpredictable in advance, and, in fact, T_E is probably the output variable of interest, as it represents the total time until the system breaks down. One goal of the simulation might be to estimate $E(T_E)$, the mean time to system failure.

EXAMPLE 11.4

A widget manufacturing process runs continuously from Monday morning until Saturday morning. The first shift of each workweek is used to load inventory buffers and chemical tanks with the components and catalysts needed to make the final product (28 varieties of widgets). These components and catalysts are made continually throughout the week, except for the last shift Friday night, which is used for cleanup and maintenance. Thus, most inventory buffers are near empty at the end of the week. During the first shift on Monday a buffer stock is built up to cover the eventuality of breakdown in some part of the process. It is desired to simulate this system during the first shift (time 0 to time $T_E = 8$ hours) to study various scheduling policies for loading inventory buffers. ◀

When simulating a terminating system, the initial conditions of the system at time 0 must be specified, and the stopping time T_E, or alternatively, the stopping event E, must be well defined. Although it is certainly true that the Shady Grove Bank in Example 11.1 will open again the next day, the simulation analyst has chosen to consider it a terminating system because the object of interest is one day's operation including startup and closedown. On the other hand, if the simulation analyst were interested in some other aspect of the bank's operations, such as the flow of money or operation of automated teller machines, then the system might be considered as a nonterminating one. Similar comments apply to the communications system of Example 11.3. If the failed component were replaced and the system continued to operate, and, most important, if the simulation analyst were interested in studying its long-run behavior, it might be considered as a nonterminating system. In Example 11.3, however, interest is in its short-run behavior, from time 0 until the first system failure at time T_E. *Therefore, whether a simulation is considered to be terminating or not depends on both the objectives of the simulation study and the nature of the system.*

Example 11.4 is a terminating system, too. It is also an example of a *transient* (or nonstationary) simulation, as the variables of interest are the in-process inventory levels which are increasing from zero or near zero (at time 0) to full or near full (at time 8 hours).

A *nonterminating system* is a system that runs continuously, or at least over a very long period of time. Examples include assembly lines which shut down infrequently, continuous production systems of many different types, telephone systems and other communications systems such as the Internet, hospital emergency rooms, police dispatching and patrolling operations, fire departments, and continuously operating computer systems.

A simulation of a nonterminating system starts at simulation time 0 under initial conditions defined by the analyst and runs for some analyst-specified period of time T_E. (Significant problems arise concerning the specification of these initial and stopping conditions, problems that have not yet been completely solved by simulation researchers.) Usually, the analyst wants to study steady-state, or long-run, properties of the system — that is, properties which are not influenced by the initial conditions of the model at time 0. *A steady-state simulation is a simulation whose objective is to study long-run, or steady-state, behavior of a nonterminating system.* The next two examples are steady-state simulations.

EXAMPLE 11.5

Consider the widget manufacturing process of Example 11.4, beginning with the second shift when the complete production process is under way. It is desired to estimate long-run production levels and production efficiencies. For the relatively long period of 13 shifts, this may be considered as a steady-state simulation. To obtain sufficiently precise estimates of production efficiency and other response variables, the analyst could decide to simulate for any length of time, T_E (even longer than 13 shifts). That is, T_E is not determined by the nature of the problem (as it was in terminating simulations); rather, it is set by the analyst as one parameter in the design of the simulation experiment. ◄

EXAMPLE 11.6

HAL Inc., a large computer service bureau, has many customers worldwide. Thus, its large computer system with many servers, workstations, and peripherals runs continuously 24 hours per day. Due to an increasing workload, HAL is considering additional CPUs, disk drives, and CD drives in various configurations. The HAL systems staff develops a simulation model of the existing system with the current workload and then explores several possibilities for expanding capacity. HAL is interested in steady-state throughput and utilization of each machine. This is an example of a steady-state simulation. The stopping time, T_E, is determined not by the nature of the problem but rather by the simulation analyst, either arbitrarily or with a certain statistical precision in mind. ◄

11.2 Stochastic Nature of Output Data

Consider one run of a simulation model over a period of time $[0, T]$. Since the model is an input-output transformation, as illustrated by Figure 10.5, and since some of the model input variables are random variables, it follows that the model output variables are random variables. This *stochastic* (or probabilistic) nature of output variables was observed in Chapter 2, Exercise 2.9, where the reader was asked to consider the effect of using a different random-number stream on the output of an inventory (newspaper) problem. Again in Tables 10.2 and 10.3, the effect of six different random-number streams for the six independent replications of the bank model of Example 10.2 was to produce six independent estimates, Y_{2i}, of average delay, which were combined into one sample mean estimator, $\bar{Y}_2$. Table 10.2 clearly exhibits the stochastic nature of the summary measure Y_2.

Three examples are now given to illustrate the nature of the output data from stochastic simulations and to give a preliminary discussion of several important properties of these data. Do not be concerned if some of these properties and the associated terminology are not entirely clear on a first reading. They will be carefully explained later in the chapter.

EXAMPLE 11.7 (Able and Baker, Revisited)

Consider the Able-Baker carhop problem (Example 2.2) which involved customers arriving according to the distribution of Table 2.11 and being served either by Able, whose service-time distribution is given in Table 2.12, or by Baker, whose service-time distribution is given in Table 2.13. The purpose of the simulation is to estimate Able's utilization, ρ, and the mean time spent in the system per customer, w, over the first 2 hours of the workday. Therefore,

Table 11.1. Results of Four Independent Runs of 2-Hour Duration of the Able-Baker Queueing Problem

Run, r	*Able's Utilization* $\hat{\rho}_r$	*Average System Time,* $\hat{w}_r$ *(Minutes)*
1	0.808	3.74
2	0.875	4.53
3	0.708	3.84
4	0.842	3.98

each run of the model is for a 2-hour period, with the system being empty and idle at time 0. Four statistically independent runs were made by using four distinct streams of random numbers to generate the interarrival and service times. Table 11.1 presents the results. The estimated utilization for run r is given by $\widehat{\rho}_r$, and the estimated average system time by $\widehat{w}_r$ (i.e., $\widehat{w}_r$ is the sample average time in system for all customers served during run r). Notice that, in this sample, the observed utilization ranges from 0.708 to 0.875, and the observed average system time ranges from 3.74 minutes to 4.53 minutes. The stochastic nature of the output data $\{\widehat{\rho}_1, \widehat{\rho}_2, \widehat{\rho}_3, \widehat{\rho}_4\}$ and $\{\widehat{w}_1, \widehat{w}_2, \widehat{w}_3, \widehat{w}_4\}$ is demonstrated by the results in Table 11.1. ◀

There are two general questions that we will address by a statistical analysis, say of the observed utilizations $\widehat{\rho}_r$, $r = 1, \ldots, 4$:

1. Estimation of the true utilization $\rho = E(\widehat{\rho}_r)$ by a single number, called a point estimate
2. Estimation of the error in our point estimate, in the form of either a standard error or a confidence interval

These questions are addressed in Section 11.4 for terminating simulations, such as Example 11.7. Classical methods of statistics may be used because $\widehat{\rho}_1, \widehat{\rho}_2, \widehat{\rho}_3$, and $\widehat{\rho}_4$ constitute a random sample; that is, they are independent and identically distributed. In addition, since $\rho = E(\widehat{\rho}_r)$ is the parameter being estimated, it follows that each $\widehat{\rho}_r$ is an unbiased estimate of the true mean utilization ρ. The analysis of Example 11.7 is considered in Example 11.10. A survey of statistical methods applicable to terminating simulations is given by Law [1980]. Additional guidance may be found in Alexopoulos and Seila [1998], Kleijnen [1987], Law and Kelton [2000], and Nelson [1992].

The next example illustrates the effects of correlation and initial conditions on the estimation of long-run mean measures of performance of a system.

Example 11.8

Consider a single-server queue with Poisson arrivals at an average rate of one every 10 minutes ($\lambda = 0.1$ per minute), and service times which are normally distributed with a mean of 9.5 minutes and a standard deviation of

1.75 minutes.[1] This is an $M/G/1$ queue, which was described and analyzed in Section 6.4.1. By Equation (6.11), the true long-run server utilization is $\rho = \lambda E(S) = (0.1)(9.5) = 0.95$. We typically would not need to simulate such a system, since we can analyze it mathematically, but we simulate it here to illustrate difficulties that occur when trying to estimate the long-run mean queue length, L_Q, defined by Equation (6.4).

Suppose we run a single simulation for a total of 5000 minutes and observe the output process $\{L_Q(t), 0 \le t \le 5000\}$, where $L_Q(t)$ is the number of customers in the waiting line at time t minutes. To make this continuous-time process a little easier to analyze, we divide the time interval [0, 5000) into five equal subintervals of 1000 minutes and compute the average number of customers in queue for each interval individually. Specifically, the average number of customers in the queue from time $(j-1)1000$ to $j(1000)$ is

$$Y_j = \frac{1}{1000}\int_{(j-1)1000}^{j(1000)} L_Q(t)\,dt, \quad j = 1, \ldots, 5 \tag{11.1}$$

Thus, $Y_1 = \int_0^{1000} L_Q(t)\,dt/1000$ is the time-weighted-average number of customers in the queue from time 0 to time 1000, $Y_2 = \int_{1000}^{2000} L_Q(t)\,dt/1000$ is the same average over [1000, 2000), and so on. Equation (11.1) is a special case of Equation (6.4). The observations $\{Y_1, Y_2, Y_3, Y_4, Y_5\}$ provide an example of "batching" raw simulation data, in this case $\{L_Q(t), 0 \le t \le 5000\}$, and the Y_j are called *batch means*. The use of batch means in analyzing output data is discussed in Section 11.5.5. But for now, notice that batching transforms the continuous-time queue-length process $\{L_Q(t), 0 \le t \le 5000\}$ into a discrete-time batch-means process $\{Y_i, i = 1, 2, 3, 4, 5\}$, where each Y_i is an estimator of L_Q.

The simulation results of three statistically independent replications are shown in Table 11.2. Each replication, or run, uses a distinct stream of random numbers. For replication 1, Y_{1j} is the batch mean for batch j (the jth interval), as defined by Equation (11.1); similarly, Y_{2j} and Y_{3j} are defined for batch j for replications 2 and 3, respectively. Table 11.2 also gives the sample mean over each replication, $\bar{Y}_{r\cdot}$, for replications $r = 1, 2, 3$.[2] That is,

$$\bar{Y}_{r\cdot} = \frac{1}{5}\sum_{j=1}^{5} Y_{rj}, \quad r = 1, 2, 3 \tag{11.2}$$

It probably will not surprise you that if we take batch averages first, then average the batch means, or just average everything together, we get the same thing. In

[1] The range of a service time is restricted to ± 5 standard deviations, which excludes the possibility of a negative service time, but which covers well over 99.999% of the normal distribution.

[2] The dot, as in the subscript $r\cdot$, indicates summation over the second subscript; and the bar, as in $\bar{Y}_{r\cdot}$, indicates an average.

Table 11.2. Batched Average Queue Length for Three Independent Replications

Batching Interval (Minutes)	*Batch, j*	*Replication* 1, Y_{1j}	2, Y_{2j}	3, Y_{3j}
[0, 1000)	1	3.61	2.91	7.67
[1000, 2000)	2	3.21	9.00	19.53
[2000, 3000)	3	2.18	16.15	20.36
[3000, 4000)	4	6.92	24.53	8.11
[4000, 5000)	5	2.82	25.19	12.62
[0, 5000)		$\bar{Y}_{1\cdot} = 3.75$	$\bar{Y}_{2\cdot} = 15.56$	$\bar{Y}_{3\cdot} = 13.66$

other words, each $\bar{Y}_{r\cdot}$ is equivalent to the time average over the entire interval [0, 5000) for replication r, as given by Equation (6.4).

Table 11.2 illustrates the inherent variability in stochastic simulations either *within* a single replication or *across* different replications. Consider the variability within replication 3, in which the average queue length over the batching intervals varies from a low of $Y_{31} = 7.67$ customers during the first 1000 minutes to a high of $Y_{33} = 20.36$ customers during the third subinterval of 1000 minutes. Table 11.2 also shows the variability across replications. Compare Y_{15} to Y_{25} to Y_{35}, the average queue lengths over the intervals 4000 to 5000 minutes across all three replications.

Suppose for the moment that a simulation analyst makes only one replication of this model and gets the result $\bar{Y}_{1\cdot} = 3.75$ customers as an estimate of mean queue length, L_Q. How precise is the estimate? This question is usually answered by attempting to estimate the standard error of $\bar{Y}_{1\cdot}$ or by forming a confidence interval. The simulation analyst may think that the five batch means $Y_{11}, Y_{12}, \ldots, Y_{15}$ could be regarded as a random sample; however, the sequence is not independent, and in fact it is autocorrelated because all of the data are obtained from within one replication. If $Y_{11}, \ldots, Y_{15}$ were mistakenly assumed to be independent observations, and their autocorrelation ignored, the usual classical methods of statistics might severely underestimate the standard error of $\bar{Y}_{1\cdot}$, possibly leading the simulation analyst to think that a high degree of precision had been achieved. On the other hand, the averages across the three replications, $\bar{Y}_{1\cdot}$, $\bar{Y}_{2\cdot}$, and $\bar{Y}_{3\cdot}$, can be regarded as independent observations because they are derived from three different replications.

Intuitively, Y_{11} and Y_{12} are correlated because in replication 1 the queue length at the end of the time interval [0,1000) is the queue length at the beginning of the interval [1000, 2000), and similarly for any two adjacent batches within a given replication. If the system is congested at the end of one interval, it will be congested for a while at the beginning of the next time interval. Similarly, periods of low congestion tend to follow each other. Within a replication, say for $Y_{r1}, Y_{r2}, \ldots, Y_{r5}$, high values of a batch mean tend to be followed by high values, and low values by low. This tendency of adjacent observations to have like values is known as positive autocorrelation. The effect of ignoring autocorrelation when it is present is discussed in more detail in Section 11.5.2. ◀

Now suppose that the purpose of the $M/G/1$ queueing simulation of Example 11.8 is to estimate "steady-state" mean queue length — that is, mean queue length under "typical operating conditions over the long run." However, each of the three replications was begun in the empty and idle state (no customers in the queue and the server available). The empty and idle initial state means that within a given replication there will be a higher than "typical" probability of the system being uncongested for times close to 0. The practical effect is that an estimator of L_Q, say $\bar{Y}_{r\cdot}$ for replication r, will be biased low [i.e., $E(\bar{Y}_{r\cdot}) < L_Q$]. The extent of the bias decreases as the run length increases, but for short-run-length simulations with atypical initial conditions this initialization bias can produce misleading results. The problem of initialization bias is discussed further in Section 11.5.1.

EXAMPLE 11.9 (Fifth National Bank of Jaspar, Revisited)

Recall Example 10.2 in which the Fifth National Bank of Jaspar planned to expand its drive-in service and wanted to investigate the impact of this change on customer delay. Reconsider the first run (replication number 1 in Table 10.2) of the bank model. The output variable $Y_{21} = 2.79$ minutes is the average delay of all customers served during the course of the simulation. Thus, Y_{21} is a sample mean of customer delays, say $D_1, D_2, \ldots, D_m$; that is,

$$Y_{21} = \frac{1}{m}\sum_{i=1}^{m} D_i$$

where D_i is the actual delay of customer i and m is the number of customers served during the 2 hours of the simulation. Chapter 10 did not work directly with the sequence $\{D_i, i = 1, 2, \ldots\}$, but rather with the summary sequence $Y_{21}, Y_{22}, \ldots, Y_{26}$ of average delay on each replication. The reason for avoiding direct statistical analysis of the within-replication output $\{D_i, i = 1, 2, \ldots\}$ is that this sequence is, in general, a nonstationary autocorrelated stochastic process. In contrast, the across-replication sequence $Y_{21}, Y_{22}, \ldots$ consists of independent and identically distributed random variables, and thus classical methods of statistical analysis can be applied for estimating standard errors and constructing confidence intervals.

To say that $\{D_i, i = 1, 2, \ldots\}$ is *nonstationary* means that $D_1, D_2, \ldots$ are not identically distributed, and in particular, that $E(D_i) \neq E(D_{i+1})$. This nonstationarity may be caused by the nature of the simulation, or more likely, by arbitrary initial conditions. For example, if at time 0 (11:00 A.M.) the teller is idle, then the first few customers will be more likely than usual to have short delays. In addition, if customer i has a long delay (D_i is large), the system will probably remain congested until customer $(i + 1)$ arrives, and thus the delay of customer $(i + 1)$, namely D_{i+1}, will also be large. Similarly, if D_i is small, D_{i+1} will tend to be relatively small. As in Example 11.8, this tendency of like values to be followed by like values causes the sequence $D_1, D_2, \ldots$ to be positively autocorrelated. The practical effect of the autocorrelation and of arbitrary initial conditions is similar to the effects in Example 11.8. ◀

11.3 Measures of Performance and Their Estimation

Consider the estimation of a performance parameter, θ (or ϕ), of a simulated system. It is desired to have a point estimate and an interval estimate of θ (or ϕ). The length of the interval estimate is a measure of the error in the point estimate. The simulation output data is of the form $\{Y_1, Y_2, \ldots, Y_n\}$ for estimating θ; we refer to such output data as *discrete-time data*, because the index n is discrete valued. The simulation output data is of the form $\{Y(t), 0 \leq t \leq T_E\}$ for estimating ϕ; we refer to such output data as *continuous-time data*, because the index t is continuous valued. For example, Y_i might be the delay of customer i, or the total cost in week i; while $Y(t)$ might be the queue length at time t, or the number of backlogged orders at time t. The parameter θ is an ordinary mean; ϕ will be referred to as a time-weighted mean. Whether we call the performance parameter θ or ϕ or μ or even "Kyle" does not really matter; we use two different symbols here simply to provide a distinction between ordinary means and time-weighted means.

11.3.1 Point Estimation

The point estimator of θ based on the data $\{Y_1, \ldots, Y_n\}$ is defined by

$$\widehat{\theta} = \frac{1}{n}\sum_{i=1}^{n} Y_i \tag{11.3}$$

where $\widehat{\theta}$ is a sample mean based on a sample of size n. Computer simulation languages may refer to this as a "discrete-time," "collect," "tally," or "observational" statistic.

The point estimator $\widehat{\theta}$ is said to be unbiased for θ if its expected value is θ — that is, if

$$E(\widehat{\theta}) = \theta \tag{11.4}$$

But in general,

$$E(\widehat{\theta}) \neq \theta \tag{11.5}$$

and $E(\widehat{\theta}) - \theta$ is called the *bias* in the point estimator $\widehat{\theta}$. It is desirable to have estimators that are unbiased or, if this is not possible, that have a small bias relative to the magnitude of θ. Examples of estimators of the form of Equation (11.3) include $\widehat{w}$ and $\widehat{w}_Q$ of Equations (6.5) and (6.7), in which case Y_i is the time spent in the (sub)system by customer i.

The point estimator of ϕ based on the data $\{Y(t), 0 \leq t \leq T_E\}$, where T_E is the simulation-run length, is defined by

$$\widehat{\phi} = \frac{1}{T_E} \int_0^{T_E} Y(t)\, dt \tag{11.6}$$

and is called a time average of $Y(t)$ over $[0, T_E]$. Simulation languages may refer to this as a "continuous-time," "discrete-change," or "time-persistent" statistic. In general,

$$E(\widehat{\phi}) \neq \phi \tag{11.7}$$

and $\widehat{\phi}$ is said to be biased for ϕ. Again, we would like to obtain unbiased, or low-bias, estimators. Examples of time averages include $\widehat{L}$ and $\widehat{L}_Q$ of Equations (6.3) and (6.4) and Y_j of Equation (11.1).

Generally, θ and ϕ are regarded as mean measures of performance of the system being simulated. Other measures usually can be put into this common framework. For example, consider estimation of the proportion of days on which sales are lost due to an out-of-stock situation. In the simulation, let

$$Y_i = \begin{cases} 1, & \text{if out of stock on day } i \\ 0, & \text{otherwise} \end{cases}$$

With n equal to the total number of days, $\widehat{\theta}$ defined by Equation (11.3) is a point estimator of θ, the proportion of out-of-stock days. For a second example, consider estimation of the proportion of time that queue length is greater than k_0 customers (for example, $k_0 = 10$). If $L_Q(t)$ represents simulated queue length at time t, then in the simulation define

$$Y(t) = \begin{cases} 1, & \text{if } L_Q(t) > k_0 \\ 0, & \text{otherwise} \end{cases}$$

Then $\widehat{\phi}$ defined by Equation (11.6) is a point estimator of ϕ, the proportion of time that the queue length is greater than k_0 customers. Thus, estimation of proportions or probabilities is a special case of the estimation of means.

A performance measure that does not fit this common framework is a quantile or percentile. Quantiles describe the level of performance that can be delivered with a given probability, p. For instance, suppose that Y represents the delay in queue that a customer experiences in a service system, measured in minutes. Then the 0.85 quantile of Y is the value θ such that

$$\Pr\{Y \leq \theta\} = p \tag{11.8}$$

where $p = 0.85$ in this case. As a percentage, θ is the $100p$th or 85th percentile of customer delay. Therefore, 85% of all customers will experience a delay of θ minutes or less. Stated differently, a customer has only a 0.15 probability of experiencing a delay of longer than θ minutes. A widely used performance measure is the median, which is the 0.5 quantile or 50th percentile.

The problem of estimating a quantile is the inverse of the problem of estimating a proportion or probability. Consider Equation (11.8). When estimating a proportion, θ is given and p is to be estimated; but when estimating a quantile, p is given and θ is to be estimated.

The easiest method for estimating a quantile is to form a histogram of the observed values of Y, then find a value $\widehat{\theta}$ such that $100p\%$ of the histogram is to the left of (smaller than) $\widehat{\theta}$. For instance, if we observe $n = 250$ customer delays $\{Y_1, \ldots, Y_{250}\}$, then an estimate of the 85th percentile of delay is a value $\widehat{\theta}$ such that $(0.85)(250) = 212.5 \approx 213$ of the observed values are less than or equal to θ. An obvious estimate is therefore to set $\widehat{\theta}$ equal to the 213th smallest value in the sample (this requires sorting the data). When the output is a continuous-time process, such as the queue-length process $\{L_Q(t), 0 \le t \le T_E\}$, then a histogram gives the *fraction of time* that the process spent at each possible level (queue length in this example). However, the method for quantile estimation remains the same: Find a value $\widehat{\theta}$ such that $100p\%$ of the histogram is to the left of $\widehat{\theta}$.

11.3.2 Interval Estimation

Valid interval estimation typically requires a method of estimating the variance of the point estimator, $\widehat{\theta}$ (everything we say in this section applies to $\widehat{\phi}$ as well). Let $\sigma^2(\widehat{\theta}) = \text{var}(\widehat{\theta})$ represent the true variance of a point estimator $\widehat{\theta}$, and let $\widehat{\sigma}^2(\widehat{\theta})$ represent an estimator of $\sigma^2(\widehat{\theta})$ based on the data $\{Y_1, \ldots, Y_n\}$. As with point estimators, it is desirable to have an unbiased variance estimator, but frequently

$$E[\widehat{\sigma}^2(\widehat{\theta})] \ne \sigma^2(\widehat{\theta}) \tag{11.9}$$

and the estimator is biased. If $\widehat{\sigma}^2(\widehat{\theta})$ is approximately unbiased, then under fairly general conditions the statistic

$$t = \frac{\widehat{\theta} - \theta}{\widehat{\sigma}(\widehat{\theta})}$$

is approximately t-distributed with some number of degrees of freedom, say f. As a shorthand we write $t \sim t_f$. Therefore, an approximate $100(1 - \alpha)\%$ confidence interval for θ is given by

$$\widehat{\theta} \pm t_{\alpha/2,f}\widehat{\sigma}(\widehat{\theta})$$

or equivalently

$$\widehat{\theta} - t_{\alpha/2,f}\widehat{\sigma}(\widehat{\theta}) \le \theta \le \widehat{\theta} + t_{\alpha/2,f}\widehat{\sigma}(\widehat{\theta}) \tag{11.10}$$

where $t_{\alpha/2,f}$ is the $100(1 - \alpha/2)$ percentage point of a t-distribution with f degrees of freedom; that is, $t_{\alpha/2,f}$ is defined by $P(t \ge t_{\alpha/2,f}) = \alpha/2$. A confidence interval of the form of (11.10) will be approximately correct, provided that the point estimator is an average of the form of Equation (11.3) or (11.6) and is relatively unbiased, and the variance estimator $\widehat{\sigma}^2(\widehat{\theta})$ is approximately unbiased for $\sigma^2(\widehat{\theta})$. (Since a quantile or percentile estimate is not an average, a different approach is required to obtain an interval estimator, as described later in the chapter.) One of the major problems in simulation output analysis is obtaining approximately unbiased estimates of $\sigma^2(\widehat{\theta})$, the variance of the point estimator.

11.4 Output Analysis for Terminating Simulations

Consider a terminating simulation that runs over a simulated time interval $[0, T_E]$ and results in observations $Y_1, \ldots, Y_n$. The sample size, n, may be a fixed number, or it may be a random variable (say, the number of observations that occur during time T_E). The goal of the simulation is to estimate

$$\theta = E\left(\frac{1}{n}\sum_{i=1}^{n} Y_i\right)$$

The method used is called the method of independent replications. The simulation is repeated a total of R times, each run using a different random-number stream and independently chosen initial conditions (which includes the case that all runs have identical initial conditions). Let Y_{ri} be the ith observation within replication r, for $i = 1, \ldots, n_r$, and $r = 1, \ldots, R$. For fixed replication r, $Y_{r1}, Y_{r2}, \ldots$ is an autocorrelated sequence *within* replication r; but *across* different replications $r \neq s$, Y_{ri} and Y_{sj} are statistically independent. Define a sample mean $\widehat{\theta}_r$, within each replication r, by Equation (11.3), which becomes

$$\widehat{\theta}_r = \frac{1}{n_r}\sum_{i=1}^{n_r} Y_{ri}, \quad r = 1, \ldots, R \tag{11.11}$$

The R sample means $\widehat{\theta}_1, \ldots, \widehat{\theta}_R$ are statistically independent and identically distributed and are unbiased estimators of θ. Thus, classical methods of confidence interval estimation can be applied.

11.4.1 Statistical Background

Suppose $\{Y_1, \ldots, Y_n\}$ are statistically independent observations. We summarize the classical method for constructing a confidence interval for the mean here, then specialize it to simulation in the following section.

In the case of independent observations, compute $\widehat{\theta}$ by Equation (11.3) and then compute the sample variance, S^2, by

$$S^2 = \sum_{i=1}^{n}\frac{(Y_i - \widehat{\theta})^2}{n-1} = \frac{\sum_{i=1}^{n} Y_i^2 - n\widehat{\theta}^2}{n-1} \tag{11.12}$$

When the Y_i are independent and identically distributed, the sample variance, S^2, is an unbiased estimator of the population variance $\sigma^2 = \text{var}(Y_i)$

(constant for all $i = 1, \ldots, n$). Since the variance of $\widehat{\theta}$ is given by

$$\sigma^2(\widehat{\theta}) = \frac{\sigma^2}{n} \tag{11.13}$$

an unbiased estimator of $\sigma^2(\widehat{\theta})$, with $f = n-1$ degrees of freedom, is provided by

$$\widehat{\sigma}^2(\widehat{\theta}) = \frac{S^2}{n} \tag{11.14}$$

Thus, the confidence interval given by Inequality (11.10) is approximately valid, provided the point estimate $\widehat{\theta}$ is unbiased. The quantity $\widehat{\sigma}(\widehat{\theta}) = S/\sqrt{n}$, sometimes denoted by s.e.$(\widehat{\theta})$, is called the *standard error* of the point estimator $\widehat{\theta}$. The standard error is also a measure of the precision of a point estimator, and can be interpreted as the *average deviation* to be expected between the point estimator and the true mean.

11.4.2 Confidence-Interval Estimation for a Fixed Number of Replications

Suppose that R independent replications are made, resulting in the independent estimates defined by Equation (11.11). Compute the overall point estimate, $\widehat{\theta}$, by

$$\widehat{\theta} = \frac{1}{R}\sum_{r=1}^{R}\widehat{\theta}_r \tag{11.15}$$

and estimate the variance of $\widehat{\theta}$ by Equation (11.14), which becomes in this context (with $n = R$),

$$\widehat{\sigma}^2(\widehat{\theta}) = \frac{1}{(R-1)R}\sum_{r=1}^{R}(\widehat{\theta}_r - \widehat{\theta})^2 \tag{11.16}$$

A $100(1-\alpha)\%$ confidence interval is given by expression (11.10) with the degrees of freedom $f = R - 1$. The quantity $\widehat{\sigma}(\widehat{\theta}) = \sqrt{\widehat{\sigma}^2(\widehat{\theta})}$ is called the standard error of the point estimator $\widehat{\theta}$. Its magnitude is an indication of the precision of $\widehat{\theta}$ as an estimator of θ. As R increases, the standard error $\widehat{\sigma}(\widehat{\theta})$ tends to become smaller and approach zero.

When the output data are of the form $\{Y_r(t), 0 \le t \le T_E\}$, for $r = 1, \ldots, R$ independent replications, then analogously to Equations (11.11), (11.15), and (11.16), compute

$$\widehat{\phi}_r = \frac{1}{T_E}\int_0^{T_E} Y_r(t)\,dt, \quad r = 1, \ldots, R, \tag{11.17}$$

$$\widehat{\phi} = \frac{1}{R}\sum_{r=1}^{R}\widehat{\phi}_r \tag{11.18}$$

and

$$\widehat{\sigma}^2(\widehat{\phi}) = \frac{1}{(R-1)R}\sum_{r=1}^{R}(\widehat{\phi}_r - \widehat{\phi})^2 \tag{11.19}$$

The confidence interval is analogous to that of (11.10). The time averages $\widehat{\phi}_1, \ldots, \widehat{\phi}_R$ are statistically independent; $\widehat{\phi}$ is an unbiased point estimator of ϕ, where by definition

$$\phi = E\left(\frac{1}{T_E}\int_0^{T_E} Y_r(t)\,dt\right)$$

and $\widehat{\sigma}^2(\widehat{\phi})$ is an unbiased estimator of $\sigma^2(\widehat{\phi}) = \mathrm{var}(\widehat{\phi})$. Therefore, the confidence interval of (11.10) is valid, again with degrees of freedom $f = R - 1$.

EXAMPLE 11.10 (The Able-Baker Carhop Problem, Continued)

Consider Example 11.7, the Able-Baker carhop problem, with the data for $R = 4$ replications given in Table 11.1. The four utilization estimates, $\widehat{\rho}_r$, are time averages of the form of $\widehat{\phi}_r$ of Equation (11.17). The simulation would produce output data of the form

$$Y_r(t) = \begin{cases} 1, & \text{if Able is busy at time } t \\ 0, & \text{otherwise} \end{cases}$$

and $\widehat{\rho}_r = \widehat{\phi}_r$ would be computed by Equation (11.17) with $T_E = 2$ hours.

The four average system times, $\widehat{w}_1, \ldots, \widehat{w}_4$, are analogous to $\widehat{\theta}_r$ of Equation (11.11), where Y_{ri} is the actual time spent in the system by customer i on replication r.

First, suppose that the analyst desires a 95% confidence interval for Able's true utilization, ρ. Using Equation (11.18), compute an overall point estimator

$$\widehat{\rho} = \frac{0.808 + 0.875 + 0.708 + 0.842}{4} = 0.808$$

Using Equation (11.19), compute the estimated variance of $\widehat{\rho}$ by

$$\widehat{\sigma}^2(\widehat{\rho}) = \frac{(0.808 - 0.808)^2 + \cdots + (0.842 - 0.808)^2}{3(4)} = (0.036)^2$$

Thus, the standard error of $\widehat{\rho} = 0.808$ is estimated by s.e.$(\widehat{\rho}) = \widehat{\sigma}(\widehat{\rho}) = 0.036$. Obtain $t_{0.025,3} = 3.18$ from Table A.5, and compute the 95% confidence interval by

$$\widehat{\rho} \pm t_{0.025,3}\,\widehat{\sigma}(\widehat{\rho})$$

$$0.808 \pm (3.18)(0.036)$$

or with 95% confidence,

$$0.694 \le \rho \le 0.922$$

In a similar fashion, compute a 95% confidence interval for mean time in system w:

$$\widehat{w} = \frac{3.74 + 4.53 + 3.84 + 3.98}{4} = 4.02 \text{ minutes}$$

$$\widehat{\sigma}^2(\widehat{w}) = \frac{(3.74 - 4.02)^2 + \cdots + (3.98 - 4.02)^2}{3(4)} = (0.176)^2$$

so that

$$\widehat{w} - t_{0.025,3}\,\widehat{\sigma}(\widehat{w}) \leq w \leq \widehat{w} + t_{0.025,3}\,\widehat{\sigma}(\widehat{w})$$

or

$$3.46 = 4.02 - (3.18)(0.176) \leq w \leq 4.02 + (3.18)(0.176) = 4.58$$

Thus, the 95% confidence interval for w is $3.46 \leq w \leq 4.58$. ◀

EXAMPLE 11.11

To show the effect of increasing the run length T_E and increasing the number of replications R, the Able-Baker carhop problem of Example 11.10 was simulated for all combinations of 2, 4, and 8 hours and 4, 8, and 16 replications. The output data for utilization are shown in Table 11.3 for all 16 independent replications. Within each replication, the 8-hour run is a continuation of the 4-hour run, which is a continuation of the 2-hour run. Table 11.4 exhibits the point estimates and their standard errors: $\widehat{\rho} \pm \text{s.e.}(\widehat{\rho})$, where $\text{s.e.}(\widehat{\rho}) = \widehat{\sigma}(\widehat{\rho})$ is an estimate of the standard deviation. Theoretically, the standard deviation of $\widehat{\rho}$ is given by

$$\sigma(\widehat{\rho}) = \sqrt{\text{var}(\widehat{\rho})} = \frac{\sigma}{\sqrt{R}}$$

where $\sigma^2 = \text{var}(\widehat{\rho}_r)$ is the variance of the population of which $\widehat{\rho}_1, \ldots, \widehat{\rho}_R$ is a random sample. Since $\text{s.e.}(\widehat{\rho})$ is an estimate of $\sigma(\widehat{\rho})$, we expect $\text{s.e.}(\widehat{\rho})$ to also decrease proportionally to $1/\sqrt{R}$. In other words, the standard error for 8 replications should be approximately $1/\sqrt{2} = 0.71$ as large as that for 4 replications; and the standard error for 16 replications should be approximately $1/\sqrt{4} = 0.5$ as large as that for 4 replications. [The relationship between sample size, R, and standard deviation, $\sigma(\widehat{\rho})$, will hold only approximately for $\text{s.e.}(\widehat{\rho})$, because a standard error is an estimator and thus it contains a certain amount of random error.] Examination of Table 11.4 shows that this relationship holds approximately for the utilization data. From the standpoint of precision, if the simulation analyst desires to reduce $\text{s.e.}(\widehat{\rho})$ by one-half, it is necessary to quadruple the number of replications (say from 4 to 16, or 8 to 32).

Increasing the run length T_E will also decrease the true standard deviation, $\sigma(\widehat{\rho})$, of the point estimator. As T_E increases in Table 11.4, the estimated standard error, $\text{s.e.}(\widehat{\rho})$, usually decreases, any exception (such as from 4 hours to 8 hours with four replications) being due to the random variation in the data. In a terminating simulation, this option is not available to the simulation analyst because, by definition, T_E is determined and fixed by the nature of the system

Table 11.3. Output Data: Able's Observed Utilization by Replication and Run Length

Replication,	*Run Length,* T_E		
r	*2 Hours*	*4 Hours*	*8 Hours*
1	0.808	0.796	0.785
2	0.875	0.825	0.833
3	0.708	0.787	0.806
4	0.842	0.837	0.833
5	0.742	0.825	0.808
6	0.767	0.775	0.800
7	0.792	0.787	0.794
8	0.950	0.867	0.827
9	0.833	0.821	0.815
10	0.717	0.750	0.821
11	0.817	0.808	0.798
12	0.842	0.746	0.817
13	0.850	0.846	0.854
14	0.850	0.846	0.848
15	0.767	0.783	0.796
16	0.817	0.804	0.813

being simulated. Nevertheless, the data are presented for illustration purposes. Notice that for equal total run length over all replications ($R \times T_E$), such as 8 replications of 4 hours each (= 32 hours) or 4 replications of 8 hours each (= 32 hours), the point estimators have approximately equal precision. ◀

Table 11.4. Able's Estimated Utilization with Standard Error $\widehat{\rho} \pm \text{s.e.}(\widehat{\rho})$ for Various Run Lengths and Number of Replications

Number of Replications	*Run Length,* T_E		
R	*2 Hours*	*4 Hours*	*8 Hours*
4	0.808 ± 0.036	0.811 ± 0.011	0.814 ± 0.011
8	0.811 ± 0.027	0.812 ± 0.011	0.810 ± 0.006
16	0.811 ± 0.015	0.806 ± 0.009	0.816 ± 0.005

11.4.3 Confidence Intervals with Specified Precision

By expression (11.10), the half-length (h.l.) of a $100(1-\alpha)\%$ confidence interval for a mean θ, based on the t distribution, is given by

$$\text{h.l.} = t_{\alpha/2,R-1}\widehat{\sigma}(\widehat{\theta})$$

where $\widehat{\sigma}(\widehat{\theta}) = S/\sqrt{R}$, S is the sample standard deviation, and R is the number of replications. Suppose that an error criterion ϵ is specified; in other words, it is desired to estimate θ by $\widehat{\theta}$ to within $\pm\epsilon$ with high probability, say at least $1 - \alpha$. Thus it is desired that a sufficiently large sample size, R, be taken to satisfy

$$P(|\widehat{\theta} - \theta| < \epsilon) \geq 1 - \alpha$$

When the sample size, R, is fixed, as in Section 11.4.2, no guarantee can be given for the resulting error. But if the sample size can be increased, an error criterion can be specified.

Assume that an initial sample of size R_0 has been observed; that is, the simulation analyst initially makes R_0 independent replications. In practice, R_0 is 2 or larger, but at least 4 or 5 is recommended with 10 or more being desirable. The R_0 replications will be used to obtain an initial estimate S_0^2 of the population variance σ^2. To meet the half-length criterion, a sample size R must be chosen such that $R \geq R_0$ and

$$\text{h.l.} = \frac{t_{\alpha/2,R-1} S_0}{\sqrt{R}} \leq \epsilon \tag{11.20}$$

Solving for R in Inequality (11.20) shows that R is the smallest integer satisfying $R \geq R_0$ and

$$R \geq \left(\frac{t_{\alpha/2,R-1} S_0}{\epsilon}\right)^2 \tag{11.21}$$

Since $t_{\alpha/2,R-1} \geq z_{\alpha/2}$, an initial estimate for R is given by

$$R \geq \left(\frac{z_{\alpha/2} S_0}{\epsilon}\right)^2 \tag{11.22}$$

where $z_{\alpha/2}$ is the $100(1-\alpha/2)$ percentage point of the standard normal distribution. And since $t_{\alpha/2,R-1} \approx z_{\alpha/2}$ for large R (say $R \geq 50$), the second inequality for R is adequate when R is large. After determining the final sample size, R, collect $R - R_0$ additional observations (i.e., make $R - R_0$ additional replications, or start over and make R total replications) and form the $100(1 - \alpha)\%$ confidence interval for θ by

$$\widehat{\theta} - t_{\alpha/2,R-1}\frac{S}{\sqrt{R}} \leq \theta \leq \widehat{\theta} + t_{\alpha/2,R-1}\frac{S}{\sqrt{R}} \tag{11.23}$$

where $\widehat{\theta}$ and S are computed based on all R replications, $\widehat{\theta}$ by Equation (11.15) and $S^2/R = \widehat{\sigma}^2(\widehat{\theta})$ by Equation (11.16). The half-length of the confidence interval given by Inequality (11.23) should be approximately ϵ or smaller; however, with the additional $R - R_0$ observations, the variance estimator S^2 may differ somewhat from the initial estimate S_0^2, possibly causing the half-length to be greater than desired. If the confidence interval (11.23) is too large, the procedure may be repeated, using Inequality (11.21) to determine an even larger sample size.

EXAMPLE 11.12

Suppose that it is desired to estimate Able's utilization in Example 11.7 to within ± 0.04 with probability 0.95. An initial sample of size $R_0 = 4$ is taken, with the results given in Table 11.1 (also, Table 11.3 with $T_E = 2$ hours). An initial estimate of the population variance is $S_0^2 = R_0\widehat{\sigma}^2(\widehat{\rho}) = 4(0.036)^2 = 0.00518$. (See Example 11.10 or Table 11.4 for the relevant data. Notice how we recover S_0^2 from the standard error by squaring it and multiplying by the number of replications.) The error criterion is $\epsilon = 0.04$ and the confidence coefficient is $1 - \alpha = 0.95$. From Inequality (11.22), the final sample size must be at least as large as

$$\frac{z_{0.025}^2 S_0^2}{\epsilon^2} = \frac{(1.96)^2(0.00518)}{(0.04)^2} = 12.44$$

Next, Inequality (11.21) can be used to test possible candidates ($R = 13, 14, \ldots$) for final sample size, as follows:

R	13	14	15
$t_{0.025,R-1}$	2.18	2.16	2.14
$\dfrac{t_{0.025,R-1}^2 S_0^2}{\epsilon^2}$	15.39	15.10	14.83

Thus, $R = 15$ is the smallest integer satisfying Inequality (11.21), so $R - R_0 = 15 - 4 = 11$ additional replications are needed. Assuming that 12 additional replications (instead of 11) were made, with the output data $\widehat{\rho}_r$ as given in Table 11.3, the half-length for a 95% confidence interval would be given by $t_{0.025,15}$ s.e.$(\widehat{\rho}) = 2.13(0.015) = 0.032$, which is well within the precision criterion $\epsilon = 0.04$. The resulting 95% confidence interval for true utilization ρ is $0.811 - 0.032 = 0.779 \le \rho \le 0.843 = 0.811 + 0.032$. Notice the considerable improvement in precision (as measured by confidence interval half-length) compared to Example 11.10, which used a fixed sample size ($R = 4$). ◀

11.4.4 *Confidence Intervals for Quantiles*

To present the interval estimator for quantiles, it is helpful to review the interval estimator for a mean in the special case when the mean represents a proportion or probability, p. In this book we have chosen to treat a proportion or probability as just a special case of a mean. However, in many statistics texts probabilities are treated separately.

When the number of independent replications $Y_1, \ldots, Y_R$ is large enough so that $t_{\alpha/2,R-1} \doteq z_{\alpha/2}$, the confidence interval for a probability p is often written as

$$\widehat{p} \pm z_{\alpha/2}\sqrt{\frac{\widehat{p}(1-\widehat{p})}{R-1}}$$

where $\widehat{p}$ is the sample proportion [tedious algebra shows that this formula is precisely equivalent to Equation (11.10) when estimating a proportion].

As mentioned in Section 11.3, the quantile estimation problem is the inverse of the probability estimation problem: Find θ such that $\Pr\{Y \leq \theta\} = p$. Thus, to estimate the p quantile, we find that value $\widehat{\theta}$ such that $100p\%$ of the data in a histogram of Y is to the left of $\widehat{\theta}$ (or stated differently, the Rpth smallest value of $Y_1, \ldots, Y_R$).

Extending this idea, an approximate $(1-\alpha)100\%$ confidence interval for θ can be obtained by finding the value θ_ℓ that cuts off $100p_\ell\%$ of the histogram, and θ_u that cuts off $100p_u\%$ of the histogram, where

$$\begin{aligned} p_\ell &= p - z_{\alpha/2}\sqrt{\frac{p(1-p)}{R-1}} \\ p_u &= p + z_{\alpha/2}\sqrt{\frac{p(1-p)}{R-1}} \end{aligned} \tag{11.24}$$

(recall that we know p). In terms of sorted values $\widehat{\theta}_\ell$ is the Rp_ℓ smallest value (rounded down), and $\widehat{\theta}_u$ is the Rp_u smallest value (rounded up), of $Y_1, \ldots, Y_R$.

Example 11.13

Suppose that we want to estimate the 0.8 quantile of the time to failure (in hours) for the communications system in Example 11.3 and form a 95% confidence interval for it. A histogram of $R = 500$ independent replications is shown in Figure 11.2.

The point estimator is $\widehat{\theta} = 4644$ hours, because 80% of the data in the histogram are to the left of 4644. Equivalently, it is the $500 \times 0.8 = 400$th smallest value of the sorted data.

To obtain the confidence interval, we first compute

$$p_\ell = p - z_{\alpha/2}\sqrt{\frac{p(1-p)}{R-1}} = 0.8 - 1.96\sqrt{\frac{0.8(0.2)}{499}} = 0.765$$

$$p_u = p + z_{\alpha/2}\sqrt{\frac{p(1-p)}{R-1}} = 0.8 + 1.96\sqrt{\frac{0.8(0.2)}{499}} = 0.835$$

The lower bound of the confidence interval is $\widehat{\theta}_\ell = 4173$ (the $500 \times p_\ell =$ 382nd smallest value, rounding down), while the upper bound of the confidence interval is $\widehat{\theta}_u = 5119$ hours (the $500 \times p_u = 418$th smallest value, rounding up). ◀

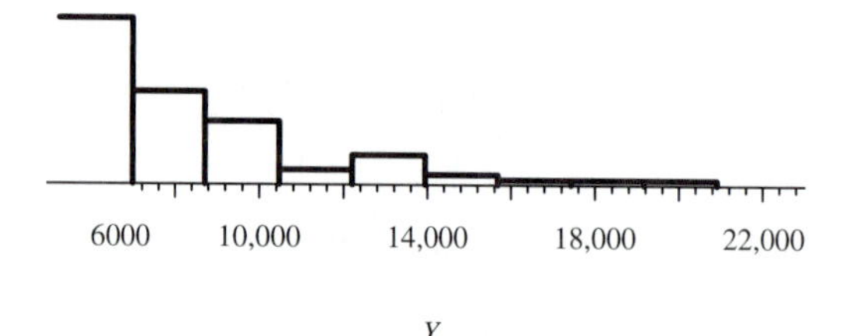

Figure 11.2. Failure data in hours for 500 replications of the communications system.

11.5 Output Analysis for Steady-State Simulations

Consider a single run of a simulation model whose purpose is to estimate a *steady-state*, or *long-run*, characteristic of the system. Suppose that the single run produces observations $Y_1, Y_2, \ldots$, which, generally, are samples of an autocorrelated time series. The steady-state (or long-run) measure of performance, θ, to be estimated is defined by

$$\theta = \lim_{n \longrightarrow \infty} \frac{1}{n} \sum_{i=1}^{n} Y_i \tag{11.25}$$

with probability 1, where the value of θ is independent of the initial conditions. (The phrase "with probability 1" means that essentially all simulations of the model, using different random numbers, will produce series $Y_i, i = 1, 2, \ldots$ whose sample average converges to θ.) For example, if Y_i was the total time job i spent in a job shop, then θ would be the long-run average time a job spends in the shop; and because θ is defined as a limit, it is independent of shop conditions at time 0.

Of course, the simulation analyst must decide to stop the simulation after some number of observations, say n, have been collected; or the simulation analyst may decide to simulate for some length of time T_E that determines n (although n may vary from run to run). The sample size n (or T_E) is a *design* choice; it is not inherently determined by the nature of the problem. The simulation analyst will choose simulation run length (n or T_E) with several considerations in mind:

1. The bias in the point estimator due to artificial or arbitrary initial conditions. The bias can be severe if run length is too short, but it generally decreases as run length increases.
2. The desired precision of the point estimator, as measured by an estimate of point-estimator variability.
3. Budget constraints on computer resources.

The next subsection discusses initialization bias and the following subsections outline two methods of estimating point-estimator variability. When discussing one replication (or run), the notation

$$Y_1, Y_2, Y_3, \ldots$$

will be used; if several replications have been made, the output data for replication r will be denoted by

$$Y_{r1}, Y_{r2}, Y_{r3}, \ldots \tag{11.26}$$

11.5.1 Initialization Bias in Steady-State Simulations

There are several methods of reducing the point-estimator bias which is caused by using artificial and unrealistic initial conditions in a steady-state simulation.

The first method is to initialize the simulation in a state that is more representative of long-run conditions. This method is sometimes called intelligent initialization. Examples include

1. Setting the inventory levels, number of backorders, and number of items on order and their arrival dates in an inventory simulation.
2. Placing customers in queue and in service in a queueing simulation.
3. Having some components failed or degraded in a reliability simulation.

There are at least two ways to intelligently specify the initial conditions. If the system exists, collect data on it and use these data to specify more typical initial conditions. This method sometimes requires a large data-collection effort. In addition, if the system being modeled does not exist—for example, if it is a variant of an existing system—this method is impossible to implement. Nevertheless, it is recommended that simulation analysts use any available data on existing systems to help initialize the simulation, as this will usually be better than assuming the system to be "completely stocked," "empty and idle," or "brand new" at time 0.

A related idea is to obtain initial conditions from a second model of the system that has been simplified enough to make it mathematically solvable. The queueing models in Chapter 6 are very useful for this purpose. The simplified model can be solved to find long-run expected or most likely conditions—such as the expected number of customers in the queue—and these conditions can be used to intialize the simulation.

A second method to reduce the impact of initial conditions, possibly used in conjunction with the first, is to divide each simulation run into two phases: first an initialization phase from time 0 to time T_0, followed by a data-collection

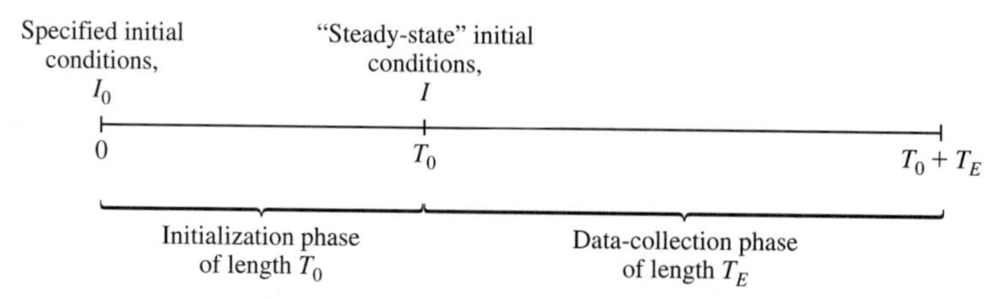

Figure 11.3. Initialization and data-collection phases of a steady-state simulation run.

phase from time T_0 to the stopping time $T_0 + T_E$. That is, the simulation begins at time 0 under specified initial conditions I_0, and runs for a specified period of time T_0. Data collection on the response variables of interest does not begin until time T_0 and continues until time $T_0 + T_E$. The choice of T_0 is quite important, as the system state at time T_0, denoted by I, should be more representative of steady-state behavior than the original initial conditions at time 0, I_0. In addition, the length T_E of the data-collection phase should be long enough to guarantee sufficiently precise estimates of steady-state behavior. Notice that the system state, I, at time T_0 is a random variable, and to say that the system has reached an approximate steady state is to say that the probability distribution of the system state at time T_0 is sufficiently close to the steady-state probability distribution as to make the bias in point estimates of response variables negligible. Figure 11.3 illustrates the two phases of a steady-state simulation. The effect of starting a simulation run of a queueing model in the empty and idle state, as well as several useful plots to aid the simulation analyst in choosing an appropriate value of T_0, are given in the following example.

Example 11.14

Consider the $M/G/1$ queue discussed in Example 11.8. Suppose that a total of 10 independent replications were made ($R = 10$), each replication beginning in the empty and idle state. The total simulation run length on each replication was $T_0 + T_E = 15{,}000$ minutes. The response variable was queue length, $L_Q(t, r)$, at time t, where the second argument, r, denotes the replication ($r = 1, \ldots, 10$). The raw output data were batched, as in Example 11.8, Equation (11.1), in batching intervals of 1000 minutes to produce the following batch means:

$$Y_{rj} = \frac{1}{1000} \int_{(j-1)1000}^{j(1000)} L_Q(t, r)dt \tag{11.27}$$

for replication $r = 1, \ldots, 10$ and for batch $j = 1, 2, \ldots, 15$. The estimator in Equation (11.27) is simply the time-weighted-average queue length over the time interval $[(j-1)1000, j(1000))$, similar to that in Equation (6.4). The 15 batch means for the 10 replications are given in Table 11.5.

Table 11.5. Individual Batch Means (y_{rj}) for $M/G/1$ Simulation with Empty and Idle Initial State

	Batch														
Replication	*1*	*2*	*3*	*4*	*5*	*6*	*7*	*8*	*9*	*10*	*11*	*12*	*13*	*14*	*15*
1	3.61	3.21	2.18	6.92	2.82	1.59	3.55	5.60	3.04	2.57	1.41	3.07	4.03	2.70	2.71
2	2.91	9.00	16.15	24.53	25.19	21.63	24.47	8.45	8.53	14.84	23.65	27.58	24.19	8.58	4.06
3	7.67	19.53	20.36	8.11	12.62	22.15	14.10	9.87	23.96	24.50	14.56	6.08	4.83	16.04	23.41
4	6.62	1.75	12.87	8.77	1.25	1.16	1.92	6.29	4.74	17.43	18.24	18.59	4.62	2.76	1.57
5	2.18	1.32	2.14	2.18	2.59	1.20	4.11	6.21	7.31	1.58	2.16	3.08	2.32	2.21	3.32
6	0.93	3.54	4.80	0.72	2.95	5.56	1.96	2.07	2.74	3.45	14.24	13.39	7.87	0.94	3.19
7	1.12	2.59	5.05	1.16	2.72	5.12	5.03	4.14	4.98	15.81	9.29	2.14	8.72	29.80	28.94
8	1.54	5.94	5.33	2.91	2.69	1.91	3.27	3.61	10.35	9.66	4.13	6.14	7.90	2.61	7.95
9	8.93	4.78	0.74	2.56	9.43	18.63	8.14	1.49	4.51	1.69	12.62	11.28	3.32	3.42	3.35
10	4.78	2.84	10.39	5.87	1.01	2.59	16.77	27.25	26.81	20.96	7.26	2.32	5.04	8.50	9.11

Normally we average all the batch means *within* each replication to obtain a replication average. However, our goal at this stage is to identify the trend in the data due to initialization bias and find out when it dissipates. To do this we will average corresponding batch means *across* replications and plot them (this idea is usually attributed to Welch [1983]). Such averages are known as *ensemble averages*. Specifically, for each batch j, define the ensemble average across all R replications to be

$$\bar{Y}_{\cdot j} = \frac{1}{R}\sum_{r=1}^{R} Y_{rj} \tag{11.28}$$

($R = 10$ here). The ensemble averages $\bar{Y}_{\cdot j}$, $j = 1, \ldots, 15$ are displayed in the third column of Table 11.6. Notice that $\bar{Y}_{\cdot 1} = 4.03$ and $\bar{Y}_{\cdot 2} = 5.45$ are estimates of mean queue length over the time periods [0, 1000) and [1000, 2000), respectively, and they are less than all other ensemble averages $\bar{Y}_{\cdot j}$ ($j = 3, \ldots, 15$). The simulation analyst may suspect that this is due to the downward bias in these estimators, which in turn is due to the queue being empty and idle at time 0. This downward bias is further illustrated in the plots that follow.

Figure 11.4 is a plot of the ensemble averages, $\bar{Y}_{\cdot j}$, versus $1000j$, for $j = 1, 2, \ldots, 15$. The actual values, $\bar{Y}_{\cdot j}$, are the discrete set of points in circles, which have been connected by straight lines as a visual aid. Figure 11.4 illustrates the

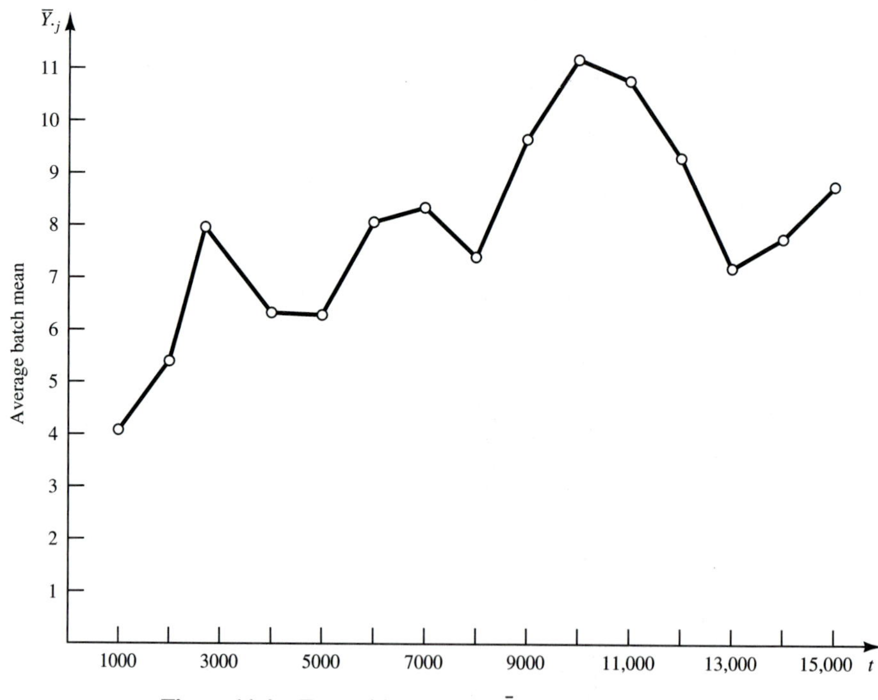

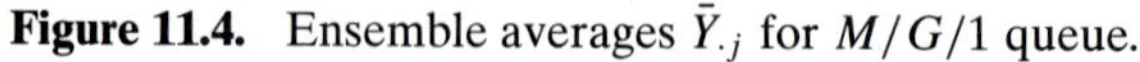
Figure 11.4. Ensemble averages $\bar{Y}_{\cdot j}$ for $M/G/1$ queue.

Table 11.6. Summary of Data for *M*/*G*/1 Simulation: Ensemble Batch Means and Cumulative Means, Averaged Over 10 Replications

Run Length, T	Batch, j	Ensemble Average Batch Mean, $\bar{Y}_{\cdot j}$	Cumulative Average (No Deletion), $\bar{Y}_{\cdot\cdot}(j,0)$	Cumulative Average (Delete 1), $\bar{Y}_{\cdot\cdot}(j,1)$	Cumulative Average (Delete 2), $\bar{Y}_{\cdot\cdot}(j,2)$
1,000	1	4.03	4.03	—	—
2,000	2	5.45	4.74	5.45	—
3,000	3	8.00	5.83	6.72	8.00
4,000	4	6.37	5.96	6.61	7.18
5,000	5	6.33	6.04	6.54	6.90
6,000	6	8.15	6.39	6.86	7.21
7,000	7	8.33	6.67	7.11	7.44
8,000	8	7.50	6.77	7.16	7.45
9,000	9	9.70	7.10	7.48	7.77
10,000	10	11.25	7.51	7.90	8.20
11,000	11	10.76	7.81	8.18	8.49
12,000	12	9.37	7.94	8.29	8.58
13,000	13	7.28	7.89	8.21	8.46
14,000	14	7.76	7.88	8.17	8.40
15,000	15	8.76	7.94	8.21	8.43

downward bias of the initial observations. As time becomes larger, the effect of the initial conditions on later observations lessens and the observations appear to vary around a common mean. When the simulation analyst feels that this point has been reached, then the data-collection phase begins.

Table 11.6 also gives the cumulative average sample mean after deleting zero, one, and two batch means from the beginning. That is, using the ensemble average batch means $\bar{Y}_{\cdot j}$, when deleting d observations out of a total of n observations, compute

$$\bar{Y}_{\cdot\cdot}(n,d) = \frac{1}{n-d}\sum_{j=d+1}^{n} \bar{Y}_{\cdot j} \tag{11.29}$$

The results in Table 11.6 for the $M/G/1$ simulation are for $d = 0, 1,$ and 2, and $n = d+1, \ldots, 15$. These cumulative averages with deletion, namely $\bar{Y}_{\cdot\cdot}(n,d)$, are plotted for comparison purposes in Figure 11.5. We do not recommend using cumulative averages to determine the initialization phase for reasons given below.

From Figures 11.4 and 11.5 it is apparent that downward bias is present, and this initialization bias in the point estimator can be reduced by deletion of one or more observations. For the 15 ensemble average batch means, it appears

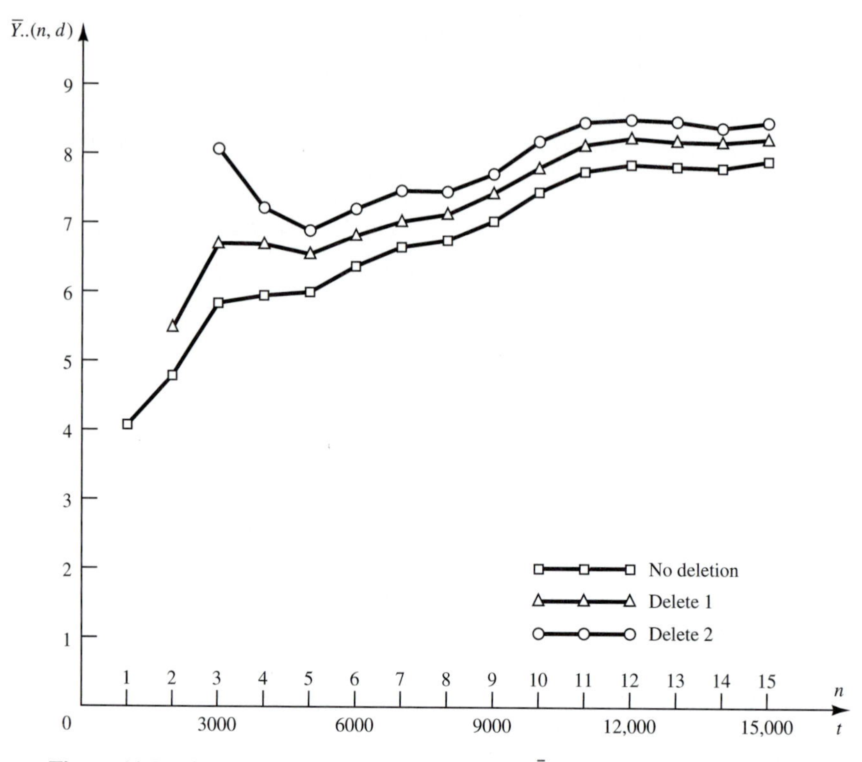

Figure 11.5. Cumulative average queue length $\bar{Y}_{\cdot\cdot}(n, d)$ versus time $1000n$.

that the first two observations have considerably more bias than any of the remaining ones. The effect of deleting first one and then two batch means is also illustrated in Table 11.6 and Figure 11.5. As expected, the estimators increase in value as more data are deleted; that is, $\bar{Y}_{\cdot\cdot}(15, 2) = 8.43$ and $\bar{Y}_{\cdot\cdot}(15, 1) = 8.21$ are larger than $\bar{Y}_{\cdot\cdot}(15, 0) = 7.94$. It also appears from Figure 11.5 that $\bar{Y}_{\cdot\cdot}(n, d)$ is increasing for $n = 5, 6, \ldots, 11$ (and all $d = 0, 1, 2$), and thus there may still be some initialization bias. It seems, however, that deletion of the first two batches removes most of the bias. ◀

Unfortunately, there is no widely accepted, objective, and proven technique to determine how much data to delete to reduce initialization bias to a negligible level. Plots can, at times, be misleading, but they are still recommended. Several points should be kept in mind:

1. Ensemble averages, such as Figure 11.4, will reveal a smoother and more precise trend as the number of replications, R, is increased. Since each ensemble average is the sample mean of i.i.d. observations, a confidence interval based on the t-distribution can be placed around each point, as shown in Figure 11.6, and these intervals can be used to judge whether or not the plot is precise enough to judge that bias has diminished. *This is the preferred method to determine a deletion point.*

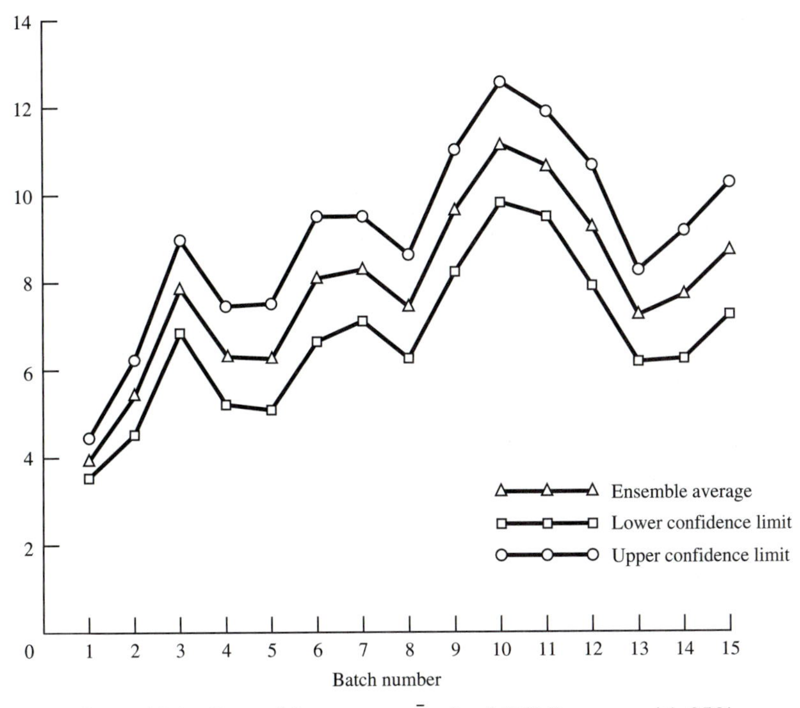

Figure 11.6. Ensemble averages $\bar{Y}_{\cdot j}$ for $M/G/1$ queue with 95% confidence intervals.

2. Ensemble averages can be smoothed further by plotting a moving average, rather than the original ensemble averages. In a moving average each plotted point is actually the average of several adjacent ensemble averages. Specifically, the jth plot point would be

$$\tilde{Y}_{\cdot j} = \frac{1}{2m+1} \sum_{i=j-m}^{j+m} \bar{Y}_{\cdot i}$$

for some $m \geq 1$, rather than the original ensemble average $\bar{Y}_{\cdot j}$. The value of m is typically chosen by trial and error until a smooth plot is obtained. See Law and Kelton [2000] or Welch [1983] for further discussion of smoothing.

3. Cumulative averages, such as in Figure 11.5, become less variable as more data are averaged. Therefore, it is expected that the left side of the curve will always be less smooth than the right side. *For this reason and others, cumulative averages should only be used if it is not feasible to compute ensemble averages*, such as when only a single replication is possible.

4. Simulation data, especially from queueing models, usually exhibits positive autocorrelation. The more correlation present, the longer it takes for

$\bar{Y}_{\cdot j}$ to approach steady state. The positive correlation between successive observations (i.e., batch means) $\bar{Y}_{\cdot 1}, \bar{Y}_{\cdot 2}, \ldots$ can be seen in Figure 11.4.

5. In most simulation studies the analyst is interested in several different output performance measures at once, such as the number in queue, customer waiting time, utilization of the servers, etc. Unfortunately, different performance measures may approach steady state at different rates. Thus, it is important to examine each performance measure individually for initialization bias and use a deletion point that is adequate for all of them.

There has been no shortage of solutions to the initialization-bias problem. Unfortunately, for every "solution" that works well in some situations, there are other situations in which it is either not applicable or performs poorly. Important ideas include testing for bias (e.g., Kelton and Law [1983], Schruben [1980], Goldsman, Schruben, and Swain [1994]); modeling the bias (e.g., Snell and Schruben [1985]); and randomly sampling the initial conditions on multiple replications (e.g., Kelton [1989]).

11.5.2 Statistical Background

If $\{Y_1, \ldots, Y_n\}$ are not statistically independent, then S^2/n given by Equation (11.14) is a biased estimator of the true variance, $\sigma^2(\widehat{\theta})$. This is almost always the case when $\{Y_1, \ldots, Y_n\}$ is a sequence of output observations from within a single replication. In this situation, $Y_1, Y_2, \ldots$ is an autocorrelated sequence, sometimes called a time series. Example 11.8 (the $M/G/1$ queue) provides an illustration of this situation.

Suppose that our point estimator for θ is the sample mean $\widehat{\theta} = \sum_{i=1}^{n} Y_i/n$. A general result from mathematical statistics is that[3]

$$\sigma^2(\widehat{\theta}) = \text{var}(\widehat{\theta}) = \frac{1}{n^2} \sum_{i=1}^{n} \sum_{j=1}^{n} \text{cov}(Y_i, Y_j) \tag{11.30}$$

where $\text{cov}(Y_i, Y_i) = \text{var}(Y_i)$. To construct a confidence interval for θ an estimate of $\sigma^2(\widehat{\theta})$ is required. But obtaining an estimate of (11.30) is pretty much hopeless, since each term $\text{cov}(Y_i, Y_j)$ may be different, in general. Fortunately, systems that have a steady state will, if simulated long enough to pass the transient phase (such as the production-line startup in Example 11.4), produce an output process that is approximately *covariance stationary*. Intuitively, stationarity implies that Y_{i+k} depends on Y_{i+1} in the same manner as Y_k depends on Y_1. In particular, the covariance between two random variables in the time series depends only on the number of observations between them, called the *lag*.

[3] This general result can be derived by recalling that for two random variables Y_1 and Y_2, the $\text{var}(Y_1 \pm Y_2) = \text{var}(Y_1) + \text{var}(Y_2) \pm 2\,\text{cov}(Y_1, Y_2)$.

For a covariance-stationary time series, $\{Y_1, Y_2, \ldots\}$, define the lag-k autocovariance by

$$\gamma_k = \text{cov}(Y_1, Y_{1+k}) = \text{cov}(Y_i, Y_{i+k}) \tag{11.31}$$

which, by definition of covariance stationarity, is not a function of i. For $k = 0$, γ_0 becomes the population variance; that is,

$$\gamma_0 = \text{cov}(Y_i, Y_{i+0}) = \text{var}(Y_i) \tag{11.32}$$

The lag-k autocorrelation is the correlation between any two observations k apart. It is defined by

$$\rho_k = \frac{\gamma_k}{\gamma_0} \tag{11.33}$$

and has the property that

$$-1 \le \rho_k \le 1, \quad k = 1, 2, \ldots$$

If a time series is covariance stationary, then Equation (11.30) can be substantially simplified. Tedious algebra shows that

$$\sigma^2(\widehat{\theta}) = \frac{\gamma_0}{n}\left[1 + 2\sum_{k=1}^{n-1}\left(1 - \frac{k}{n}\right)\rho_k\right] \tag{11.34}$$

where ρ_k is the lag-k autocorrelation given by Equation (11.33), and γ_0 is the population variance defined in Equation (11.32).

When $\rho_k > 0$ for all k (or most k), the time series is said to be positively autocorrelated. In this case large observations tend to be followed by large observations, and small observations by small ones. Such a series will tend to drift slowly above and then below its mean. Figure 11.7(a) is an example of a stationary time series exhibiting positive autocorrelation. The output data from most queueing simulations are positively autocorrelated, as are the successive customer delays $D_1, D_2, \ldots$ in the bank model discussed in Example 11.9.

On the other hand, if some of the $\rho_k < 0$, the series $Y_1, Y_2, \ldots$ will display the characteristics of negative autocorrelation. In this case large observations tend to be followed by small observations, and vice versa. Figure 11.7(b) is an example of a stationary time series exhibiting negative autocorrelation. The output of certain inventory simulations may be negatively autocorrelated.

Figure 11.7(c) also shows an example of a time series with an upward trend. Such a time series is not stationary, since the probability distribution of Y_i is changing with the index i.

Why does autocorrelation make it difficult to estimate $\sigma^2(\widehat{\theta})$? Recall that the standard estimator for the variance of a sample mean is S^2/n from Equation (11.14). Using Equation (11.34), it can be shown [Law, 1977] that the expected value of the variance estimator S^2/n is

$$E\left(\frac{S^2}{n}\right) = B\sigma^2(\widehat{\theta}) \tag{11.35}$$

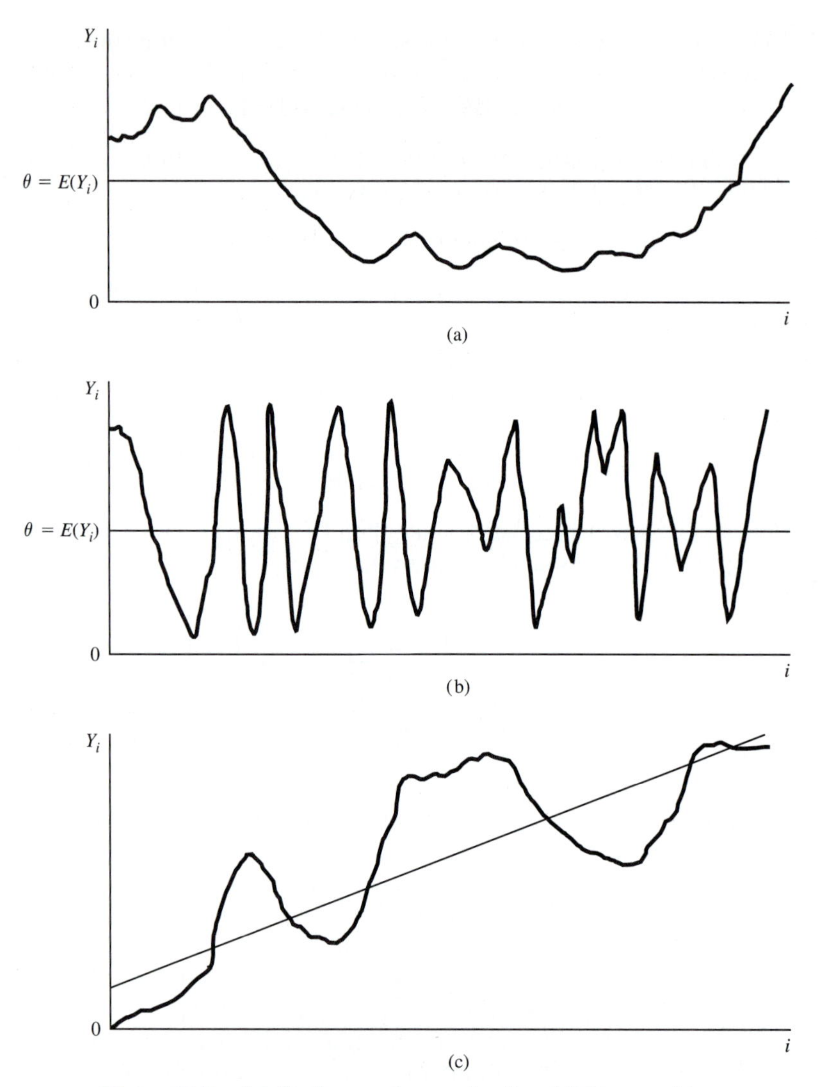

Figure 11.7. (a) Stationary time series Y_i exhibiting positive autocorrelation; (b) stationary time series Y_i exhibiting negative autocorrelation; (c) nonstationary time series with an upward trend.

where

$$B = \frac{n/c - 1}{n - 1} \tag{11.36}$$

and c is the quantity in brackets in Equation (11.34). The effect of the autocorrelation on the estimator S^2/n is derived by an examination of Equations (11.34) and (11.36). There are essentially three possibilities:

Case 1

If the Y_i are independent, then $\rho_k = 0$ for $k = 1, 2, 3, \ldots$. Therefore, $c = 1+2\sum_{k=1}^{n-1}(1-k/n)\rho_k = 1$, and Equation (11.34) reduces to the familiar γ_0/n of Equation (11.13) (notice that $\gamma_0 = \sigma^2$). Notice also that $B = 1$, so S^2/n is an unbiased estimator of $\sigma^2(\widehat{\theta})$. The Y_i will always be independent when they are obtained from different replications, which is the primary reason that we prefer experiment designs calling for multiple replications.

Case 2

If the correlations ρ_k are primarily positive, then $c = 1+2\sum_{k=1}^{n-1}(1-k/n)\rho_k > 1$, so that $n/c < n$, and hence $B < 1$. Therefore, S^2/n is biased low as an estimator of $\sigma^2(\widehat{\theta})$. If this correlation were ignored, the nominal $100(1-\alpha)\%$ confidence interval given by expression (11.10) would be too short and its true confidence coefficient would be less than $1-\alpha$. The practical effect would be that the simulation analyst would have unjustified confidence in the apparent precision of the point estimator due to the shortness of the confidence interval. If the correlations ρ_k are large, B could be quite small, implying a significant underestimation.

Case 3

If the correlations ρ_k are substantially negative, then $0 \le c < 1$, and it follows that $B > 1$ and S^2/n is biased high for $\sigma^2(\widehat{\theta})$. In other words, the true precision of the point estimator $\widehat{\theta}$ would be greater than what is indicated by its variance estimator S^2/n, because

$$\sigma^2(\widehat{\theta}) < E\left(\frac{S^2}{n}\right)$$

As a result, the nominal $100(1-\alpha)\%$ confidence interval of expression (11.10) would have true confidence coefficient greater than $1-\alpha$. This error is less serious than Case 2, because we are unlikely to make incorrect decisions if our estimate is actually more precise than we think it is.

A simple example demonstrates why we are especially concerned about positive correlation: Suppose you want to know how students on a university campus will vote in an upcoming election. To estimate their preferences you plan to solicit 100 responses. The standard experiment is to randomly select 100 students to poll; call this experiment A. An alternative is to randomly select 20 students and ask each of them to state their preference 5 times; call this experiment B. Both experiments obtain 100 responses, but clearly an estimate based on experiment B will be less precise (will have larger variance) than an estimate based on experiment A. Experiment A obtains 100 independent responses, while experiment B obtains only 20 independent responses and 80 dependent ones. The five opinions from any one student are perfectly positively correlated (assuming a student names the same candidate all five times). Although this is an extreme example, it illustrates that estimates based on positively correlated data are more variable than estimates based on independent data. Therefore, a confidence interval or other measure of error should correctly account for dependent data, but S^2/n does not.

Two methods for eliminating or reducing the deleterious effects of autocorrelation when estimating a mean are given in the following sections. Unfortunately, some simulation languages either use or facilitate the use of S^2/n as an estimator of $\sigma^2(\widehat{\theta})$, the variance of the sample mean, in all situations. If used uncritically in a simulation with positively autocorrelated output data, the downward bias in S^2/n and the resulting shortness of a confidence interval for θ may convey the impression of much greater precision than actually is the case. When such positive autocorrelation is present in the output data, the true variance of the point estimator, $\widehat{\theta}$, can be many times greater than is indicated by S^2/n.

11.5.3 Replication Method for Steady-State Simulations

If initialization bias in the point estimator has been reduced to a negligible level (through some combination of intelligent initialization and deletion), then the method of independent replications can be used to estimate point-estimator variability and to construct a confidence interval. The basic idea is simple: Make R replications, initializing and deleting from each one the same way.

If, however, significant bias remains in the point estimator and a large number of replications are used to reduce point estimator variability, the resulting confidence interval can be misleading. This happens because *bias is not affected by the number of replications* (R); it is affected only by deleting more data (i.e., increasing T_0) or extending the length of each run (i.e., increasing T_E). Thus, increasing the number of replications (R) may produce shorter confidence intervals around the "wrong point." Therefore, it is important to do a thorough job of investigating the initial-condition bias.

If the simulation analyst decides to delete d observations of the total of n observations in a replication, then the point estimator of θ is $\bar{Y}_{..}(n, d)$, defined by Equation (11.29); that is, the point estimator is the average of the remaining data. The basic raw output data, $\{Y_{rj}, r = 1, \ldots, R; j = 1, \ldots, n\}$, are exhibited in Table 11.7. Each Y_{rj} is derived in one of the following ways:

Case 1

Y_{rj} is an individual observation from within replication r; for example, Y_{rj} could be the delay of customer j in a queue, or the response time to job j in a job shop.

Case 2

Y_{rj} is a batch mean from within replication r of some number of discrete-time observations. (Batch means are discussed further in Section 11.5.5.)

Case 3

Y_{rj} is a batch mean of a continuous-time process over interval j; for instance, as in Example 11.14, Equation (11.27) defines Y_{rj} as the time-average (batch mean) number in queue over the interval $[1000(j-1), 1000j)$.

In case 1, the number d of deleted observations and the total number of observations n may vary from one replication to the next, in which case replace d by d_r, and n by n_r. For simplicity, assume that d and n are constant over replications. In cases 2 and 3, d and n will be constant.

When using the replication method, each replication is regarded as a single sample for the purpose of estimating θ. For replication r, define

$$\bar{Y}_{r\cdot}(n, d) = \frac{1}{n-d} \sum_{j=d+1}^{n} Y_{rj} \tag{11.37}$$

as the sample mean of all (nondeleted) observations in replication r. Since all replications use different random-number streams and all are initialized at time 0 by the same set of initial conditions (I_0), the replication averages

$$\bar{Y}_{1\cdot}(n, d), \ldots, \bar{Y}_{R\cdot}(n, d)$$

are independent and identically distributed random variables; that is, they constitute a random sample from some underlying population having unknown mean

$$\theta_{n,d} = \mathrm{E}[\bar{Y}_{r\cdot}(n, d)] \tag{11.38}$$

The overall point estimator, given in Equation (11.29), is also given by

$$\bar{Y}_{\cdot\cdot}(n, d) = \frac{1}{R} \sum_{r=1}^{R} \bar{Y}_{r\cdot}(n, d) \tag{11.39}$$

as can be seen from Table 11.7 or from using Equation (11.28). Thus, it follows that

$$\mathrm{E}[\bar{Y}_{\cdot\cdot}(n, d)] = \theta_{n,d}$$

also. If d and n are chosen sufficiently large, then $\theta_{n,d} \approx \theta$, and $\bar{Y}_{\cdot\cdot}(n, d)$ is an approximately unbiased estimator of θ. The bias in $\bar{Y}_{\cdot\cdot}(n, d)$ is $\theta_{n,d} - \theta$.

For convenience, when the value of n and d are understood, abbreviate $\bar{Y}_{r\cdot}(n, d)$ (the mean of the undeleted observations from the rth replication) and

Table 11.7. Raw Output Data from a Steady-State Simulation

	Observations						*Replication*
Replication	*1*	...	*d*	*d+ 1*	...	*n*	*Averages*
1	$Y_{1,1}$	...	$Y_{1,d}$	$Y_{1,d+1}$	...	$Y_{1,n}$	$\bar{Y}_{1\cdot}(n, d)$
2	$Y_{2,1}$	...	$Y_{2,d}$	$Y_{2,d+1}$	...	$Y_{2,n}$	$\bar{Y}_{2\cdot}(n, d)$
.	.		.	.		.	.
.	.		.	.		.	.
.	.		.	.		.	.
R	$Y_{R,1}$	...	$Y_{R,d}$	$Y_{R,d+1}$	...	$Y_{R,n}$	$\bar{Y}_{R\cdot}(n, d)$
	$\bar{Y}_{\cdot 1}$	...	$\bar{Y}_{\cdot d}$	$\bar{Y}_{\cdot,d+1}$	...	$\bar{Y}_{\cdot,n}$	$\bar{Y}_{\cdot\cdot}(n, d)$

$\bar{Y}_{..}(n, d)$ (the mean of $\bar{Y}_{1.}(n, d), \ldots, \bar{Y}_{R.}(n, d)$) by $\bar{Y}_{r.}$ and $\bar{Y}_{..}$, respectively. To estimate the standard error of $\bar{Y}_{..}$, first compute the sample variance

$$S^2 = \frac{1}{R-1}\sum_{r=1}^{R}(\bar{Y}_{r.} - \bar{Y}_{..})^2 = \frac{1}{R-1}\left(\sum_{r=1}^{R}\bar{Y}_{r.}^2 - R\bar{Y}_{..}^2\right) \tag{11.40}$$

The standard error of $\bar{Y}_{..}$ is given by

$$\text{s.e.}(\bar{Y}_{..}) = \frac{S}{\sqrt{R}} \tag{11.41}$$

A $100(1-\alpha)\%$ confidence interval for θ, based on the t-distribution, is given by

$$\bar{Y}_{..} - t_{\alpha/2,R-1}\frac{S}{\sqrt{R}} \leq \theta \leq \bar{Y}_{..} + t_{\alpha/2,R-1}\frac{S}{\sqrt{R}} \tag{11.42}$$

where $t_{\alpha/2,R-1}$ is the $100(1-\alpha/2)$ percentage point of a t-distribution with $R-1$ degrees of freedom. This confidence interval is valid only if the bias of $\bar{Y}_{..}$ is approximately zero.

As a rough rule, the length of each replication, beyond the deletion point, should be at least ten times the amount of data deleted. In other words, $(n-d)$ should at least $10d$ (or more generally, T_E should be at least $10T_0$). Given this run length, the number of replications should be as many as time permits, up to about 25 replications. Kelton [1986] established that there is little value in dividing the available time into more than 25 replications, so if time permits making more than 25 replications of length T_0+10T_0, then make 25 replications of longer than $T_0 + 10T_0$, instead.

EXAMPLE 11.15

Consider again the $M/G/1$ queueing simulation of Examples 11.8 and 11.14. Suppose that the simulation analyst decides to make $R = 10$ replications, each of length $T_E = 15{,}000$ minutes, each starting at time 0 in the empty and idle state, and each initialized for $T_0 = 2000$ minutes before data collection begins. The raw output data consist of the batch means defined by Equation (11.27); recall that each batch mean is simply the average number of customers in queue for a 1000-minute interval. The first two batch means are deleted ($d = 2$). The purpose of the simulation is to estimate, by a 95% confidence interval, the long-run time average queue length, denoted by L_Q (or θ).

The replication averages $\bar{Y}_{r.}(15, 2), r = 1, 2, \ldots, 10$, are shown in Table 11.8 in the rightmost column. The point estimator is computed by Equation (11.39) as

$$\bar{Y}_{..}(15, 2) = 8.43$$

Its standard error is given by Equation (11.41) as

$$\text{s.e.}(\bar{Y}_{..}(15, 2)) = 1.59$$

and using $\alpha = 0.05$ and $t_{0.025,9} = 2.26$, the 95% confidence interval for long-run mean queue length is given by Inequality (11.42) as

$$8.43 - 2.26(1.59) \leq L_Q \leq 8.43 + 2.26(1.59)$$

or

$$4.84 \leq L_Q \leq 12.02$$

The simulation analyst may conclude with a high degree of confidence that the long-run mean queue length is between 4.84 and 12.02 customers. The confidence interval computed here as given by Inequality (11.42) should be used with caution, because a key assumption behind its validity is that enough data have been deleted to remove any significant bias due to initial conditions—that is, d and n are sufficiently large so that the bias $\theta_{n,d} - \theta$ is negligible. ◀

Table 11.8. Data Summary for $M/G/1$ Simulation by Replication

Replication, r	*Sample Mean for Replication r* (No Deletion) $\bar{Y}_{r\cdot}(15,0)$	(Delete 1) $\bar{Y}_{r\cdot}(15,1)$	(Delete 2) $\bar{Y}_{r\cdot}(15,2)$
1	3.27	3.24	3.25
2	16.25	17.20	17.83
3	15.19	15.72	15.43
4	7.24	7.28	7.71
5	2.93	2.98	3.11
6	4.56	4.82	4.91
7	8.44	8.96	9.45
8	5.06	5.32	5.27
9	6.33	6.14	6.24
10	10.10	10.48	11.07
$\bar{Y}_{\cdot\cdot}(15,d)$	7.94	8.21	8.43
$\sum_{r=1}^{R} \bar{Y}_{r\cdot}^2$	826.20	894.68	938.34
S^2	21.75	24.52	25.30
S	4.66	4.95	5.03
$S/\sqrt{10} = \text{s.e.}(\bar{Y}_{\cdot\cdot})$	1.47	1.57	1.59

Example 11.16

Suppose that in Example 11.15, the simulation analyst had decided to delete one batch ($d = 1$), or no batches ($d = 0$). The quantities needed to compute 95% confidence intervals are shown in Table 11.8. The resulting 95% confidence intervals are computed by Inequality (11.42) as follows:

$$(d = 1)\ 4.66 = 8.21 - 2.26(1.57) \leq L_Q \leq 8.21 + 2.26(1.57) = 11.76$$

$$(d = 0)\ 4.62 = 7.94 - 2.26(1.47) \leq L_Q \leq 7.94 + 2.26(1.47) = 11.26$$

Notice that, for a fixed total sample size, n, two things happen as fewer data are deleted:

1. The confidence interval shifts downward, reflecting the greater downward bias in $\bar{Y}_{..}(15, d)$ as d decreases.
2. The standard error of $\bar{Y}_{..}(n, d)$, namely $S/\sqrt{R}$, decreases as d decreases.

In this example, $\bar{Y}_{..}(n, d)$ is based on a run length of $T_E = 1000(n - d) = 15{,}000 - 1000d$ minutes. Thus, as d decreases, T_E increases, and in effect the sample mean $\bar{Y}_{..}$ is based on a larger "sample size" (i.e., longer run length). In general, the larger the sample size, the smaller the standard error of the point estimator. This larger sample size can be due to a longer run length (T_E) per replication, or to more replications (R).

Therefore, there is a trade-off between reducing bias and increasing the variance of a point estimator, when the total sample size (R and T_0+T_E) is fixed. The more deletion (i.e., the larger T_0 is and the smaller T_E is, keeping $T_0 + T_E$ fixed), the less bias but greater variance there is in the point estimator. ◀

Recall that each batch in Examples 11.15 and 11.16 consists of 1000 minutes of simulated time. Therefore, discarding $d = 2$ batches really means discarding 2000 minutes of data, a substantial amount. It is not uncommon for very large deletions to be necessary to overcome the initial conditions.

11.5.4 Sample Size in Steady-State Simulations

Suppose it is desired to estimate a long-run performance measure, θ, within $\pm\epsilon$ with confidence $100(1 - \alpha)\%$. In a steady-state simulation, a specified precision may be achieved either by increasing the number of replications (R) or by increasing the run length (T_E). The first solution, controlling R, is carried out as given in Section 11.4.3 for terminating simulations.

EXAMPLE 11.17

Consider the data in Table 11.8 for the $M/G/1$ queueing simulation as an initial sample of size $R_0 = 10$. Assuming that $d = 2$ observations were deleted, the initial estimate of variance is $S_0^2 = 25.30$. Suppose that it is desired to estimate long-run mean queue length, L_Q, within $\epsilon = 2$ customers with 90% confidence. The final sample size needed must satisfy Inequality (11.21). Using $\alpha = 0.10$ in Inequality (11.22) yields an initial estimate:

$$R \geq \left(\frac{z_{0.05} S_0}{\epsilon}\right)^2 = \frac{1.645^2(25.30)}{2^2} = 17.1$$

Thus, at least 18 replications will be needed. Proceeding as in Example 11.12, next try $R = 18$, $R = 19, \ldots$ as follows:

R	18	19
$t_{0.05,R-1}$	1.74	1.73
$\left(\dfrac{t_{0.05,R-1} S_0}{\epsilon}\right)^2$	19.15	18.93

Since $R = 19 \geq (t_{0.05,18}\, S_0/\epsilon)^2 = 18.93$ is the smallest integer R satisfying inequality (11.21), a total sample size of $R = 19$ replications are needed to estimate L_Q to within ± 2 customers. Therefore, $R - R_0 = 19 - 10 = 9$ additional replications are needed to achieve the specified error. ◀

An alternative to increasing R is to increase total run length $T_0 + T_E$ within each replication. If the calculations in Section 11.4.3, as illustrated in Example 11.17, indicate that $R - R_0$ additional replications are needed beyond the initial number, R_0, then an alternative is to increase run length $(T_0 + T_E)$ in the same proportion (R/R_0) to a new run length $(R/R_0)(T_0 + T_E)$. Thus, additional data will be deleted, from time 0 to time $(R/R_0)T_0$, and more data will be used to compute the point estimates, as illustrated by Figure 11.8. However, the total amount of simulation effort is the same as if we had simply increased the number of replications but maintained the same run length. The advantage of increasing total run length per replication and deleting a fixed proportion $[T_0/(T_0 + T_E)]$ of the total run length is that any residual bias in the point estimator should be further reduced by the additional deletion of data at the beginning of the run. A possible disadvantage of the method is that in order to continue the simulation of all R replications [from time $T_0 + T_E$ to time $(R/R_0)(T_0 + T_E)$] it is necessary to have saved the state of the model at time $T_0 + T_E$ and to be able to restart the model and run it for the additional required time. Otherwise, the simulations would have to be rerun from time 0, which could be time consuming for a complex model. Most simulation languages have a restart capability to facilitate intermittent stopping and restarting of a model.

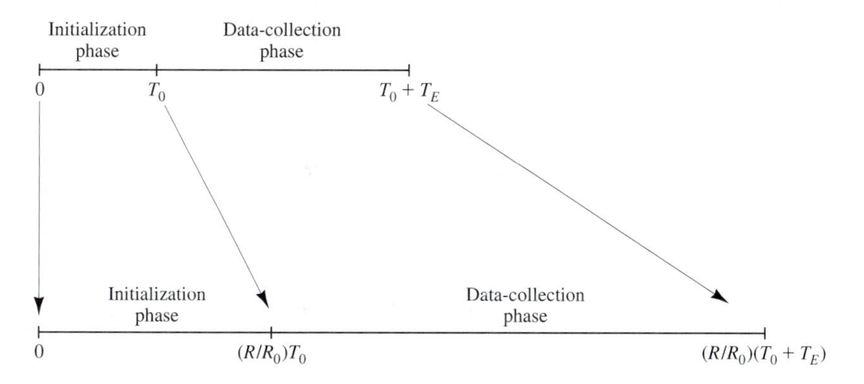

Figure 11.8. Increasing run-length to achieve specified accuracy.

EXAMPLE 11.18

In Example 11.17, suppose that run length was to be increased to achieve the desired error of ±2 customers. Since $R/R_0 = 19/10 = 1.9$, the run length should be almost doubled to $(R/R_0)(T_0+T_E) = 1.9(15{,}000) = 28{,}500$ minutes. The data collected from time 0 to time $(R/R_0)T_0 = 1.9(2000) = 3800$ minutes would be deleted, and the data from time 3800 to time 28,500 used to compute new point estimates and confidence intervals. ◀

11.5.5 Batch Means for Interval Estimation in Steady-State Simulations

One disadvantage of the replication method is that data must be deleted on each replication, and, in one sense, deleted data are wasted data, or at least lost information. This suggests that there may be merit in using an experiment design that is based on a single, long replication. The disadvantage of a single-replication design arises when we try to compute the standard error of the sample mean. Since we only have data from within one replication, the data are dependent and the usual estimator is biased.

The method of *batch means* attempts to solve this problem by dividing the output data from one replication (after appropriate deletion) into a few large batches, and then treating the means of these batches as if they were independent. When the raw output data after deletion form a continuous-time process, $\{Y(t), T_0 \le t \le T_0 + T_E\}$, such as the length of a queue or the level of inventory, then we form k batches of size $m = T_E/k$ and compute the batch means as

$$\bar{Y}_j = \frac{1}{m} \int_{(j-1)m}^{jm} Y(t + T_0)\, dt$$

for $j = 1, 2, \ldots, k$. In other words, the jth batch mean is just the time-weighted average of the process over the time interval $[T_0 + (j-1)m, T_0 + jm)$, exactly as in Example 11.8.

When the raw output data after deletion form a discrete-time process, $\{Y_i, i = d+1, d+2, \ldots, n\}$, such as the customer delays in a queue or the cost per period of an inventory system, then we form k batches of size $m = (n-d)/k$ and compute the batch means as

$$\bar{Y}_j = \frac{1}{m} \sum_{i=(j-1)m+1}^{jm} Y_{i+d}$$

for $j = 1, 2, \ldots, k$ (assuming k divides $n - d$ evenly, otherwise round down to the nearest integer). That is, the batch means are formed as shown below:

$$\underbrace{Y_1, \ldots, Y_d}_{\text{deleted}}, \underbrace{Y_{d+1}, \ldots, Y_{d+m}}_{\bar{Y}_1}, \underbrace{Y_{d+m+1}, \ldots, Y_{d+2m}}_{\bar{Y}_2}, \ldots, \underbrace{Y_{d+(k-1)m+1}, \ldots, Y_{d+km}}_{\bar{Y}_k}$$

Starting with either continuous-time or discrete-time data, the variance of the sample mean is estimated by

$$\frac{S^2}{k} = \frac{1}{k}\sum_{j=1}^{k}\frac{(\bar{Y}_j - \bar{Y})^2}{k-1} = \frac{\sum_{j=1}^{k}\bar{Y}_j^2 - k\bar{Y}^2}{k(k-1)} \tag{11.43}$$

where $\bar{Y}$ is the overall sample mean of the data after deletion. As discussed in Section 11.2, the batch means $\bar{Y}_1, \bar{Y}_2, \ldots, \bar{Y}_k$ are not independent; however, if the batch size is sufficiently large, successive batch means will be approximately independent, and the variance estimator will be approximately unbiased.

Unfortunately, there is no widely accepted and relatively simple method for choosing an acceptable batch size m (or equivalently choosing a number of batches k). But some general guidelines can be culled from the research literature:

- Schmeiser [1982] found that for a *fixed total sample size* there is little benefit from dividing it into more than $k = 30$ batches, even if we could do so and still retain independence between the batch means. Therefore, there is no reason to consider numbers of batches much greater than 30, no matter how many raw data are available. He also found that the performance of the confidence interval, in terms of its width and the variability of its width, is poor for fewer than 10 batches. Therefore, a number of batches between 10 and 30 should be used in most applications.
- Although there is typically autocorrelation between batch means at all lags, the lag–1 autocorrelation $\rho_1 = \text{corr}(\bar{Y}_j, \bar{Y}_{j+1})$ is usually studied to assess the dependence between batch means. When the lag–1 autocorrelation is nearly 0, then the batch means are treated as independent. This approach is based on the observation that the autocorrelation in many stochastic processes decreases as the lag increases. Therefore, all lag autocorrelations should be smaller (in absolute value) than the lag–1 autocorrelation.
- The lag–1 autocorrelation between batch means can be estimated as described below. However, the autocorrelation should not be estimated from a small number of batch means (such as the $10 \le k \le 30$ recommended above), due to bias in the autocorrelation estimator. Law and Carson [1979] suggest estimating the lag–1 autocorrelation from a large number of batch means based on a smaller batch size (perhaps $100 \le k \le 400$). When the autocorrelation between these batch means is approximately 0, then the autocorrelation will be even smaller if we rebatch the data to between 10 and 30 batch means based on a larger batch size. Hypothesis tests for 0 autocorrelation are available, as described below.
- If the *total sample size is to be chosen sequentially*, say to attain a specified precision, then it is helpful to allow the batch size and number of batches to grow as the run length increases. It can be shown that a good strategy

is to allow the number of batches to increase as the square root of the sample size after first finding a batch size at which the lag–1 autocorrelation is approximately 0. Although we will not discuss this point further, an algorithm based on it can be found in Fishman and Yarberry [1997].

Based on these insights, we recommend the following general strategy:

1. Obtain output data from a single replication and delete as appropriate. Recall our guideline of collecting at least 10 times as much data as is deleted.
2. Form up to $k = 400$ batches (but at least 100 batches) with the retained data, and compute the batch means. Estimate the sample lag–1 autocorrelation of the batch means as

$$\widehat{\rho}_1 = \frac{\sum_{j=1}^{k-1}(\bar{Y}_j - \bar{Y})(\bar{Y}_{j+1} - \bar{Y})}{\sum_{j=1}^{k}(\bar{Y}_j - \bar{Y})^2}$$

3. Check the correlation to see if it is sufficiently small.
   (a) If $\widehat{\rho}_1 \leq 0.2$, then rebatch the data into $30 \leq k \leq 40$ batches, and form a confidence interval using $k - 1$ degrees of freedom for the t-distribution and Equation (11.43) to estimate the variance of $\bar{Y}$.
   (b) If $\widehat{\rho}_1 > 0.2$, then extend the replication by 50% to 100% and go to step 2. If it is not possible to extend the replication, then rebatch the data into approximately $k = 10$ batches, and form the confidence interval using $k - 1$ degrees of freedom for the t-distribution and Equation (11.43) to estimate the variance of $\bar{Y}$.
4. As an additional check on the confidence interval, examine the batch means (at the larger batch size) for independence using the following test (see, for instance, Alexopoulos and Seila [1998]). Compute the test statistic

$$C = \sqrt{\frac{k^2 - 1}{k - 2}}\left(\widehat{\rho}_1 + \frac{(\bar{Y}_1 - \bar{Y})^2 + (\bar{Y}_k - \bar{Y})^2}{2\sum_{j=1}^{k}(\bar{Y}_j - \bar{Y})^2}\right)$$

   If $C < z_\beta$, then accept the independence of the batch means, where β is the Type I error level of the test (such as $0.1, 0.05, 0.01$). Otherwise, extend the replication by 50% to 100% and go to step 2. If it is not possible to extend the replication, then rebatch the data into approximately $k = 10$ batches, and form the confidence interval using $k - 1$ degrees of freedom for the t-distribution and Equation (11.43) to estimate the variance of $\bar{Y}$.

This procedure, including the final check, is conservative in several respects. First, if the lag–1 autocorrelation is substantially negative, then we proceed to form the confidence interval anyway. A dominant negative correlation tends to make the confidence interval wider than necessary, which is an error, but not one that will cause us to make incorrect decisions. The requirement that $\widehat{\rho}_1 < 0.2$

at $100 \leq k \leq 400$ batches is pretty stringent, and will tend to force us to get more data (and therefore create larger batches) if there is any hint of positive dependence. Any finally, the hypothesis test at the end has a probability of β of forcing us to get more data when none is really needed. But this conservatism is by design, since the cost of an incorrect decision is typically much greater than the cost of some additional computer runtime.

The batch means approach to confidence interval estimation is illustrated in the next example.

EXAMPLE 11.19

Reconsider the $M/G/1$ simulation of Example 11.8, except that the mean service time is changed from 9.5 minutes to 7 minutes (implying a long-run server utilization of 0.7). Suppose that we want to estimate the steady-state expected delay in queue, w_Q, by a 95% confidence interval. To illustrate the method of batch means, assume that one run of the model has been made simulating 3000 customers after the deletion point. We then form batch means from $k = 100$ batches of size $m = 30$, and estimate the lag–1 autocorrelation to be $\widehat{\rho}_1 = 0.346 > 0.2$. Thus, we decide to extend the simulation to 6000 customers after the deletion point, and again estimate the lag–1 autocorrelation. The estimate, based on $k = 100$ batches of size $m = 60$, is $\widehat{\rho}_1 = 0.004 < 0.2$.

Having passed the correlation check, we rebatch the data into $k = 30$ batches of size $m = 200$. The point estimate is the overall mean

$$\bar{Y} = \frac{1}{6000} \sum_{j=1}^{6000} \bar{Y}_j = 9.04$$

minutes. The variance of $\bar{Y}$, computed from the 30 batch means, is

$$\frac{S^2}{k} = \frac{\sum_{j=1}^{30} \bar{Y}_j^2 - 30\bar{Y}^2}{30(29)} = 0.604$$

Thus, a 95% confidence interval is given by

$$\bar{Y} - t_{0.025,29}\sqrt{0.604} \leq w_Q \leq \bar{Y} + t_{0.025,29}\sqrt{0.604}$$

or

$$7.45 = 9.04 - 2.04(0.777) \leq w_Q \leq 9.04 + 2.04(0.777) = 10.63$$

Thus, we assert with 95% confidence that true mean delay in queue, w_Q, is between 7.45 and 10.63 minutes. If these results are not sufficiently precise for practical use, the run length should be increased to achieve greater precision.

As a further check on the validity of the confidence interval, we can apply the correlation hypothesis test. To do so, we compute the test statistic from the $k = 30$ batches of size $m = 200$ used to form the confidence interval. This gives

$$C = -0.31 < 1.96 = z_{0.05}$$

confirming the lack of correlation at the 0.05 significance level. Notice that at this small number of batches the estimated lag–1 autocorrelation appears to be slightly negative, illustrating our point about the difficulty of estimating correlation with small numbers of observations. ◀

11.5.6 Confidence Intervals for Quantiles

Constructing confidence intervals for quantile estimates in a steady-state simulation can be tricky, especially if the output process of interest is a continuous-time process, such as $L_Q(t)$, the number of customers in queue at time t. In this section we outline the main issues.

Taking the easier case first, suppose that the output process from a single replication (after appropriate deletion of initial data) is $Y_{d+1}, \ldots, Y_n$. To be concrete, Y_i might be the delay in queue of the ith customer. Then the point estimate of the pth quantile can be obtained as before, either from the histogram of the data or the sorted values. Of course, only the data after the deletion point are used. Suppose we make R replications, and let $\widehat{\theta}_r$ be the quantile estimate from the rth. Then the R quantile estimates $\widehat{\theta}_1, \ldots, \widehat{\theta}_R$ are independent and identically distributed. Their average

$$\widehat{\theta}. = \frac{1}{R}\sum_{i=1}^{R}\widehat{\theta}_i$$

can be used as the point estimate of θ, and an approximate confidence interval is

$$\widehat{\theta}. \pm t_{\alpha/2,R-1}\frac{S}{\sqrt{R}}$$

where S^2 is the usual sample variance of $\widehat{\theta}_1, \ldots, \widehat{\theta}_R$.

What if only a single replication is obtained? Then the same reasoning applies if we let $\widehat{\theta}_i$ be the quantile estimate from *within* the ith batch of data. This requires sorting the data, or forming a histogram, within each batch. If the batches are large enough, then these within-batch quantile estimates will also be approximately i.i.d.

When we have a continuous-time output process, then, in principle, the same methods apply. However, we must be careful not to transform the data in a way that changes the problem. In particular, we cannot form batch means first—as we have done throughout this chapter—and then estimate the quantile from these batch means. The p quantile of the batch means of $L_Q(t)$ is not the same as the p quantile of $L_Q(t)$ itself. Thus, the quantile point estimate must be formed from the histogram of the raw data, either from each run if we make replications or within each batch if we make a single replication.

11.6 Summary

This chapter emphasized the idea that a stochastic discrete-event simulation is a statistical experiment. Therefore, before sound conclusions can be drawn on the basis of the simulation-generated output data, a proper statistical analysis is required. The purpose of the simulation experiment is to obtain estimates of the performance measures of the system under study. The purpose of the statistical analysis is to acquire some assurance that these estimates are sufficiently precise for the proposed use of the model.

A distinction was made between terminating simulations and steady-state simulations. Steady-state simulation output data are more difficult to analyze, because the simulation analyst must address the problem of initial conditions and the choice of run length. Some suggestions were given regarding these problems, but unfortunately no simple, complete, and satisfactory solution exists. Nevertheless, simulation analysts should be aware of the potential problems and of the possible solutions — namely, deletion of data and increasing the run length. More advanced statistical techniques (not discussed in this text) are given in Alexopoulos and Seila [1998], Bratley, Fox, and Schrage [1987], and Law and Kelton [2000].

The statistical precision of point estimators can be measured by a standard error estimate, or a confidence interval. The method of independent replications was emphasized. With this method, the simulation analyst generates statistically independent observations, and thus standard statistical methods can be employed. For steady-state simulations, the method of batch means was also discussed.

The main point is that simulation output data contain some amount of random variability, and without some assessment of its magnitude, the point estimates cannot be used with any degree of reliability.

REFERENCES

ALEXOPOULOS, C., AND A. F. SEILA [1998], "Output Data Analysis," Chapter 7 in *Handbook of Simulation*, John Wiley, New York.

BRATLEY, P., B. L. FOX, AND L. E. SCHRAGE [1987], *A Guide to Simulation*, 2d ed., Springer-Verlag, New York.

FISHMAN, G. S. [1973], *Concepts and Methods in Discrete Event Digital Simulation*, John Wiley, New York.

FISHMAN, G. S. [1978], "Grouping Observations in Digital Simulation," *Management Science*, Vol. 24, pp. 510–521.

FISHMAN, G. S., AND L. S. YARBERRY [1997], "An Implementation of the Batch Means Method," *INFORMS Journal on Computing*, Vol. 9, pp. 296–310.

GOLDSMAN, D., L. SCHRUBEN, AND J. J. SWAIN [1994], "Tests for Transient Means in Simulated Time Series," *Naval Research Logistics*, Vol. 41, pp. 171–187.

KELTON, W. D. [1986], "Replication Splitting and Variance for Simulating Discrete-Parameter Stochastic Processes," *Operations Research Letters*, Vol. 4, pp. 275–279.

KELTON, W. D. [1989], "Random Initialization Methods in Simulation," *IIE Transactions*, Vol. 21, pp. 355–367.

KELTON, W. D., and A. M. LAW [1983], "A New Approach for Dealing with the Startup Problem in Discrete Event Simulation," *Naval Research Logistics Quarterly*, Vol. 30, pp. 641–658.

KLEIJNEN, J. P. C. [1987], *Statistical Tools for Simulation Practitioners*, Dekker, New York.

LAW, A. M. [1977], "Confidence Intervals in Discrete Event Simulation: A Comparison of Replication and Batch Means," *Naval Research Logistics Quarterly*, Vol. 24, pp. 667–78.

LAW, A. M. [1980], "Statistical Analysis of the Output Data from Terminating Simulations," *Naval Research Logistics Quarterly*, Vol. 27, pp. 131–43.

LAW, A. M., AND J. S. CARSON [1979], "A Sequential Procedure for Determining the Length of a Steady-State Simulation," *Operations Research*, Vol. 27, pp. 1011–25.

LAW, A. M., AND W. D. KELTON [1982] "Confidence Intervals for Steady-State Simulations, II: A Survey of Sequential Procedures," *Management Science*, Vol. 28, No. 5, pp. 550–662.

LAW, A. M., AND W. D. KELTON [2000], *Simulation Modeling and Analysis*, 3d ed., McGraw-Hill, New York.

NELSON, B. L. [1992], "Statistical Analysis of Simulation Results," Chapter 102 in *Handbook of Industrial Engineering*, 2d ed., John Wiley, New York.

SCHMEISER, B. [1982], "Batch Size Effects in the Analysis of Simulation Output," *Operations Research*, Vol. 30, pp. 556–568.

SCHRUBEN, L. [1980], "Detecting Initialization Bias in Simulation Output," *Operations Research*, Vol. 30, pp. 569–90.

SNELL, M. AND L. SCHRUBEN [1985], "Weighting Simulation Data to Reduce Initialization Effects," *IIE Transactions*, Vol. 17, pp. 354–363.

WELCH, P. D. [1983], "The Statistical Analysis of Simulation Results," in *The Computer Performance Modeling Handbook*, ed. S. Lavenberg, Academic Press, New York, pp. 268–328.

EXERCISES

1. **(a)** Consider Example 2.1. Under what circumstances would it be appropriate to use a terminating simulation, versus a steady-state simulation, to analyze this system?

(b) Repeat part (a) for Example 2.2 (the Able-Baker carhop problem).

(c) Repeat part (a) for Example 2.3.

(d) Repeat part (a) for Example 2.4.

(e) Repeat part (a) for Example 2.5.

(f) Repeat part (a) for Example 2.6.

(g) Repeat part (a) for Example 3.5 (the dump truck problem).

2. Suppose that in Example 11.14 the simulation analyst decided to investigate the bias using batch means over a batching interval of 2000 minutes. By definition, a batch mean for the interval $[(j-1)2000, j(2000))$ is defined by

$$Y_j = \frac{1}{2000} \int_{(j-1)2000}^{j(2000)} L_Q(t)\,dt$$

(a) Show algebraically that such a batch mean can be obtained from two adjacent batch means over the two halves of the interval.

(b) Compute the seven averaged batch means for the intervals $[0, 2000)$, $[2000, 4000)$, ... for the $M/G/1$ simulation. Use the data $(\bar{Y}_{\cdot j})$ in Table 11.6 (ignoring $\bar{Y}_{\cdot 15} = 8.76$).

(c) Draw plots of the type in Figures 11.4 and 11.5. Does it still appear that deletion of the data over $[0, 2000)$ (the first "new" batch mean) is sufficient to remove most of the point-estimator bias?

3. Suppose in Example 11.14 that the simulation analyst could afford to run only 5 independent replications (instead of 10). Use the batch means in Table 11.5 for replications 1 to 5 to compute a 95% confidence interval for mean queue length L_Q. Investigate deletion of initial data. Compare the results using 5 replications to those using 10 replications.

4. In Example 11.7, suppose that management desired 95% confidence in the estimate of mean system time w, and the error allowed was $\epsilon = 0.4$ minute.

(a) Using the same initial sample of size $R_0 = 4$ (given in Table 11.1), determine the required total sample size.

(b) Assume that the additional $R - R_0$ replications were made. Use the results in the first column of Table 11.3 ($r = 1, 2, \ldots, R_0, \ldots, R$; $T_E = 2$ hours) to compute a 95% confidence interval for w. Compare the half-length to the criteria $\epsilon = 0.4$ minute.

5. Simulate the dump truck problem in Example 3.5. At first make the run length $T_E =$ 40 hours. Make four independent replications. Compute a 90% confidence interval for mean cycle time, where a cycle time for a given truck is the time between its successive arrivals to the loader. Investigate the effect of different initial conditions (all trucks initially at the loader queue, versus all at the scale, versus all traveling, versus the trucks distributed throughout the system in some manner).

6. Consider an (M, L) inventory system, in which the procurement quantity, Q, is defined by

$$Q = \begin{cases} M - I, & \text{if } I < L \\ 0, & \text{if } I \geq L \end{cases}$$

where I is the level of inventory on hand plus on order at the end of a month, M is the maximum inventory level, and L is the reorder point. Since M and L are under management control, the pair (M, L) is called the inventory policy. Under certain conditions, the analytical solution of such a model is possible, but the computational effort may be prohibitive. Use simulation to investigate an (M, L) inventory system with the following properties. The inventory status is checked at the end of each month. Backordering is allowed at a cost of \$4 per item short per month. When an order arrives, it will first be used to relieve the backorder. The lead time is given by a uniform distribution on the interval (0.25, 1.25) months. Let the beginning inventory level stand at 50 units, with no orders outstanding. Let the holding cost be \$1 per unit in inventory per month. Assume that the inventory position is reviewed each month. If an order is placed, its cost is \$60 + \$5Q, where \$60 is the ordering cost and \$5 is the cost of each item. The time between demands is exponentially distributed with a mean of 1/15 month. The sizes of the demands follow the distribution:

Demand	*Probability*
1	1/2
2	1/4
3	1/8
4	1/8

(a) Make four independent replications, each of run length 100 months preceded by a 12-month initialization period, for the $(M, L) = (50, 30)$ policy. Estimate long-run mean monthly cost with a 90% confidence interval.

(b) Using the results of part (a), estimate the total number of replications needed to estimate mean monthly cost within \$5.

7. Reconsider Exercise 6, except that if the inventory level at a monthly review is zero or negative, a rush order for Q units is placed. The cost for a rush order is \$120 + \$12Q, where \$120 is the ordering cost and \$12 is the cost of each item. The lead time for a rush order is given by a uniform distribution on the inteval (0.10, 0.25) months.

(a) Make four independent replications for the (M, L) policy, and estimate long-run mean monthly cost with a 90% confidence interval.

(b) Using the results of part (a), estimate the total number of replications needed to estimate mean monthly cost within \$5.

8. Suppose that the items in Exercise 6 are perishable, with a selling price given by the following data:

On the Shelf (Months)	*Selling Price*
0–1	\$10
1–2	5
>2	0

Thus, any item that has been on the shelf more than 2 months cannot be sold. The age is determined at the time the demand occurs. If an item is outdated, it is discarded and the next item is brought forward. Simulate the system for 100 months.

(a) Make four independent replications for the $(M, L) = (50, 30)$ policy, and estimate long-run mean monthly cost with a 90% confidence interval.

(b) Using the results of part (a), estimate the total number of replications needed to estimate mean monthly cost within \$5.

At first, assume that all the items in the beginning inventory are fresh. Is this a good assumption? What effect does this "all-fresh" assumption have on the estimates of long-run mean monthly cost? What can be done to improve these estimates? Carry out a complete analysis.

9. Consider the following inventory system:

(a) Whenever the inventory level falls to or below 10 units, an order is placed. Only one order can be outstanding at a time.

(b) The size of each order is Q. Maintaining an inventory costs \$0.50 per day per item in inventory. Placing an order results in a fixed cost of \$10.00.

(c) Lead time is distributed in accordance with a discrete uniform distribution between zero and 5 days.

(d) If a demand occurs during a period when the inventory level is zero, the sale is lost at a cost of \$2.00 per unit.

(e) The number of customers each day is given by the following distribution:

Number of Customers per Day	*Probability*
1	0.23
2	0.41
3	0.22
4	0.14

(f) The demand on the part of each customer is Poisson distributed with a mean of 3 units.

(g) For simplicity, assume that all demands occur at 12 noon and that all orders are placed immediately thereafter.

Assume further that orders are received at 5:00 P.M., or after the demand that occurred on that day. Consider the policy having $Q = 20$. Make five independent replications, each of length 100 days, and compute a 90% confidence interval for long-run mean daily cost. Investigate the effect of initial inventory level and existence of an outstanding order on the estimate of mean daily cost. Begin with an initial inventory of $Q + 10$, and no outstanding orders.

10. A store selling Mother's Day cards must decide 6 months in advance on the number of cards to stock. Reordering is not allowed. Cards cost \$0.25 and sell for \$0.60. Any cards not sold by Mother's Day go on sale for \$0.20 for 2 weeks. However, sales of the remaining cards are probabilistic in nature according to the following distribution:

32% of the time, all cards remaining get sold.
40% of the time, 80% of all cards remaining are sold.
28% of the time, 60% of all cards remaining are sold.

Any cards left after 2 weeks are sold for \$0.30. The card shop owner is not sure how many cards can be sold but thinks it is somewhere (i.e., uniformly distributed) between 200 and 400. Suppose that the card shop owner decides to order 300 cards. Estimate the expected total profit with an error of at most \$5.00. [*Hint*: Make three or four initial replications. Use these data to estimate the total sample size needed. Each replication consists of one Mother's Day.]

11. A very large mining operation has decided to control the inventory of high-pressure piping by a periodic-review, order-up-to-M policy, where M is a target level. The annual demand for this piping is normally distributed with a mean of 600 and a variance of 800. This demand occurs fairly uniformly over the year. The lead time for resupply is Erlang distributed of order $k = 2$ with a mean of 2 months. The cost of each unit is $400. The inventory carrying charge, as a proportion of item cost on an annual basis, is expected to fluctuate normally about a mean of 0.25 (simple interest) with a standard deviation of 0.01. The cost of making a review and placing an order is $200, and the cost of a backorder is estimated to be $100 per unit backordered. Suppose that the inventory level is reviewed every 2 months, and let $M = 337$.

(a) Make five independent replications, each of run length 100 months, to estimate long-run mean monthly cost by means of a 90% confidence interval.

(b) Investigate the effects of initial conditions. Determine an appropriate number of monthly observations to delete to reduce initialization bias to a negligible level.

12. Consider some number, say N, of $M/M/1$ queues in series. The $M/M/1$ queue, described in Section 6.4, has Poisson arrivals at some rate λ customers per hour, exponentially distributed service times with mean $1/\mu$, and a single server. (Recall that "Poisson arrivals" means that interarrival times are exponentially distributed.) By $M/M/1$ queues in series, it is meant that upon completion of service at a given server, a customer joins a waiting line for the next server. The system can be shown as follows:

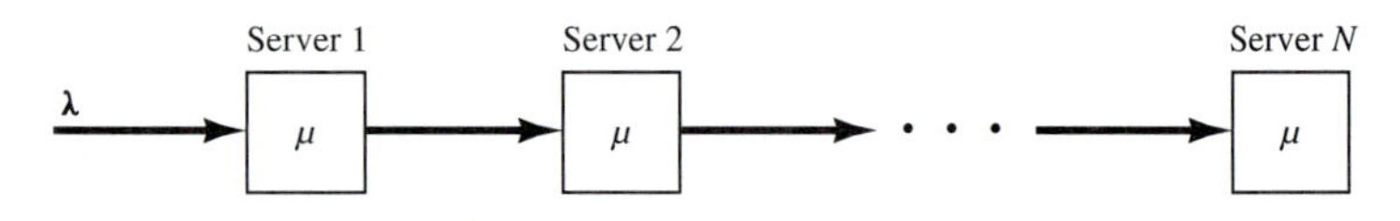

All service times are exponentially distributed with mean $1/\mu$, and the capacity of each waiting line is assumed to be unlimited. Assume that $\lambda = 8$ customers per hour, and $1/\mu = 0.1$ hour. The measure of performance is response time, which is defined to be the total time a customer is in the system.

(a) By making appropriate simulation runs, compare the initialization bias for $N = 1$ (i.e., one $M/M/1$ queue) to $N = 2$ (i.e., two $M/M/1$ queues in series). Start each system with all servers idle and no customers present. The purpose of the simulation is to estimate mean response time.

(b) Investigate the initialization bias as a function of N, for $N = 1, 2, 3, 4$, and 5.

(c) Draw some general conclusions concerning initialization bias for "large" queueing systems when at time 0 the system is assumed to be empty and idle.

13. Jobs enter a job shop in random fashion according to a Poisson process at overall rate 2 every 8-hour day. The jobs are of four types. They flow from workstation to workstation in a fixed order depending on type, as shown below. The proportions of each type are also shown.

Type	*Flow through Stations*	*Proportion*
1	1, 2, 3, 4	0.4
2	1, 3, 4	0.3
3	2, 4, 3	0.2
4	1, 4	0.1

Processing times per job at each station depend on type, but all times are (approximately) normally distributed with mean and s.d. (in hours) as follows:

	Station			
Type	*1*	*2*	*3*	*4*
1	(20, 3)	(30, 5)	(75, 4)	(20, 3)
2	(18, 2)		(60, 5)	(10, 1)
3		(20, 2)	(50, 8)	(10, 1)
4	(30, 5)			(15, 2)

Station i will have c_i workers ($i = 1, 2, 3, 4$). Each job occupies one worker at a station for the duration of a processing time. All jobs are processed on a first-in, first-out basis, and all queues for waiting jobs are assumed to have unlimited capacity. Simulate the system for 800 hours, preceded by a 200-hour initialization period. Assume that $c_1 = 8$, $c_2 = 8$, $c_3 = 20$, $c_4 = 7$. Based on $R = 5$ replications, compute a 97.5% confidence interval for average worker utilization at each of the four stations. Also compute a 95% confidence interval for mean total response time for each job type, where a total response time is the total time that a job spends in the shop.

14. Change Exercise 13 to give priority at each station to the jobs by type. Type 1 jobs have priority over type 2, type 2 over type 3, and type 3 over type 4. Use a run length of 800 hours, an initialization period of 200 hours, and $R = 5$ replications. Compute four 97.5% confidence intervals for mean total response time by type. Also run the model without priorities and compute the same confidence intervals. Discuss the trade-offs when using first-in, first-out versus a priority system.

15. Consider a single-server queue with Poisson arrivals at rate $\lambda = 10.82$ per minute, and normally distributed service times with a mean of 5.1 seconds and a variance of 0.98^2 seconds. It is desired to estimate the mean time in the system for a customer who upon arrival finds i other customers in the system; i.e., to estimate

$$w_i = E(W|N = i) \quad \text{for } i = 0, 1, 2, \ldots$$

where W is a typical system time, and N is the number of customers found by an arrival. For example, w_0 is the mean system time for those customers who find the system empty, w_1 is the mean system time for those customers who find one other customer present upon arrival, and so on. The estimate $\widehat{w}_i$ of w_i will be a sample mean of system times taken over all arrivals who find i in the system. Plot $\widehat{w}_i$ vs i. Hypothesize and attempt to verify a relation between w_i and i.

(a) Simulate for a 10-hour period with empty and idle initial conditions.

(b) Simulate for a 10-hour period after an initialization of one hour. Are there observable differences in the results of (a) and (b)?

(c) Repeat parts (a) and (b) with service times exponentially distributed with mean 5.1 seconds.

(d) Repeat parts (a) and (b) with deterministic service times equal to 5.1 seconds.

(e) Determine the number of replications needed to estimate $w_0, w_1, \ldots, w_6$ with a standard error for each of at most 3 seconds. Repeat parts (a)–(d) using this number of replications.

16. At Smalltown U. there is one graphics workstation for student use located across campus from the computer center. At 2:00 A.M. one night six students arrive at the workstation to complete an assignment. A student uses the workstation for 10 ± 8 minutes, then leaves to go to the computer center to pick up graphics output. There is a 25% chance that the run will be OK and the student will go to sleep. If it is not OK, the student returns to the workstation and waits until it becomes free. The round trip from workstation to computer center and back takes 30 ± 5 minutes. The computer becomes inaccessible at 5:00 A.M. Estimate the probability, p, that at least five of the six students will finish their assignment in the 3-hour period. First, make $R = 10$ replications and compute a 95% confidence interval for p. Next determine the number of replications needed to estimate p within $\pm.02$ and make these replications. Recompute the 95% confidence interval for p.

17. Four workers are evenly spaced along a conveyor belt. Items needing processing arrive according to a Poisson process at rate of 2 per minute. Processing time is exponentially distributed with a mean of 1.6 minutes. If a worker becomes idle, then he or she takes the first item to come by on the conveyor. If a worker is busy when an item comes by, that item moves down the conveyor to the next worker, taking 20 seconds between two successive workers. When a worker finishes processing an item, the item leaves the system. If an item passes by the last worker, it is recirculated on a loop conveyor and will return to the first worker after 5 minutes.

Management is interested in having a balanced workload; that is, management would like worker utilizations to be equal. Let ρ_i be the long-run utilization of worker i, and let ρ be the average utilization of all workers. Thus, $\rho = (\rho_1+\rho_2+\rho_3+\rho_4)/4$. Using queueing theory, ρ can be estimated by $\rho = \lambda/c\mu$, where $\lambda = 2$ arrivals per minute, $c = 4$ servers, and $1/\mu = 1.6$ minutes is the mean service time. Thus, $\rho = \lambda/c\mu = (2/4)1.6 = 0.8$, so on the average a worker will be busy 80% of the time.

(a) Make 5 independent replications, each of run length 40 hours preceded by a one-hour initialization period. Compute 95% confidence intervals for ρ_1 and ρ_4. Draw conclusions concerning workload balance.

(b) Based on the same 5 replications, test the hypothesis H_0: $\rho_1 = 0.8$ at a level of significance $\alpha = 0.05$. If a difference of $\pm.05$ is important to detect, determine the probability that such a deviation is detected. In addition, if it is desired to detect such a deviation with probability at least 0.9, determine the sample size needed to do so. (See any basic statistics textbook for guidance on hypothesis testing.)

(c) Repeat (b) for H_0: $\rho_4 = 0.8$.

(d) Based on the results from (a)–(c), draw conclusions for management about the balancing of workloads.

18. At a small rock quarry, a single power shovel dumps a scoopful of rocks at the loading area approximately every 10 minutes, with the actual time between scoops being well modeled as exponentially distributed with mean 10 minutes. Three scoops of rocks make a pile, and whenever one pile of rocks is completed, the shovel starts a new pile.

The quarry has a single truck that can carry one pile (three scoops) at a time. It takes approximately 27 minutes for a pile of rocks to be loaded into the truck, driven to the processing plant, unloaded, and for the truck to return back to the loading area. The actual time to do these things (altogether) is well modeled as being normally distributed with mean 27 minutes and standard deviation 12 minutes.

When the truck returns to the loading area, it will load and transport another pile if one is waiting to be loaded; otherwise it stays idle until another pile is ready. For safety reasons, no loading of the truck occurs until a complete pile (all three scoops) is waiting.

The quarry operates in this manner for an 8-hour day. We are interested in estimating the utilization of the trucks and the expected number of piles waiting to be transported if an additional truck is purchased.

19. Big Bruin, Inc., plans to open a small grocery store in Juneberry, NC. They expect to have two checkout lanes, with one lane being reserved for customers paying with cash. The question they want to answer is: how many grocery carts do they need?

During business hours (6 A.M.–8 P.M.) cash-paying customers are expected to arrive at a rate of 8 per hour. All other customers are expected to arrive at a rate of 9 per hour. The time between arrivals of each type can be modeled as exponentially distributed random variables.

The time spent shopping is modeled as normally distributed with mean 40 minutes and standard deviation 10 minutes. The time required to check out after shopping can be modeled as lognormally distributed with (a) mean 4 minutes and standard deviation 1 minute for cash-paying customers; (b) mean 6 minutes and standard deviation 1 minute for all other customers.

We will assume that every customer uses a shopping cart, and that after finishing shopping, the customer leaves the cart in the store so that it is immediately available for another customer. We will also assume that if a cart is not available, the customer immediately leaves the store disgusted.

The primary performance measures of interest to Big Bruin are the expected number of shopping carts in use, and the expected number of customers lost per day. Recommend a number of carts for the store, remembering that carts are expensive, but so are lost customers.

20. Develop a simulation model of the total time in the system for an $M/M/1$ queue with service rate $\mu = 1$ (therefore, the traffic intensity is $\rho = \lambda/\mu = \lambda$, the arrival rate). Use the simulation, in conjunction with the technique of plotting ensemble averages, to study the effect of traffic intensity on initialization bias when the queue starts empty. Specifically, see how the initialization phase T_0 changes for $\rho = 0.5, 0.7, 0.8, 0.9, 0.95$.

12

Comparison and Evaluation of Alternative System Designs

Chapter 11 dealt with the precise estimation of a measure of performance for one system. This chapter discusses a few of the many statistical methods that can be used to compare two or more system designs on the basis of some performance measure. One of the most important uses of simulation is the comparison of alternative system designs. Since the observations of the response variables contain random variation, statistical analysis is needed to determine whether any observed differences are due to differences in design or merely to the random fluctuation inherent in the models.

The comparison of two system designs is computationally easier than the simultaneous comparison of multiple (more than two) system designs. Section 12.1 discusses the case of two system designs, using two possible statistical techniques: *independent sampling* and *correlated sampling*. Correlated sampling is also known as the common random-numbers technique; simply put, the same random numbers are used to simulate both alternative system designs. If implemented correctly, correlated sampling usually reduces the variance of the estimated difference of the performance measures and thus can provide, for a given sample size, more precise estimates of the mean difference than can independent sampling. Section 12.2 extends the statistical techniques of Section 12.1 to the comparison of multiple (more than two) system designs, using the Bonferroni approach to confidence-interval estimation and selecting the best. Since the Bonferroni approach is limited to ten or fewer system designs, Section 12.3 describes how a large number of complex system designs can sometimes be represented by a simpler metamodel. Finally, for comparison and evaluation of a very large number of system designs that are related in a less structured way, Section 12.4 presents optimization via simulation.

12.1 Comparison of Two System Designs

Suppose that a simulation analyst desires to compare two possible configurations of a system. In a queueing system, perhaps two possible queue disciplines, or two possible sets of servers, are to be compared. In a supply-chain inventory system, perhaps two possible ordering policies will be compared. A job shop may have many possible scheduling rules; a production system may have in-process inventory buffers of various capacities. Many other examples of alternative system designs can be provided.

The method of replications will be used to analyze the output data. The mean performance measure for system i will be denoted by $\theta_i (i = 1, 2)$. If it is a steady-state simulation, it will be assumed that deletion of data, or other appropriate techniques, have been used to assure that the point estimators are approximately unbiased estimators of the mean performance measures, θ_i. The goal of the simulation experiment is to obtain point and interval estimates of the difference in mean performance, namely $\theta_1 - \theta_2$. Three methods of computing a confidence interval for $\theta_1 - \theta_2$ will be discussed. But first an example and a general framework will be given.

EXAMPLE 12.1

A vehicle safety inspection station performs three jobs: (1) brake check, (2) headlight check, and (3) steering check. The present system has three stalls in parallel; that is, a vehicle enters a stall, where one attendant makes all three inspections. The present system is illustrated in Figure 12.1(a). Based on data from the existing system, it has been assumed that arrivals occur completely at random (i.e., according to a Poisson process) at an average rate of 9.5 per hour, and that the time for a brake check, a headlight check, and a steering check

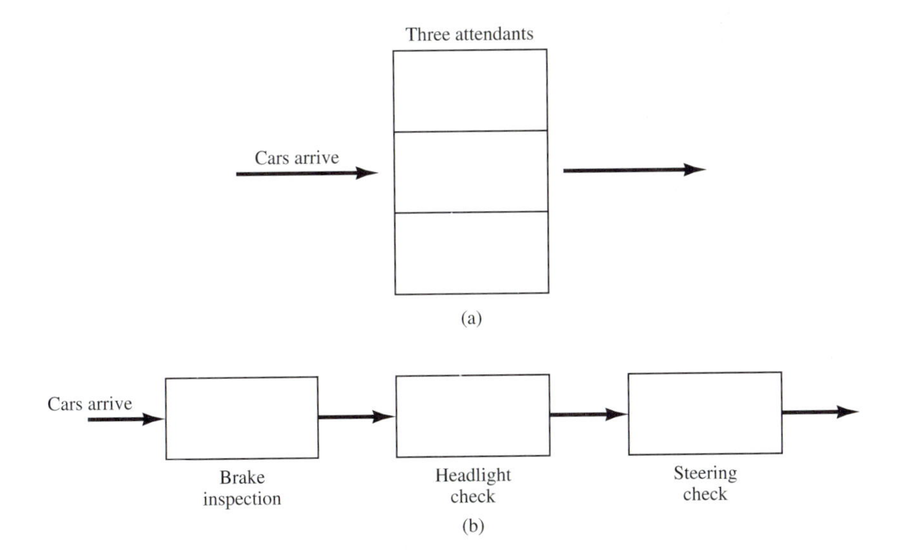

Figure 12.1. Vehicle safety inspection station and a possible alternative design.

are normally distributed with means of 6.5, 6, and 5.5 minutes, respectively, all having standard deviations of approximately 0.5 minute. There is no limit on the queue of waiting vehicles.

An alternative system design is shown in Figure 12.1(b). Each attendant will specialize in a single task, and each vehicle will pass through three workstations in series. No space is allowed for vehicles between the brake and headlight checks or between the headlight and steering checks. Therefore, a vehicle in the brake or headlight check must move to the next attendant, while a vehicle in the steering check must exit before the next vehicle can move ahead. Due to the increased specialization of the inspectors, it is anticipated that mean inspection times for each type of check will decrease by 10% to 5.85, 5.4, and 4.95 minutes, respectively, for the brake, headlight, and steering inspections. The Safety Inspection Council has decided to compare the two systems on the basis of mean response time per vehicle, where a response time is defined as the total time from a vehicle's arrival until its departure from the system. ◀

When comparing two systems, such as those in Example 12.1, the simulation analyst must decide on a run length $T_E^{(i)}$ for each model $(i = 1, 2)$, and a number of replications R_i to be made of each model. From replication r of system i, the simulation analyst obtains an estimate Y_{ri} of the mean performance measure, θ_i. In Example 12.1, Y_{ri} would be the average response time observed during replication r for system i $(r = 1, \ldots, R_i; i = 1, 2)$. The data, together with the two summary measures, the sample means $\bar{Y}_{\cdot i}$, and the sample variances S_i^2, are exhibited in Table 12.1. Assuming that the estimators Y_{ri} are (at least approximately) unbiased, it follows that

$$\theta_1 = E(Y_{r1}),\ r = 1, \ldots, R_1; \quad \theta_2 = E(Y_{r2}),\ r = 1, \ldots, R_2$$

In Example 12.1, since the Safety Inspection Council is interested in a comparison of the two system designs, the simulation analyst decides to compute a confidence interval for $\theta_1 - \theta_2$, the difference between the two mean performance measures. The confidence interval is used to answer two questions: (1) How large is the mean difference, and how precise is the estimator of mean difference? (2) Is there a significant difference between the two systems? This second question will lead to one of three possible conclusions:

1. If the confidence interval (c.i.) for $\theta_1 - \theta_2$ is totally to the left of zero, as shown in Figure 12.2(a), then there is strong evidence for the hypothesis that $\theta_1 - \theta_2 < 0$, or equivalently $\theta_1 < \theta_2$.

In Example 12.1, $\theta_1 < \theta_2$ implies that the mean response time for system 1 (the original system) is smaller than for system 2 (the alternative system).

2. If the c.i. for $\theta_1 - \theta_2$ is totally to the right of zero, as shown in Figure 12.2(b), then there is strong evidence that $\theta_1 - \theta_2 > 0$, or equivalently, $\theta_1 > \theta_2$.

In Example 12.1, $\theta_1 > \theta_2$ can be interpreted as system 2 being better than system 1, in the sense that system 2 has smaller mean response time.

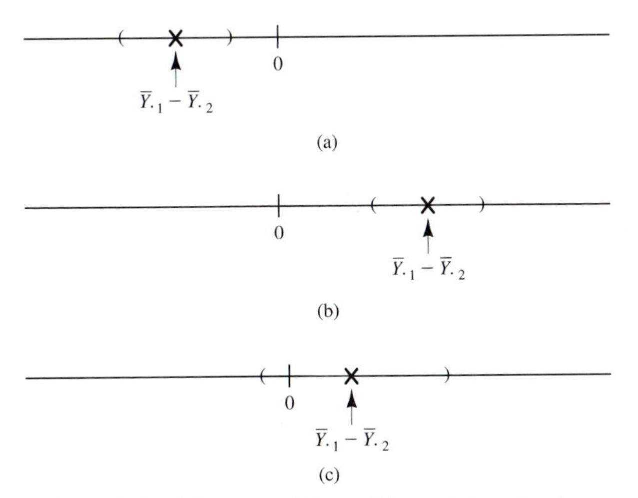

Figure 12.2. Three possible confidence intervals when comparing two systems.

3. If the c.i. for $\theta_1 - \theta_2$ contains zero, then, based on the data at hand, there is no strong statistical evidence that one system design is better than the other.

Some statistics textbooks say that the weak conclusion $\theta_1 = \theta_2$ can be drawn, but such statements can be misleading. A "weak" conclusion is often no conclusion at all. Most likely, if enough additional data were collected (i.e., R_i increased), the c.i. would possibly shift, and definitely shrink in length, until conclusion 1 or 2 would be drawn. In addition to one of these three conclusions, the confidence interval provides a measure of the precision of the estimator of $\theta_1 - \theta_2$.

A two-sided $100(1-\alpha)\%$ c.i. for $\theta_1 - \theta_2$ will always be of the form

$$(\bar{Y}_{\cdot 1} - \bar{Y}_{\cdot 2}) \pm t_{\alpha/2,\nu}\text{s.e.}(\bar{Y}_{\cdot 1} - \bar{Y}_{\cdot 2}) \tag{12.1}$$

where $\bar{Y}_{\cdot i}$ is the sample mean performance measure for system i over all replications

$$\bar{Y}_{\cdot i} = \frac{1}{R_i}\sum_{r=1}^{R_i} Y_{ri} \tag{12.2}$$

and ν is the degrees of freedom associated with the variance estimator, $t_{\alpha/2,\nu}$ is the $100(1-\alpha/2)$ percentage point of a t-distribution with ν degrees of freedom, and s.e.$(\cdot)$ represents the standard error of the specified point estimator. To obtain the standard error and the degrees of freedom, the analyst uses one of three statistical techniques. All three techniques assume that the basic data, Y_{ri} of Table 12.1, are approximately normally distributed. This assumption is reasonable, provided that each Y_{ri} is itself a sample mean of observations from replication r (which is indeed the situation in Example 12.1). [Note the similarity between Equations (12.1) and (11.10).]

Table 12.1. Simulation Output Data and Summary Measures When Comparing Two Systems

System	*Replication*				*Sample Mean*	*Sample Variance*
	1	*2*	...	R_i		
1	Y_{11}	Y_{21}	...	$Y_{R_1 1}$	$\bar{Y}_{\cdot 1}$	S_1^2
2	Y_{12}	Y_{22}	...	$Y_{R_2 2}$	$\bar{Y}_{\cdot 2}$	S_2^2

By design of the simulation experiment, $Y_{r1} (r = 1, \ldots, R_1)$ are independently and identically distributed (i.i.d.) with mean θ_1 and variance σ_1^2 (say). Similarly, $Y_{r2} (r = 1, \ldots, R_2)$ are i.i.d. with mean θ_2 and variance σ_2^2 (say). The three techniques for computing the confidence interval in Equation (12.1), which are based on three different sets of assumptions, are discussed in the following subsections.

There is an important distinction between *statistically significant* differences and *practically significant* differences in systems performance. Statistical significance answers the question: Is the observed difference $\bar{Y}_{\cdot 1} - \bar{Y}_{\cdot 2}$ larger than the variability in $\bar{Y}_{\cdot 1} - \bar{Y}_{\cdot 2}$? This question can be restated as: Have we collected enough data to be confident that the difference we observed is real, or just chance? Conclusions 1 and 2 above imply a statistically significant difference, while conclusion 3 implies that the observed difference is not statistically significant (even though the systems may indeed be different). Statistical significance is a function of the simulation experiment and the output data.

Practical significance answers the question: Is the true difference $\theta_1 - \theta_2$ large enough to matter for the decision we need to make? In Example 12.1, we may reach the conclusion that $\theta_1 > \theta_2$ and decide that system 2 is better (smaller expected response time). However, if the actual difference $\theta_1 - \theta_2$ is very small—say small enough that a customer would not notice the improvement—then it may not be worth the cost to replace system 1 with system 2. Practical significance is a function of the actual difference between the systems and is independent of the simulation experiment.

Confidence intervals do not answer the question of practical significance directly. Instead, they bound (with probability $1-\alpha$) the true difference $\theta_1 - \theta_2$ within the range

$$\bar{Y}_{\cdot 1} - \bar{Y}_{\cdot 2} - t_{\alpha/2,\nu}\text{s.e.}(\bar{Y}_{\cdot 1} - \bar{Y}_{\cdot 2}) \leq \theta_1 - \theta_2 \leq \bar{Y}_{\cdot 1} - \bar{Y}_{\cdot 2} + t_{\alpha/2,\nu}\text{s.e.}(\bar{Y}_{\cdot 1} - \bar{Y}_{\cdot 2})$$

Whether or not a difference within these bounds is practically significant depends on the particular problem.

12.1.1 Independent Sampling with Equal Variances

Independent sampling means that different and independent random-number streams will be used to simulate the two systems. This implies that all the observations of simulated system 1, namely $\{Y_{r1}, r = 1, \ldots, R_1\}$, are statistically independent of all the observations of simulated system 2, namely

$\{Y_{r2}, r = 1, \ldots, R_2\}$. By Equation (12.2) and the independence of the replications, the variance of the sample mean, $\bar{Y}_{.i}$, is given by

$$\text{var}(\bar{Y}_{.i}) = \frac{\text{var}(Y_{ri})}{R_i} = \frac{\sigma_i^2}{R_i}, \quad i = 1, 2$$

When using independent sampling, $\bar{Y}_{.1}$ and $\bar{Y}_{.2}$ are statistically independent; hence,

$$\begin{aligned} \text{var}(\bar{Y}_{.1} - \bar{Y}_{.2}) &= \text{var}(\bar{Y}_{.1}) + \text{var}(\bar{Y}_{.2}) \\ &= \frac{\sigma_1^2}{R_1} + \frac{\sigma_2^2}{R_2} \end{aligned} \tag{12.3}$$

In some cases it is reasonable to assume that the two variances are equal (but unknown in value); that is, $\sigma_1^2 = \sigma_2^2$. The data can be used to test the hypothesis of equal variances; if rejected, the method of Section 12.1.2 must be used. In a steady-state simulation, the variance σ_i^2 decreases as the run length $T_E^{(i)}$ increases; therefore, it may be possible to adjust the two run lengths, $T_E^{(1)}$ and $T_E^{(2)}$, to achieve at least approximate equality of σ_1^2 and σ_2^2.

If it is reasonable to assume that $\sigma_1^2 = \sigma_2^2$ (approximately), a two-sample t confidence-interval approach can be used. The point estimate of the mean performance difference is

$$\widehat{\theta_1} - \widehat{\theta_2} = \bar{Y}_{.1} - \bar{Y}_{.2} \tag{12.4}$$

with $\bar{Y}_{.i}$ given by Equation (12.2). Next compute the sample variance for system i by

$$\begin{aligned} S_i^2 &= \frac{1}{R_i - 1} \sum_{r=1}^{R_i} (Y_{ri} - \bar{Y}_{.i})^2 \\ &= \frac{1}{R_i - 1} \left(\sum_{r=1}^{R_i} Y_{ri}^2 - R_i \bar{Y}_{.i}^2 \right) \end{aligned} \tag{12.5}$$

Note that S_i^2 is an unbiased estimator of the variance σ_i^2. Since by assumption $\sigma_1^2 = \sigma_2^2 = \sigma^2$ (say), a pooled estimate of σ^2 is obtained by

$$S_p^2 = \frac{(R_1 - 1)S_1^2 + (R_2 - 1)S_2^2}{R_1 + R_2 - 2}$$

which has $\nu = R_1 + R_2 - 2$ degrees of freedom. The c.i. for $\theta_1 - \theta_2$ is then given by expression (12.1) with the standard error computed by

$$\text{s.e.}(\bar{Y}_{.1} - \bar{Y}_{.2}) = S_p \sqrt{\frac{1}{R_1} + \frac{1}{R_2}} \tag{12.6}$$

This standard error is an estimate of the standard deviation of the point estimate, which, by Equation (12.3), is given by $\sigma\sqrt{1/R_1 + 1/R_2}$.

In some cases, the simulation analyst may have $R_1 = R_2$, in which case it is safe to use the c.i. in expression (12.1) with the standard error taken from Equation (12.6) even if the variances (σ_1^2 and σ_2^2) are not equal. However, if the variances are unequal and the sample sizes differ, it has been shown that use of the two-sample t c.i. may yield invalid confidence intervals whose true probability of containing $\theta_1 - \theta_2$ is much less than $1 - \alpha$. Thus, if there is no evidence that $\sigma_1^2 = \sigma_2^2$, and if $R_1 \neq R_2$, the approximate procedure in the next subsection is recommended.

12.1.2 Independent Sampling with Unequal Variances

If the assumption of equal variances cannot safely be made, an approximate $100(1-\alpha)\%$ c.i. for $\theta_1 - \theta_2$ can be computed as follows. The point estimate and sample variances are computed by Equations (12.4) and (12.5). The standard error of the point estimate is given by

$$\text{s.e.}(\bar{Y}_{\cdot 1} - \bar{Y}_{\cdot 2}) = \sqrt{\frac{S_1^2}{R_1} + \frac{S_2^2}{R_2}} \tag{12.7}$$

with degrees of freedom, ν, approximated by the expression

$$\nu = \frac{(S_1^2/R_1 + S_2^2/R_2)^2}{[(S_1^2/R_1)^2/(R_1 - 1)] + [(S_2^2/R_2)^2/(R_2 - 1)]} \tag{12.8}$$

rounded to an integer. The confidence interval is then given by expression (12.1) using the standard error of Equation (12.7). A minimum number of replications $R_1 \geq 6$ and $R_2 \geq 6$ is recommended for this procedure.

12.1.3 Correlated Sampling, or Common Random Numbers

Correlated sampling means that, for each replication, the same random numbers are used to simulate both systems. Therefore, R_1 and R_2 must be equal, say $R_1 = R_2 = R$. Thus, for each replication r, the two estimates, Y_{r1} and Y_{r2}, are no longer independent but rather are correlated. Since independent streams of random numbers are used on different replications, the pairs (Y_{r1}, Y_{s2}) are mutually independent when $r \neq s$, however. (For example, in Table 12.1, the observation Y_{11} is correlated with Y_{12}, but Y_{11} is independent of all other observations.) The purpose of using correlated sampling is to induce a positive correlation between Y_{r1} and Y_{r2} (for each r), and thus to achieve a variance reduction in the point estimator of mean difference, $\bar{Y}_{\cdot 1} - \bar{Y}_{\cdot 2}$. In general, this variance is given by

$$\begin{aligned} \text{var}(\bar{Y}_{\cdot 1} - \bar{Y}_{\cdot 2}) &= \text{var}(\bar{Y}_{\cdot 1}) + \text{var}(\bar{Y}_{\cdot 2}) - 2\text{cov}(\bar{Y}_{\cdot 1}, \bar{Y}_{\cdot 2}) \\ &= \frac{\sigma_1^2}{R} + \frac{\sigma_2^2}{R} - \frac{2\rho_{12}\sigma_1\sigma_2}{R} \end{aligned} \tag{12.9}$$

where ρ_{12} is the correlation between Y_{r1} and Y_{r2}. [By definition, $\rho_{12} = \text{cov}(Y_{r1}, Y_{r2})/\sigma_1\sigma_2$, which does not depend on r.]

Now compare the variance of $\bar{Y}_{.1} - \bar{Y}_{.2}$ when using correlated sampling [Equation (12.9), call it V_{CORR}] to the variance when using independent sampling with equal sample sizes [Equation (12.3) with $R_1 = R_2 = R$, call it V_{IND}]. Notice that

$$V_{CORR} = V_{IND} - \frac{2\rho_{12}\sigma_1\sigma_2}{R} \tag{12.10}$$

If correlated sampling works as intended, the correlation ρ_{12} will be positive; hence, the second term on the right side of Equation (12.10) will be positive and therefore

$$V_{CORR} < V_{IND}$$

That is, the variance of the point estimator will be smaller when using correlated sampling than when using independent sampling. A smaller variance (for the same sample size R) implies that the estimator based on correlated sampling is more precise.

To compute a $100(1-\alpha)\%$ c.i. with correlated data, first compute the differences

$$D_r = Y_{r1} - Y_{r2} \tag{12.11}$$

which, by the definition of correlated sampling, are i.i.d.; then compute the sample mean difference by

$$\bar{D} = \frac{1}{R}\sum_{r=1}^{R} D_r \tag{12.12}$$

(Thus, $\bar{D} = \bar{Y}_{.1} - \bar{Y}_{.2}$) The sample variance of the differences $\{D_r\}$ is computed by

$$\begin{aligned} S_D^2 &= \frac{1}{R-1}\sum_{r=1}^{R}(D_r - \bar{D})^2 \\ &= \frac{1}{R-1}\left(\sum_{r=1}^{R} D_r^2 - R\bar{D}^2\right) \end{aligned} \tag{12.13}$$

which has degrees of freedom $\nu = R - 1$. The $100(1-\alpha)\%$ c.i. for $\theta_1 - \theta_2$ is given by expression (12.1) with standard error of $\bar{Y}_{.1} - \bar{Y}_{.2} = \bar{D}$ estimated by

$$\text{s.e.}(\bar{D}) = \text{s.e.}(\bar{Y}_{.1} - \bar{Y}_{.2}) = \frac{S_D}{\sqrt{R}} \tag{12.14}$$

Since $S_D/\sqrt{R}$ of Equation (12.14) is an estimate of $\sqrt{V_{CORR}}$, and expression (12.6) or (12.7) is an estimate of $\sqrt{V_{IND}}$, correlated sampling will typically produce a c.i. which is shorter for a given sample size than the c.i. produced by independent sampling if $\rho_{12} > 0$. In fact, the expected length of the c.i. will be shorter when using correlated sampling if $\rho_{12} > 0.1$, provided $R > 10$. And the larger R is, the smaller ρ_{12} can be and still yield a shorter expected length [Nelson 1987].

For any problem, there are always many ways of implementing common random numbers. It is never enough to simply use the same seed on the random-number generator(s). Each random number used in one model for some purpose must be used for the same purpose in the second model; that is, the use of the random numbers must be synchronized. For example, if the ith random number is used to generate a service time at workstation 2 for the jth arrival in model 1, the ith random number should be used for the very same purpose in model 2. For queueing systems or service facilities, synchronization of the common random numbers guarantees that both systems face identical workloads: both systems face arrivals at the same instants of time, and these arrivals demand equal amounts of service. (The actual service times of a given arrival in the two models may not be equal, but may be proportional, if the server in one model is faster than the server in the other model.) For an inventory system when comparing two different ordering policies, synchronization guarantees that the two systems face identical demand for a given product. For production or reliability systems, synchronization guarantees that downtimes for a given machine will occur at exactly the same times, and will have identical durations, in the two models. On the other hand, if some aspect of one of the systems is totally different from that in the other system, synchronization may be inappropriate, or even impossible to achieve. In summary, those aspects of the two system designs which are sufficiently similar should be simulated with common random numbers in such a way that the two models "behave" similarly; but those aspects that are totally different should be simulated with independent random numbers.

Implementation of common random numbers is model dependent, but certain guidelines can be given that will make correlated sampling more likely to yield a positive correlation. The purpose of the guidelines is to ensure that synchronization occurs:

1. Dedicate a random-number stream to a specific purpose, and use as many different streams as needed. (Different random-number generators, or widely spaced seeds on the same generator, can be used to get two different, nonoverlapping streams. Some simulation languages allow the number of random numbers, or offset, between seeds to be specified rather than having to specify the seeds themselves.) It is not sufficient to assign seeds at the beginning of the first replication and then let the random-number generator merely continue for the second and subsequent replications. If conducted in this manner, the first replication will be synchronized but subsequent replications may not be.
2. For systems (or subsystems) with external arrivals: As each entity enters the system, the next interarrival time is generated, and then immediately all random variables (such as service times, order sizes, etc.) needed by the arriving entity and identical in both models are generated in a fixed order and stored as attributes of the entity, to be used later as needed. Apply guideline 1; that is, dedicate one random-number stream to these

external arrivals and all their attributes.

3. For systems having an entity performing given activities in a cyclic or repeating fashion, assign a random-number stream to this entity. (Example: a machine that cycles between two states: up-down-up-down-.... Use a dedicated random-number stream to generate the uptimes and downtimes.)
4. If synchronization is not possible or it is inappropriate for some part of the two models, use independent streams of random numbers for this subset of random variates.

Unfortunately, there is no guarantee that correlated sampling will always induce a positive correlation between comparable runs of the two models. It is known that if, for each input random variate X, the estimators Y_{r1} and Y_{r2} are increasing functions of the random variate X (or both are decreasing functions of X), then ρ_{12} will be positive. The intuitive idea is that both models (i.e., both Y_{r1} and Y_{r2}) respond in the same direction to each input random variate, and this results in positive correlation. This increasing or decreasing nature of the response variables (called *monotonicity*) with respect to the input random variables is known to hold for certain queueing systems (such as the $GI/G/c$ queues), when the response variable is customer delay, so some evidence exists that common random numbers is a worthwhile technique for queueing simulations. (For simple queues, customer delay is an increasing function of service times and a decreasing function of interarrival times.) Wright and Ramsay [1979] reported a negative correlation for certain inventory simulations, however. In summary, the guidelines above should be followed, and some reasonable notion that the response variable of interest is a monotonic function of the random input variables should be evident.

EXAMPLE 12.1 (Continued)

The two inspection systems shown in Figure 12.1 will be compared using both independent sampling and correlated sampling, in order to illustrate the greater precision of correlated sampling when it works.

Each vehicle arriving to be inspected has four input random variables associated with it:

$$A_n = \text{interarrival time between vehicles } n \text{ and } n+1$$

$$S_n^{(1)} = \text{brake inspection time for vehicle } n \text{ in model 1}$$

$$S_n^{(2)} = \text{headlight inspection time for vehicle } n \text{ in model 1}$$

$$S_n^{(3)} = \text{steering inspection time for vehicle } n \text{ in model 1}$$

For model 2 (of the proposed system), mean service times are decreased by 10%. When using independent sampling, different values of service (and interarrival) times would be generated for model 2. But when using correlated sampling, the random-number generator must be used in such a way that exactly the same values are generated for $A_1, A_2, A_3, \ldots$. For service times, $S_n^{(i)}$ $(i = 1, 2, 3)$ could be generated for model 1, and $S_n^{(i)} - 0.1E(S_n^{(i)})$ used in model 2. Alternatively, since normal random variates are usually generated by first generating a standard normal variate and then using Equation (8.26), the service times for a brake inspection could be generated by

$$E(S_n^{(1)}) + \sigma Z_n^{(1)} \tag{12.15}$$

where $Z_n^{(1)}$ is a standard normal variate, $\sigma = 0.5$ minute, $E(S_n^{(1)}) = 6.5$ minutes for model 1, and $E(S_n^{(1)}) = 5.85$ minutes for model 2; the other two inspection times would be generated in a similar fashion. To implement (synchronized) common random numbers, the simulation analyst would generate identical $Z_n^{(i)}$ sequences $(i = 1, 2, 3; n = 1, 2, \ldots)$ in both models and then use the appropriate version of Equation (12.15) to generate the inspection times. For the synchronized runs, the service times for a vehicle were generated at the instant of arrival (by guideline 2) and stored as an attribute of the vehicle, to be used as needed. Runs were also made with nonsynchronized common random numbers, in which case one random-number stream was used as needed.

Table 12.2 gives the average response time for each of $R = 10$ replications, each of run length $T_E = 16$ hours. It was assumed that two cars were present at time 0, waiting to be inspected. Column 1 gives the outputs from model 1. Model 2 was run using independent random numbers (column 2I) and common random numbers without synchronization (column 2C*) and with synchronization (column 2C). The purpose of the simulation is to estimate mean difference in response times for the two systems.

Table 12.2. Comparison of System Designs for the Vehicle Safety Inspection System

	Average Response Time for Model				*Observed Differences*	
Replication	*1*	*2I*	*2C**	*2C*	$D_{1,2C^*}$	$D_{1,2C}$
1	29.59	51.62	56.47	29.55	−26.88	0.04
2	23.49	51.91	33.34	24.26	−9.85	−0.77
3	25.68	45.27	35.82	26.03	−10.14	−0.35
4	41.09	30.85	34.29	42.64	6.80	−1.55
5	33.84	56.15	39.07	32.45	−5.23	1.39
6	39.57	28.82	32.07	37.91	7.50	1.66
7	37.04	41.30	51.64	36.48	−14.60	0.56
8	40.20	73.06	41.41	41.24	−1.21	−1.04
9	61.82	23.00	48.29	60.59	13.53	1.23
10	44.00	28.44	22.44	41.49	21.56	2.51
Sample mean	37.63	43.04			−1.85	0.37
Sample variance	118.90	244.33			208.94	1.74
Standard error	6.03				4.57	0.42

For the two independent runs (1 and 2I), it was assumed that the variances were not necessarily equal, so the method of Section 12.1.2 was applied. Sample variances and the standard error were computed by Equations (12.5) and (12.7), yielding

$$S_1^2 = 118.9, \quad S_{2I}^2 = 244.3$$

and

$$\text{s.e.}(\bar{Y}_{.1} - \bar{Y}_{.2I}) = \sqrt{\frac{118.9}{10} + \frac{244.3}{10}} = 6.03$$

with degrees of freedom, ν, equal to 17, as given by Equation (12.8). The point estimate is $\bar{Y}_{.1} - \bar{Y}_{.2I} = -5.4$ minutes, and a 95% c.i. [expression (12.1)] is given by

$$-5.4 \pm 2.11(6.03)$$

or

$$-18.1 \leq \theta_1 - \theta_2 \leq 7.3 \tag{12.16}$$

The 95% confidence interval in Inequality (12.16) contains zero, which indicates that there is no strong evidence that the observed difference of -5.4 minutes is due to anything other than random variation in the output data. In other words, it is not statistically significant. Thus, if the simulation analyst had decided to use independent sampling, no strong conclusion is possible because the estimate of $\theta_1 - \theta_2$ is quite imprecise.

For the two sets of correlated runs (1 and 2C*, and 1 and 2C), the observations are paired and analyzed as given in Equations (12.11) through (12.14). The point estimate when not synchronizing the random numbers is given by Equation (12.12) as

$$\bar{D} = -1.9 \text{ minutes}$$

the sample variance by $S_D^2 = 208.9$ (with $\nu = 9$ degrees of freedom), and the standard error by s.e.$(\bar{D}) = 4.6$. Thus, a 95% c.i. for the true mean difference in response times, as given by expression (12.1), is

$$-1.9 \pm 2.26(4.6)$$

or

$$-12.3 < \theta_1 - \theta_2 < 8.5 \tag{12.17}$$

Again no strong conclusion is possible, since the confidence interval contains zero. Notice, however, that the estimate of $\theta_1 - \theta_2$ is slightly more precise than that in Inequality (12.16), since the length of the interval is smaller.

When complete synchronization of the random numbers was used in run 2C, the point estimate of the mean difference in response times was

$$\bar{D} = 0.4 \text{ minute}$$

the sample variance was $S_D^2 = 1.7$ (with $\nu = 9$ degrees of freedom), and the standard error was s.e.$(\bar{D}) = 0.4$. A 95% c.i. for the true mean difference is given by

$$-0.50 < \theta_1 - \theta_2 < 1.30 \tag{12.18}$$

The confidence interval in Inequality (12.18) again contains zero, but it is considerably shorter than the previous two intervals. This greater precision in the estimation of $\theta_1 - \theta_2$ is due to the use of synchronized common random numbers. The short length of the interval in Inequality (12.18) suggests that the true difference, $\theta_1 - \theta_2$, is close to zero. In fact, the upper bound of 1.30 indicates that system 2 is at most 1.30 minutes faster, in expectation. If such a small difference is not practically significant, then there is no need to determine which system is truly better.

As seen by comparing the confidence intervals in Inequalities (12.16), (12.17), and (12.18), the width of the confidence interval is reduced by 18% when using nonsynchronized common random numbers, and by 93% when using common random numbers with full synchronization. Comparing the estimated variance of $\bar{D}$ using synchronized common random numbers to the variance of $\bar{Y}_{.1} - \bar{Y}_{.2}$ using independent sampling shows a variance reduction of 99.5%, which means that to achieve precision comparable to that achieved by correlated sampling, a total of approximately $R = 2090$ independent replications would have to be made.

The next few examples show how common random numbers can be implemented in other contexts. ◀

Example 12.2 (The Dump Truck Problem, Revisited)

Consider Example 3.5 (the dump truck problem), shown in Figure 3.7. Each of the trucks repeatedly goes through three activities: loading, weighing, and traveling. Assume that there are eight trucks, and at time 0 all eight are at the loaders. Weighing time per truck on the single scale is uniformly distributed between 1 and 9 minutes; and travel time per truck is exponentially distributed with mean 85 minutes. An unlimited queue is allowed before the loader(s) and before the scale. All trucks can be traveling at the same time. Management desires to compare one fast loader to the two slower loaders currently being used. Each of the slow loaders can fill a truck in 1 to 27 minutes, uniformly distributed. The new fast loader can fill a truck in 1 to 19 minutes, uniformly distributed. The basis for comparison is mean system response time, where a response time is defined as the duration of time from a truck arrival at the loader queue to that truck's departure from the scale.

To implement synchronized common random numbers, a separate and distinct random-number stream was assigned to each of the eight trucks. At the beginning of each replication (i.e., at time 0), a new and independently chosen set of eight seeds was specified, one seed for each random-number stream. Thus, weighing times and travel times for each truck were identical in both models, and the loading time for a given truck's ith visit to the fast loader was

Table 12.3. Comparison of System Designs for the Dump Truck Problem

	Average System Response Time for Model			
1	*2I*	*2C*	*Differences,*	
Replication	*(2 Loaders)*	*(1 Loader)*	*(1 Loader)*	$D_{1,2C}$
1	21.38	29.01	24.30	−2.92
2	24.06	24.70	27.13	−3.07
3	21.39	26.85	23.04	−1.65
4	21.90	24.49	23.15	−1.25
5	23.55	27.18	26.75	−3.20
6	22.36	26.91	25.62	−3.26
Sample mean	22.44	26.52		−2.56
Sample variance	1.28	2.86		0.767
Sample standard deviation	1.13	1.69		0.876

proportional to the loading time in the original system (with two slow loaders). Implementation of common random numbers without synchronization (e.g., using one random-number stream to generate all loading, weighing, and travel times as needed) would likely lead to a given random number being used to generate a loading time in model 1 but a travel time in model 2, or vice versa, and from that point on the use of a random number would most likely be different in the two models.

Six replications of each model were run, each of run length $T_E = 40$ hours. The results are shown in Table 12.3. Both independent sampling and correlated sampling were used, to illustrate the advantage of correlated sampling. The first column (labeled model 1) contains the observed average system response time for the existing system with two loaders. The columns labeled 2I and 2C are for the alternative design having one loader; the independent sampling results are in 2I, and the correlated sampling results are in the column labeled 2C. The rightmost column, labeled $D_{1,2C}$, lists the observed differences between the runs of model 1 and model 2C.

For independent sampling assuming unequal variances, the following summary statistics were computed using Equations (12.2), (12.5), (12.7), (12.8), and (12.1) and the data (in columns 1 and 2I) in Table 12.3:

$$\text{Point estimate}: \quad \bar{Y}_{.1} - \bar{Y}_{.2I} = 22.44 - 26.52 = -4.08 \text{ minutes}$$

$$\text{Sample variances}: \quad S_1^2 = 1.28, \; S_{2I}^2 = 2.86$$

$$\text{Standard error}: \quad \text{s.e.}(\bar{Y}_{.1} - \bar{Y}_{.2}) = (S_1^2/R_1 + S_{2I}^2/R_2)^{1/2} = 0.831$$

$$\text{Degrees of freedom}: \quad \nu = 8.73 \approx 9$$

$$95\% \text{ c.i. for } \theta_1 - \theta_2: \quad -4.08 \pm 2.26(0.831) \quad \text{or} \quad -4.08 \pm 1.878$$

$$-5.96 \le \theta_1 - \theta_2 \le -2.20$$

For correlated sampling, implemented by the use of synchronized common random numbers, the following summary statistics were computed using Equations (12.12), (12.13), (12.14), and (12.1), plus the data (in columns 1 and 2C) in Table 12.3:

Point estimate: $\bar{D} = -2.56$ minutes

Sample variance: $S_D^2 = 0.767$

Standard error: $\text{s.e.}(\bar{D}) = S_D/\sqrt{R} = 0.876/\sqrt{6} = 0.358$

Degrees of freedom: $\nu = R - 1 = 5$

95% c.i. for $\theta_1 - \theta_2$: $-2.56 \pm 2.57(0.358)$ or -2.56 ± 0.919

$$-3.48 \leq \theta_1 - \theta_2 \leq -1.641$$

By comparing the c.i. widths, we see that the use of correlated sampling with synchronization reduced c.i. width by 50%. This reduction could be important if, say, a difference of as much as 5.96 is considered practically significant, but a difference of at most 3.48 is not. Equivalently, if equal precision were desired, independent sampling would require approximately four times as many observations as would correlated sampling, or approximately 24 replications of each model instead of six. ◀

To illustrate how correlated sampling can fail when not implemented correctly, consider the dump truck model again. There were eight trucks, and each was assigned its own random-number stream. For each of the six replications, eight seeds were randomly chosen, one seed for each random-number stream. Therefore, a total of 48 (6 times 8) seeds were specified for the correct implementation of common random numbers. When the authors first developed and ran this example, eight seeds were specified at the beginning of the first replication only; on the remaining five replications the random numbers were generated by continuing down the eight original streams. Since comparable replications with one and two loaders required different numbers of random variables, only the first replications of the two models were synchronized. The remaining five were not synchronized. The resulting confidence interval for $\theta_1 - \theta_2$ when using correlated sampling was approximately the same length, or only slightly shorter, than the confidence interval when using independent sampling. Therefore, correlated sampling is quite likely to fail in reducing the standard error of the estimated difference unless proper care is taken to guarantee synchronization of the random-number streams on all replications.

Example 12.3

In Example 2.5, two policies for replacing bearings in a milling machine were compared. In the "pencil-and-paper" simulation of Tables 2.24 and 2.25, common random numbers were used for bearing lifetimes but independent random numbers were used for repairperson delay times. The bearing-life distribution, assumed discrete in Example 2.5 (Table 2.22), is now more realistically assumed to be continuous on the range 950 to 1950 hours, with the first column of Table 2.22 giving the midpoint of 10 intervals of width 100 hours. The repairperson delay-time distribution of Table 2.23 is also assumed continuous, in the range 2.5 to 17.5 minutes, with interval midpoints as given in the first column. The probabilities of each interval are given in the second columns of Tables 2.22 and 2.23.

The two models were run using correlated sampling and, for illustrative purposes, independent sampling, each for $R = 10$ replications with a run length of $T_E = 3$ years. The purpose was to estimate the difference in mean total costs per year, with the cost data given in Example 2.5. The estimated total cost over the 3-year period for the two policies is given in Table 12.4.

Table 12.4. Total Costs for Alternative Designs of Bearing-Replacement Problem

Replication	*Total Cost over 3 Years for Policy*			*Difference in Total Cost*
r	2	*1I*	*1C*	$D_{1C,2}$
1	6,670	8,505	8,778	2,108
2	6,380	8,764	8,580	2,200
3	6,501	8,978	8,904	2,403
4	6,762	8,960	9,006	2,244
5	6,877	9,440	9,100	2,223
6	6,659	8,764	8,968	2,310
7	6,716	8,787	9,175	2,459
8	7,104	8,977	9,699	2,595
9	6,612	9,145	8,806	2,194
10	6,589	8,680	8,978	2,388
Sample mean	6,687	8,900		2,312
Sample variance	40,178	69,047		21,820

Policy 1 was to replace each bearing as it failed. Policy 2 was to replace all three bearings whenever one bearing failed. Policy 2 was run first, and then policy 1 was run using independent sampling (column 1I), and using correlated sampling (column 1C). The 95% confidence intervals for mean cost difference are as follows:

$$\text{Independent sampling: } \$2213 \pm 219$$

$$\text{Correlated sampling: } \$2312 \pm 106$$

(The computation of these confidence intervals is left as an exercise for the reader.)

Notice that the confidence interval for mean cost difference when using correlated sampling is approximately 50% of the length of the confidence interval based on independent sampling. Therefore, for the same computer costs (i.e., for $R = 10$ replications), correlated sampling produces estimates which are twice as precise in this example. If correlated sampling were used, the simulation analyst could conclude with 95% confidence that the mean cost difference between the two policies is between $2206 and $2418 over a 3-year period. ◀

12.1.4 Confidence Intervals with Specified Precision

Section 11.4.3 described a procedure for obtaining confidence intervals with specified precision. Confidence intervals for the *difference* between two systems' performance may be obtained in an analogous manner.

Suppose that we want the error in our estimate of $\theta_1 - \theta_2$ to be less than $\pm\epsilon$ (the quantity ϵ might be a practically significant difference). Therefore, our goal is to find a number of replications R such that

$$\text{h.l.} = t_{\alpha/2,\nu}\text{s.e.}(\bar{Y}_{\cdot 1} - \bar{Y}_{\cdot 2}) \leq \epsilon \tag{12.19}$$

As in Section 11.4.3, we begin by making $R_0 \geq 2$ replications of each system to obtain an initial estimate of $\text{s.e.}(\bar{Y}_{\cdot 1} - \bar{Y}_{\cdot 2})$. We then solve for the total number of replications $R \geq R_0$ needed to achieve the half-length criterion (12.19). Finally, we make an additional $R - R_0$ replications (or a fresh R replications) of each system, compute the confidence interval, and check that the half-length criterion has been attained.

EXAMPLE 12.1 (Continued)

Recall that $R_0 = 10$ replications and complete synchronization of the random numbers yielded the 95% confidence interval for the difference in expected response time of the two vehicle inspection stations in inequality (12.18); this interval can be rewritten as 0.4 ± 0.90 minutes. Although system 2 appears to have the smaller expected response time, the difference is not statistically significant, since the confidence interval contains 0. Suppose that a difference larger than ± 0.5 minute is considered to be practically significant. We therefore want to make enough replications to obtain a $\text{h.l.} \leq \epsilon = 0.5$.

The confidence interval used in Example 12.1 was $\bar{D} \pm t_{\alpha/2,R_0-1}S_D/\sqrt{R_0}$, with the specific values $\bar{D} = 0.4$, $R_0 = 10$, $t_{0.025,9} = 2.26$ and $S_D^2 = 1.7$. To obtain the desired precision we need to find R such that

$$\frac{t_{\alpha/2,R-1}S_D}{\sqrt{R}} \leq \epsilon$$

Therefore, R is the smallest integer satisfying $R \geq R_0$ and

$$R \geq \left(\frac{t_{\alpha/2,R-1}S_D}{\epsilon}\right)^2$$

Since $t_{\alpha/2,R-1} \leq t_{\alpha/2,R_0-1}$, a conservative estimate for R is given by

$$R \geq \left(\frac{t_{\alpha/2,R_0-1}S_D}{\epsilon}\right)^2$$

Substituting $t_{0.025,9} = 2.26$ and $S_D^2 = 1.7$ we obtain

$$R \geq \frac{(2.26)^2(1.7)}{(0.5)^2} = 34.73$$

implying that a total of 35 replications are needed, 25 more than in the initial experiment. ◀

12.2 Comparison of Several System Designs

Suppose that a simulation analyst desires to compare K alternative system designs. The comparison will be made on the basis of some specified performance measure, θ_i, of system i, for $i = 1, 2, \ldots, K$. Many different statistical procedures have been developed which can be used to analyze simulation data and draw statistically sound inferences concerning the parameters θ_i. These procedures can be classified as being fixed-sample-size procedures, or sequential sampling (or multistage) procedures. In the first type, a predetermined sample size (i.e., run length and number of replications) is used to draw inferences via hypothesis tests or confidence intervals. Examples of fixed-sample-size procedures include the interval estimation of a mean performance measure [as in expression (11.10), Section 11.3], and the interval estimation of the difference in mean performance measures of two systems [as by expression (12.1) in Section 12.1]. Advantages of fixed-sample-size procedures include the known or easily estimated cost in terms of computer time before running the experiments. When computer time is limited, or when conducting a pilot study, a fixed-sample-size procedure may be appropriate. In some cases, clearly inferior system designs may be ruled out at this early stage. A major disadvantage is that no strong conclusion may be possible. For example, the confidence interval may be too wide for practical use, since the width is an indication of the precision of the point estimator. A hypothesis test may lead to a failure to reject the null hypothesis, a weak conclusion in general, meaning that there is no strong evidence one way or the other about the truth or falsity of the null hypothesis.

A sequential sampling scheme is one in which more and more data are collected until an estimator with a prespecified precision is achieved, or until one of several alternative hypotheses is selected, with the probability of correct selection being larger than a prespecified value. A two-stage (or multistage) procedure is one in which an initial sample is used to estimate how many additional observations are needed to draw conclusions with a specified precision. An example of a two-stage procedure for estimating the performance measure of a single system was given in Section 11.4.3.

The proper procedure to use depends on the goal of the simulation analyst. Some possible goals include:

1. Estimation of each parameter, θ_i
2. Comparison of each performance measure, θ_i, to a control, θ_1, (where θ_1 may represent the mean performance of an existing system)

3. All possible comparisons, $\theta_i - \theta_j$, for $i \neq j$
4. Selection of the best θ_i (largest or smallest)

The first three goals will be achieved by the construction of confidence intervals. The number of such confidence intervals is $C = K, C = K - 1$, and $C = K(K - 1)/2$, respectively. Hochberg and Tamhane [1987] and Hsu [1996] are comprehensive references for such multiple-comparison procedures. The fourth goal requires the use of a type of statistical procedure known as a multiple ranking and selection procedure. Procedures to achieve these and other goals are discussed by Kleijnen [1975, Chaps. II and V], who also discusses their relative merit and disadvantages. Goldsman and Nelson [1998] and Law and Kelton [2000] discuss those selection procedures most relevant to simulation. A comprehensive reference is Bechhofer, Santner, and Goldsman [1995]. The next subsection presents a fixed-sample-size procedure which can be used to meet goals 1, 2 and 3 and which is applicable in a wide range of circumstances. Subsection 12.2.2 presents a related procedure to achieve goal 4.

12.2.1 Bonferroni Approach to Multiple Comparisons

Suppose that a total of C confidence intervals are computed, and that the ith interval has confidence coefficient $1 - \alpha_i$. Let S_i be the statement that the ith confidence interval contains the parameter (or difference of two parameters) being estimated. This statement may be true or false for a given set of data, but the procedure leading to the interval is designed so that statement S_i will be true with probability $1 - \alpha_i$. When it is desired to make statements about several parameters simultaneously, as in goals 1, 2 and 3, the analyst would like to have high confidence that *all* statements are true simultaneously. The Bonferroni inequality states that

$$P(\text{all statements } S_i \text{ are true}, i = 1, \ldots, C) \geq 1 - \sum_{j=1}^{C} \alpha_j = 1 - \alpha_E \quad (12.20)$$

where $\alpha_E = \sum_{j=1}^{C} \alpha_j$ is called the overall error probability. Expression (12.20) can be restated as

$$P(\text{one or more statements } S_i \text{ is false}, i = 1, \ldots, C) \leq \alpha_E$$

or equivalently,

$$P(\text{one or more of the } C \text{ confidence intervals does not contain the parameter being estimated}) \leq \alpha_E$$

Thus, α_E provides an upper bound on the probability of a false conclusion. When conducting an experiment making C comparisons, first select the overall error probability, say $\alpha_E = 0.05$ or 0.10. The individual α_j may be chosen to be equal ($\alpha_j = \alpha_E/C$), or unequal, as desired. The smaller the value of α_j, the wider the jth confidence interval will be. For example, if two 95% c.i.'s

($\alpha_1 = \alpha_2 = 0.05$) are constructed, the overall confidence level will be 90% or greater ($\alpha_E = \alpha_1 + \alpha_2 = 0.10$). If ten 95% c.i.'s are constructed ($\alpha_i = 0.05$, $i = 1, \ldots, 10$), the resulting overall confidence level could be as low as 50% ($\alpha_E = \sum_{i=1}^{10} \alpha_i = 0.50$), which is far too low for practical use. To guarantee an overall confidence level of 95%, when ten comparisons are being made, then one solution is to construct ten 99.5% confidence intervals for the parameters (or differences) of interest.

The Bonferroni approach to multiple confidence intervals is based on expression (12.20). A major advantage is that it holds whether the models for the alternative designs are run with independent sampling or with common random numbers.

The major disadvantage of the Bonferroni approach when making a large number of comparisons is the increased width of each individual interval. For example, for a given set of data and a large sample size, a 99.5% c.i. will be $z_{0.0025}/z_{0.025} = 2.807/1.96 = 1.43$ times longer than a 95% c.i. For small sample sizes, say for a sample size of 5, a 99.5% c.i. will be $t_{0.0025,4}/t_{0.025,4} = 5.598/2.776 = 1.99$ times longer than an individual 95% c.i. The width of a c.i. is a measure of the precision of the estimate. For these reasons, it is recommended that the Bonferroni approach only be used when a small number of comparisons are being made. Ten or so comparisons appears to be the practical upper limit.

Corresponding to goals 1, 2, and 3 above, there are at least three possible ways of using the Bonferroni inequality (12.20) when comparing K alternative system designs:

1. (*Individual c.i.'s*): Construct a $100(1 - \alpha_i)\%$ c.i. for parameter θ_i by expression (11.10), in which case the number of intervals is $C = K$. If independent sampling were used, the K c.i.'s would be mutually independent, and thus the overall confidence level would be $(1-\alpha_1) \times (1-\alpha_2) \times \cdots \times (1-\alpha_C)$, which is larger (but not much larger) than the right side of expression (12.20). This type of procedure is most often used to estimate multiple parameters of a single system, rather than to compare systems. And since multiple parameter estimates from the same system are likely to be dependent, the Bonferroni inequality is typically needed.
2. (*Comparison to an existing system*): Compare all designs to one specific design, usually to an existing system. That is, construct a $100(1 - \alpha_i)\%$ c.i. for $\theta_i - \theta_1 (i = 2, 3, \ldots, K)$ using expression (12.1). (System 1 with performance measure θ_1 is assumed to be the existing system.) In this case, the number of intervals is $C = K - 1$. This type of procedure is most often used to compare several competitors to the present system in order to determine which are better.
3. (*All possible comparisons*): Compare all designs to each other. That is, for any two system designs $i \neq j$, construct a $100(1 - \alpha_{ij})\%$ c.i. for $\theta_i - \theta_j$. With K designs, the number of confidence intervals computed is $C = \binom{K}{2} = K(K-1)/2$. The overall confidence coefficient would be

bounded below by $1 - \alpha_E = 1 - \sum\sum_{i \neq j} \alpha_{ij}$ [which follows by expression (12.20)]. It is generally believed that correlated sampling (common random numbers) will make the true overall confidence level larger than the right side of expression (12.20), and usually larger than when using independent sampling. The right side of expression (12.20) can be thought of as giving the worst case (i.e., the lowest possible overall confidence level).

Example 12.4

Reconsider the vehicle inspection station of Example 12.1. Suppose that the construction of additional space to hold one waiting car is being considered. The alternative system designs are:

1. Existing system (parallel stations)
2. No space between stations in series
3. One space between brake and headlight inspection only
4. One space between headlight and steering inspection only

Design 2 was compared to the existing set-up in Example 12.1. Designs 2, 3, and 4 are series queues as shown in Figure 12.1(b), the only difference being the number or location of a waiting space between two successive inspections. The arrival process and the inspection times are as given in Example 12.1. The basis for comparison will be mean response time, θ_i, for system i, where a response time is the total time it takes for a car to get through the system. Confidence intervals for $\theta_2 - \theta_1$, $\theta_3 - \theta_1$, and $\theta_4 - \theta_1$ will be constructed having an overall confidence level of 95%. The run length T_E has now been set at 40 hours (instead of the 16 hours used in Example 12.1), and the number of replications R of each model is 10. Common random numbers will be used in all models, but this does not affect the overall confidence level, because, as mentioned, the Bonferroni inequality (12.20) holds regardless of the statistical independence or dependence of the data.

Since the overall error probability is $\alpha_E = 0.05$ and $C = 3$ confidence intervals are to be constructed, let $\alpha_i = 0.05/3 = 0.0167$ for $i = 2, 3, 4$. Then use expression (12.1) (with proper modifications) to construct $C = 3$ confidence intervals with $\alpha = \alpha_i = 0.0167$ and degrees of freedom $\nu = 10 - 1 = 9$. The standard error is computed by Equation (12.14), since common random numbers are being used. The output data Y_{ri} are displayed in Table 12.5; Y_{ri} is the sample mean response time for replication r on system i ($r = 1, \ldots, 10$; $i = 1, 2, 3, 4$). The differences $D_{ri} = Y_{r1} - Y_{ri}$ are also shown, together with the sample mean differences, $\bar{D}_{\cdot i}$, averaged over all replications as in Equation (12.12), the sample variances $S^2_{D_i}$ and the standard error. By expression (12.1), the three confidence intervals, with overall confidence coefficient at least $1 - \alpha_E$, are given by

$$\bar{D}_{\cdot i} - t_{\alpha_i/2, R-1}\text{s.e.}(\bar{D}_{\cdot i}) \leq \theta_1 - \theta_i \leq \bar{D}_{\cdot i} + t_{\alpha_i/2, R-1}\text{s.e.}(\bar{D}_{\cdot i}), \quad i = 2, 3, 4$$

Table 12.5. Analysis of Output Data for Vehicle Inspection System When Using Correlated Sampling

Replication,	*Average Response Time for System Design*				*Observed Difference*		
	1,	*2,*	*3,*	*4,*			
r	Y_{r1}	Y_{r2}	Y_{r3}	Y_{r4}	D_{r2}	D_{r3}	D_{r4}
1	63.72	63.06	57.74	62.63	0.66	5.98	1.09
2	32.24	31.78	29.65	31.56	0.46	2.59	0.68
3	40.28	40.32	36.52	39.87	−0.04	3.76	0.41
4	36.94	37.71	35.71	37.35	−0.77	1.23	−0.41
5	36.29	36.79	33.81	36.65	−0.50	2.48	−0.36
6	56.94	57.93	51.54	57.15	−0.99	5.40	−0.21
7	34.10	33.39	31.39	33.30	0.71	2.71	0.80
8	63.36	62.92	57.24	62.21	0.44	6.12	1.15
9	49.29	47.67	42.63	47.46	1.62	6.66	1.83
10	87.20	80.79	67.27	79.60	6.41	19.93	7.60
Sample mean, $\bar{D}_{.i}$					0.80	5.686	1.258
Sample standard deviation, S_{D_i}					2.12	5.338	2.340
Sample variance, $S^2_{D_i}$					4.498	28.498	5.489
Standard error, $S_{D_i}/\sqrt{R}$					0.671	1.688	0.741

The value of $t_{\alpha_i/2,R-1} = t_{0.0083,9} = 2.97$ is obtained from Table A.5 by interpolation. For these data, with 95% confidence, it is stated that

$$-1.19 \le \theta_1 - \theta_2 \le 2.79$$
$$0.67 \le \theta_1 - \theta_3 \le 10.71$$
$$-0.94 \le \theta_1 - \theta_4 \le 3.46$$

The simulation analyst has high confidence (at least 95%) that all three confidence statements are correct. Notice that the c.i. for $\theta_1 - \theta_2$ again contains zero; thus, there is no statistically significant difference between design 1 and design 2, a conclusion which confirms the previous results in Example 12.1. The c.i. for $\theta_1 - \theta_3$ lies completely above zero, which provides strong evidence that $\theta_1 - \theta_3 > 0$ — that is, that design 3 is better than design 1, because its mean response time is smaller. Since the c.i. for $\theta_1 - \theta_4$ contains zero, there is no statistically significant difference between designs 1 and 4.

If the simulation analyst now decides that it would be desirable to compare designs 3 and 4, more simulation runs would be necessary, because it is not formally correct to decide which confidence intervals to compute after the data have been examined. On the other hand, if the simulation analyst had decided to compute all possible confidence intervals (and had made this decision before collecting the data, Y_{ri}), the number of confidence intervals would have been $C = 6$ and the three c.i.'s above would have been $t_{0.0042,9}/t_{0.0083,9} \approx 3.32/2.97 = 1.12$ times (or 12%) longer. There is always a trade-off between the number

Table 12.6. Analysis of Output Data for Vehicle Inspection System When Using Independent Sampling

	Average Response Time for System Design			
Replication, r	*1,* Y_{r1}	*2,* Y_{r2}	*3,* Y_{r3}	*4,* Y_{r4}
1	63.72	59.37	52.00	59.03
2	32.24	50.06	47.04	49.97
3	40.28	60.63	53.21	60.18
4	36.94	46.36	40.88	45.44
5	36.29	68.87	50.85	66.65
6	56.94	66.44	60.42	66.03
7	34.10	27.51	26.70	27.45
8	63.36	47.98	40.12	47.50
9	49.29	29.92	28.59	29.84
10	87.20	47.14	41.62	46.44
Sample mean $\bar{Y}_{\cdot i}$	50.04	50.43	44.14	49.85
S_i	17.70	13.98	10.76	13.64
S_i^2	313.38	195.54	115.74	185.98
$\bar{Y}_{\cdot 1} - \bar{Y}_{\cdot i}$		−0.39	5.89	0.18
s.e.$(\bar{Y}_{\cdot 1} - \bar{Y}_{\cdot i})$		7.13	6.55	7.07

of intervals (C) and the width of each interval. The simulation analyst should carefully consider the possible conclusions before running the simulation experiments, and decide on those runs and analyses that will provide the most useful information. In particular, the number of confidence intervals computed should be as small as possible, preferably 10 or less.

For purposes of illustration, 10 replications of each of the four designs were run using independent sampling (i.e., different random numbers for all runs). The results are presented in Table 12.6, together with sample means $(\bar{Y}_{\cdot i})$, sample standard deviations (S_i), and sample variances (S_i^2), plus the observed difference of sample means $(\bar{Y}_{\cdot 1} - \bar{Y}_{\cdot i})$ and the standard error (s.e.) of the observed difference. It is observed that all three confidence intervals for $\theta_1 - \theta_i (i = 2, 3, 4)$ contain zero. Therefore, no strong conclusion is possible based on these data and this sample size. By contrast, a sample size of ten was sufficient, when using correlated sampling, to provide strong evidence that design 3 is superior to design 1.

Notice the large increase in standard error of the estimated difference when using independent sampling versus using common random numbers. These standard errors are compared in Table 12.7. In addition, a careful examination of Tables 12.5 and 12.6 illustrates the superiority of correlated sampling. In Table 12.5, in all 10 replications, system design 3 has a smaller average response time than that for system design 1. Comparing replications 1 and 2 in Table 12.5, it can be seen that a random-number stream that leads to high congestion and large response times in system design 1, as in the first replication, produces results of similar magnitude across all four system designs. Similarly, when system design 1 exhibits relatively low congestion and low response times, as in the second replication, all system designs produce relatively low average response times. This similarity of results on each replication is due, of course, to the use of common random numbers across systems. By contrast, for independent sampling, Table 12.6 shows no such similarity across system designs. In only 5 of the 10 replications is the average response time for system design 3 smaller than that for system design 1, although the average difference in response times across all 10 replications is of approximately the same magnitude in each case: 5.69 minutes when using correlated sampling, and 5.89 minutes when using independent sampling. The greater variability of independent sampling is reflected also in the standard errors of the point estimates: ±1.69 minutes for correlated sampling versus ±6.55 minutes for independent sampling, an increase of 388%, as seen in Table 12.7. This example illustrates again the advantage of correlated sampling. ◀

Table 12.7. Comparison of Standard Errors for Correlated Sampling and Independent Sampling for the Vehicle Inspection Problem

	Standard Error When Using:		
Difference in Sample Means	*Correlated Sampling*	*Independent Sampling*	*Percentage Increase*
$\bar{Y}_{\cdot 1} - \bar{Y}_{\cdot 2}$	0.67	7.13	1064%
$\bar{Y}_{\cdot 1} - \bar{Y}_{\cdot 3}$	1.69	6.55	388%
$\bar{Y}_{\cdot 1} - \bar{Y}_{\cdot 4}$	0.74	7.07	955%

As stated previously, correlated sampling does not yield a variance reduction in all simulation models. It is recommended that a pilot study be undertaken and variances estimated to confirm (or possibly deny) the assumption that correlated sampling will reduce the variance (or standard error) of an estimated difference. The reader is referred to the discussion in Section 12.1.3.

Some of the exercises at the end of this chapter provide an opportunity to compare correlated and independent sampling, and to compute simultaneous confidence intervals using the Bonferroni approach.

12.2.2 Bonferroni Approach to Selecting the Best

Suppose that there are K system designs, and the ith design has expected performance θ_i. At a gross level we are interested in which system is best, where "best" is defined to be having maximum expected performance.[1] At a

[1] If "best" is defined to be having minimum expected performance, then the procedure in this section is easily modified, as we illustrate in the example.

more refined level we may also be interested in how much better the best is relative to each alternative, since secondary criteria that are not reflected in the performance measure θ_i (such as ease of installation, cost to maintain, etc.) may tempt us to choose an inferior system if it is not deficient by much.

If system design i is the best, then $\theta_i - \max_{j \neq i} \theta_j$ is equal to the difference in performance between the best and the second best. If system design i is not the best, then $\theta_i - \max_{j \neq i} \theta_j$ is equal to the difference between system i and the best. The selection procedure we describe in this section focuses on the parameters $\theta_i - \max_{j \neq i} \theta_j$ for $i = 1, 2, \ldots, K$.

Let i^* denote the (unknown) index of the best system. As a general rule, the smaller the true difference $\theta_{i^*} - \max_{j \neq i^*} \theta_j$ is, and the more certain we want to be that we find the best system, the more replications are required to achieve our goal. Therefore, instead of demanding that we find i^*, we can compromise and ask to find i^* with high probability whenever the difference between system i^* and the others is at least some practically significant amount. More precisely, we want the probability that we select the best system to be at least $1 - \alpha$ whenever $\theta_{i^*} - \max_{j \neq i^*} \theta_j \geq \epsilon$. If there are one or more systems that are within ϵ of the best, then we will be satisfied to select either the best or any one of the near best. Both the probability of correct selection, $1 - \alpha$, and the practically significant difference, ϵ, will be under our control.

The following procedure achieves the desired probability of correct selection (Nelson and Matejcik [1995]). And because we are also interested in how much each system differs from the best, it also forms $100(1 - \alpha)\%$ confidence intervals for $\theta_i - \max_{j \neq i} \theta_j$ for $i = 1, 2, \ldots, K$. The procedure is valid for normally distributed data using either correlated or independent sampling.

Two-Stage Bonferroni Procedure

1. Specify the practically signficant difference ϵ, the probability of correct selection $1 - \alpha$, and the first-stage sample size $R_0 \geq 10$. Let $t = t_{\alpha/(K-1), R_0 - 1}$.
2. Make R_0 replications of system i to obtain $Y_{1i}, Y_{2i}, \ldots, Y_{R_0,i}$, for systems $i = 1, 2, \ldots, K$.
3. Calculate the first-stage sample means $\bar{Y}_{\cdot i}$, $i = 1, 2, \ldots, K$. For all $i \neq j$ calculate the sample variance of the difference[2]

$$S_{ij}^2 = \frac{1}{R_0 - 1} \sum_{r=1}^{R_0} \left(Y_{ri} - Y_{rj} - (\bar{Y}_{\cdot i} - \bar{Y}_{\cdot j})\right)^2$$

Let $\widehat{S}^2 = \max_{i \neq j} S_{ij}^2$, the largest sample variance.

[2] Notice that S_{ij}^2 is algebraically equivalent to S_D^2, the sample variance of $D_r = Y_{ri} - Y_{rj}$, for $r = 1, 2, \ldots, R_0$.

4. Calculate the second-stage sample size

$$R = \max\left\{R_0, \left\lceil \frac{t^2\widehat{S}^2}{\epsilon^2} \right\rceil\right\}$$

where $\lceil\cdot\rceil$ means to round up.

5. Make $R - R_0$ additional replications of system i to obtain the output data $Y_{R_0+1,i}, Y_{R_0+2,i}, \ldots, Y_{R,i}$, for $i = 1, 2, \ldots, K$.[3]

6. Calculate the overall sample means

$$\bar{\bar{Y}}_i = \frac{1}{R}\sum_{r=1}^{R} Y_{ri}$$

for $i = 1, 2, \ldots, K$.

7. Select the system with largest $\bar{\bar{Y}}_i$ as the best. Also form the confidence intervals

$$\min\{0, \bar{\bar{Y}}_i - \max_{j\neq i} \bar{\bar{Y}}_j - \epsilon\} \le \theta_i - \max_{j\neq i}\theta_j \le \max\{0, \bar{\bar{Y}}_i - \max_{j\neq i} \bar{\bar{Y}}_j + \epsilon\}$$

for $i = 1, 2, \ldots, K$.

The confidence intervals in Step 7 are not like the usual $\pm$ intervals presented elsewhere in this chapter. Perhaps the most useful interpretation of them is as follows. Let $\hat{i}$ be the index of the system selected as best. Then for each of the other systems i, we make one of the declarations below:

- If $\bar{\bar{Y}}_i - \bar{\bar{Y}}_{\hat{i}} + \epsilon \le 0$, then declare system i to be inferior to the best.
- If $\bar{\bar{Y}}_i - \bar{\bar{Y}}_{\hat{i}} + \epsilon > 0$, then declare system i to be statistically indistinguishable from the best, and therefore system i may be the best.

EXAMPLE 12.4 (Continued)

Recall that in Example 12.4 we considered $K = 4$ different designs for the vehicle inspection station. Suppose that we would like a 0.95 probability of selecting the best (smallest expected response time) system design when the best differs from the second best by at least two minutes. Since this is a minimization problem, we focus on the differences $\theta_i - \min_{j\neq i}\theta_j$ for $i = 1, 2, 3, 4$. Then we can apply the two-stage Bonferroni procedure as follows:

1. $\epsilon = 2$ minutes, $1 - \alpha = 0.95$, $R_0 = 10$, and $t = t_{0.0167,9} = 2.508$.
2. The data in Table 12.5, which was obtained using correlated sampling.
3. From Table 12.5 we get $S_{12}^2 = S_{D_2}^2 = 4.498$, $S_{13}^2 = S_{D_3}^2 = 28.498$, and $S_{14}^2 = S_{D_4}^2 = 5.489$. By similar calculations we obtain $S_{23}^2 = 11.857$, $S_{24}^2 = 0.106$, and $S_{34}^2 = 9.957$.

---

[3] If it is more convenient, a total of R replications can be generated from system i by restarting the entire experiment.

4. Since $\widehat{S}^2 = S_{13}^2 = 28.498$ is the largest sample variance,

$$R = \max\left\{10, \left\lceil \frac{(2.508)^2(28.498)}{2^2} \right\rceil\right\} = \max\{10, \lceil 44.8 \rceil\} = 45$$

5. Make $45 - 10 = 35$ additional replications of each system.
6. Calculate the overall sample means

$$\bar{\bar{Y}}_i = \frac{1}{45}\sum_{r=1}^{45} Y_{ri}$$

for $i = 1, 2, 3, 4$.

7. Select the system with smallest $\bar{\bar{Y}}_i$ as the best. Also form the confidence intervals

$$\min\{0, \bar{\bar{Y}}_i - \min_{j \neq i} \bar{\bar{Y}}_j - 2\} \leq \theta_i - \min_{j \neq i} \theta_j \leq \max\{0, \bar{\bar{Y}}_i - \min_{j \neq i} \bar{\bar{Y}}_j + 2\}$$

for $i = 1, 2, 3, 4$. ◀

12.3 Metamodeling

Suppose that there is a simulation output response variable, Y, that is related to k independent variables, say $x_1, x_2, \ldots, x_k$. The dependent variable, Y, is a random variable, while the independent variables $x_1, x_2, \ldots, x_k$ are called design variables and are usually subject to control. The true relationship between the variables Y and x is represented by the (often complex) simulation model. Our goal is to approximate this relationship by a simpler mathematical function called a metamodel. In some cases the analyst will know the exact form of the functional relationship between Y and $x_1, x_2, \ldots, x_k$, say $Y = f(x_1, x_2, \ldots, x_k)$. However, in most cases, the functional relationship is unknown, and the analyst must select an appropriate f containing unknown parameters, and then estimate those parameters from a set of data (Y, x). Regression analysis is one method for estimating the parameters.

Example 12.5

An insurance company promises to process all claims it receives each day by the end of the next day. They have developed a simulation model of their proposed claims-processing system to evaluate how hard it will be to meet this promise. The actual number and types of claims that will need to be processed each day will vary, and the number may grow over time. Therefore, the company would like to have a model that predicts the total processing time as a function of the number of claims received. ◀

The primary value of a metamodel is to make it easy to answer "what-if" questions, such as what the processing time will be if there are x claims. Evaluating a function f, or perhaps its derivatives, at a number of values of x is typically much easier than running a simulation experiment for each value.

12.3.1 Simple Linear Regression

Suppose that it is desired to estimate the relationship between a single independent variable x and a dependent variable Y, and suppose that the true relationship between Y and x is suspected to be linear. Mathematically, the expected value of Y for a given value of x is assumed to be

$$E(Y|x) = \beta_0 + \beta_1 x \tag{12.21}$$

where β_0 is the intercept on the Y axis, an unknown constant; and β_1 is the slope, or change in Y for a unit change in x, also an unknown constant. It is further assumed that each observation of Y can be described by the model

$$Y = \beta_0 + \beta_1 x + \epsilon \tag{12.22}$$

where ϵ is a random error with mean zero and constant variance σ^2. The regression model given by Equation (12.22) involves a single variable x and is commonly called a simple linear regression model.

Suppose that there are n pairs of observations $(Y_1, x_1), (Y_2, x_2), \ldots, (Y_n, x_n)$. These observations may be used to estimate β_0 and β_1 in Equation (12.22). The method of least squares is commonly used to form the estimates. In the method of least squares, β_0 and β_1 are estimated such that the sum of the squares of the deviations between the observations and the regression line is minimized. The individual observations in Equation (12.22) may be written as

$$Y_i = \beta_0 + \beta_1 x_i + \epsilon_i, \quad i = 1, 2, \ldots, n \tag{12.23}$$

where $\epsilon_1, \epsilon_2, \ldots$ are assumed to be uncorrelated random variables.

Each ϵ_i in Equation (12.23) is given by

$$\epsilon_i = Y_i - \beta_0 - \beta_1 x_i \tag{12.24}$$

and represents the difference between the observed response, Y_i, and the expected response, $\beta_0 + \beta_1 x_i$, predicted by the model in Equation (12.21). Figure 12.3 shows how ϵ_i is related to x_i, Y_i, and $E(Y_i|x_i)$.

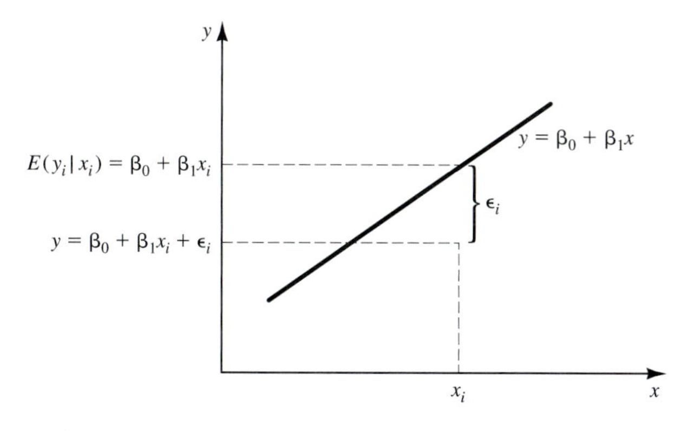

Figure 12.3. Relationship of ϵ_i to x_i, Y_i and $E(Y_i|x_i)$.

The sum of squares of the deviations given in Equation (12.24) is given by

$$L = \sum_{i=1}^{n} \epsilon_i^2 = \sum_{i=1}^{n} (Y_i - \beta_0 - \beta_1 x_i)^2 \tag{12.25}$$

where L is called the least-squares function. It is convenient to rewrite Y_i as follows:

$$Y_i = \beta_0' + \beta_1(x_i - \bar{x}) + \epsilon_i \tag{12.26}$$

where $\beta_0' = \beta_0 + \beta_1 \bar{x}$ and $\bar{x} = \sum_{i=1}^{n} x_i / n$. Equation (12.26) is often called the transformed linear regression model. Using Equation (12.26), Equation (12.25) becomes

$$L = \sum_{i=1}^{n} [Y_i - \beta_0' - \beta_1(x_i - \bar{x})]^2$$

To minimize L, find $\partial L / \partial \beta_0'$ and $\partial L / \partial \beta_1$, set each to zero, and solve for $\widehat{\beta}_0'$ and $\widehat{\beta}_1$. Taking the partial derivatives and setting each to zero yields

$$n\widehat{\beta}_0' = \sum_{i=1}^{n} Y_i$$

$$\widehat{\beta}_1 \sum_{i=1}^{n} (x_i - \bar{x})^2 = \sum_{i=1}^{n} Y_i (x_i - \bar{x}) \tag{12.27}$$

Equations (12.27) are often called the "normal equations," which have the solutions

$$\widehat{\beta}_0' = \bar{Y} = \sum_{i=1}^{n} \frac{Y_i}{n} \tag{12.28}$$

and

$$\widehat{\beta}_1 = \frac{\sum_{i=1}^{n} Y_i (x_i - \bar{x})}{\sum_{i=1}^{n} (x_i - \bar{x})^2} \tag{12.29}$$

The numerator in Equation (12.29) is rewritten for computational purposes as

$$S_{xy} = \sum_{i=1}^{n} Y_i (x_i - \bar{x}) = \sum_{i=1}^{n} x_i Y_i - \frac{\left(\sum_{i=1}^{n} x_i\right)\left(\sum_{i=1}^{n} Y_i\right)}{n} \tag{12.30}$$

where S_{xy} denotes the corrected sum of cross products of x and Y. The denominator of Equation (12.29) is rewritten for computational purposes as

$$S_{xx} = \sum_{i=1}^{n} (x_i - \bar{x})^2 = \sum_{i=1}^{n} x_i^2 - \frac{\left(\sum_{i=1}^{n} x_i\right)^2}{n} \tag{12.31}$$

where S_{xx} denotes the corrected sum of squares of x. The value of $\widehat{\beta}_0$ can be retrieved easily as

$$\widehat{\beta}_0 = \widehat{\beta}_0' - \widehat{\beta}_1 \bar{x} \tag{12.32}$$

EXAMPLE 12.6 (Calculating $\widehat{\beta}_0$ and $\widehat{\beta}_1$)

The simulation model of the claims-processing system in Example 12.5 was executed with initial conditions $x = 100, 150, 200, 250$, and 300 claims received the previous day. Three replications were obtained at each setting. The response Y is the number of hours required to process x claims. The results are shown in Table 12.8. The graphical relationship between the number of claims received and total processing time is shown in Figure 12.4. Such a display is called a scatter diagram. Examination of this scatter diagram indicates that there is a strong relationship between number of claims and processing time. The tentative assumption of the linear model given by Equation (12.22) appears to be reasonable.

With the processing times as the Y_i values (the dependent variables) and the number of claims as the x_i values (the independent variables), $\widehat{\beta}_0$ and $\widehat{\beta}_1$ can be determined by the following computations: $n = 15$, $\sum_{i=1}^{15} x_i = 3000$, $\sum_{i=1}^{15} Y_i = 178$, $\sum_{i=1}^{15} x_i^2 = 675{,}000$, $\sum_{i=1}^{15} x_i Y_i = 39080$, and $\bar{x} = 3000/15 = 200$.

Table 12.8. Simulation Results for Processing Time Given x Claims

Number of Claims x	*Hours of Processing Time Y*
100	8.1
100	7.8
100	7.0
150	9.6
150	8.5
150	9.0
200	10.9
200	13.3
200	11.6
250	12.7
250	14.5
250	14.7
300	16.5
300	17.5
300	16.3

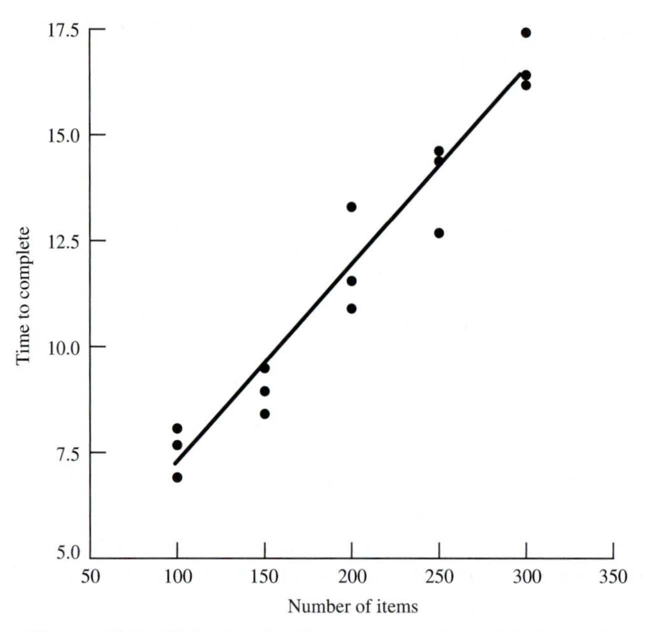

Figure 12.4. Relationship between number of claims and hours of processing time.

From Equation (12.30) S_{xy} is calculated as

$$S_{xy} = 39{,}080 - \frac{(3000)(178)}{15} = 3480$$

From Equation (12.31), S_{xx} is calculated as

$$S_{xx} = 675{,}000 - \frac{(3000)^2}{15} = 75{,}000$$

Then, $\widehat{\beta}_1$ is calculated from Equation (12.29) as

$$\widehat{\beta}_1 = \frac{S_{xy}}{S_{xx}} = \frac{3480}{75{,}000} = 0.0464$$

As shown in Equation (12.28), $\widehat{\beta}'_0$ is just $\bar{Y}$, or

$$\widehat{\beta}'_0 = \frac{178}{15} \approx 11.8667$$

To express the model in the original terms, compute $\widehat{\beta}_0$ from Equation (12.32) as

$$\widehat{\beta}_0 = 11.8667 - 0.0464(200) = 2.5867$$

Then an estimate of the mean of Y given x, $E(Y|x)$, is given by

$$\widehat{y} = \widehat{\beta}_0 + \widehat{\beta}_1 x = 2.5867 + 0.0464x \tag{12.33}$$

For a given number of claims, x, this model can be used to predict the number of hours required to process them. And the coefficient $\widehat{\beta}_1$ has the interpretation that each additional claim received adds an expected 0.0464 hours, or 2.8 minutes, to the expected total processing time. ◀

Regression analysis is widely used and frequently misused. Several of the common abuses are briefly mentioned here. Relationships derived in the manner of Equation (12.33) are valid for values of the independent variable within the range of the original data. The linear relationship that has been tentatively assumed may not be valid outside the original range. Therefore, Equation (12.33) can only be considered as valid for $100 \leq x \leq 300$. Regression models are not advised for extrapolation purposes.

Care should be taken in selecting variables that have a plausible causal relationship with each other. It is quite possible to develop statistical relationships that are unrelated in a practical sense. For example, an attempt might be made to relate monthly output of a steel mill to the weight of computer printouts appearing on a manager's desk during the month. A straight line may appear to provide a good model for the data, but the relationship between the two variables is tenuous. A strong observed relationship does not imply that a causal relationship exists between the variables. Causality can be inferred only when analysis uncovers some plausible reasons for its existence. In Example 12.5 it is clear that starting with more claims implies that more time is needed to process them. Therefore, a relationship of the form of Equation (12.33) is at least plausible.

12.3.2 Testing for Significance of Regression

In Section 12.3.1 it was assumed that a linear relationship existed between Y and x. In Example 12.5 a scatter diagram, shown in Figure 12.4, relating number of claims and processing time was prepared to determine whether a linear model was a reasonable tentative assumption prior to the calculation of $\widehat{\beta}_0$ and $\widehat{\beta}_1$. However, the adequacy of the simple linear relationship should be tested prior to using the model for predicting the response, Y_i, given an independent variable, x_i. There are several tests which may be conducted to aid in determining model adequacy. Testing whether the order of the model tentatively assumed is correct, commonly called the "lack-of-fit test," is suggested. The procedure is explained by Box and Draper [1987], Hines and Montgomery [1990], and Montgomery [1991].

Testing for the significance of regression provides another means for assessing the adequacy of the model. The hypothesis test described below requires the additional assumption that the error component ϵ_i is normally distributed. Thus, the complete assumptions are that the errors are NID$(0, \sigma^2)$, that is, normally and independently distributed with mean zero and constant variance σ^2. The adequacy of the assumptions can and should be checked by residual analysis, discussed by Box and Draper [1987], Hines and Montgomery [1990], and Montgomery [1991].

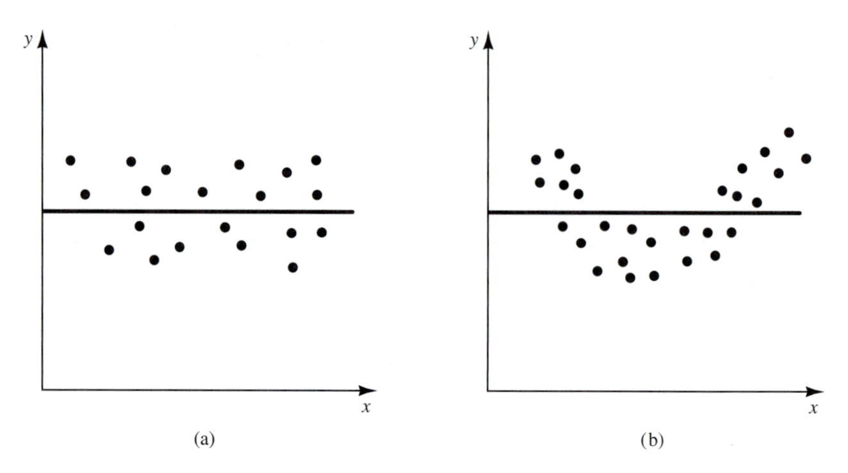

Figure 12.5. Failure to reject $H_0: \beta_1 = 0$.

Testing for significance of regression is one of many hypothesis tests that can be developed from the variance properties of $\widehat{\beta}_0$ and $\widehat{\beta}_1$. The interested reader is referred to the references cited above for extensive discussion of hypothesis testing in regression. Just the highlights of testing for significance of regression are given in this section.

Suppose that the alternative hypotheses are

$$H_0\colon \beta_1 = 0$$
$$H_1\colon \beta_1 \neq 0$$

Failure to reject H_0 indicates that there is no linear relationship between x and Y. This situation is illustrated in Figure 12.5. Notice that two possibilities exist. In Figure 12.5(a), the implication is that x is of little value in explaining the variability in Y, and that $\widehat{y} = \bar{Y}$ is the best estimator. In Figure 12.5(b), the implication is that the true relationship is not linear.

Alternatively, if H_0 is rejected, the implication is that x is of value in explaining the variability in Y. This situation is illustrated in Figure 12.6. Here, also, two possibilities exist. In Figure 12.6(a), the straight-line model is adequate. However, in Figure 12.6(b), even though there is a linear effect of x, a model with higher-order terms (such as $x^2, x^3, \ldots$) is necessary. Thus, even though there may be significance of regression, testing of the residuals and testing for lack of fit are needed to assure the adequacy of the model.

The appropriate test statistic for significance of regression is given by

$$t_0 = \frac{\widehat{\beta}_1}{\sqrt{\text{MS}_E / S_{xx}}} \tag{12.34}$$

where MS_E is the mean squared error. The error is the difference between the observed value, Y_i, and the predicted value, $\widehat{y}_i$, at x_i, or $e_i = Y_i - \widehat{y}_i$. The

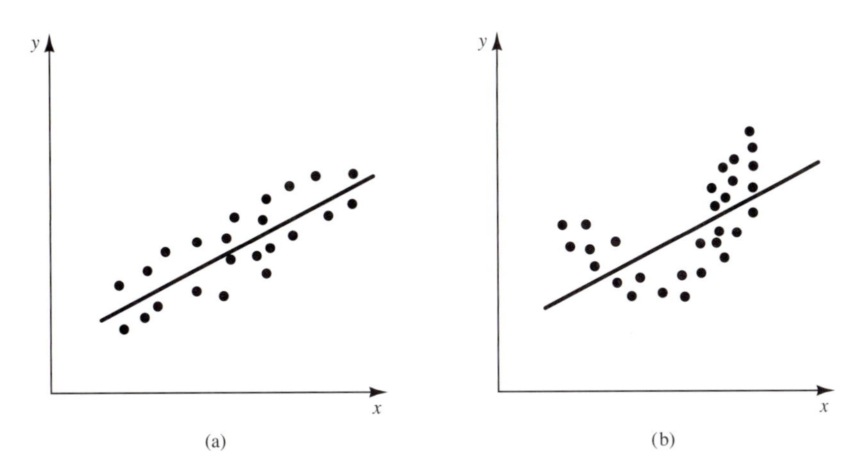

Figure 12.6. $H_0: \beta_1 = 0$ is rejected.

squared error is given by $\sum_{i=1}^{n} e_i^2$, and the mean squared error, given by

$$\mathrm{MS}_E = \sum_{i=1}^{n} \frac{e_i^2}{n-2} \tag{12.35}$$

is an unbiased estimator of $\sigma^2 = \mathrm{var}(\epsilon_i)$. The direct method can be used to calculate $\sum_{i=1}^{n} e_i^2$; that is, calculate each $\widehat{y}_i$, compute e_i^2, and sum all the e_i^2 values, $i = 1, 2, \ldots, n$. However, it can be shown that

$$\sum_{i=1}^{n} e_i^2 = S_{yy} - \widehat{\beta}_1 S_{xy} \tag{12.36}$$

where S_{yy}, the corrected sum of squares of Y, is given by

$$S_{yy} = \sum_{i=1}^{n} Y_i^2 - \frac{\left(\sum_{i=1}^{n} Y_i\right)^2}{n} \tag{12.37}$$

and S_{xy} is given by Equation (12.30). Equation (12.36) may be easier to use than the direct method.

The statistic defined by Equation (12.34) has the t-distribution with $n-2$ degrees of freedom. The null hypothesis H_0 is rejected if $|t_0| > t_{\alpha/2,n-2}$.

Example 12.7 (Testing for Significance of Regression)

Given the results in Example 12.6, the test for the significance of regression is conducted. One more computation is needed prior to conducting the test. That is, $\sum_{i=1}^{n} Y_i^2 = 2282.94$. Using Equation (12.37) yields

$$S_{yy} = 2282.94 - \frac{(178)^2}{15} = 170.6734$$

Then $\sum_{i=1}^{15} e_i^2$ is computed according to Equation (12.36) as

$$\sum_{i=1}^{15} e_i^2 = 170.6734 - 0.0464(3480) = 9.2014$$

Now, the value of MS_E is calculated from Equation (12.35) as

$$MS_E = \frac{9.2014}{13} = 0.7078$$

The value of t_0 can be calculated using Equation (12.34) as

$$t_0 = \frac{0.0464}{\sqrt{0.7078/75000}} = 15.13$$

Since $t_{0.025,13} = 2.16$ from Table A.5, we reject the hypothesis that $\beta_1 = 0$. Thus, there is significant evidence that x and Y are related. ◀

12.3.3 Multiple Linear Regression

If the simple linear regression model of Section 12.3.1 is inadequate, several other possibilities exist. There may be several independent variables, so that the relationship is of the form

$$Y = \beta_0 + \beta_1 x_1 + \beta_2 x_2 + \cdots + \beta_m x_m + \epsilon \tag{12.38}$$

Notice that this model is still linear, but has more than one independent variable. Regression models having the form shown in Equation (12.38) are called multiple linear regression models. Another possibility is that the model is of a quadratic form, such as

$$Y = \beta_0 + \beta_1 x + \beta_2 x^2 + \epsilon \tag{12.39}$$

Equation (12.39) is also a linear model which may be transformed to the form of Equation (12.38) by letting $x_1 = x$ and $x_2 = x^2$.

Yet another possibility is a model of the form such as

$$Y = \beta_0 + \beta_1 x_1 + \beta_2 x_2 + \beta_3 x_1 x_2 + \epsilon$$

which is also a linear model. The analysis of these three models with the forms shown above, and related models, can be found in Box and Draper [1987], Hines and Montgomery [1990], Montgomery [1991], and other applied statistics texts; and also Kleijnen [1987, 1998] which is primarily concerned with the application of these models in simulation.

12.3.4 Random-Number Assignment for Regression

The assignment of random-number seeds or streams is part of the design of a simulation experiment.[4] Assigning a different seed or stream to different

[4] This section is based on Nelson [1992].

design points (settings for $x_1, x_2, \ldots, x_m$ in a multiple linear regression) guarantees that the responses Y from different design points will be statistically independent. Similarly, assigning the same seed or stream to different design points induces dependence among the corresponding responses, since they all have the same source of randomness.

Many textbook experiment designs assume independent responses across design points. To conform to this assumption we must assign different seeds or streams to each design point. However, it is often useful to assign the same random-number seeds or streams to all of the design points—in other words, to use common random numbers.

The intuition behind common random numbers for metamodels is that a fairer comparison among design points is achieved if the design points are subjected to the same experimental conditions, specifically the same source of randomness. The mathematical justification is as follows: suppose we fit the simple linear regression $Y_i = \beta_0 + \beta_1 x_i + \epsilon_i$ and obtain least-squares estimates $\widehat{\beta}_0$ and $\widehat{\beta}_1$. Then an estimator of the expected difference in performance between design points i and j is

$$\widehat{\beta}_0 + \widehat{\beta}_1 x_i - (\widehat{\beta}_0 + \widehat{\beta}_1 x_j) = \widehat{\beta}_1 (x_i - x_j)$$

Since x_i and x_j are fixed design points, $\widehat{\beta}_1$ determines the estimated difference between design points i and j, or for that matter any other two values of x. Therefore, common random numbers can be expected to reduce the variance of $\widehat{\beta}_1$, and more generally reduce the variance of all of the slope terms in a multiple linear regression. Common random numbers do not typically reduce the variance of the intercept term, $\widehat{\beta}_0$.

The least-squares estimators $\widehat{\beta}_0$ and $\widehat{\beta}_1$ are appropriate whether or not we use common random numbers, but the associated statistical analysis is affected by it. For statistical analysis of a metamodel under common random numbers see Kleijnen [1988] and Nelson [1992].

12.4 Optimization via Simulation

Consider the following examples.[5]

Example 12.8 [Materials Handling System (MHS)]

Engineers need to design an MHS consisting of a large automated storage and retrieval device, automated guided vehicles (AGVs), AGV stations, lifters and conveyors. Among the design variables they can control are the number of AGVs, the load per AGV, and the routing algorithm used to dispatch the AGVs. Alternative designs will be evaluated based on AGV utilization, transportation delay for material that needs to be moved, and overall investment and operation costs. ◀

[5] Some of these descriptions are based on Boesel, Nelson, and Ishii [2000].

EXAMPLE 12.9 [Liquefied Natural Gas (LNG) Transportation]

An LNG transporation system will consist of LNG tankers and loading, unloading and storage facilities. In order to minimize cost, designers can control tanker size, number of tankers in use, number of jetties at the loading and unloading facilities, and capacity of the storage tanks. ◀

EXAMPLE 12.10 [Automobile Engine Assembly]

In an assembly line, a large buffer (queue) between workstations can increase station utilization—since there will tend to be something waiting to be processed —but drive up space requirements and work-in-process inventory. An allocation of buffer capacity that minimizes these competing costs is desired. ◀

EXAMPLE 12.11 [Traffic-Signal Sequencing]

Civil engineers want to sequence the traffic signals along a busy section of road to reduce driver delay and the congestion occurring along narrow cross streets. For each traffic signal, the length of the red, green, and green-turn-arrow cycles can be individually set. ◀

EXAMPLE 12.12 [On-Line Services]

A company offering on-line information services over the Internet is changing its computer architecture from central mainframe computers to distributed workstation computing. The numbers and types of CPUs, the network structure, and the allocation of processing tasks all need to be determined. Response time to customer queries is the key performance measure. ◀

What do these design problems have in common? Clearly a simulation model could be useful in each, and all have an implied goal of finding the best design relative to some performance measures (cost, delay, etc.). In each example there are potentially a very large number of alternative designs, ranging from tens to thousands, and certainly more than the 2 to 10 we considered in Section 12.2.2. Some of the examples contain a diverse collection of decision variables: discrete (number of AGVs, number of CPUs), continuous (tanker size, red-cycle length), and qualitative (routing strategy, algorithm for allocating processing tasks). This makes developing a metamodel, as described in Section 12.3, difficult.

All of these problems fall under the general topic of "optimization via simulation," where the goal is to minimize or maximize some measures of system performance, and system performance can be evaluated only by running a computer simulation. Optimization via simulation is a relatively new, but already vast, topic, and commercial software has just recently become widely available. In this section we describe the key issues that should be considered in undertaking optimization via simulation, provide some pointers to the available literature, and give one example algorithm.

12.4.1 What Does "Optimization via Simulation" Mean?

Optimization is a key tool used by operations researchers and management scientists, and there are well-developed algorithms for many classes of problems, the most famous being linear programming. Much of work on optimization deals with problems in which all aspects of the system are treated as being known with certainty; most critically, the performance of any design (cost, profit, makespan, etc.) can be evaluated exactly.

In stochastic, discrete-event simulation, the result of any simulation run is a random variable. For notation, let $x_1, x_2, \ldots, x_m$ be the m controllable design variables, and let $Y(x_1, x_2, \ldots, x_m)$ be the observed simulation output performance on one run. To be concrete, x_1, x_2, x_3 might denote the number of AGVs, the load per AGV, and the routing algorithm used to dispatch the AGVs, respectively, in Example 12.8, while $Y(x_1, x_2, x_3)$ could be total MHS acquisition and operation cost.

What does it mean to "optimize" $Y(x_1, x_2, \ldots, x_m)$ with respect to $x_1, x_2, \ldots, x_m$? Since Y is a random variable, we cannot optimize the *actual* value of Y. The most common definition of optimization is

$$\text{maximize or minimize } E\left(Y(x_1, x_2, \ldots, x_m)\right) \tag{12.40}$$

In other words, the mathematical expectation, or long-run average, of performance is maximized or minimized. *This is the default definition of optimization used in all commercial packages of which we are aware.* In our example $E\left(Y(x_1, x_2, x_3)\right)$ is the expected, or long-run average, cost of operating the MHS with x_1 AGVs, x_2 load per AGV, and routing algorithm x_3.

It is important to note that (12.40) is not the only possible definition, however. For instance, we might want to select the MHS design that has the best chance of costing less than $\$D$ to purchase and operate, changing the objective to

$$\text{maximize } \Pr\left(Y(x_1, x_2, x_3) \le D\right)$$

We can fit this objective into formulation (12.40) by defining a new performance measure

$$Y'(x_1, x_2, x_3) = \begin{cases} 1, & \text{if } Y(x_1, x_2, x_3) \le D \\ 0, & \text{otherwise} \end{cases}$$

and maximizing the $E\left(Y'(x_1, x_2, x_3)\right)$ instead.

A more complex optimization problem occurs when we want to select the system design that is *most likely* to be the best. Such an objective is relevant when one-shot, rather than long-run average, performance matters. Examples include a Space Shuttle launch, or the delivery of a unique, large order of products. Bechhofer, Santner, and Goldsman [1995] address this problem under the topic of "multinomial selection."

We have been assuming that a system design $x_1, x_2, \ldots, x_m$ can be evaluated in terms of a single performance measure, Y, such as cost. Obviously, this may not always be the case. In the MHS example we may also be interested in some measure of system productivity, such as throughput or cycle time. At present, multiple objective optimization via simulation is not well developed. Therefore, one of two strategies is typically employed:

1. Combine all of the performance measures into a single measure, the most common being cost. For instance, the revenue generated by each completed product in the MHS could represent productivity and be included as a negative cost.
2. Optimize with respect to one key performance measure, but then evaluate the top solutions with respect to secondary performance measures. For instance, the MHS could be optimized with respect to expected cost, and then the cycle time could be compared for the top 5 designs. *This approach requires that information on more than just the best solution be maintained.*

12.4.2 Why Is Optimization via Simulation Difficult?

Even when there is no uncertainly, optimization can be very difficult if the number of design variables is large, the problem contains a diverse collection of design variable types, and little is known about the structure of the performance function. Optimization via simulation adds an additional complication because the performance of a particular design cannot be evaluated exactly, but instead must be estimated. Because we have estimates, it may not be possible to conclusively determine if one design is better than another, frustrating optimization algorithms that try to move in improving directions. In principle, one can eliminate this complication by making so many replications, or such long runs, at each design point that the performance estimate has essentially no variance. In practice, this could mean that very few alternative designs will be explored due to the time required to simulate each one.

The existence of sampling variability forces optimization via simulation to make compromises. The following are the standard ones:

- **Guarantee a prespecified probability of correct selection.** The two-stage Bonferroni procedure in Section 12.2.2 is an example of this approach, which allows the analyst to specify the desired chance of being right. Such algorithms typically require either that every possible design is simulated or that a strong functional relationship among the designs (such as a metamodel) applies. Other algorithms can be found in Goldsman and Nelson [1998].
- **Guarantee asymptotic convergence.** There are many algorithms that guarantee convergence to the global or a local optimal solution as the simulation effort (number of replications, length of replications) becomes

infinite. These guarantees are useful because they indicate that the algorithm tends to get to where the analyst wants it to go. However, convergence can be slow, and there is often no guarantee as to how good the reported solution is when the algorithm is terminated in finite time (as it must be in practice). See Andradóttir [1988] for specific algorithms that apply to discrete- or continuous-variable problems.

- **Optimal for deterministic counterpart.** The idea here is to use an algorithm that would find the optimal solution *if the performance of each design could be evaluated with certainty.* An example might be applying a standard nonlinear programming algorithm to the simulation optimization problem. It is typically up to the analyst to make sure that enough simulation effort is expended (replications or run length) to insure that such an algorithm is not misled by sampling variability. Direct application of an algorithm that assumes deterministic evaluation to a stochastic simulation is not recommended.
- **Robust heuristics.** Many heuristics have been developed for deterministic optimization problems that do not guarantee finding the optimal solution, but nevertheless have been shown to be very effective on difficult, practical problems. Some of these heuristics use randomness as part of their search strategy, so one might argue that they are less sensitive to sampling variability than other types of algorithms. Nevertheless, it is still important to make sure that enough simulation effort is expended (replications or run length) to insure that such an algorithm is not misled by sampling variability.

Since robust heuristics are the most common algorithms found in commercial optimization via simulation software, we provide some guidance on their use in the next section.

12.4.3 Using Robust Heuristics

By a "robust heuristic" we mean a procedure that does not depend on strong problem structure—such as continuity or convexity of $E(Y(x_1, \ldots, x_m))$—to be effective, can be applied to problems with mixed types of decision variables, and—ideally—is tolerant of some sampling variability. Genetic algorithms (GA) and tabu search (TS) are two prominent examples, but there are many others and many variations of them. Such heuristics form the core of most commercial implementations. To give a sense of these heuristics, we describe GA and TS below. We caution the reader that only a high-level description of the simplest version of each procedure is provided. The commercial implementations are much more sophisticated.

Suppose that there are k possible solutions to the optimization-via-simulation problem. Let $\mathbf{X} = \{\mathbf{x}_1, \mathbf{x}_2, \ldots, \mathbf{x}_k\}$ denote the solutions, where the ith solution $\mathbf{x}_i = (x_{i1}, x_{i2}, \ldots, x_{im})$ provides specific settings for the m decision variables. The simulation output at solution $\mathbf{x}_i$ is denoted $Y(\mathbf{x}_i)$; this could be the output of a single replication, or the average of several replications. Our goal is to find the solution $\mathbf{x}^*$ that minimizes $E(Y(\mathbf{x}))$.

On each iteration (known as a "generation"), a GA operates on a "population" of p solutions. Denote the population of solutions on the jth iteration as $\mathbf{P}(j) = \{\mathbf{x}_1(j), \mathbf{x}_2(j), \ldots, \mathbf{x}_p(j)\}$. There may be multiple copies of the same solution in $\mathbf{P}(j)$, and $\mathbf{P}(j)$ may contain solutions that were discovered on previous iterations. From iteration to iteration this population evolves in such a way that good solutions tend to survive and give birth to new, and hopefully better, solutions, while inferior solutions tend to be removed from the population. The basic GA is given below:

Basic GA

1. Set the iteration counter $j = 0$, and select (perhaps randomly) an initial population of p solutions $\mathbf{P}(0) = \{\mathbf{x}_1(0), \ldots, \mathbf{x}_p(0)\}$.
2. Run simulation experiments to obtain performance estimates $Y(\mathbf{x})$ for all p solutions $\mathbf{x}(j)$ in $\mathbf{P}(j)$.
3. Select a population of p solutions from those in $\mathbf{P}(j)$ in such a way that those with smaller $Y(\mathbf{x})$ values are more likely, but not certain, to be selected. Denote this population of solutions as $\mathbf{P}(j+1)$.
4. Recombine the solutions in $\mathbf{P}(j+1)$ via crossover (which joins parts of two solutions $\mathbf{x}_i(j+1)$ and $\mathbf{x}_\ell(j+1)$ to form a new solution) and mutation (which randomly changes a part of a solution $\mathbf{x}_i(j+1)$).
5. Set $j = j+1$ and go to step 2.

The GA can be terminated after a specified number of iterations, when little or no improvement is noted in the population, or when the population contains p copies of the same solution. At termination, the solution $\mathbf{x}^*$ that has the smallest $Y(\mathbf{x})$ value in the last population is chosen as best (or alternatively, the solution with the smallest $Y(\mathbf{x})$ over all iterations could be chosen).

GAs are applicable to almost any optimization problem, because the operations of selection, crossover, and mutation can be defined in a very generic way that does not depend on specifics of the problem. However, when these operations are not tuned to the specific problem, a GA's progress can be very slow. Commercial versions are often self-tuning, meaning that they update selection, crossover, and mutation parameters during the course of the search. There is some evidence that GAs are tolerant of sampling variability in $Y(\mathbf{x})$ because they maintain a population of solutions, rather than focusing on improving a current best solution. In other words, it is not critical that the GA perfectly rank the solutions in a population of solutions, because the next iteration depends on the entire population, not a single solution.

TS, on the other hand, identifies a current best solution on each iteration and then tries to improve it. Improvements occur by changing the solution via "moves." For example, the solution (x_1, x_2, x_3) could be changed to the solution

$(x_1 + 1, x_2, x_3)$ by the move of adding 1 to the first decision variable (perhaps x_1 represents the number of AGVs in Example 12.8, so the move would add one more AGV). The "neighbors" of solution $\mathbf{x}$ are all of those solutions that can be reached by legal moves. TS finds the best neighbor solution and moves to it. However, to avoid making moves that return the search to a previously visited solution, moves may become "taboo" (not usable) for some number of iterations. Conceptually, think about how you would find your way through a maze: If you took a path that led to a dead end, then you would avoid taking that path again (it would be taboo).

The basic TS algorithm is given below. The description is based on Glover [1989].

Basic TS

1. Set the iteration counter $j = 0$ and the list of taboo moves to empty. Select an initial solution $\mathbf{x}^*$ in $\mathbf{X}$ (perhaps randomly).
2. Find the solution $\mathbf{x}'$ that minimizes $Y(\mathbf{x})$ over all of the neighbors of $\mathbf{x}^*$ that are not reached by taboo moves, running whatever simulations are needed to do the optimization.
3. If $Y(\mathbf{x}') < Y(\mathbf{x}^*)$, then $\mathbf{x}^* = \mathbf{x}'$ (move the current best solution to $\mathbf{x}'$).
4. Update the list of taboo moves and go to step 2.

The TS can be terminated when a specified number of iterations have been completed, when some number of iterations has passed without changing $\mathbf{x}^*$, or when there are no more feasible moves. At termination, the solution $\mathbf{x}^*$ is chosen as best.

TS is fundamentally a discrete-decision-variable optimizer, but continuous decision variables can be discretized as described in Section 12.4.4 below. TS aggressively pursues improving solutions, and therefore tends to make rapid progress. However, it is more sensitive to random variability in $Y(\mathbf{x})$, since $\mathbf{x}^*$ is taken to be the *true* best solution so far and attempts are made to improve it. There are probabilistic versions of TS that should be less sensitive, however. An important feature of commercial implementations of TS, which is not present in the Basic TS above, is a mechanism for overiding the taboo list when it is advantageous.

We offer two suggestions for using commercial products that employ a GA, TS, or other robust heuristic:

Control Sampling Variability

In many cases it will up to the user to determine how much sampling (replications or run length) will be undertaken at each potential solution. This is a difficult problem in general. Ideally, sampling should increase as the heuristic closes in on the better solutions, simply because it is much more difficult to distinguish solutions that are close in expected performance from those that differ widely. Early in the search it may be easy for the heuristic to determine good solutions and search directions because clearly inferior solutions are being compared to much better ones, but late in the search this may not be the case.

If the analyst must specify a fixed number of replications per solution that will be used through the search, then a preliminary experiment should be conducted. Simulate several designs, some at the extremes of the solution space and some nearer the center. Compare the apparent best and apparent worst of these designs, using the approaches in Section 12.1. Using the technique described in Section 12.1.4, determine the minimum number of replications required to declare these designs to be statistically significantly different. This is a minimum number of replications that should be used.

After the optimization run has completed, perform a second set of experiments on the top 5 to 10 designs identified by the heuristic. Use the comparison techniques in Section 12.2 to rigorously determine which are the best or near best of these designs.

Restarting

Because robust heuristics provide no guarantees that they converge to the optimal solution for optimization via simulation, it makes sense to run the optimization two or more times to see which run yields the best solution. Each optimization run should use different random-number seeds or streams, and ideally should start from different initial solutions. Try starting the optimization at solutions on the extremes of the solution space, in the center of the space, and at randomly generated solutions. If people familiar with the system suspect that certain designs will be good, be sure to include them as possible starting solutions for the heuristic.

12.4.4 An Illustration: Random Search

In this section we present an algorithm for optimization via simulation known as random search. The specific implementation is based on Algorithm 2 in Andradóttir [1998], which has guaranteed asymptotic convergence. Thus, it will find the true optimal solution if permitted to run long enough. However, in practice convergence can be slow, and the memory requirements of this particular version of random search can be quite large. Even though random search is not a "robust heuristic," we will also use it to demonstrate some strategies we would employ in conjunction with such heuristics, and to demonstrate why optimization via simulation is tricky even with what appears to be an uncomplicated algorithm.

The random-search algorithm that we present requires that there be a finite number of possible system designs (although that number may be quite large). This might seem to rule out problems with continuous decision variables, such as conveyor speed. In practice, however, apparently continuous decision variables can often be discretized in a reasonable way. For instance, if conveyor speed can be anything from 60 to 120 feet per minute, little may be lost by treating the possible conveyor speeds as 60, 61, 62, ..., 120 feet per minute (61 possible values). Note, however, that there are algorithms designed specifically for continuous-variable problems (Andradóttir [1998]).

Again let the k possible solutions to the optimization via simulation problem be denoted $\{\mathbf{x}_1, \mathbf{x}_2, \ldots, \mathbf{x}_k\}$, where the ith solution $\mathbf{x}_i = (x_{i1}, x_{i2}, \ldots, x_{im})$ provides specific settings for the m decision variables. The simulation output at solution $\mathbf{x}_i$ is denoted $Y(\mathbf{x}_i)$; this could be the output of a single replication, or the average of several replications. Our goal is to find the solution $\mathbf{x}^*$ that minimizes $E(Y(\mathbf{x}))$.

On each iteration of the random-search algorithm we compare a current good solution to a randomly chosen competitor. If the competitor is better, then it becomes the current good solution. When we terminate the search, the solution we choose is the one that has been visited most often (which means that we expect to revisit solutions many times).

Random-Search Algorithm

1. Initialize counter variables $C(i) = 0$ for $i = 1, 2, \ldots, k$. Select an initial solution i^0, and set $C(i^0) = 1$. ($C(i)$ counts the number of times we visit solution i.)
2. Choose another solution i' from the set of all solutions *except* i^0 in such a way that each solution has an equal chance of being selected.
3. Run simulation experiments at the two solutions i^0 and i' to obtain outputs $Y(i^0)$ and $Y(i')$. If $Y(i') < Y(i^0)$, then set $i^0 = i'$.
4. Set $C(i^0) = C(i^0) + 1$. If not done, then go to step 2. If done, then select as the estimated optimal solution x_{i^*} such that $C(i^*)$ is the largest count.

Note that if the problem is a maximization problem, then replace step 3 with

3. Run simulation experiments at the two solutions i^0 and i' to obtain outputs $Y(i^0)$ and $Y(i')$. If $Y(i') > Y(i^0)$, then set $i^0 = i'$.

One of the difficult problems with many optimization-via-simulation algorithms is knowing when to stop (exceptions include algorithms that guarantee a probability of correct selection). Typical rules might be to stop after a certain number of iterations, when the best solution has not changed much in several iterations, or when all time available to solve the problem has been exhausted. Whatever rule is used, we recommend applying a statistical selection procedure, such as the two-stage Bonferroni procedure in Section 12.2.2, to the 5 to 10 apparent best solutions. This is done to determine which among them is the true best with guaranteed confidence. If the raw data from the search have been saved, then this data can be used as the first-stage sample for a two-stage selection procedure (Boesel, Nelson, and Ishii [2000]).

Example 12.13 (Implementing Random Search)

Suppose that a manufacturing system consists of 4 stations in series. The zeroth station always has raw material available. When the zeroth station completes work on a part, it passes the part along to the first station, then the first passes the part to the second, and so on. Buffer space between stations 0 and 1, 1 and 2, and 2 and 3 is limited to 50 parts total. If, say, station 2 finishes a part but there is no buffer space available in front of station 3, then station 2 is blocked, meaning that it cannot do any further work. The question is how to allocate these 50 spaces to minimize the expected cycle time per part over one shift.

Let x_i be the number of buffer spaces in front of station i. Then the decision variables are x_1, x_2, x_3 with the constraint that $x_1 + x_2 + x_3 = 50$ (it makes no sense to allocate fewer buffer spaces than we have available). This implies a total of 1326 possible designs (can you figure out how this number is determined?).

To simplify the presentation of the random-search algorithm, let the counter for solution (x_1, x_2, x_3) be denoted as $C(x_1, x_2, x_3)$.

Random-Search Algorithm

1. Initialize 1326 counter variables $C(x_1, x_2, x_3) = 0$ for all possible solutions (x_1, x_2, x_3). Select an initial solution, say $(x_1 = 20, x_2 = 15, x_3 = 15)$ and set $C(20, 15, 15) = 1$.
2. Choose another solution from the set of all solutions *except* (20, 15, 15) in such a way that each solution has an equal chance of being selected. Suppose (11, 35, 4) is chosen.
3. Run simulation experiments at the two solutions to obtain estimates of the expected cycle time $Y(20, 15, 15)$ and $Y(11, 35, 4)$. Suppose that $Y(20, 15, 15) < Y(11, 35, 4)$. Then (20, 15, 15) remains as the current good solution.
4. Set $C(20, 15, 15) = C(20, 15, 15) + 1$.

2. Choose another solution from the set of all solutions *except* (20, 15, 15) in such a way that each solution has an equal chance of being selected. Suppose (28, 12, 10) is chosen.
3. Run simulation experiments at the two solutions to obtain estimates of the expected cycle time $Y(20, 15, 15)$ and $Y(28, 12, 10)$. Suppose that $Y(28, 12, 10) < Y(20, 15, 15)$. Then (28, 12, 10) becomes the current good solution.
4. Set $C(28, 12, 10) = C(28, 12, 10) + 1$.

2. Choose another solution from the set of all solutions *except* (28, 12, 10) in such a way that each solution has an equal chance of being selected. Suppose (0, 14, 36) is chosen.
3. Continue ....

When the search is terminated, we select the solution (x_1, x_2, x_3) that gives the largest $C(x_1, x_2, x_3)$ count. As we discussed earlier, the top 5 to 10 solutions should then be subjected to a separate statistical analysis to determine which among them is the true best (with high confidence). In this case, the solutions with the largest counts would receive the second analysis. ◀

Despite the apparent simplicity of the random-search algorithm, we have glossed over a subtle issue that often arises in algorithms with provable performance. In step 2 the algorithm must randomly choose a solution such that all are equally likely to be selected (except the current one). How can this be accomplished in Example 12.13? The constraint that $x_1 + x_2 + x_3 = 50$ means that x_1, x_2, and x_3 cannot be sampled independently. One might be tempted to sample x_1 as a discrete uniform random variable on 0 to 50, then sample x_2 as a discrete uniform on 0 to $50 - x_1$, and finally set $x_3 = 50 - x_1 - x_2$. But this method does not make all solutions equally likely, as the following illustration shows: Suppose that x_1 is randomly sampled to be 50. Then the trial solution must be (50, 0, 0); there is only one choice. But if $x_1 = 49$, then both (49, 1, 0) and (49, 0, 1) are possible. Thus, $x_1 = 49$ should be more likely than $x_1 = 50$ if all solutions with $x_1 + x_2 + x_3 = 50$ are to be equally likely.

12.5 Summary

This chapter provided a basic introduction to the comparative evaluation of alternative system designs based on data collected from simulation runs. It was assumed that a fixed set of alternative system designs had been selected for consideration. Comparisons based on confidence intervals and the use of common random numbers were emphasized. A brief introduction to metamodels (whose purpose is to describe the relationship between design variables and the output response) and optimization via simulation (whose purpose is to select the best from among a large and diverse collection of system designs) was also provided. Although beyond the scope of this text, there are many additional topics of potential interest in the realm of statistical analysis techniques relevant to simulation. Some of these topics include:

1. Experimental design models, whose purpose is to determine which factors have a significant impact on the performance of system alternatives
2. Output analysis methods other than the methods of replication and batch means
3. Variance-reduction techniques, which are methods to improve the statistical efficiency of simulation experiments (common random numbers being an important example)

The reader is referred to Banks [1998] and Law and Kelton [2000] for discussions of these and other topics relevant to simulation.

The most important idea in Chapters 11 and 12 is that simulation output data require a statistical analysis in order to be interpreted correctly. In particular, a statistical analysis can provide a measure of the precision of the results produced by a simulation, and can provide techniques for achieving a specified precision.

REFERENCES

ANDRADÓTTIR, S. [1998], "Simulation Optimization," Chapter 9 in *Handbook of Simulation*, John Wiley, New York.

BANKS, J., ed. [1998], *Handbook of Simulation*, John Wiley, New York.

BECHHOFER R. E., T. J. SANTNER, AND D. GOLDSMAN [1995], *Design and Analysis for Statistical Selection, Screening and Multiple Comparisons*, John Wiley, New York.

BOESEL, J., B. L. NELSON, AND N. ISHII [2000], "A Framework for Simulation-Optimization Software," *IIE Transactions*, forthcoming.

BOX, G. E. P., AND N. R. DRAPER [1987], *Empirical Model-Building and Response Surfaces*, John Wiley, New York.

GLOVER, F. [1989], "Tabu Search—Part I," *ORSA Journal on Computing*, Vol. 1, pp. 190–206.

GOLDSMAN, D., AND B. L. NELSON [1998], "Comparing Systems via Simulation," Chapter 8 in *Handbook of Simulation*, John Wiley, New York.

HINES, W. W., AND D. C. MONTGOMERY [1990], *Probability and Statistics in Engineering and Management Science*, 3d ed., John Wiley, New York.

HOCHBERG Y., AND A. C. TAMHANE [1987], *Multiple Comparison Procedures*, John Wiley, New York.

HSU, J. C. [1996], *Multiple Comparisons: Theory and Methods*, Chapman & Hall, New York.

KLEIJNEN, J. P. C. [1975], *Statistical Techniques in Simulation, Parts I and II*, Dekker, New York.

KLEIJNEN, J. P. C. [1987], *Statistical Tools for Simulation Practitioners*, Dekker, New York.

KLEIJNEN, J. P. C. [1988], "Analyzing Simulation Experiments with Common Random Numbers," *Management Science*, Vol. 34, pp. 65–74.

KLEIJNEN, J. P. C. [1998], "Experimental Design for Sensitivity Analysis, Optimization, and Validation of Simulation Models," Chapter 6 in *Handbook of Simulation*, John Wiley, New York.

LAW, A. M., AND W. D. KELTON [2000], *Simulation Modeling and Analysis*, 3d ed., McGraw-Hill, New York.

MONTGOMERY, D. C. [1991], *Design and Analysis of Experiments*, 3d ed., John Wiley, New York.

NELSON, B. L., AND F. J. MATEJCIK [1995], "Using Common Random Numbers for Indifference-Zone Selection and Multiple Comparisons in Simulation," *Management Science*, Vol. 41, pp. 1935–1945.

NELSON, B. L. [1987], "Some Properties of Simulation Interval Estimators Under Dependence Induction," *Operations Research Letters*, Vol. 6, pp. 169–176.

NELSON, B. L. [1992], "Designing Efficient Simulation Experiments," *1992 Winter Simulation Conference Proceedings*, Washington, D.C.

WRIGHT, R. D., AND T. E. RAMSAY, JR. [1979], "On the Effectiveness of Common Random Numbers," *Management Science*, Vol. 25, pp. 649–56.

EXERCISES

1. Reconsider the dump truck problem of Example 3.5, which was also analyzed in Example 12.2. As business expands, the company buys new trucks, making the total number of trucks now equal to 16. The company desires to have a sufficient number of loaders and scales so that the average number of trucks waiting at the loader queue plus the average number at the weigh queue is no more than three. Investigate the following combinations of number of loaders and number of scales:

Number of Scales	*Number of Loaders* 2	3	4
1	–	–	–
2	–	–	–

   The loaders being considered are the "slow" loaders in Example 12.2. Loading time, weighing time, and travel time for each truck are as previously defined in Example 12.2. Use common random numbers to the greatest extent possible when comparing alternative systems designs. The goal is to find the smallest number of loaders and scales to meet the company's objective of an average total queue length of no more than three trucks. In your solution, take into account the initialization conditions, run length, and number of replications needed to achieve a reasonable likelihood of valid conclusions.

2. In Exercise 11.6, consider the following alternative (M, L) policies:

			L	
			Low 30	*High* 40
M	Low	50	(50,30)	(50, 40)
	High	100	(100,30)	(100, 40)

   Investigate the relative costs of these policies using suitable modifications of the simulation model developed in Exercise 11.6. Compare the four system designs on the basis of long-run mean monthly cost. First make four replications of each (M, L) policy, using common random numbers to the greatest extent possible. Each replication should have a 12-month initialization phase followed by a 100-month data-collection phase. Compute confidence intervals having an overall confidence level of 90% for mean monthly cost for each policy. Then estimate the additional replications needed to achieve confidence intervals which do not overlap. Draw conclusions as to which is the best policy.

3. Reconsider Exercise 11.7. Compare the four inventory policies studied in Exercise 2, taking the cost of rush orders into account when computing monthly cost.

4. Reconsider Exercise 11.8. Compare the four monthly inventory policies studied in Exercise 2, taking into account the selling price of the perishable items.

5. In Exercise 11.9, investigate the effect of the order quantity on long-run mean daily cost. Since each order arrives on a pallet on a delivery truck, the permissible order quantities, Q, are multiples of 10 (i.e., Q may equal 10, or 20, or 30, ...). In Exercise 11.9, the policy $Q = 20$ was investigated.

(a) First, investigate the two policies $Q = 10$ and $Q = 50$. Use the run lengths, and so on, suggested in Exercise 11.9. On the basis of these runs, decide whether the optimal Q, say Q^*, is between 10 and 50, or greater than 50. (The cost curve as a function of Q should have what kind of shape?)

(b) Based on the results in part (a), suggest two additional values for Q and simulate the two policies. Draw conclusions. Include an analysis of the strength of your conclusions.

6. In Exercise 11.10, determine the number of cards Q that the card shop owner should purchase to maximize the profit within an error (for total profit) of approximately \$5.00 at most. First, make runs for the policy of ordering $Q = 250$ and $Q = 350$ cards. With these results, plus the results of Exercise 11.10 for ordering $Q = 300$ cards, decide on a range of Q worth considering further (e.g., $200 \leq Q \leq 250$, or $250 \leq Q \leq 300$, etc.). Then investigate this restricted range for two additional policies (e.g., $Q = 200$ and $Q = 225$, or $Q = 265$ and $Q = 285$, etc.).

7. In Exercise 11.11, investigate the effect of target level M and review period N on mean monthly cost. Consider two target levels, M, determined by ± 10 from the target level used in Exercise 11.11, and consider review periods N of 1 month and 3 months. Which (N, M) pair is best, based on these simulations?

8. Reconsider Exercises 11.13 and 11.14, which involved the scheduling rules (or queue disciplines) of first-in, first-out (FIFO) and priority by type (PR) in a job shop. In addition to these two rules, consider a shortest-imminent-operation (SIO) scheduling rule. For a given station, all jobs of the type with the smallest mean processing time are given highest priority. For example, when using an SIO rule at station 1, jobs are processed in the following order: type 2 first, then type 1, and type 3 last. Two jobs of the same type are processed on a FIFO basis. Develop a simulation experiment to compare the FIFO, PR, and SIO rules on the basis of mean total response time over all jobs.

9. In Exercise 11.13 (the job shop with FIFO rule), determine the minimum number of workers needed at each station to avoid bottlenecks. A bottleneck occurs when average queue lengths at a station steadily increase over time. (Do not confuse increasing average queue length due to an inadequate number of servers with increasing average queue length due to initialization bias. In the former case, average queue length continues to increase indefinitely and server utilization is 1.0. In the latter case, average queue length eventually levels off and server utilization is less than 1.) Report on utilization of workers and total time it takes for a job to get through the job shop, by type and over all types. [*Hint*: If server utilization at a workstation is 1.0, and if average queue length tends to increase linearly as simulation run length increases, it is a good possibility that the workstation is unstable and therefore is a bottleneck. In this case, at least one additional worker is needed at the work station. Use queueing theory, namely $\lambda / c_i \mu < 1$, to suggest the minimum number of workers needed at station 1. Recall that λ is the arrival rate, $1/\mu$ is the overall mean service time for one job with one worker, and c_i is the number of workers at station i. Attempt to use the same basic condition, $\lambda/c_i\mu < 1$, to determine an initial number of servers at station i for $i = 2, 3, 4$.]

10. **(a)** Repeat Exercise 9 for the PR scheduling rule (see Exercise 11.14).

(b) Repeat Exercise 9 for the SIO scheduling rule (see Exercise 8).

(c) Compare the minimum required number of workers for each scheduling rule: FIFO, versus PR, versus SIO.

11. With the minimum number of workers determined by Exercises 9 and 10 for the job shop of Exercise 11.13, consider adding one worker to the entire shop. This worker can be trained to handle the processing at only one station. At which station should this worker be placed? How does this additional worker affect mean total response time over all jobs? Over type 1 jobs? Investigate the job shop with and without the additional worker for each scheduling rule: FIFO, PR, SIO.

12. In Exercise 11.17, suppose that a buffer of capacity one item is constructed in front of each worker. Design an experiment to determine if this change in system design has a significant impact upon individual worker utilizations (ρ_1, ρ_2, ρ_3, and ρ_4). At the very least, compute confidence intervals for $\rho_1^0 - \rho_1^1$ and $\rho_4^0 - \rho_4^1$, where ρ_i^s, is utilization for worker i when the buffer has capacity s.

13. A clerk in the admissions office at Small State University processes requests for admissions materials. The time to process requests depends on the program of interest (e.g., industrial engineering, management science, computer science, etc.) and the level of the program (Bachelors, Masters, Ph.D.). Suppose that the processing time is well modeled as normally distributed with mean 7 minutes and standard deviation 2 minutes. At the beginning of the day it takes the clerk some time to get set to begin working on requests; suppose that this time is well modeled as exponentially distributed with mean 20 minutes. The admissions office typically receives between 40 and 60 requests per day.

Let x be the number of applications received on a day, and let Y be the time required to process them (including the set-up time). Fit a metamodel for $E(Y|x)$ by making n replications at the design points $x = 40, 50, 60$. Notice that in this case we know that the correct model is

$$E(Y|x) = \beta_0 + \beta_1 x = 20 + 7x$$

(why?). Begin with $n = 2$ replications at each design point and estimate β_0 and β_1. Gradually increase the number of replications and observe how many are required for the estimates to be close to the true values.

14. Repeat the previous exercise using correlated sampling. How do the results change?

15. The usual statistical analysis used to test for $\beta_1 \neq 0$ does not hold if we use correlated sampling. Where does it break down?

16. Riches and Associates retains its cash reserves primarily in the form of certificates of deposit (CDs), which earn interest at an annual rate of 8%. Periodically, however, withdrawals must be made from these CDs in order to pay suppliers, etc. These cash outflows are made through a checking account that earns no interest. The need for cash cannot be predicted with certainty. Transfers from CDs to checking can be made instantaneously, but there is a "substantial penalty" for early withdrawal from CDs. Therefore, it may make sense for R&A to make use of the overdraft protection on their checking account, which charges interest at a rate of \$0.00033 per dollar per day (i.e., 12% per year) for overdrafts.

R&A likes simple policies in which they transfer a fixed amount, a fixed number of times, per year. Currently they make 6 transfers per year of \$18,250 each time. Your job is to find a policy that reduces their long-run cost per day.

Based on historical patterns, demands for cash arrive a rate of about 1 per day, with the arrivals being well modeled as a Poisson process. The amount of cash needed to satisfy each demand is reasonably represented by a lognormally distributed random variable with mean \$300 and standard deviation \$150.

The penalty for early withdrawal is different for different CDs. It averages \$150 for each withdrawal (regardless of size), but the actual penalty can be modeled as a uniformly distributed random variable with range \$100 to \$200.

Use cash level in checking to determine the length of the initialization phase. Make enough replications so that your confidence interval for the difference in long-run cost per day does not contain zero. Be sure to use correlated sampling in your experiment design.

17. If you have access to commercial optimization-via-simulation software, test how well it works as the variability of the simulation outputs increases. Use a simple model such as $Y = x^2 + \varepsilon$, where ε is a random variable with a $N(0, \sigma^2)$ distribution, and for which the optimal solution is known ($x = 0$ for minimization, in this case). See how quickly, or whether, the software can find the true optimal solution as σ^2 increases. Next try more complex models with more than one design variable.

18. For Example 12.13, show why there are 1326 solutions. Then derive a way to sample x_1, x_2, and x_3 such that $x_1 + x_2 + x_3 = 50$ and all outcomes are equally likely.

19. An electronic component with mean time to failure of x years can be purchased for $2x$ thousand dollars (thus, the more reliable the component, the more expensive it is). The value of x is restricted to be between 1 to 10 years, and the actual time to failure is modeled as exponentially distributed. The mission for which the component is to be used lasts one year; if the component fails in less than one year, then there is a cost of \$20,000 for early failure. What value of x should be chosen to minimize the expected total cost (purchase plus early failure)?

To solve this problem, develop a simulation that generates a total cost for a component with mean time to failure of x years. This requires sampling an exponentially distributed random variable with mean x, and then computing the total cost as $2000x$ plus 20,000 if the failure time is less than 1. Fit a quadratic metamodel in x and use it to determine the value of x that minimizes the fitted model. [*Hints:* Select several values of x between 1 and 10 as design points. At each value of x, let the response variable $Y(x)$ be the average of at least 30 observations of total cost.]

20. Use optimization-via-simulation software to solve Exercise 19. If you do not have access to such software, use the random-search algorithm and let the possible values of x be $\{1.00, 1.25, 1.50, \ldots, 10.00\}$.

21. Suppose that demand for a certain product has a Poisson distribution with mean 10 units. Use optimization-via-simulation software (or the random-search algorithm, if you do not have access to such software) to determine the order size x that maximizes the probability that demand equals x units, assuming that $0 \leq x \leq 100$, and x is an integer.[6] [*Hint:* On each replication of the simulation for a trial value of x, a random demand is generated from the Poisson distribution with mean 10. If this random demand equals x, then the response is $Y(x) = 1$; otherwise it is $Y(x) = 0$.]

[6] This problem is based on Example 5.1 in S. Andradóttir [2000], "Accelerating the Convergence of Random Search Methods for Discrete Stochastic Optimization," *ACM TOMACS*, forthcoming.

13

Simulation of Manufacturing and Material Handling Systems

Manufacturing and material handling systems provide one of the most important applications of simulation. Simulation has been used successfully as an aid in the design of new production facilities, warehouses, and distribution centers. It has also been used to evaluate suggested improvements to existing systems. Engineers and analysts using simulation have found it valuable for evaluating the impact of capital investments in equipment and physical facility, and of proposed changes to material handling and layout. They have also found it useful to evaluate staffing and operating rules, and proposed rules and algorithms to be incorporated into warehouse management control software and production control systems. Managers have found simulation useful in providing a "test drive" before making capital investments, without disrupting the existing system with untried changes.

Section 13.1 provides an introduction and discusses some of the features of simulation models of manufacturing and material handling systems. Section 13.2 discusses the goals of manufacturing simulation and the most common measures of system performance. Section 13.3 discusses a number of the issues common to many manufacturing and material handling simulations, including the treatment of downtimes and failure, and trace-driven simulations using actual historical data or historical order files. Section 13.4 provides a number of short case studies, with references for additional reading. For an overview of simulation software for manufacturing and material handling applications, see Section 4.7.

13.1 Manufacturing and Material Handling Simulations

As do all modeling projects, manufacturing and material handling simulation projects need to address the issues of scope and level of detail. Consider scope as analogous to breadth and level of detail as analogous to depth. Scope determines the boundaries of the project, what's in the model and what's not. For a subsystem, process, machine, or other component, the project scope determines whether the object is in the model or not. Given that a component or subsystem is part of a model, it often can be simulated at many different levels of detail.

The proper scope and level of detail should be determined by the objectives of the study and the questions being asked. On the other hand, level of detail may be constrained by the availability of input data and the knowledge of how system components work. For new nonexistent systems, data availability may be limited and system knowledge may be based on assumptions.

While some guidelines can be provided, the judgment of experienced simulation analysts, working with the customer to determine early in the project the questions the model is being designed to address, provide the basis for selecting a proper scope and level of detail.

Should the model simulate each conveyor section or vehicle movement, or can some be replaced by a simple time delay? Should the model simulate auxiliary parts, or the handling of purchased parts, or can the model assume that such parts are always available at the right location when needed for assembly?

At what level of detail does the control system need to be simulated? Many modern manufacturing facilities, distribution centers, baggage handling systems, and other material handling systems are computer controlled by a management control software system. The algorithms built into such control software play a key role in system performance. Simulation is often used to evaluate and compare the effectiveness of competing control schemes and to evaluate suggested improvements. It can be used to debug and fine-tune the logic of a control system before it is installed.

These questions are representative of the issues that need to be addressed when deciding the correct level of model detail and scope of a project. In turn, the scope and level of model detail determine the type of questions that can be addressed by the model. In addition, models can be developed in an iterative fashion, adding detail for peripheral operations at later stages if such operations are later judged to significantly impact the main operation. It is good advice to start as simple as possible, adding detail only as needed.

13.1.1 Models of Manufacturing Systems

Models of manufacturing systems may have to take into account a number of characteristics of such systems, some of which are:

- Physical layout
- Labor
    - Shift schedules
    - Job duties and certification

- Equipment
  - Rates and capacities
  - Breakdowns
    - Time to failure
    - Time to repair
    - Resources needed for repair
- Maintenance
  - PM schedule
  - Time and resources required
  - Tooling and fixtures
- Workcenters
  - Processing
  - Assembly
  - Disassembly
- Product
  - Product flow, routing, and resources needed
  - Bill of materials
- Production schedules
  - Made-to-stock
  - Made-to-order
    - Customer orders
    - Line items and quantities
- Production control
  - Assignment of jobs to work areas
  - Task selection at workcenters
  - Routing decisions
- Supplies
  - Ordering
  - Receipt and storage
  - Delivery to workcenters
- Storage
  - Supplies
  - Spare parts
  - Work-in-process (WIP)
  - Finished goods
- Packing and shipping
  - Order consolidation
  - Paperwork
  - Loading trailers

13.1.2 Models of Material Handling

In manufacturing systems, it is not unusual for 80% to 85% of an item's total time in system to be expended in material handling or waiting for material handling to occur. This work-in-process (WIP) represents a vast investment, and reductions in WIP and associated delays can result in large cost savings. Therefore, for some studies, detailed material handling simulations are cost effective.

In some production lines, the material handling system is an essential component. For example, automotive paint shops typically consist of a power-and-free conveyor system that transports automobile bodies or body parts through the paint booths.

In warehouses, distribution centers, flow-through and cross-docking operations, material handling is clearly a key component of any material-flow model. Manual warehouses typically use manual fork trucks to move pallets from receiving dock to storage and storage to shipping dock. More automated distribution centers may use extensive conveyor systems to support putaway, order picking, order sortation, and consolidation.

Models of material handling systems may have to contain some of the following types of subsystems:

- Conveyors
  - Accumulating
  - Nonaccumulating
  - Indexing and other special purpose
  - Fixed window or random spacing
  - Power and free
- Transporters
  - Unconstrained vehicles (e.g., manually guided fork trucks)
  - Guided vehicles (automated or operator controlled, wire-guided, chemical paths, rail-guided)
  - Bridge cranes and other overhead lifts
- Storage systems
  - Pallet storage
  - Case storage
  - Small part storage (totes)
  - Oversize items
  - Rack storage or block stacked
  - Automated storage and retrieval systems (AS/RS) with storage-retrieval machines (SRM)

13.1.3 Some Common Material Handling Equipment

There are numerous types of material handling devices common to manufacturing, warehousing, and distribution operations. They include unconstrained transporters such as carts, manually driven fork lift trucks, and pallet jacks; guided-path transporters such as AGVs (automated guided vehicles); and fixed-path devices such as various types of conveyor.

The class of unconstrained transporters, sometimes called free-path transporters, includes carts, fork lift trucks, pallet jacks, and other manually driven vehicles that are free to travel throughout a facility unconstrained by a guide path of any kind. Unconstrained transporters are not constrained to a network of paths and may choose an alternate path or move around an obstruction. In contrast, the guided-path transporters move along a fixed path, such as chemical trails on the floor, wires imbedded in the floor, infrared lights placed strategically, self-guidance using radio communications, laser guidance and dead reckoning, and rail. Guided-path transporters may contend with each other for space along that path and usually have limited options upon meeting obstacles and congestion. Examples of guided-path transporters include the automated guided vehicle (AGV); a rail-guided turret truck for storage and retrievals of pallets in rack storage; and a crane in an AS/RS (automated storage and retrieval system).

The conveyor is a fixed-path device for moving entities from point to point, following a fixed path with specific load points, stopping or processing points, and unload points. A conveyor system may consist of numerous connected sections with merges and diverts. Each section may be one of a number of different types. Examples of conveyor types include belt, powered and gravity roller, bucket, chain, tilt-tray, and power-and-free, each with its own characteristics that must be modeled accurately.

Most conveyor sections can be classified as either accumulating or nonaccumulating. An accumulating conveyor section runs continuously. If the forward progress of an item is halted while on the accumulating conveyor, slippage occurs allowing the item to remain stationary and items behind it to continue moving until they reach the stationary item. Some belt and most roller conveyors operate in this manner. Only items that will not be damaged by bumping into each other can be placed on an accumulating conveyor.

In contrast, after an item is on a nonaccumulating conveyor section, its spacing relative to other items does not change. If one item stops moving, the entire section stops moving and hence all items on the section stop. For example, a nonaccumulating conveyor is used for moving televisions not yet in cartons, for they must be kept a safe distance from each other while moving from one assembly or testing station to the next. Bucket conveyors, tilt-tray conveyors, most belt conveyors, and conveyors designed to carry heavy loads (usually, pallets) are nonaccumulating conveyors.

Conveyors can also be classified as fixed-window or random spacing. In fixed-window spacing, items on the conveyor must always be within zones of equal length, which can be pictured as lines drawn on a belt conveyor or trays pulled by a chain. For example, in a tilt-tray conveyor, continuously moving trays of fixed size are used to move items. The control system is designed to induct items in such a way that each item is in a separate tray; thus it is a nonaccumulating fixed-window conveyor. In contrast, with random spacing, items can be anywhere on the conveyor section relative to other items. To be inducted, they simply require sufficient space.

Besides these basic types, there are innumerable types of specialized conveyors for special purposes. For example, a specialized indexing conveyor may move forward in increments, always maintaining a fixed distance between the trailing edge of the load ahead and the leading edge of the load behind. Its purpose is to form a "slug" of items, equally spaced apart, to be inducted all together onto a transport conveyor. For the local behavior of some systems, that is, the performance at a particular workstation or induction point, a detailed understanding and accurate model of the physical workings plus the control logic are essential for accurate results.

13.2 Goals and Performance Measures

The purpose of simulation is insight, not numbers. Those who purchase and use simulation software and services want to gain insight and understanding into how a new or modified system will work. Will it meet throughput expectations? What happens to response time at peak periods? Is the system resilient to short-term surges? What is the recovery time when short-term surges cause congestion and queueing? What are the staffing requirements? What problems occur? If problems occur, what is their cause and how do they arise? What is the system capacity? What conditions and loads cause a system to reach its capacity?

While simulations are expected to provide numeric measures of performance, such as throughput under a given set of conditions, the major benefit of simulation comes from the insight and understanding gained regarding system operations. Visualization through animation and graphics provides major assistance in the communication of model assumptions, system operations, and model results. Often, visualization is the major contributor to a model's credibility, which in turn leads to acceptance of the model's numeric outputs. Of course, a proper experimental design that includes the right range of experimental conditions plus a rigorous analysis and, for stochastic simulation models, a proper statistical analysis is of utmost importance for the simulation analyst to draw correct conclusions from simulation outputs.

The major goals of manufacturing simulation models are to identify problem areas and quantify system performance. Some common measures of system performance include:

- Throughput under average and peak loads
- System cycle time (how long it takes to produce one part)

- Utilization of resources, labor and machines
- Bottlenecks and choke points
- Queueing at work locations
- Queueing and delays caused by material handling devices and systems
- WIP storage needs
- Staffing requirements
- Effectiveness of scheduling systems
- Effectiveness of control systems

Often, material handling is an important part of a manufacturing system and its performance. Nonmanufacturing material handling systems include warehouses, distribution centers, cross-docking operations, and baggage handling systems at airports. The major goals of these nonmanufacturing material handling systems are similar to those identified for manufacturing systems. Some additional considerations are:

- How long it takes to process one day of customer orders
- Effect of changes in order profiles (for distribution centers)
- Truck/trailer queueing and delays at receiving and shipping docks
- Effectiveness of material handling systems for peak loads
- Recovery time from short-term surges (for example, with baggage handling)

13.3 Issues in Manufacturing and Material Handling Simulations

There are a number of modeling issues especially important for the achievement of accurate and valid simulation models of manufacturing and material handling systems. Two of these issues are the proper modeling of downtimes, and whether to use actual system data for some inputs versus a statistical model of those inputs.

13.3.1 Modeling Downtimes and Failures

Unscheduled random downtimes can have a major effect on the performance of manufacturing systems. Many authors have discussed the proper modeling of downtime data [Williams, 1994; Clark, 1994; Law and Kelton, 1991]. This section discusses the problems that can arise when downtime is modeled incorrectly, and suggests a number of ways to correctly model machine and system downtimes.

Scheduled downtime, such as for preventive maintenance, or periodic downtime, such as for tool replacement, also can have a major effect on system performance. But these downtimes usually are (or should be) predictable and can be scheduled to minimize disruptions. In addition, engineering efforts or new technology may be able to reduce their duration.

There are a number of alternatives for modeling random unscheduled downtimes, some better than others:

1. Ignore it.
2. Do not model it explicitly but adjust processing times appropriately.
3. Use constant values for time to failure and time to repair.
4. Use statistical distributions for time to failure and time to repair.

Of course, alternative 1 generally is not the suggested approach. This is certainly an irresponsible modeling decision if downtimes have an impact on the results, as they do in almost all situations. One situation in which ignoring downtimes may be appropriate, with the full knowledge of the customer, is to leave out catastrophic downtimes that occur rarely. In other words, the model may incorporate normal downtimes but ignore those catastrophic downtimes, such as general power failures, snow storms, cyclones and hurricanes, that occur rarely but stop all production when they do occur. The documented scope of the project should clearly state the assumed operating conditions and those conditions that are not included in the model. If it is generally known that a plant will be closed for some number of snow days per year, then the simulation need not take these downtimes into account, for the effect of any given number of days can easily be factored into the simulation results when making annual projections.

The second possibility, to factor into the model the effect of downtimes by adjusting processing times applied to each job or part, may be an acceptable approximation under limited circumstances. If each job or part is subject to a large number of small delays associated with downtime of equipment or tools, then the total of such delays may be added to the pure processing time to arrive at an adjusted processing time. If total delay time and pure processing time are random in nature, then an appropriate statistical distribution should be used for the total adjusted processing time. If the pure processing time is constant while the total delay time in one cycle is random and variable, it is almost never accurate to adjust the processing time by a constant factor. For example, if processing time is usually 10 minutes but the equipment is subject to downtimes that cause about a 10% loss in capacity, it is not appropriate to merely change the processing time to a constant 11 minutes. While such a deterministic adjustment may provide reasonably accurate estimates of overall system throughput, it will not provide accurate estimates of local behavior such as queue and buffer space needed at peak times. Queueing and short-term congestion are strongly influenced by randomness and variability.

The third possibility, using constant durations for time to failure and time to repair, may be appropriate when, for example, the downtime is actually due to preventative maintenance that is on a fixed schedule. In almost all other circumstances, the fourth possibility, modeling time to failure and time to repair by appropriate statistical distributions, is the appropriate technique. This requires either actual data for choosing a statistical distribution based on the techniques in Chapter 9, or when data is lacking a reasonable assumption based on the physical nature of the causes of downtimes.

The nature of time to failure is also important. Are times to failure completely random in nature, typically due to a large number of possible causes of failure? In this case, the exponential distribution may provide a good statistical model. Or, rather, are times to failure more regular, typically due to some major component, say a tool, wearing out? In this case, a uniform or (truncated) normal distribution may be more appropriate. If so, the mean of the distribution represents the average time to failure and the distribution places a plus or minus around the mean.

Time to failure can be measured in a number of different ways:

1. by wall-clock time
2. by machine or equipment busy time
3. by number of cycle times
4. by number of items produced

Breakdowns or failures can be based on clock time, actual usage, or cycles. Note that the word breakdown or failure is used, even though product jams or preventive maintenance may be the reason for a downtime. As mentioned, breakdowns or failures can be probabilistic or deterministic in duration.

Actual usage breakdowns are based on the time during which the resource is used. For example, wear on a machine tool occurs only when the machine is in use. Time to failure is measured against machine busy time and not against wall-clock time. If the time to failure is 90 hours, then the model keeps track of total busy time since the last downtime ended, and when 90 hours is reached, processing is interrupted and a downtime occurs.

Clock-time breakdowns might be associated with scheduled maintenance —for example, changes of fluids every three months when a complete lubrication is required. Downtimes based on wall-clock time may also be used for equipment that is always busy or equipment that "runs" whether or not it is processing parts.

Cycle breakdowns or failures are based on the number of times the resource is used. For example, after every 50 uses of a tool, it needs to be sharpened. Downtimes based on number of cycle times or number of items produced are implemented by generating the number of times or items, and in the model simply counting until this number is reached. Typical uses of downtimes based on busy time or cycle times may be for maintenance or tool replacement.

Another issue is what happens to a part at a machine when the jam, breakdown, or failure occurs. Possibilities include scrapping the part, reworking it, or simply continuing processing after repair. In some cases, for example when preventive maintenance is due, the part in the machine may complete processing before the repair (or maintenance activity) begins.

Time to repair can also be modeled in two fundamentally different ways:

1. as a pure time delay (no resources required)
2. as a wait time for a resource (e.g., maintenance person) plus a time delay for actual repair

Of course, there are many variations on these methods in actual modeling situations. When a repair or maintenance person is a limited resource, the second approach will be a more accurate model and provide more information.

The next example illustrates the importance of using the proper approach for modeling downtimes, and the consequences and inaccurate results that may occur when making incorrect assumptions.

Example 13.1

Consider a single machine that processes a wide variety of parts that arrive in random mixes at random times. Data analysis has shown that an exponentially distributed processing time with a mean of 7.5 minutes provides a fairly accurate representation. Parts arrive at random, and time between arrivals is exponentially distributed with a mean of 10 minutes. The machine fails at random times. Downtime studies have shown that time to failure can be reasonably approximated by an exponential distribution with a mean time of 1000 minutes. The time to repair the resource is also exponentially distributed with a mean time of 50 minutes. When a failure occurs, the current part in the machine is removed from the machine; when the repair has been completed, the part completes its processing.

When a part arrives, it queues and waits its turn at the machine. It is desired to estimate the size of this queue. An experiment was designed to estimate the average number of parts in the queue. To illustrate the effect of an accurate treatment of downtimes, the model was run under a number of different assumptions. For each case and replication, the simulation runlength was 100,000 minutes.

Table 13.1 shows the average number of parts in the queue for six different treatments of the time between breakdowns. For each treatment that involves randomness, five replications of those treatments and the average for those five replications are shown.

Case A ignores the breakdowns. The average number in the queue is 2.31 parts. Across the 5 independent replications, the averages range from 2.05 to 2.70 parts. This treatment of breakdowns is not recommended.

Case B increases the average service time from 7.5 minutes to 8.0 minutes in an attempt to approximate the effect of downtimes. On average, each downtime and repair cycle is 1050 minutes, with the machine down for 50 minutes. Thus the machine is down, on the average in the long run, $50/1050 = 4.8\%$ of total time. Thus, some have argued that downtime has approximately the same effect as increasing the processing time of each part by 4.8%, which is about 7.86 minutes. Therefore, an assumed constant 8 minutes per part should

Table 13.1. Simulation of Breakdowns

Case	*1st Rep*	*2nd Rep*	*3rd Rep*	*4th Rep*	*5th Rep*	*Average*
A. Ignore the breakdowns	2.36	2.05	2.38	2.05	2.70	2.31
B. Increase service time to 8.0	3.32	2.82	3.32	2.81	4.03	3.26
C. All random	4.05	3.77	4.36	3.95	4.43	4.11
D. Random processing, deterministic breakdowns	3.24	2.85	3.28	3.05	3.79	3.24
E. All deterministic						0.52
F. Deterministic processing, random breakdowns	1.06	1.04	1.10	1.32	1.16	1.13

be (it is argued) a conservative approach. For this treatment of downtimes, the average number of parts in the queue, over the five replications, is about 3.26 parts. Across the 5 replications, the range is from 2.81 to 4.03 parts. While the treatment in Case B may be appropriate under some limited circumstances, as discussed in a previous section it is not appropriate under the assumptions of this example.

The proper treatment, shown as Case C, treats the randomness in processing and breakdowns properly with the assumed correct exponential distributions. The average value is about 4.11 parts waiting for the machine. Across the 5 replications, the average queue length ranges from 3.77 to 4.43 parts. The average number waiting differs from that of Case B by almost one part.

Case D is a simplification that treats the processing randomly, but treats the breakdowns as deterministic. The results average about 3.24 parts in the queue. The range of averages is from 2.85 to 3.79 parts.

Case E treats all of the times as deterministic. Only one replication is needed, since additional replications will only provide the same result. The average value in the queue is 0.52 parts, well below the value in Case C, or any other case for that matter. The conclusion: Ignoring randomness is dangerous and leads to totally unrealistic results.

Case F treats arrivals and processing as deterministic, but breakdowns are random. The average number of parts in the queue at the machine is about 1.13. The range is from 1.04 to 1.32 parts. For some machines and processing in manufacturing environments, Case F is the realistic situation; that is, processing times are constant and arrivals are regulated, that is, are also constant. The reader is left to consider the inaccuracies that would result from making faulty assumptions regarding the nature of time to failure and time to repair.

In conclusion, there may be a significant difference in the average number in a queue based on the treatment of randomness. The results using the correct treatment of randomness may be far different than the other alternatives. Often it is tempting, due to the unavailability of detailed data but the availability of averages, to want to use average time to failure as if it were a constant. Example 13.1 illustrates the dangers of inappropriate assumptions. The appropriate technique to use, as well as the appropriate statistical distribution, depend on the available data and the situation at hand.

As discussed by Williams [1994], the accurate treatment of downtimes is essential for achieving valid models of manufacturing systems. Some of the essential ingredients are:

- avoidance of oversimplified and inaccurate assumptions
- careful collection of downtime data
- accurate representation of time to failure and time to repair by statistical distributions
- accurate modeling of system logic when a downtime occurs, in terms of both repair time logic and what happens to the part currently processing ◄

13.3.2 Trace-Driven Models

Consider a model of a distribution center that receives customer orders that must be processed and shipped in one day. One modeling question is how to represent the day's set of orders. A typical order may contain one or more line items, and each line item may have a quantity of one or more pieces. For example, when you buy a new stereo, you may purchase an amplifier, a tuner, and a CD player (all separate line items with a quantity of one piece), and 4 identical speakers (another line item with a quantity of 4 pieces). The overall order profile can have a major impact on the performance of a particular system design. A system designed to handle large orders going to a small number of customers may not perform well if order profiles shift toward a larger number of customers (or larger number of separate shipments) with one or two items per order.

One approach is to characterize the order profile by using a discrete statistical distribution for each variable in an order:

1. the number of line items
2. for each line item, the number of pieces

If these two variables are statistically independent, then this approach may provide a valid model of the order profile. For many applications, however, these two variables may be highly correlated in ways that could be difficult to characterize statistically. For example, an apparel and shoe company has six large customers (the large department stores and discount chains), representing 50% of unit sales volume, that typically order dozens or hundreds of line items and large quantities of many of the items. At the opposite pole, on any given day approximately 70% of the orders are for one or two pairs of shoes (just-in-time with a vengeance!). For this company, the number of line items in an order is highly positively correlated with the quantity ordered; that is, large orders with a large number of line items also usually have large quantities of many of the line items. And small orders with only a few line items typically order small quantities of each item.

What would happen if the two variables, number of line items and quantity per line item, were modeled by independent statistical distributions? When an order began processing, the model would make two uncorrelated random draws, which could result in order profiles quite different from those found in practice. Such an erroneous assumption could result, for example, in far too large a proportion of orders having one or two line items with large unrealistic quantities.

Another common but more serious error is to assume that there is an average order and to simulate only the number of orders in a day with each being the typical order. In the author's experience, analysis of many order profiles has shown (1) that there is no such thing as a typical order and (2) that there is no such thing as a typical order profile.

An alternative approach, and one that has proven successful in many studies, is for the company to provide the actual orders for a sample of days over the previous year. Usually, it is desirable to simulate peak days. A model driven by actual historical data is called a trace-driven model.

A trace-driven model eliminates all possibility of error due to ignoring or misestimating correlations in the data. One apparent limitation may be a customer's desire, at times, to be able to simulate hypothesized changes to the order profile, such as a higher proportion of smaller orders in terms of both line items and quantities. In practice, this limitation can be removed by adding "dials" to the order-profile portion of the model, so that a simulation analyst can "dial up" more or less of certain characteristics, as desired. One approach is to treat the day's orders as a statistical population from which the model draws samples in a random fashion. This approach makes it easy to change overall order volume but not to modify the profile. A second related approach would be to subdivide a day's orders into subgroups based on number of line items, quantities, or other numeric parameters, and then sample in a specified proportion from each subgroup. By changing the proportion of each subgroup, different order profiles can be "dialed up" and fed into the model. A third approach is to use factors to adjust the number of daily orders, the number of line items, and/or the quantities. In practice, one of these approaches may be as accurate as can be expected for hypothesized future order profiles, and may provide a cost-effective and reasonably accurate model, especially for testing the robustness of a system design for assumed changes in order characteristics.

Other examples of trace-driven models include:

- Orders to a custom job shop using actual historical orders
- Product mix and quantities, and production sequencing, for an assembly line making 100 styles and sizes of hot water heaters
- Time to failure and downtime using actual maintenance records
- Truck arrival times to a warehouse using gate records

Whether to make an input variable trace-driven or to characterize it as a statistical distribution depends on a number of issues, including the nature of the variable itself and whether it is correlated with or independent of other variables, the availability of accurate data, and the questions being addressed. In one situation, a trace driven model is a requirement. For example, a frozen food warehouse uses a simulation model on a daily basis as a decision support tool to assist with staffing levels and assignments and prestaging of large orders. The model must be driven by that day's actual (known) orders.

13.4 Case Studies of the Simulation of Manufacturing and Material Handling Systems

The *Winter Simulation Conference Proceedings, IIE Magazine, Modern Material Handling*, and other periodicals are excellent sources of information for short cases in the simulation of manufacturing and material handling systems.

An abstract of some of the papers from the *1994 Winter Simulation Conference Proceedings* will provide some insight into the types of problems that can be addressed by simulation. These abstracts have been paraphrased and shortened where appropriate; our goal is to provide an indication of the breadth of real-world applications of simulation.

Session: Semiconductor Wafer Manufacturing

Paper: Modeling and Simulation of Material Handling for Semiconductor Wafer Manufacturing

Authors: Neal G. Pierce and Richard Stafford

Abstract: This paper presents the results of a design study to analyze the interbay material handling systems for semiconductor wafer manufacturing. The authors developed discrete-event simulation models of the performance of conventional cleanroom material handling including manual and automated systems. The components of a conventional cleanroom material handling system include an overhead monorail system for interbay (bay-to-bay) transport, work-in-process stockers for lot storage, and manual systems for intrabay movement. The authors constructed models and experiments that assisted with analyzing cleanroom material handling issues such as designing conventional automated material handling systems and specifying requirements for transport vehicles.

Session: Simulation in Aerospace Manufacturing

Paper: Modeling Aircraft Assembly Operations

Authors: Harold A. Scott

Abstract: A simulation model is used to aid in the understanding of complex interactions of aircraft assembly operations. Simulation helps to identify the effects of resource constraints on dynamic process capacity and cycle time. To analyze these effects, the model must capture job and crew interactions at the control-code level. This paper explores five aspects of developing simulation models to analyze crew operations on aircraft assembly lines:

Representing job-precedence relationships
Simulating crew members with different skill and job-proficiency levels
Reallocating crew members to assist ongoing jobs
Depicting shifts and overtime
Modeling spatial constraints and crew movements in the production area.

Session: Control of Manufacturing Systems

Paper: Discrete Event Simulation for Shop Floor Control

Authors: J. S. Smith, R. A. Wysk, D. T. Sturrock, S. E. Ramaswamy, G. D. Smith, S. B. Joshi

Abstract: This paper describes an application of simulation to shop floor control of a flexible manufacturing system. In this application, the simulation is used not only as an analysis and evaluation tool, but also as a "task generator" for the specification of shop floor control tasks. Using this approach, the effort applied to the development of the simulation is not duplicated in the development of the control system. Instead, the same control logic is used for the control system as was used for the simulation. Additionally, since the simulation implements the control, it provides very high fidelity performance predictions. Implementation experience in two flexible manufacturing laboratories is described.

Session: Flexible Manufacturing

Paper: Developing and Analyzing Flexible Cell Systems Using Simulation

Authors: Edward F. Watson and Randall P. Sadowski

Abstract: This paper develops and evaluates flexible cell alternatives to support an agile production environment at a mid-sized manufacturer of industrial equipment. Three work-cell alternatives were developed based on traditional flow analysis studies, past experience, and common sense. To support the analysis of each option, a simulation model was developed and validated for the current manufacturing environment. The simulation model allowed the analyst to evaluate each cell alternative under current conditions as well as anticipated future conditions that included changes to product demand, product mix, and process technology.

Session: Modeling of Production Systems

Paper: Inventory Cost Model for Just-in-Time Production

Authors: Mahesh Mathur

Abstract: This paper presents the design and operation of a simulation model used to compare set-up and inventory carrying costs with varying lot sizes.

While reduction of lot sizes is a necessary step toward implementation of just-in-time (JIT) in a job shop environment, a careful cost study is required to determine the optimum lot size under the present set-up conditions. A simulation model can be designed to graphically display the fluctuation of carrying costs and accumulation of set-up costs on a time scale in a dynamic manner. The selection of the optimum lot size then can be based on realistic cost figures.

Session: Analysis of Manufacturing Systems

Paper: Modeling Strain of Manual Work in Manufacturing Systems

Authors: I. Ehrhardt, H. Herper, and H. Gebhardt

Abstract: Simulation allows complex logistical issues to be handled properly. Current research concentrates on technological aspects in the improvement of system behavior. Even though there is ever-increasing automation, there are vital tasks in production and logistics that are still assigned to humans. Present simulation modeling efforts rarely concentrate on the manual activities assigned to humans. This paper describes a simulation model that considers manual operations for increasing the effectiveness of planning logistic systems.

13.5 Summary

This chapter introduced some of the ideas and concepts most relevant to manufacturing and material handling simulation. Some of the key points are the importance of accurately modeling downtimes, the advantages of trace-driven simulations with respect to some of the inputs, and the need in some models for accurate modeling of material handling equipment and the control system.

REFERENCES

BANKS, J. [1994], "Software for Simulation," *1994 Winter Simulation Conference Proceedings*, ed. J. D. Tew, S. Manivannan, D. A. Sadowski, and A. F. Seila, Association for Computing Machinery, New York, pp. 26–33.

CLARK, G. M. [1994], "Introduction to Manufacturing Applications," *1994 Winter Simulation Conference Proceedings*, ed. J. D. Tew, S. Manivannan, D. A. Sadowski, and A. F. Seila, Association for Computing Machinery, New York, pp. 15–21.

EHRHARDT, I., H. HERPER, AND H. GEBHARDT [1994], "Modelling Strain of Manual Work in Manufacturing Systems," *1994 Winter Simulation Conference Proceedings*, ed. J. D. Tew, S. Manivannan, D. A. Sadowski, and A. F. Seila, Association for Computing Machinery, New York, pp. 1044–1049.

LAW, A. M. AND W. D. KELTON [1999], *Simulation Modeling and Analysis*, 3d ed., McGraw-Hill, New York.

PIERCE, N. G., AND R. STAFFORD [1994], "Modeling and Simulation of Material Handling for Semiconductor Wafer Fabrication," *1994 Winter Simulation Conference Proceedings*, ed. J. D. Tew, S. Manivannan, D. A. Sadowski, and A. F. Seila, Association for Computing Machinery, New York, pp. 900–906.

SCOTT, H. A. [1994], "Modeling Aircraft Assembly Operations," *1994 Winter Simulation Conference Proceedings*, ed. J. D. Tew, S. Manivannan, D. A. Sadowski, and A. F. Seila, Association for Computing Machinery, New York, pp. 920–927.

SMITH J. S. et al. [1994], "Discrete Event Simulation for Shop Floor Control," *1994 Winter Simulation Conference Proceedings*, ed. J. D. Tew, S. Manivannan, D. A. Sadowski, A. F. Seila, Association for Computing Machinery, New York, pp. 962–969.

WATSON, E. F., AND R. P. SADOWSKI [1994], "Developing and Analyzing Flexible Cell Systems Using Simulation," *1994 Winter Simulation Conference Proceedings*, ed. J. D. Tew, S. Manivannan, D. A. Sadowski, and A. F. Seila, Association for Computing Machinery, New York, pp. 978–985.

WILLIAMS, E. J. [1994], "Downtime Data — Its Collection, Analysis, and Importance," *1994 Winter Simulation Conference Proceedings*, ed. J. D. Tew, S. Manivannan, D. A. Sadowski, and A. F. Seila, Association for Computing Machinery, New York, pp. 1040–1043.

EXERCISES

Instructions to the student: Many of the following exercises involve material handling equipment such as conveyors and vehicles. The student is expected to use any simulation language or simulator that supports modeling conveyors and vehicles at a high level.

Some of the following exercises use the uniform, exponential, normal, or triangular distributions. Virtually all simulation languages and simulators support these plus other distributions. The use of the first three distributions was explained in the note to the exercises in Chapter 4; the use of the triangular is explained in the exercise that requires it. For reference, the properties of these distributions plus others used in simulation are given in Chapter 5, and random-variate generation is covered in Chapter 8.

1. A case sortation system consists of one infeed conveyor and 12 sortation lanes, as shown in the accompanying schematic (not to scale).

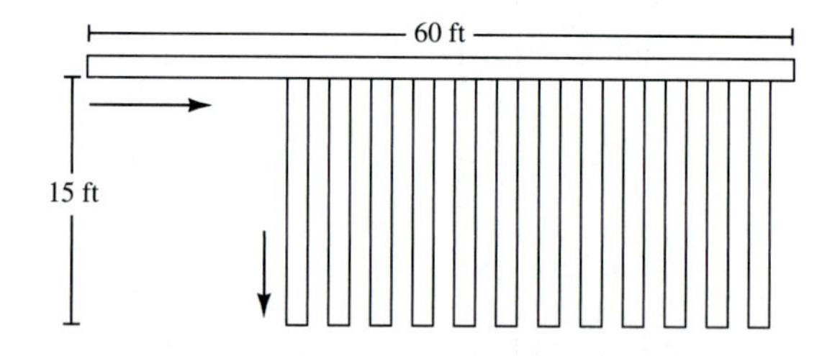

Cases enter the system from the left at a rate of 50 per minute at random times. All cases are 18 by 12 inches and travel along the 18-inch dimension. The incoming mainline conveyor is 20 inches wide and 60 feet in length (as shown). The sortation lanes are numbered 1 to 12 from left to right, and are 18 inches wide and 15 feet in length with 2 feet spacing between adjacent lanes. (Estimate any other dimensions that may be needed.) The infeed conveyor runs at 180 feet/minute and the sortation lanes at 90 feet/minute. All conveyor sections are accumulating, but on entrance at the left, incoming cases are at least 2 feet apart from leading edge to leading edge. On the sortation lanes the cases accumulate with no gap between them.

Incoming cases are distributed to the 12 lanes in the following proportions:

1	6%	7	11%
2	6%	8	6%
3	5%	9	5%
4	24%	10	5%
5	15%	11	3%
6	14%	12	0%

The 12th lane is an overflow lane; it is only used if one of the other lanes fills and a divert is not possible.

At the end of the sortation lanes is a group of operators who scan each case with a bar-code scanner, apply a label, and then place it on a pallet. Operators move from lane to lane as necessary to avoid having a lane fill. There is one pallet per lane, each holding 40 cases. When a pallet is full, assume a new empty one is immediately available. If a lane fills to 10 cases and another case arrives at the divert point, this last case continues to move down the 60-foot mainline conveyor and is diverted into lane 12, the overflow lane.

Assume that one operator can handle 8.5 cases per minute, on the average. Ignore walking time and assignment of an operator to a particular lane; in other words, assume the operators work as a group and can immediately service any lane.

(a) Set up an experiment that varies the number of operators and addresses the question: How many operators are needed? The objective is to have the minimum number of operators but also to avoid overflow.

(b) For each experiment in part (a), report the following output statistics:

Operator utilization
Total number of cases palletized
Number of cases palletized by lane
Number of cases to the overflow lane

(c) For each experiment in part (a), verify that all cases are being palletized. In other words, verify that the system can handle 50 cases per minute, or explain why it cannot.

2. Re-do Exercise 1 to a greater level of detail by modeling operator walking time and operator assignment to lanes. Assume that operators walk at 200 feet per minute and that the walking distance from one lane to the next is 5 feet. Handling time per case is now assumed to be 7.5 cases per minute. Devise a set of rules that can be used by operators for lane changing. (For example, change lanes to that lane with the greatest number of cases only when the current lane is empty or the other lane reaches a certain level.) Assume that each operator is assigned to a certain number of adjacent lanes and handles only those lanes. However, if necessary, two operators (but no more) may be assigned to one lane—that is, operator assignments may overlap.

(a) If your lane-changing rule has any numeric parameters, experiment to find the best settings. Under these circumstances, how many operators are needed? What is the average operator utilization?

(b) Does a model that has more detail, as does Exercise 2(a) when compared to Exercise 1, always have greater accuracy? How about this particular model? Compare the results of Exercise 2(a) to the results for Exercise 1. Are the same or different conclusions drawn?

(c) Devise a second lane-changing rule. Compare results between the two rules. Compare total walking time or percent of time spent walking between the two rules.

Suggestion: A lane-changing rule could have one or two "triggers." A one-trigger rule might state that if a lane reaches a certain level, the operator moves to that lane. (Without modification, such a rule could lead to excessive operator movement, if two lanes had about the same number of cases near the trigger level.) A two-trigger rule might state that if a lane reaches a certain level and the operator's current lane becomes empty, then change to the new lane; but if a lane reaches a specified higher "critical" level, then the operator immediately changes lanes.

(d) Compare your results with those of other students who may have used a different lane-changing rule.

3. Re-do Exercise 2 with a different operator-assignment rule. Basically, operators can be assigned to any lane as conditions warrant, but as before no more than two operators can be assigned to a lane at the same time. Address questions (a)–(d) as in Exercise 2.

4. A package sortation system consists of one in-feed conveyor, 12 sortation lanes (or chutes), and a takeaway conveyor, as shown in the accompanying schematic (not to scale).

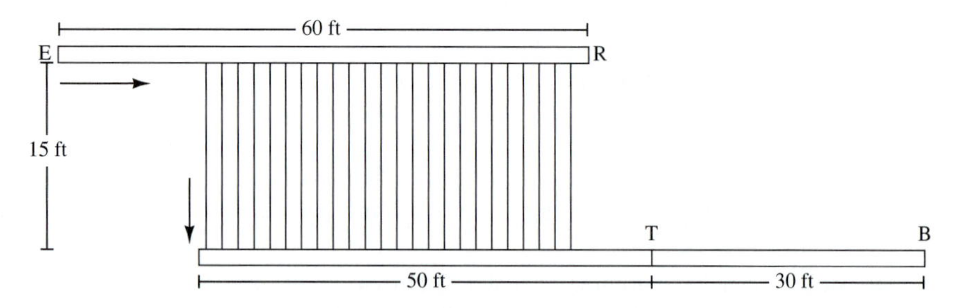

Packages enter the system from the left (at point E) at a rate of 100 per minute at random times. A package is diverted to a sortation lane based on its shipping destination; in the model, a given percentage of the packages chosen randomly according to the proportions in the following table are assigned to each lane. The percentage for each lane/destination is:

1	6%	7	10%
2	6%	8	6%
3	5%	9	4%
4	24%	10	4%
5	15%	11	3%
6	14%	12	3%

After a slug of 8 packages has accumulated, they are all released together onto a 50-foot section of takeaway conveyor and head toward the point labeled T, where they transfer (at conveyor speed) onto a second section of takeaway conveyor, at the end of which is a bar-code scanner (at the point labeled B). A slug cannot be released until the previous slug has cleared the point T. In this model, packages disappear at point B.

All packages are 18 by 12 inches and travel along their 18-inch dimension. The incoming mainline conveyor is 20 inches wide and 60 feet in length (as shown). The 12 sortation lanes are 18 inches wide and 15 feet in length with 2 feet spacing between adjacent lanes. (Estimate any other dimensions that may be needed.) The infeed conveyor runs at 240 feet/minute and the sortation lanes at 90 feet/minute. The infeed and sortation conveyor sections are accumulating, but on entrance at the left incoming packages are at least 2 feet apart from leading edge to leading edge. On the sortation lanes the packages accumulate with no gap between them. The first 50-foot section of takeaway conveyor is a nonaccumulating belt-type conveyor, and the second 30-foot section is accumulating.

When a lane contains 8 packages, no more diverts are allowed until the lane is released and the slug completely clears the bottom of the chute. When a lane is full or before the slug has cleared, and another package arrives to divert, the affected package does not divert but rather moves to the end of the infeed conveyor (the point labeled R) and then recirculates to the start of the infeed conveyor on a conveyor that is not shown in the schematic. Simulate recirculation as a time delay of 45 seconds without explicitly modeling the recirculation conveyor itself.

(a) Set up an experiment to determine the minimum required speed of the take-away conveyor to minimize or eliminate recirculation. Available conveyor from vendors has speeds starting at 60 feet/minute and increases in increments of 30 feet/minute. (A faster conveyor is more expensive, so minimizing the speed will reduce costs, provided recirculation is not a problem.)

(b) Could destination assignment to available lane be modified in such a way as to improve system performance? Devise a better assignment rule and test it using the simulation model.

5. A fleet of AGVs (automated guided vehicles) services 5 workstations in series by transporting parts on fixtures from one workstation to the next, remaining with a part from the pickup point through all workstations to the offload point. The incoming staging conveyor, the AGV guidepath, the five workstations (A–E), and the offload conveyor are shown in the accompanying schematic (not to scale).

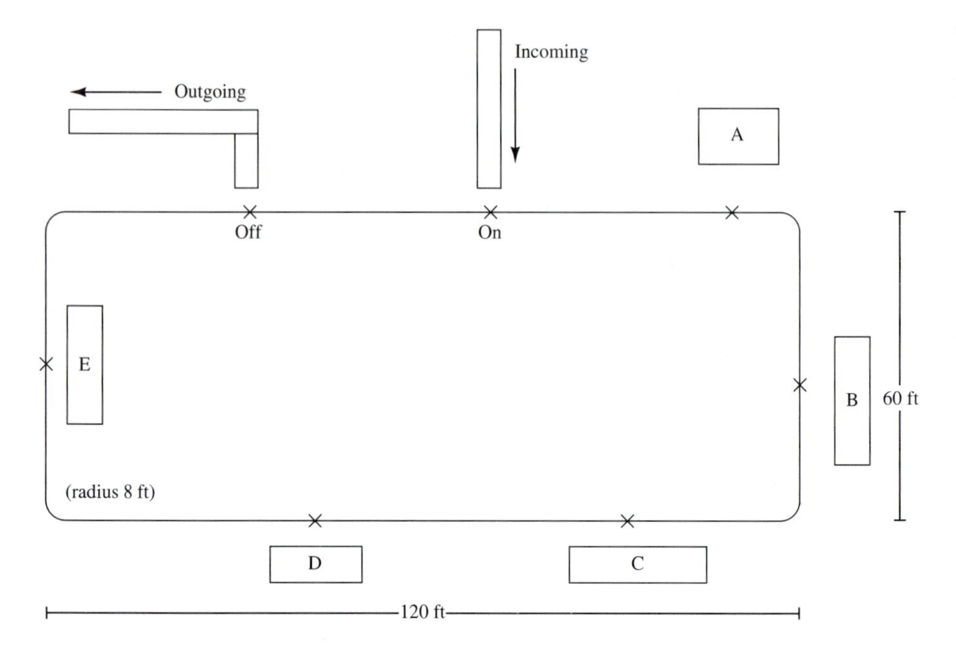

AGVs travel in a clockwise direction. Initially, all are empty and queued at the incoming conveyor at the "On" point. The workstations are labeled A–E. Parts on fixtures are staged on the incoming conveyor and picked up by an AGV at the point labeled "On." The AGV then travels to machine A and stops at the point marked on the guidepath. The part and fixture remain on the AGV during processing, then travel together to the next workstation. After processing is completed at station E, the AGV travels to the point labeled "Off," the part with fixture is automatically unloaded onto the offload conveyor, and the empty AGV queues at the point labeled "On" for the next part.

Assume that the incoming conveyor remains full. It is 15 feet long and can hold 3 parts with fixtures, each taking 5 feet of conveyor space. Therefore, when one part with fixture is picked up by an AGV, another part and fixture are placed on the end of the conveyor. Pickup time is 30 seconds. The incoming conveyor runs at 30 feet per minute.

When a part has completed processing at all stations and has been transported to the offload point labeled "Off," the offload operation takes 45 seconds. The offload conveyor itself runs at 30 feet per minute. The parts and fixtures, although taking a space of 3 feet by 3 feet, require 5 feet on the conveyor.

Initial workstation placements are shown in approximate fashion with no exact locations given. Constraints are that each workstation must be placed along the straight section of guidepath on which they are shown on the schematic, with the AGV stopping point no closer than 5 feet from the entrance to or exit from a curve. The incoming and offload conveyor must be placed as shown, with the offload point 20 feet from the end of the preceding curve and the pickup point 40 additional feet to the right. Otherwise, AGV stopping points at workstations can be placed at any desired location on the straight section of guidepath.

AGVs travel at 15 feet per second and accelerate and decelerate at 3 ft/sec/sec. They are 7 feet in length and queue 8 feet apart (1 foot gap between two AGVs) when waiting for the processing position at a workstation or the pickup/offload position at a conveyor.

At each of the workstations, processing time per part, mean time to failure (MTTF), and mean time to repair (MTTR) are as follows:

Workstation	*Processing time (sec.)*	*MTTF (min.)*	*MTTR (min.)*
A	120	90	3.1
B	100	82	4.0
C	140	110	4.5
D	45	20	5.0
E	100	240	9.0

As detailed data is not available, certain assumptions regarding time to failure and time to repair are made, as follows: Assume that actual time to failure is exponentially distributed with the specified mean (MTTF). Assume that repair time requires a maintenance person and there is only one available. Actual repair times are uniformly distributed with the mean as shown and a half-width of one-half the mean. Time to failure is measured against wall-clock time, not against station busy time. After a repair, any remaining processing time is taken and the part continues.

There are three basic questions:

What is maximum system throughput?

How many AGVs are needed?

What is the best workstation placement (within the stated constraints)?

For each set of experiments in parts (a) and (b) following, consider the following issues:

Should there be a "warmup" period for loading the system during which no statistics are collected? How long should the simulation run? (One suggestion: The warmup period should last at least until several parts have completed processing. The runlength should allow for at least 10 downtimes on each machine. Simulate with as many AGVs as will reasonably fit on the guidepath and then try to reduce the number without reducing throughput.)

Your experiments should vary the number of AGVs and workstation placement. Before simulating, make your best estimate regarding workstation placement and justify your decision. Also, make a reasonable estimate regarding the likely range for number of AGVs.

(a) In the first set of experiments, use a rough-cut model that does not model the AGVs explicitly. Compute the approximate travel time between adjacent workstations and use this value as a time delay in the model. Assume there are sufficient AGVs. (Use this model to address the question of maximum throughput but not the number of AGVs needed.) What are the advantages and limitations of this approach?

(b) In the second set of experiments, model the AGVs and guidepath explicitly. Design an experiment to determine maximum system throughput (parts per hour) with the smallest number of AGVs.

6. More accurate information has been obtained on failure times for the workstations in Exercise 5. Re-do Exercise 5 with the new information. Failures appear to be more closely related to the number of parts processed than to the amount of time passed. The new data is as follows:

Workstation	*MTTF (parts)*	*MTTR (min.)*
A	45	3.1
B	49	4.0
C	47	4.5
D	27	5.0
E	144	9.0

Note that MTTF (mean time to failure) is measured in terms of number of parts, but assumptions on time to repair are unchanged from Exercise 5. For example, for workstation A, if time to failure happened to be 45 parts, then after a failure, the 45th part would experience the next failure. In reality, failure occurs during processing, but for modeling purposes, assume the failure occurs just as processing begins. (Does this make any difference?)

For the distribution of time to failure, assume a discrete uniform distribution with mean (MTTF) over the range from 1 to 2*MTTF − 1. For example, for workstation A, the actual times to failure are equally likely to be any value from 1 to 89 parts.

Address the same questions as in Exercise 5. Do these new assumptions make any difference? How many failures are there per 8-hour shift for the first set of assumptions as compared to the revised assumptions?

7. For the problem in Exercises 5 and 6, a study was done to determine the repair times more accurately for each workstation. It was found that a triangular distribution provided a closer fit to the actual repair times. Assume the following parameters for the triangular distribution:

Workstation	*Repair time (triangular distribution)*			
	Minimum (min.)	*Most likely (min.)*	*Maximum (min.)*	*Average (min.)*
A	1.0	2.3	6.0	3.1
B	2.0	3.0	7.0	4.0
C	2.5	3.5	7.5	4.5
D	2.0	4.0	9.0	5.0
E	4.0	7.0	16.0	9.0

Note that the average repair times are identical to the assumed averages (MTTR) in Exercises 5 and 6, but the distribution is triangular instead of uniform. (Most simulation languages or simulators have a built-in triangular distribution that requires specification of the minimum value, most likely value or mode, and maximum value. The average value is given for informational purposes only; it is the sum of the other three parameters divided by 3. The triangular distribution is discussed in Chapter 5.)

(a) Re-do Exercise 5 with the new assumptions for repair times.

(b) Re-do Exercise 6 with the new assumptions for repair times.

8. One approach to modeling downtimes discussed in this chapter, but not recommended, was to replace the processing time with a larger adjusted processing time and to leave out the downtimes. The adjusted processing time is meant to account for downtimes, at least in the long run and on the average. How would this method have affected the results for Exercise 5?

(a) For Exercise 5, compute the adjusted processing times for each workstation.

(b) For Exercise 5, use the adjusted processing times and leave out all explicit downtimes due to failure. How does this change the results? Is this a good approach for this model?

9. Suppose that improved maintenance and better machines could totally eliminate downtime due to failure for the system in Exercise 5. By how much would throughput improve?

10. Develop a model for Example 5.1 and attempt to reproduce qualitatively the results found in the text regarding different assumptions for simulating downtimes. Do not attempt to get exactly the same numerical results, but rather to show the same qualitative results.

(a) Do your models support the conclusions discussed in the text? Provide a discussion and conclusions.

(b) Make a plot of the number of entities in the queue versus time. Can you tell when failures occurred? After a repair, about how long does it take for the queue to get back to "normal"?

11. In Example 13.1, the failures occurred at low frequency compared to the processing time of an entity. Time to failure was 1000 minutes and interarrival time was 10 minutes, implying that few entities would experience a failure. But when an entity did experience a failure of 50 minutes on average, it was several times larger than the processing time of 7.5 minutes.

Re-do the model for Example 13.1, assuming high-frequency failures. Specifically, assume that the time to failure is exponentially distributed with a mean of 2 minutes and the time to repair is exponentially distributed with a mean of 0.1 minute or 6 seconds. Compared to the low-frequency case, entities will tend to experience a number of short downtimes.

For low-frequency versus high-frequency downtimes, compare the average number of downtimes experienced per entity, the average duration of downtime experienced, the average time-to-complete service (including downtime, if any), and the percent of time down.

Note that the percentage of time the machine is down for repair should be the same in both cases:

$$50/(1000 + 50) = 4.76\%$$

$$6 \text{ sec}/(2 \text{ min} + 6 \text{ sec}) = 4.76\%$$

Verify percentage downtime from the simulation results. Are the results identical? Are they close? Should they be identical, or just close? As the simulation runlength increases, what should happen to percentage of time down?

With high-frequency failures, do you come to the same conclusions as were drawn in the text regarding the different ways to simulate downtimes? Make recommendations regarding how to model low-frequency versus high-frequency failures.

12. Re-do Exercise 11 (based on Example 13.1) with one change. When an entity experiences a downtime, it must be reprocessed from the beginning. If service time is random, take a new draw from the assumed distribution. If service time is constant, it starts over again. How does this assumption affect the results?

13. Re-do Exercise 11 (based on Example 13.1) with one change. When an entity experiences a downtime, it is scrapped. How does scrapping entities on failure affect the results in the low-frequency and in the high-frequency situations? What are your recommendations regarding the handling of low- versus high-frequency downtimes when parts are scrapped?

14. Sheets of metal pass sequentially through 4 presses: shear, punch, form, and bend. Each machine is subject to downtime and die change. The parameters for each machine are as follows:

Press	*Process rate (per min.)*	*Time to failure (min.)*	*Time to repair (min.)*	*No. sheets to a die change (no. sheets)*	*Time to change die (min.)*
Shear	4.5	100	8	500	25
Punch	5.5	90	10	400	25
Form	3.8	180	9	750	25
Bend	3.2	240	20	600	25

Note that processing time is given as a rate, for example, the shear press works at a rate of 4.5 sheets per minute. Assume that processing time is constant. Because of automated equipment, the time to change a die is fairly constant, so it is assumed to be always 25 minutes. Die changes occur between the stamping of two sheets after the number shown in the table have gone through a machine. Time to failure is assumed to be exponentially distributed with the mean given in the table. Time to repair is assumed to be uniformly distributed with the mean taken from the table and a half-width of 5 minutes. When a failure occurs, 20% of the sheets are scrapped. The remaining 80% are reprocessed at the failed machine after the repair.

Assume that an unlimited supply of material is available in front of the shear press, which processes one sheet after the next as long as there is space available between itself and the next machine, the punch press. In general, one machine processes one sheet after another continuously, stopping only for a downtime, a die change, or because the available buffer space between itself and the next machine becomes full. Assume that sheets are taken away after bending at the bend press. Buffer space is divided into 3 separate areas, one between the shear and the punch presses, the second between the punch and form presses, and the last between form and bend.

(a) Assume that there is an unlimited amount of space between machines. Run the simulation for 480 hours (about 1 month with 24-hour days, 5 days per week). Where do backups occur? If the total buffer space for all three buffers is limited to 15 sheets (not counting before shear or after bend), how would you recommend dividing this space among the three adjacent pairs of machines? Does this simulation provide enough information to make a reasonable decision.

(b) Modify the model so that there is a finite buffer between adjacent machines. When the buffer becomes full and the machine feeding the buffer completes a sheet, the sheet is not able to exit the machine. It remains in the machine, blocking additional work. Assume that total buffer space is 15 sheets for the 3 buffers.

Use the recommendation from part (a) as a starting point for each buffer size. Attempt to minimize the number of runs. You are allowed to experiment with a maximum of 3 buffer sizes for each buffer. (How many runs does this make?) Run a set of experiments to determine the allocation of buffer space that maximizes production. Simulate each alternative for at least 1000 hours.

Report total production per hour on the average, press utilization broken down by percentage of time busy, down, changing dies, and idle, and average number of sheets in each buffer.

14

Simulation of Computer Systems

It is only natural that simulation is used extensively to simulate computer systems, because of their great importance to the everyday operations of business, industry, government and universities. In this chapter we look at the motivations for simulating computer systems, the different types of approaches used, and the interplay between characteristics of the model and implementation strategies. We begin by looking at general characteristics of computer system simulations. Next we lay the groundwork for investigating simulation of computer systems by looking at various types of simulation tools used to perform those simulations. In Section 14.3 we describe different ways that input is presented or generated for these simulations. We next work through an example of a high-level computer system one might simulate, paying attention to problems of model construction and output analysis. In Section 14.5 we turn to the central processing unit (CPU), and point out what is generally simulated, and how. Following this, in Section 14.6, we consider simulation of memory systems.

14.1 Introduction

Computer systems are incredibly complex. At the time of this writing, a single chip contains nearly ten million transistors. A computer system exhibits complicated behavior at time scales from the time it takes to "flip" a transistor's state (on the order of 10^{-11} seconds) to the time it takes a human to interact with it (on the order of seconds or minutes). Computer systems are designed hierarchically, in an effort to manage this complexity. Figure 14.1 illustrates the point. At a high level of abstraction (the system level) one might view computational activity in terms of tasks circulating among servers, queueing

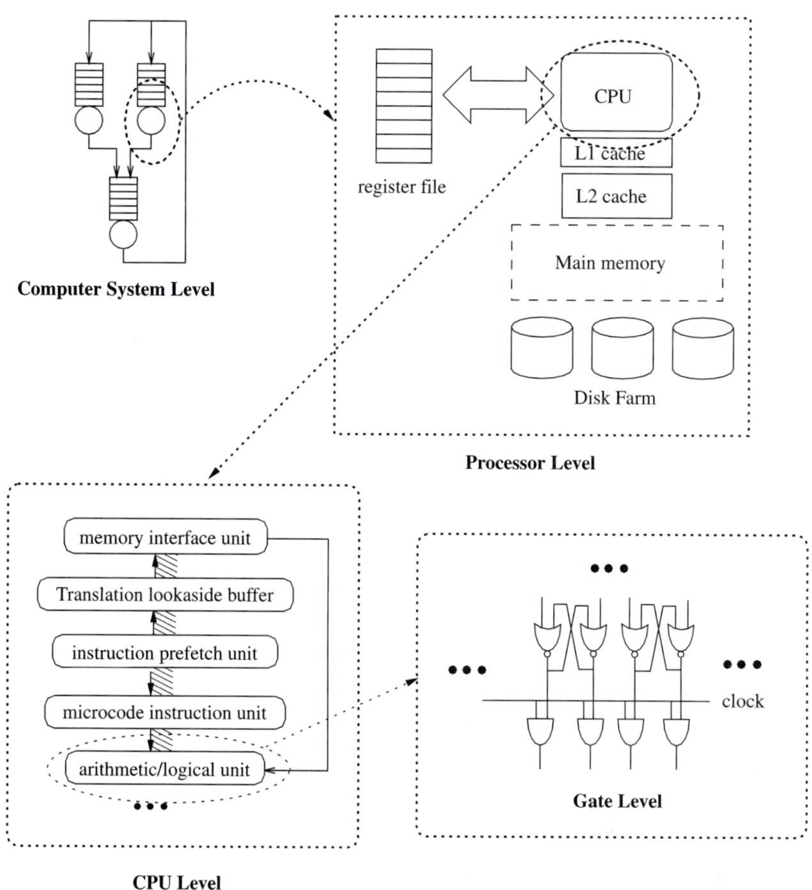

Figure 14.1. Different levels of abstraction in computer systems.

for service when a server is busy. A lower level in the hierarchy can view the activity among components of a given processor (its registers, its memory hierarchy). At a lower level still one views the activity of functional units that together make up a central processing unit, and at an even lower level one can view the logical circuitry that makes it all happen.

Simulation is used extensively at every level of this hierarchy, with results from one level being used at another. For instance, engineers working on designing a new chip will begin by partitioning the chip functionally (e.g., the subsystem that does arithmetic, the subsystem that interacts with memory, and so on), establish interfaces between the subsystems, then design and test the subsystems individually. Given a subsystem design, the electrical properties of the circuit are first studied, using a circuit simulator that solves differential equations describing electrical behavior. At this level engineers work to ensure that signal timing is correct throughout the circuit, and that the electrical properties fall within the parameters intended by the design. Once this level of validation has been achieved, the electrical behavior is abstracted into logical behavior; e.g., signals formerly thought of as electrical waveforms are now thought of as logical 1's or 0's. A different type of simulator is next used to test the correctness of the circuit's logical behavior. A common testing technique is to present the design with many different sets of logical inputs ("test vectors") for which the desired logical outputs are known. Discrete-event simulation is used to evaluate the logical response of the circuit to each test vector, and is also used to evaluate timing (e.g., the time required to load a register with a datum from the main memory). Once a chip's subsystems are designed and tested, the designs are integrated, and then the whole system is subjected to testing, again by simulation.

At a higher level one simulates using functional abstractions. For instance, a memory chip could be modeled simply as an array of numbers, and a reference to memory as just an indexing operation. A special type of description language exists for this level, called "register-transfer language" (see, for instance, Mano [1993]). This is like a programming language, with preassigned names for registers and other hardware-specific entities, and with assignment statements used to indicate data transfer between hardware entities. For example, the sequence below loads into register `r3` the data whose memory address is in register `r6`, subtracts one from it, and writes the result into the memory location that is word-adjacent (a word in this example is 4 bytes in size) to the location first read.

```
r3 = M[r6];
r3 = r3-1;
r6 = r6+4;
M[r6] = r3;
```

A simulator of such a language might ascribe deterministic time constants to the execution of each of these statements. This is a useful level of abstraction to use when one needs to express sequencing of data transfers at a low level, but not so low as the gates themselves. The abstraction makes sense when one is content to assume that the memory works, and that the time to put a datum in or out is a known constant. The "known constant" is a value resulting from analysis at a lower level of abstraction. Functional abstraction is also commonly used to simulate subsystems of a central processing unit (CPU), when studying how an executing program exercises special architectural features of the CPU.

At a higher level still one might study how an input-output (I/O) system behaves in response to execution of a computer program. The program's behavior may be abstracted to the point of being *modeled* but with some detailed description of I/O demands, e.g., with a Markov-chain that with some specificity describes an I/O operation as the Markov chain transitions. The behavior of the I/O devices may be abstracted to the point that all that is considered is how long it takes to complete a specified I/O operation. Because of these abstractions one can simulate larger systems, and simulate them more quickly. Continuing in this vein, at a higher level of abstraction still, one dispenses with specificity altogether. The execution of a program is modeled with a randomly sampled CPU service interval; its I/O demand is modeled as a randomly sampled service time on a randomly sampled I/O device.

Different levels of abstraction serve to answer different sorts of questions about a computer system, and different simulation tools exist for each level. Highly abstract models rely on stochastically modeled behavior to estimate high-level system performance, such as throughput (average number of "jobs" processed per unit time) and mean response time (per job). Such models may also incorporate system failure and repair, and estimate metrics such as mean time to failure and availability. Less abstract models are used to evaluate specific system components. A study of an advanced CPU design might be aimed at estimating the throughput (instructions executed per unit time); a study of a hierarchical memory system might seek to estimate the fraction of time that a sought memory reference was immediately found in the examined memory. As we have already seen, more detailed models are used to evaluate functional correctness of circuit design.

14.2 Simulation Tools

Hand-in-hand with different abstraction levels one finds different tools used to perform and evaluate simulations. We next examine different types of tools, and identify important characteristics about their function and their use.

An important characteristic of a tool is how it supports model building. In many tools one constructs networks of components whose local behavior is already known and already programmed into the tool. This is a powerful paradigm for complex model construction. At the low end of the abstraction hierarchy, electrical circuit simulators and gate-level simulators are driven by network descriptions. Likewise, at the high end of the abstraction hierarchy, tools that simulate queueing networks and Petri nets are driven by network descriptions, as are sophisticated commercial communication-system simulators that have extensive libraries of preprogrammed protocol behaviors. Some of these tools allow one to incorporate user-programmed behavior, but it appears this is not the normal usage pattern.

A very significant player in computer-system design at lower levels of abstraction is the VHDL language (see Ashenden [1996]). VHDL is the result of a U.S. effort in the 1980s to standardize the languages used to build electronic systems for the government. It has since undergone the IEEE standardization process and is widely used throughout the industry. As a language for describing digital electronic systems, VHDL serves both as a design specification and as a simulation specification. VHDL is a rich language, full of constructs specific to digital systems, as well as the normal constructs one finds in a procedural programming language. It achieves its dual role by imposing a clear separation between system topology and system behavior. Design specification is a matter of topology, whereas simulation specification is a matter of behavior. Libraries of predefined subsystems and behaviors are widely available, but the language itself very much promotes user-defined programmed behavior. VHDL is also innovative in its use of abstract interfaces (e.g., to a functional unit) to which different "architectures" at different levels of abstraction may be attached. For instance, the interface to the arithmetic-logical-unit (ALU) would be VHDL "signals" that identify the input operands, the operation to be applied to them, and the output. One could attach to this interface an architecture that in a few lines of code just performs the operation—if an addition is specified, just one VHDL statement assigns the output signal to be the sum (using the VHDL addition operator) of the two input signals. An alternative architecture could completely specify the gate-level logical design of the ALU. Models that interact with the ALU interface cannot tell how the semantics of the interface are implemented. This separation of interface and architecture supports modular construction of models, allowing one to validate a new submodel architecture by comparing the results it gives the interface with those provided by a different architecture on the same inputs. A substantive treatment of VHDL is well beyond the scope of this book. VHDL is widely used in the electrical engineering community but is hardly used outside of it.

Drawbacks to VHDL are that it is a big language, requiring a substantive VHDL compiler, and vendors typically target the commercial market at prices that exclude academic research. Of course, other simulation languages exist, and this text describes several in Chapter 4. Such languages are good for modeling certain types of computer systems at a high level, but are not designed or suited for expression of computer systems modeling at lower levels of the abstraction hierarchy. As a result, when computer scientists need to simulate specialized model behavior, they will often write a simulation (or a simulator) from scratch. For example, if a new policy for moving data between memories in a hierarchy is considered, an existing language will not have that policy preprogrammed; when a new architectural feature in a CPU is designed, the modeler will have to describe that feature and its interaction with the rest of the CPU using a general programming language. A class of tools exist that use a general programming language to express simulation-model behavior, e.g., SimPack (Fishwick [1992]), C++SIM (Little and McCue [1993]), CSIM (Schwetman [1986]), Awesime (Grunwald [1995]), SSF (Cowie et al. [1999]). This type of tool defines objects and libraries for use with languages like C, C++, or Java. Model behavior is expressed as a computer program that manipulates these predefined objects. The technique is especially powerful when used with object-oriented languages, because the tool can define base-class objects whose behavior is extended by the modeler.

Some commercial simulation languages do support interaction with general programming languages; however, simulation languages are not frequently used in the academic computer science world. Cost is a partial explanation; while commercial packages are developed with commercial needs and commercial budgets in mind, computer scientists can usually develop what they need relatively quickly, themselves. Another explanation is a matter of emphasis: simulation languages tend to include a rich number of predefined simulation objects and actions, and allow access to a programming language to express object behavior; a simulation model is expressed primarily in the constructs of the simulation language, and the model is evaluated by either compiling the model (using a simulation-language-specific compiler) and running it, or using a simulation-language-specific interpreter.

One of the many advantages to such an approach is that the relative rigidity of the programming model makes possible graphical model building, thereby raising the whole model-building endeavor up to a higher level of abstraction. Some tools have so much preprogrammed functionality that it is possible to design and run a model without writing a single line of computer code.

By contrast, programming languages with simulation constructs tend to define a few elemental simulation objects; a simulation model is expressed principally using the notions and control flow of the general programming language, with references to simulation objects interspersed. To evaluate the model one compiles or interprets the program using a compiler or interpreter associated with the general programming language, as opposed to one associated with the simulation language. The former approach supports more rapid model development in contexts where the language is tuned to the application; the latter approach supports much greater generality in the sorts of models that can be expressed.

Among tools supporting user-programmed behavior, a fundamental characteristic is the world view which is supported. In the following two subsections we look closely at a process-oriented tool called CSIM, and then at an event-oriented approach using a C++ base framework.

14.2.1 Process Orientation

A process-oriented view (see Chapter 3) implies that the tool must support separately schedulable "user threads." Threading is a fundamental concept in programming, and a discussion of its capabilities and implementation serves to highlight important issues in simulation modeling. Fundamentally, a "user thread" is a separately schedulable thread of execution control, implemented as part of a single executing process (as seen by the operating system; see Nutt [2000]). An operating system has the notion of separate processes (which might interact) which typically have their own separate and independent memory space. A group of user threads operate in the same process memory space, with each thread having allocated to it a relatively small portion of that space for its own use. The private space is used to store variables that are local to the thread, and information needed to support the suspension and resumption of that thread.

The CSIM code we discussed in Chapter 4 (Figure 4.14) illustrates some of these points. Recall that this code models a single server with exponentially distributed interarrival times and positive normal service times. A cursory glance shows the model to be legitimate C++ code. CSIM defines a number of objects in file `cpp.h`—`event` and `facility` are two such used in this model. However, both have methods (i.e., associated procedures) with specific thread-oriented semantic behavior that is beyond C++. In this model, procedures `sim` and `customer` define the execution body of threads. In this and any other user threading system, at thread creation one specifies where the thread begins to execute. In CSIM the `create` call as the first executable statement of a procedure creates a CSIM thread whose thread body is that procedure. It is certainly possible for multiple threads to exist simultaneously using the same body of code; we will see this when we step through the semantics of the CSIM model.

In a user-level threading system, a scheduling mechanism selects a thread known to be ready to execute, and gives execution control over to it. The thread itself is responsible for yielding control back to the scheduler. In process-oriented simulations, a thread is executed at a particular point in simulation time, say t, as a result of some event being scheduled (explicitly or implicitly) at time t; the thread execution is in some sense the event handler. All of the execution of the thread, up until the point it yields, is associated with time t; simulation time does not advance. The thread yields by executing a statement whose semantics mean "suspend this thread until the following condition occurs." The `hold` statement is a classic example of this, it specifies how long in simulated time the thread suspends. Other statements may cause the thread to suspend until it is specifically signaled by some other thread, or until some resource the thread is enqueued upon is granted to it. The CSIM model has examples of all of these. Procedure `sim` has a `hold` statement used to model interarrival times between customers; it also has a `wait` statement associated with CSIM event variable `done` that causes the thread to suspend until some other thread explicitly signals it through that variable. Procedure `customer` suspends through the `use` method, associated with CSIM facility variable `f`. A facility is a queue; calling `use` with a floating-point argument suspends the thread. The thread is then treated by the CSIM scheduling logic as a customer to a FCFS queue, and the passed argument to the `use` call as the customer's service time. At the simulation time corresponding to that customer finishing service, the thread is scheduled to execute again, at the statement following the call to `f.use`.

Let's now step through how the CSIM model executes. First, the model shown is compiled and linked with CSIM libraries, creating an executable which, when run, simulates the model. The `main` procedure associated with any C++ program—where program execution begins—is buried inside the CSIM kernel. Beginning execution, control enters inside the CSIM kernel where initialization is done; then the kernel calls procedure `sim`. Because the kernel is already compiled to make a call to a procedure so named, CSIM requires a modeler to declare a procedure called `sim`, and that `sim` call `create` to turn the call into a thread. This call creates the first thread. In the specific example of interest that thread generates customers, 1000 of them. It enters a loop where it first suspends for an exponentially distributed period of time (an interarrival delay), and after that delay simply creates a new customer, and returns to the top of the loop. The call to `customer` creates a new thread. Specifically, `sim` calls `customer`, which calls `create`. `create` allocates the internal data structures for the new thread, *schedules* that new thread to execute at the current simulation time, but before executing the thread further, control is returned to `sim`. Note that these procedure call semantics are different from C++'s, and must be understood by a CSIM modeler in order to correctly read and write models. On receiving back control, `sim` immediately suspends through a `hold` call, and the scheduler selects the thread whose invocation time is least among all executable threads. Almost always this will be the `customer` thread just created. This thread begins execution at the call to `use`, which suspends the thread until all customers generated before it have finished their executions, and then an implicit hold is performed on behalf of the thread, modeling its service delay.

When these queueing and service delays have passed, the thread resumes execution. It increments a counter of the total number of customer threads that have received their service, and if that counter is equal to the total number of customers, applies a "signal" to event variable `done`. In any case, following this test and possible signaling, the thread destroys itself by leaving the execution body of the `customer` procedure. Exiting a thread procedure body is another place where execution semantics differ from C++. It is important to observe that at any given instant of the simulation there may be multiple instances of the `customer` thread in existence, one for each customer in queue at that instant. After the main thread has generated all customers, it suspends on event variable `done`. Eventually the last customer created receives its simulated service, recognizes it is last, and signals the main thread. `sim` resumes execution, and calls a CSIM routine to print statistics about the simulation run. At several points the runtime semantics of the CSIM model differ from single-thread-of-control C++, owing to threading. The implementation of these semantics is very relevant to simulation modelers, as the implementation has a dramatic impact on the execution cost. To understand these costs we next consider how threads are typically implemented.

The CSIM example illustrates that threads access variables that are globally visible to all threads, and variables that are local to a specific thread. Variables `TotalCustomers` and `NumberOfCustomers` are examples of global variables; local variables `i` (in `sim`) and `service` (in `customer`) are examples of local variables. Every instance of `customer` has its own copy of `service`. Local variables, the location of where a thread resumes execution, and the contents of machine registers at the time a thread suspended are all part of the thread *state*. To start a thread's execution, the scheduler must restore all of a thread's state to the appropriate places so that, as the thread executes, its execution environment is not perceptibly different from what it would be if it had not suspended. There are different ways to store and restore a thread's state. The most common involve the program's runtime stack (see Hennessy and Patterson [1994]). When a program makes a procedure call, the runtime

```
void procA(int a) {
create("thread");
  int x;
  procB(x=2*a);
  procB(2*x);
}

void procB(int b) {
 int x;
 procC(x=2*b);
 procC(2*x);
}

void procC(int c) {
 hold(c);
}
```

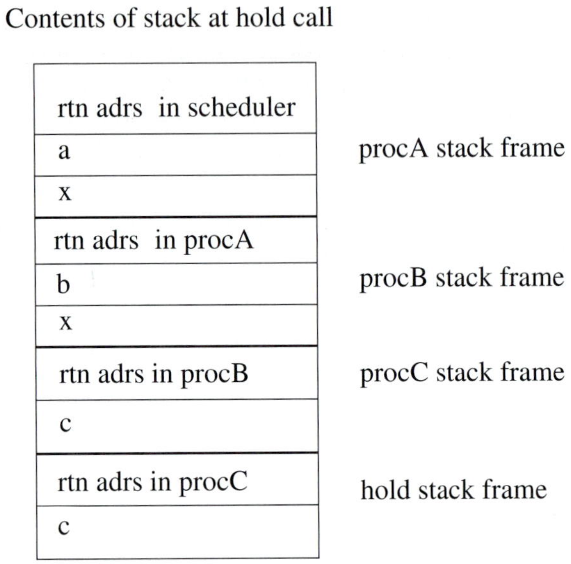

Contents of stack at hold call

rtn adrs in scheduler	procA stack frame
a	
x	
rtn adrs in procA	procB stack frame
b	
x	
rtn adrs in procB	procC stack frame
c	
rtn adrs in procC	hold stack frame
c	

Figure 14.2. Runtime stack in threaded simulation.

system pushes a new "stack frame" onto the stack. A stack frame contains the return address of the calling routine, copies of input arguments to the procedure, and space for local variables used by the called procedure. A machine register called the stack-frame pointer contains the address of the first byte of the frame; machine code that references local variables does so using addressing modes that specify an offset from the stack-frame pointer and cause the hardware to compute the actual address by adding the offset to the contents of the stack-frame pointer. When the called procedure returns control to the caller, the runtime system pops its frame off the stack.

Figure 14.2 illustrates a sequence of procedures and the stack of frames present on the first call to `hold`. The call to `hold` suspends the thread, and when the thread executes once again it will require all the information shown on the stack. Consequently the threading mechanism has to save this portion of the stack somehow, and restore it just before the thread is executed again. CSIM's approach is to have all threads use the same stack space as they execute, with the frame for the first procedure in the thread body always starting at the same memory location. When the thread suspends, the CSIM kernel copies the used stack into separate storage associated with the thread. To restore a thread, the CSIM kernel first copies the previously stored stack into the fixed stack position, then lets the normal runtime control take over. The CSIM kernel also saves and restores machine registers. An alternative approach to thread stack management avoids the copying costs by preallocating memory for the stack of each different thread. Machine registers must still be saved and restored, but a switch between threads is faster because only the stack pointer must be changed, not the stack itself. The key limitation of such an approach is that one must declare ahead of time the maximum stack size a thread may require, and there is always the danger of the thread's pushing the stack beyond its allocation. CSIM's approach admits more generality.

Process-oriented simulators are known to be slower than event-oriented simulators, stack management being a key contributing factor. Depending on the threading implementation, there may be additional factors. CSIM's threads are tailored to simulation, and the internal scheduling mechanisms are purely temporal ones. However, threading is a concept that extends well beyond simulation, and threads exist in more general forms. The Java language has threads built in (e.g., Grand [1997]); for C and C++ there is a Posix standard for thread libraries (see Nichols et al. [1996]); and a Microsoft Windows NT definition exists, to name but a few. A simulator built using such threads has to build its temporal scheduling policy using only the scheduling interface provided by the language or thread library. Any such system will make a distinction between runnable and blocked threads. The thread scheduler will select the next thread to run from among all threads in the pool of runnable threads, possibly based on priorities.

The problem for a simulator is that *it* wants to select the next thread to run, based on simulation time. A finesse then is to arrange things so that at any time there is exactly one thread in the runnable pool, and to have the simulation scheduler be in control of unblocking the thread it next wishes to run. Most thread systems have ways of explicitly making a blocked thread runnable, and of having a running thread block itself. The mechanisms are not so important as the idea. The simulation scheduler can itself be a thread. When it runs, it identifies the next thread to run, as a function of an ordinary event-list (that identifies associated threads), puts that thread in the runnable set, and blocks itself. The one thread that can now be run does, and as a part of its suspension sequence causes the scheduler to be runnable, and blocks itself.

There is a certain attraction to using a standard language or package to implement threads, but we see that there is an additional cost of implementing the simulation scheduler in a somewhat indirect way. Add that cost to the overhead of a thread context switch, realize that all this work is expended on the execution of each and every event in a process-oriented simulation, and so come to understand why process-view simulators are slower than event-view simulators.

14.2.2 Event Orientation

An event-oriented simulator is much simpler to design and implement using a general programming language. The semantics of the program expressing a model are the semantics of the programming language; there is no hidden activity akin to stack management or thread scheduling. It is natural to use object-oriented techniques and express a simulation in terms of messages (events) passed between simulation objects. Languages like C++ and Java are outstanding for developing abstract base classes to express the general structure of the simulator. In such an approach model building entails development of concrete classes derived from the base classes, which includes expression of their methods. These methods express the model's event handlers, and so explicitly include scheduling of future events. To illustrate these ideas, we work through a simple example in C++.

For our example we suppose a simulation system has an abstract base class for all simulation objects, and for all events communicated between such objects. Figure 14.3 displays a simple example of such base classes, in C++. Without worrying over much about C++ syntax, the sense of the declarations can be made out. A class contains declarations of data and functions (known as "methods"). The class definition identifies all relevant types (for data, for return values, for arguments). In this system simulation time will be represented by abstract type `stime`. This avoids the issue of whether the time-stamp ought to be in floating-point or integer format. The `SimEvt` class defines an interface for all event data structures a user would define. The `SimEvt` base class contains a "private" data member that stores the time-stamp of a delivered event, and contains a "public" method that allows any code to see what that time-stamp is. Class `SimScheduler` is declared to be a "friend class," which means that its methods have the same access privileges to an event's data as does the `SimEvt` class. The last bit of mystery is the functions `~SimEvt` and `~SimObj` prefaced by the keyword `virtual`. To appreciate these and the power of the object-oriented approach, we must first talk about derivation. A modeler may tailor to the application, by definition, new event classes that include the entire interface of the `SimEvt` class. `PktEvt` is an example of this. The user wishes for the event to carry a data packet of general length, and so endows the class with the infrastructure necessary to store and access that packet. The user creates a "derived" instance of the class by calling the `PktEvt` constructor, which initializes its specialized data structures. The critical feature is that a pointer to an `PktEvt` instance can be passed to any routine with a type declaration of `SimEvt`, and that routine can call the `SimEvt` methods and deal with the event in those terms. As we will see, this is exactly what happens when a simulation object schedules an event.

Class `SimScheduler` gives the interface for a scheduler object. Normally there would be one of these in the simulation.

Class `SimObj` is the abstract base class for a simulation object. Its interface specifies a "virtual" method `acceptEvt`; any schedulable object the modeler may wish to define must include this method as part of its interface. `acceptEvt` is called by the simulation scheduler to pass an event to the object. Different types of objects define different handlers—an event delivered to a CPU object is dealt with differently from one delivered to a Memory object. The action where the object's `acceptEvt` method is called is the simulation of the event; a pointer to the event that causes the

```
class SimEvt {
  friend bool operator<(const SimEvt &e1, const SimEvt &e2);
  friend bool operator==(const SimEvt &e1, const SimEvt &e2);
  friend class SimScheduler;   // scheduler has access
 private;
   stime _time;
 public:
  SimEvt();                  // default constructor
  virtual ~SimEvt();         // destructor is virtual
  void set_time(stime when) { _time = when; }
  stime get_time() { return _time; }
};

class SimObj {
  protected:
  public:
    SimObj();
    virtual ~SimObj();
    virtual void acceptEvt(SimEvt *)=0;
};

class SimScheduler {
 public:
   SimScheduler();
   int ScheduleEvt(SimObj *, SimEvt *, stime delay);
   stime now();
};

// User defined derived class
class PktEvt : public SimEvt {
  private:
   int pkt_length;
   void *packet;
  public:
   PktEvt();
   PktEvt(int len, void* pkt )  {
       packet = (void *)new char[len];
       memcpy(packet, pkt, len);
       pkt_length = len;
   }
   ~PktEvt() { delete packet; }
   void* get_pkt()      { return pkt; }
   int   get_pkt_length { return pkt_length; }
};
```

Figure 14.3 Sample classes for simulator.

execution is passed as a parameter in that call. This event was scheduled in the first place by a call to the `ScheduleEvt` method associated with an instance of the `SimScheduler` class. The code scheduling the event has first created the event, acquired a pointer to the `SimObj` instance to receive the event, and decided on the delay into the future when the event ought to be received and acted upon.

We illustrate these points with the code fragment below.

```
class Packet {
  public:
    int _src, _dest, _data;
    Packet(int src, dest, data) : _src(src),
      _dest(dest), _data(data) {}
};
class Node : public SimObj {
  void acceptEvt(SimObj *simob);
   ...
};
SimScheduler *sched;
Node *node;
Packet p(12,15,96);
   ...
sched->ScheduleEvt( node,
  new PktEvt ( (int)sizeof(Packet), (void *)&p), 1);
```

Here we identify the existence of a class `Node` derived from the `SimObj` base class, and a class `Packet` that defines the payload of a `PktEvt` event. The definition of local variable `p` creates a `Packet` with specified fields. An event that delivers a copy of `p` one time unit into the future is to be scheduled. The first two arguments of method `ScheduleEvt` are declared to be a pointer to a `SimObj`, and a pointer to a `SimEvt`, respectively. However, in the code fragment we pass a pointer to a `Node`, and a pointer to `Packet`. C++ accepts this, because its interpretation is that if class A derives from class B, then A "is-a" B. In practice this means that wherever a pointer to a `SimObj` is called for, we can supply a pointer to an object whose class ultimately derives from the `SimObj` class. The question is, if the simulation system thinks it is being given a pointer to a `SimObj` when in fact it is given a pointer to a `Node`, how will it happen that the `Node` version of `acceptEvt` is called when the event happens? This is where the keyword "virtual" plays a role. The C++ compiler knows to "tag" a `SimObj` instance with information that both identifies what the derived type is and allows it to find a pointer to the corresponding `acceptEvt` (through a lookup table). So, in code unseen by the modeler, a call like

```
simObj *objptr;
simEvt *evtptr;
...
objptr->acceptEvt( evtptr );
```

is implemented by looking up the class of the the object pointed to by `objptr`, and finding the location of that class's implementation of `acceptEvt`. Now `Node`'s version of `acceptEvt` must be declared to accept a pointer to a `SimEvt`, even though it knows (by convention) that what is pointed to is a `PktEvt`. It deals with this by a "cast," i.e.

```
void Node::acceptEvt( SimEvt *simevt ) {
  if( PktEvt* mysimevt  = dynamic_cast<PktEvt *>simevt ) {
   // fetch presumed packet, cast through void* declaration
   Packet* p = (Packet *)(mysimevt->get_pkt())
   int src  = p->_src;
   int dest = p->_dest;
   ...
  } else {
    // report an error
    ...
 }
}
```

It should be remembered that C++ cannot guarantee that `simevt` points to an instance of `PktEvt`—it is the responsibility of the modeler to check that all calls to `ScheduleEvt` which pass a pointer to a `Node` object also pass a pointer to a `PktEvt` object; this can also be checked at runtime. The "dynamic cast" operation in this code snippet returns a pointer if `simevt` does point to a `PktEvt`, and zero otherwise.

This example illustrates the power of object-oriented modeling. The object pointers are formally to base classes, not the specific classes created by the modeler, yet the right things happen. In practical terms this means that one can build the simulation kernel in terms of the base class types, but have it put events on an event-list and have them delivered to the specified objects at the specified times without its ever knowing precisely the type of the events or the objects! To make this all work, a methodology must be defined and observed by the modeler for interpreting the types of the events passed between objects. For example, one might insist that simulation objects and events to be delivered to them be codesigned; if one defines `NewSimObj`, then one also defines `NewSimObjEvt` and insists that any event scheduled for a `NewSimObj` instance be of type `NewSimObjEvt`. Adherence to this protocol means that the `acceptEvt` method for `NewSimObj` can safely treat the pointer it is passed as one to a `NewSimObjEvt` event, and so extract whatever specialized information is defined there.

A last bit of explanation is in order. Class `SimObj` contains method `AcceptEvt`, which is declared to be virtual, and whose declaration ends in the cryptic designation "= 0." This syntax means that nowhere will the base class actually define its own version of `AcceptEvt`, and that any instantiated class that derives from `SimObj` *must* provide this method (or have an ancestor closer to `SimObj` that does). Finally, the destructor methods `~SimObj` and `~SimEvt` are declared to be virtual. Lacking this, if `delete` is applied to a pointer of the base class, then only the base class destructor will be called. This could cause a memory leak. For instance, `PktEvt` allocates dynamic memory to store the packet when it is constructed. The destructor must be called to release that memory. Making the destructor virtual ensures that it will.

Table 14.1. Decreasing Abstraction and Model Results

Typical System	*Model Results*	*Tools*
CPU network	Job throughput, Job response time	Queueing network, Petri net simulators, scratch
Processor	Instruction throughput, Time/instruction	VHDL, scratch
Memory system	Miss rates, response time	VHDL, scratch
ALU	Timing, correctness	VHDL, scratch
Logic network	Timing, correctness	VHDL, scratch

To conclude this discussion on tools, we remark that *flexibility* is the key requirement in computer-systems simulation. Flexibility in most contexts means the ability to use the full power of a general programming language. This requires a level of programming expertise that is not needed for the use of commercial graphically oriented modeling packages. The implementation requirements of an object-oriented event-oriented approach are much less delicate than for a threaded simulator, and the amount of simulator overhead involved in delivering an event to an object is considerably less than the cost of a context switch in a threaded system. For these reasons, most of the simulators written from scratch take the event-oriented view. The choice of using a process-oriented or event-oriented simulator—or writing one's own—is a function of the level of modeling ease, versus execution speed. It also happens that the process-oriented view is more naturally employed at the higher levels of the abstraction hierarchy.

To summarize this section we present a table (Table 14.1) that lists different levels of abstraction in computer-systems simulation, the sorts of questions whose answers are sought from the models, and the sorts of tools typically used for modeling. The level of abstraction decreases as one descends through the table.

14.3 Model Input

Just as there are different levels of abstraction in computer-systems simulation, there are different means of providing input to a model. The model might be driven by stochastically generated input, using either simple or complicated operational assumptions, or it might be given trace input, measured from actual systems. Simulations at the high end of the abstraction hierarchy most typically use stochastic input, while simulations at lower levels commonly employ trace input. Stochastic input models are particularly useful when one wishes to study system behavior over a range of scenarios; all that may be required is to adjust an input model parameter and rerun the simulation. Of course, using randomly generated input raises the question of how real or representative the input is, which is why systems people frequently prefer trace data on lower-level simulations. While using a trace means that one cannot explore different input scenarios, traces are useful when directly comparing two different implementations of some policy or some mechanism on the same input. The realism of the input gives the simulation added authority.

In all cases, the data used to drive the simulation is intended to exercise whatever facet of the computer system is of interest. High-level systems simulations accept a stream of job descriptions; CPU simulations accept a stream of instruction descriptions, memory simulations accept a stream of memory references; and gate-level simulations accept a stream of logical signals.

Computer systems modeled as queueing networks (recall Chapter 6) typically interpret "customers" as computer programs; servers typically represent services such as attention by the CPU or an input/output system. Random sampling generates customer interarrival times, and it may also be used to govern routing and time-in-service. However, it is common in computer systems contexts to have routing and service times be state dependent; e.g., the next server visited is already specified in the customer's description, or may be the attached server with least queue length.

Interarrival processes have historically been modeled using Poisson processes (where times between successive arrivals have an exponential distribution). However, this assumption has fallen from favor as a result of empirical observations that significantly contradict Poisson assumptions in current computer and communication systems. The real value of Poisson assumptions lies in their tractability for mathematical analysis, and so as simulationists we can discard them with little loss.

In the subsections to follow we look at the mathematical formulation of common input models, stochastic input models for virtual memory, and direct-execution techniques.

14.3.1 Modulated Poisson Process

Stochastic input models ought to reflect the real-life phenomena of burstiness — that is, brief periods when traffic intensity is much higher than normal. An input model that is sometimes used to support this but which retains a level of tractability is a *modulated Poisson process*, or MPP (see Fischer and Meier-Hellstern [1993]). The underlying framework is a continuous-time Markov chain (CTMC), whose details we sketch so as to employ the concept later. A CTMC is always in some *state*; for descriptive purposes states are named by the integers: $1, 2, \ldots$. The CTMC remains in a state for a random period of time, transitions randomly to another state, stays there for a random period of time, transitions again, and so on. The CTMC behavior is completely determined by its *generator matrix*, $Q = \{q_{i,j}\}$. For states $i \neq j$, entry $q_{i,j}$ describes the rate at which the chain transitions from state i into state j (this is the total transition rate out of state i, times the probability that it transitions then into state j). The rate describes how quickly the transition is made; its units are transitions per unit simulation time. Diagonal element $q_{i,i}$ is the negated sum of all rates out of state i: $q_{i,i} = -\sum_{j \neq i} q_{i,j}$. An operational view of the CTMC is that upon entering a state i, it remains in that state for an exponentially distributed period of time, the exponential having rate $-q_{i,i}$. Making the transition, it chooses state j with probability $-q_{i,j}/q_{i,i}$. Many CTMCs are *ergodic*, meaning that, left to run forever, every state is visited infinitely often. In an ergodic chain π_i denotes state i's *stationary probability*, which we can interpret as the long-term average fraction of time the CTMC is in state i. A critical relationship exists between stationary probabilities and transition rates: for every state i,

$$\pi_i \sum_{j \neq i} q_{i,j} = \sum_{j \neq i} \pi_j q_{j,i}$$

If we think of $q_{i,j}$ as describing a probability "flow" that is enabled when the CTMC is in state i, then these equations say that in the long term, the sum of all flows out of state i is the same as the sum of all flows into the state. We will see in the example below that we can use the balance equations to build a stochastic input with desired characteristics. To complete the definition of an MPP, it remains only to associate a customer arrival rate λ_i with state i. When in state i, customers are generated as a Poisson process with rate λ_i.

To illustrate, let us consider an input process that is either OFF, ON, or BURSTY. We wish for the process to be OFF one-half of the time, on average for 1 second, and when not OFF, we wish for it to be BURSTY for 10% of the time. We will assume that the CTMC transitions into BURSTY only from the ON state, and transitions out of BURSTY only into the ON state. We will say that state 0 corresponds to OFF, 1 to ON, and 2 to BURSTY. Our problem statement implies that $\pi_0 = 0.5$, $\pi_1 = 0.45$, and $\pi_2 = 0.05$. Since the only transition from OFF is to ON, and the mean OFF time is 1, we infer that $q_{0,1} = 1$. The balance equation for state 0 can be rewritten as

$$0.5 = 0.45 q_{1,0}$$

so that $q_{1,0} = (0.5/0.45)$. The balance equation for state 1 can be rewritten as

$$0.45((0.5/0.45) + q_{1,2}) = 0.5 + 0.05 q_{2,1}$$

and the balance equation for state 2 is

$$0.05 q_{2,1} = 0.45 q_{1,2}$$

The equations for states 1 and 2 are identical; mathematically we don't have enough conditions to force a unique solution. If we add the constraint that a BURSTY period lasts on average 1/10 of a second, we thus define $q_{2,1} = 10$ and hence $q_{1,2} = (0.5/0.45)$. Operationally, the simulation of this CTMC is simple. In state 0 one samples an exponential with mean 1 to determine the state's holding time. Following this period, the CTMC transitions into state 1, and samples a holding time from an exponential with mean 0.45, after which it transitions to OFF or BURSTY with equal probability. In the BURSTY state it samples an exponential holding time with mean 0.1. Now all that remains is for us to define the state-dependent customer arrival rates. Obviously $\lambda_0 = 0$; for illustration we choose $\lambda_1 = 10$ and $\lambda_2 = 500$.

Figure 14.4 illustrates a snippet of code used to generate times of arrivals in this process. Transitions between states are sampled using the inverse transform technique, described in Chapter 8 (the variable `acc` accumulates the cumulative probability function in the distribution described by the row vector `P[state]`).

```
double Finish;            // sim termination
double time = 0.0;        // current clock
double htime, etime;      // transition times
double **P;               // trans.  prob.  matrix
int state = 0;            // current state id
int total = 0;            // total pkts emitted
 ...
while( time < Finish ) {
 // generate exponential holding time, state-dependent mean
 htime = time+expntl( hold[state] );
 // emit packets until state transition time.  State-dependent
 // rate.  Note assignment made to etime in while condition test
 while( (etime = time+expntl( 1.0/rate[state])
                  < min( htime, Finish) ) ) {
       cout << etime << '' '' << total << endl;
    total++;
    time = etime;   // advance to packet issue time
 }
 time = htime;
 // select next state
 double trans = u01();
 double acc = P[state][0];
 int i = 0;
 while( acc < trans ) acc += P[state][++i];
 state = i;
 }
```

Figure 14.4 C++ code generating MPP trace.

Figure 14.5 plots total customers generated as a function of time for a short period of a sample run, and for a longer period. In the shorter run we see regions where the graph increases sharply; this corresponds to periods in the BURSTY state. While not in this state, a mixture of OFF and ON periods moves the accumulated packet count up at a much more gradual rate. The MPP model can clearly describe burstiness, but the burstiness is limited in time scale. The longer run views the data at a time scale that is two orders of magnitude larger, and we see that the irregularities are largely smoothed.

To contrast the MPP model, consider a traffic source that remains OFF for an exponentially distributed period of time with mean 1.0, but when it comes ON it remains on for a period of time sampled from a Pareto distribution. While ON, packets arrive as a Poisson process. The cumulative distribution function

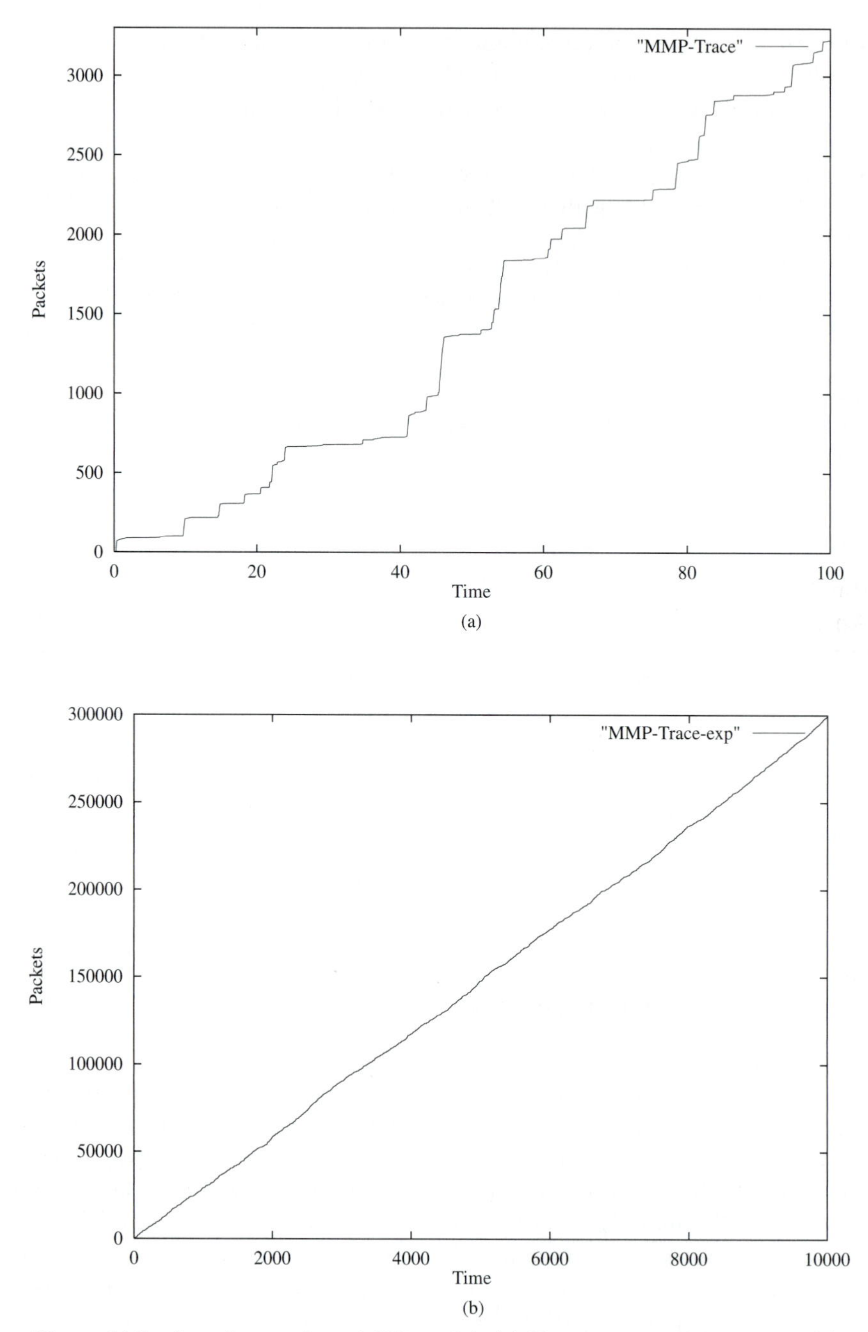

Figure 14.5. Sample runs from MPP model. (a) Short run, small time scale. (b) Long run, large time scale.

for a Pareto with parameters k and λ is

$$F(t) = 1 - \left(\frac{\kappa}{t}\right)^{\alpha}$$

for $0 < \kappa \leq t$ and $0 < \alpha$. The Pareto distribution is of particular interest in network modeling because it gives rise to "self-similarity," which informally means preservation of irregularities at multiple time scales. Figure 14.6 parallels the MPP data, illustrating accumulated packet counts as a function of time, illustrating behavior for the first 1000 units of time and for the first 100,000 units of time. Here, despite two orders of magnitude difference in run length, the visual impression of behavior is much the same between the two traces. This sort of behavior is frequently seen in computer and communication systems; the long lengths have been seen to reflect things such as burstiness of packets, file lengths, and demand on a server.

14.3.2 Virtual Memory Referencing

Randomness can also be used to drive models in the middle levels of abstraction. An example is a model of program execution behavior in a computer with virtual memory (see Nutt [2000]). In such a system, the data and instructions used by the program are organized in units called *pages*. All pages are the same size, typically 2^{10} to 2^{12} bytes in size. The physical memory of a computer is divided into *page frames*, each capable of holding exactly one page. The decision of which page to map to which frame is made by the operating system. As the program executes, it makes memory references to the "virtual memory," as though it occupied a very large memory starting at address 0, and is the only occupant of the memory. On every memory reference made by the program, the hardware looks up the identity of the page frame containing the reference, and translates the virtual address into a physical address. The hardware may well discover that the referenced page is not present in the main memory; this is called a *page fault*. When a page fault occurs, the hardware alerts the operating system, which then takes over to bring in the referenced page from a disk and decides which page frame should contain it. The operating system may need to evict a page from a page frame to make room for the new one. The policy the operating system uses to decide which page to evict is called the "replacement policy." The quality of a replacement policy is often measured in terms of the fraction of references made whose page frames are found immediately, the *hit ratio*.

Virtual memory systems are used in computers that support concurrent execution of multiple programs. In order to study different replacement policies one could simulate the memory-referencing behavior of several different programs, simulate the replacement policy, and count the number of references that page-fault. For this simulation to be meaningful it is necessary that the stochastically generated references capture essential characteristics of program behavior. Virtual memory works well precisely because programs do

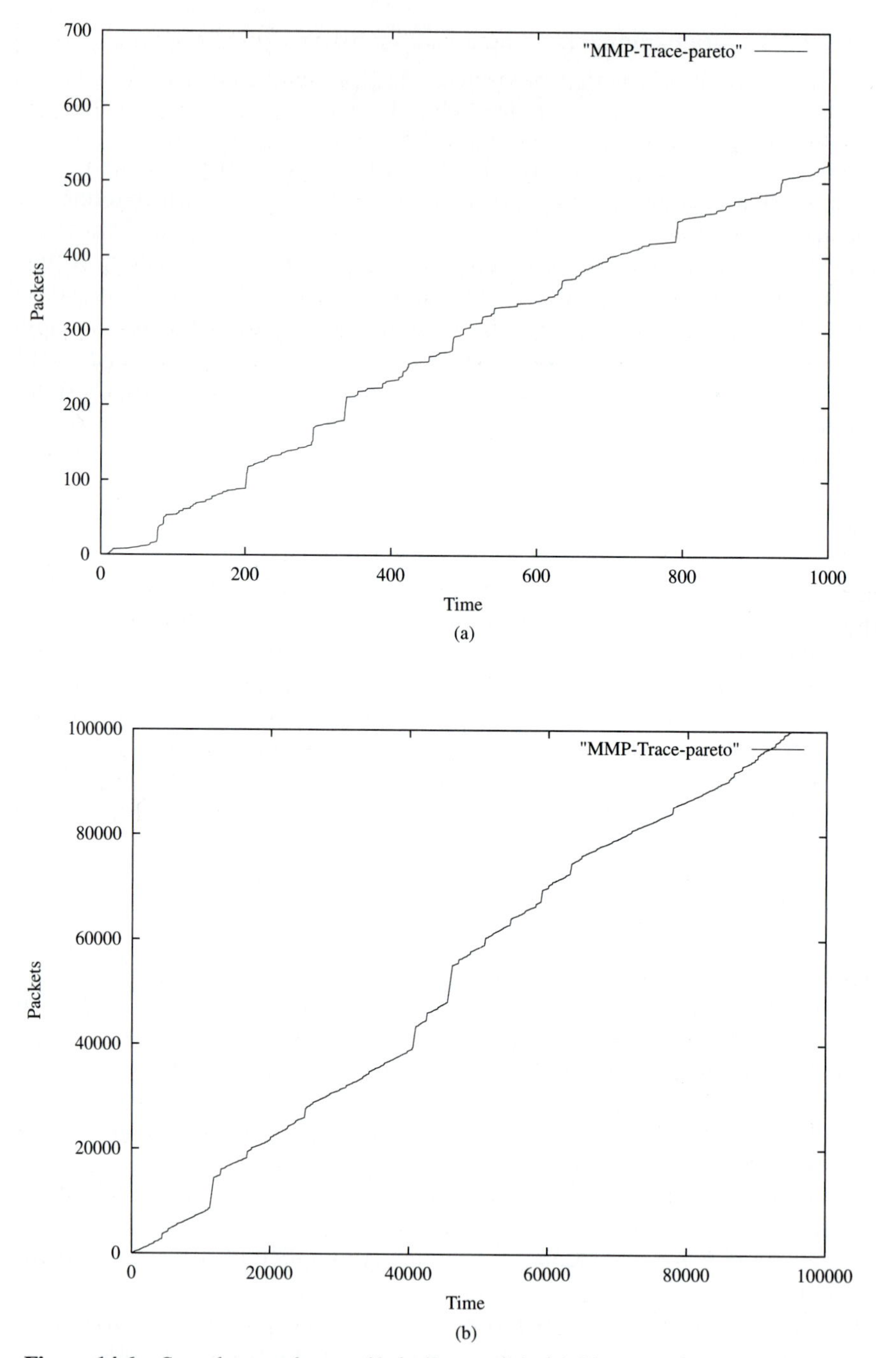

Figure 14.6. Sample runs from self-similar model. (a) Short run, small time scale. (b) Long run, large time scale.

tend to exhibit a certain type of behavior, so-called locality of reference. What this means intuitively is that program references tend to cluster in time and space, that when a reference to a new page is made and the page is brought in from the disk, it is likely that the other data or instructions on the page will also soon be referenced. In this way the overhead of bringing in the page is amortized over all the references made to that page before it is eventually evicted. A program's referencing behavior can usually be separated into a sequence of "phases," where during each phase the program makes references to a relatively small collection of pages called its *working set.* Phase transitions essentially change the program's working set. The challenge for the operating system is to recognize when the pages used by a program are no longer in its working set, for these are the pages it can safely evict to make room for pages that *are* in some program's working set.

Figure 14.7 illustrates a stream of memory references taken from a execution of the commonly used `gcc` compiler. One graph gives a global picture, the other cuts out references to pages over number 100 and shows more fine detail. Each graph depicts points of the form (i, p_i), where p_i is the page number of the ith reference made by the program (arithmetically shifted so that the smallest page number referenced is 10). The phases are clearly seen; each member of the working set of a phase is seen clearly as lines (which are really just a concatenation of many points). One striking facet of this graph is how certain pages remain in almost all working sets. However, other kinds of programs exhibit other behaviors. A common characteristic of scientific programs is that the execution is dominated by an inner loop that sweeps over arrays of data; the pages containing the instructions are in the working set throughout the loop, but data pages migrate in and out.

Despite various differences, a near-invariant among program executions is phaselike behavior, and working sets. When building a stochastic reference generator it makes sense therefore to focus modeling effort on phase and working-set definition. As a simple starting point we might, with every reference generated, randomly choose (with some small probability) whether to start a new phase, by changing the working set. Given a working set, we would choose to reference some page in the working set with high probability, and if choosing to stay in the set, choose the same page as last referenced in the working set, with high probability. The inner loop of a program that generates references in this manner appears in Figure 14.8. Details of working-set definition are hidden inside of routine `new_wrkset`, and might vary with the type of program being modeled. For the purposes of illustration here we wrote a version that defined a working set by randomly choosing a working-set size between 2 and 8, and a maximum page number of 100. A working set of size n is constructed by randomly choosing a "center" page c from among all pages, randomly choosing an integer dispersion factor d from 2 to 6, and then randomly selecting a working set from among all pages within distance $d \times n$ from center page c (with appropriate wraparound of page numbers at the endpoints 0 and 100). In order to model the referencing pattern of a scientific program's

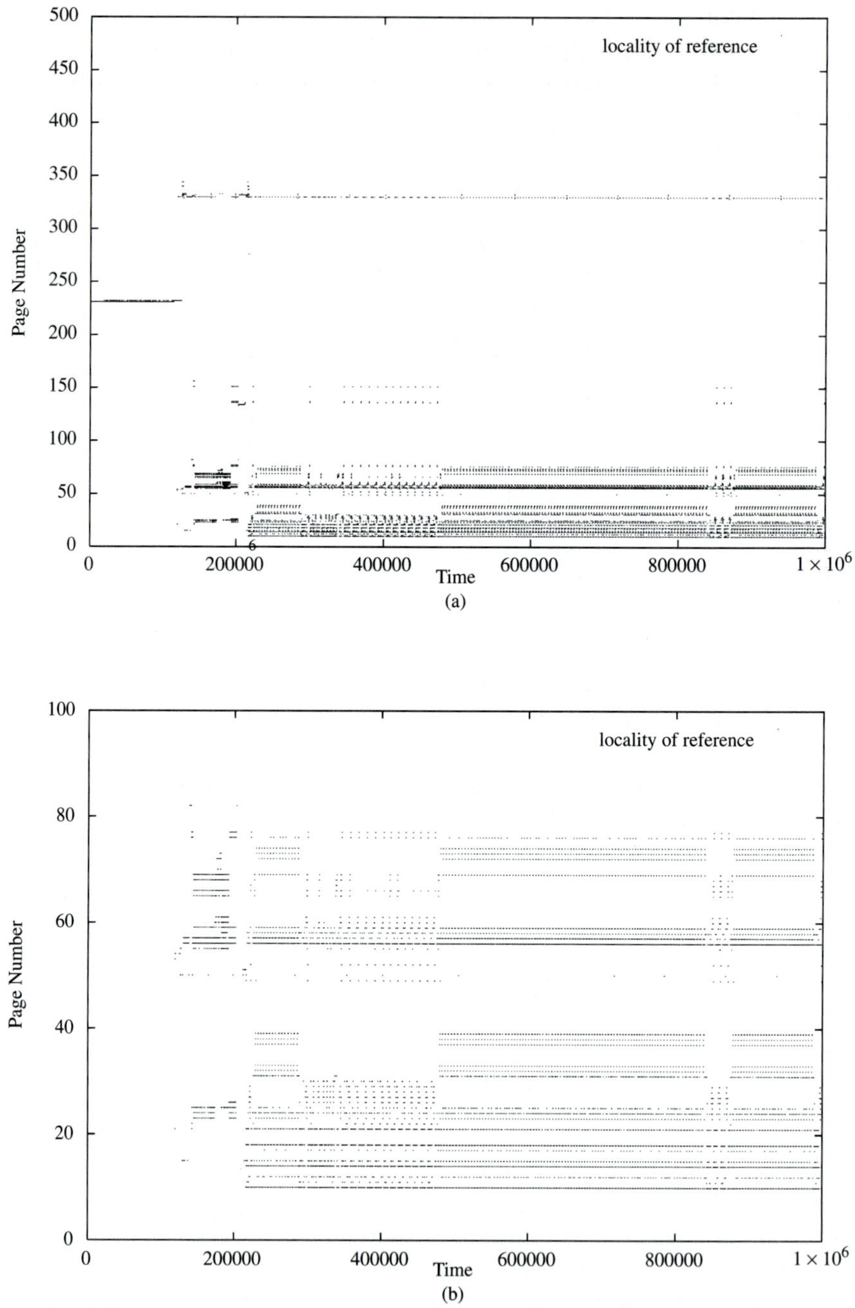

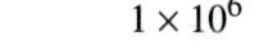

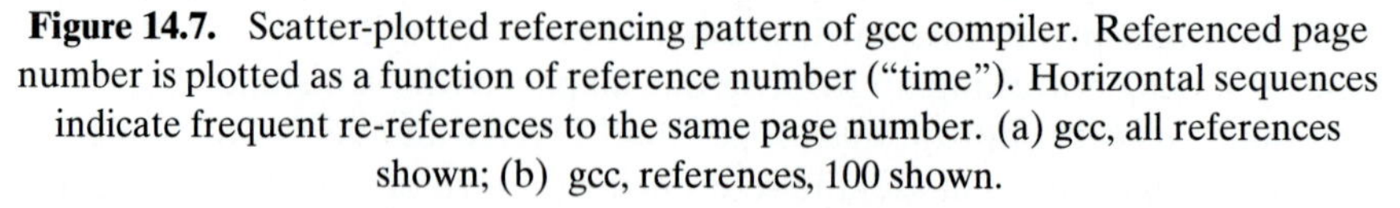

Figure 14.7. Scatter-plotted referencing pattern of gcc compiler. Referenced page number is plotted as a function of reference number ("time"). Horizontal sequences indicate frequent re-references to the same page number. (a) gcc, all references shown; (b) gcc, references, 100 shown.

instruction stream we manipulated the logic illustrated above to "lock down" a working set for a long time in the middle of the program execution. Figure 14.9 illustrates the result. As designed, phases and working sets are clearly defined.

```
double ppt = 0.0001;  // Pr{phase transition}
double psw = 0.999;   // Pr{ref in WS}
double psp = 0.9;     // Pr{reference same page}

extern void new_wrkset();  // create new WS
extern int  from_wrkset(); // sample from WS
extern int  not_from_wrkset(); // sample outside WS

int ref;              // last page referenced
int sv_ref;           // save ref

extern double u01();  // uniform (0,1) RNG

for(int i=0; i<length; i++) {
   if( u01() <  ppt ) new_wrkset();  // phase transition
   if( u01() < psw ) {               // stay in working set?
    if( psp < u01() )                // change page, in wrkset
       ref = sv_ref = from_wrkset();
   } else ref = not_from_wrkset();   // step outside of wrkset
 cout << i << '' '' << ref << endl;
 ref = sv_ref;
 }
```

Figure 14.8 Pseudocode for generating reference trace.

While the example above illustrates how one can in principle stochastically generate an execution path, simulations at the middle level of abstraction also commonly use traces. Studies of CPU design will use a measured trace of instructions executed by a running program, while studies of memory systems will use a measured trace of the addresses referenced by an executing program.

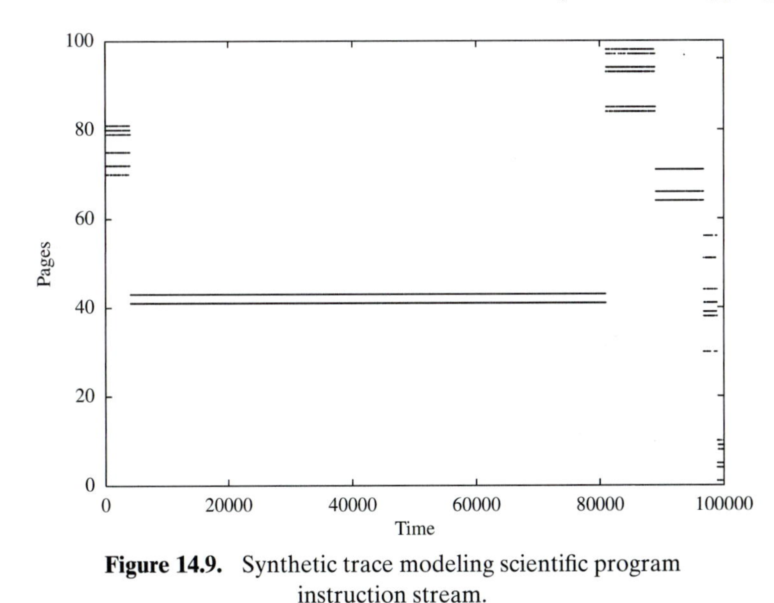

Figure 14.9. Synthetic trace modeling scientific program instruction stream.

Considering that a modern CPU can execute tens to hundreds of millions of instructions each second, such traces get to be lengthy. A small piece of a typical trace of memory references is shown below.

```
2     430d70
2     430d74
2     415130
0   1000acac
2     414134
1   7fff00ac
2     414138
```

The first number is a code describing the type of access; 2 represents an instruction fetch, 0 a data read, 1 a data write. The second number represents a memory address, in hexadecimal. If the trace were also to describe the instruction stream, a hexadecimal word giving the machine code of the instruction fetched could follow the memory address on every instruction fetch line. Two or three words of memory are needed to represent one reference, even when the information is efficiently packed (not as characters, as shown, which take much more space!). Consider also the amount of computation needed to simulate a CPU or memory for the execution of a significantly long run of a nontrivial program. These observations help us understand the motivation for techniques that compress the address trace, and for techniques that allow one to infer information about multiple systems from a single pass through a long trace. We will say more about these techniques later in this chapter.

Another method of generating input is called "direct-execution" simulation (for examples, see Covington et al. [1991], Lebeck and Wood [1997], and Dickens et al. [1996]), one approach to which is illustrated in Figure 14.10. Direct-execution is like generating a trace, and driving the simulation with that trace, all at once. Computer programs are "instrumented" with additional code that observes the instructions the program executes and the memory and I/O references the program makes as it executes. The instrumented program is compiled and linked with a simulation kernel library. Execution control rests with the simulation kernel, which calls the instrumented program to provide the next instruction or reference that the program generates. The simulation kernel uses the returned information to drive the model for the next step. The simulation model driven by the program's execution can be of an entirely different CPU design, or a memory system, or even (given multiple instrumented programs) the internals of a communications network. Direct-execution simulation solves the problem of storing very large traces—the trace is consumed as it is being generated. However, it is tricky to modify computer programs to get at the trace information and to coordinate the trace generator with the discrete-event simulator. The only practical way an ordinary simulation practitioner can use such methods is when the system has a software tool for making such modifications, and this is not common.

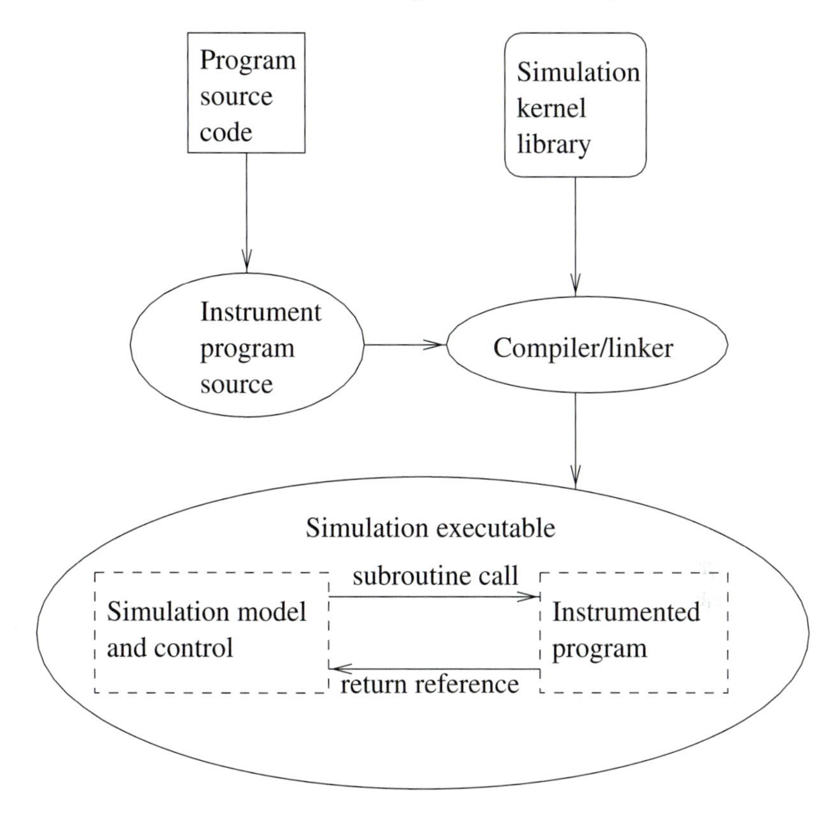

Figure 14.10. Direct-execution simulation.

14.4 High-Level Computer-System Simulation

In this section we illustrate concepts typical of high-level computer simulations, by sketching a simulation model of a computer system that services requests from the World Wide Web.

Example 14.1

A company that provides a major web site for searching and links to sites for travel, commerce, entertainment, and the like wishes to conduct a capacity planning study. The overall architecture of their system is shown in Figure 14.11. At the back end one finds data servers responsible for all aspects of handling specific queries and updating databases. Data servers receive requests for service from application servers—machines dedicated to running specific applications (e.g., a search engine) supported by the site. In front of applications are web servers that manage the interaction of applications with the World Wide Web, and the portal to the whole system is a load-balancing router that distributes requests directed to the web site among web servers.

The goal of the study is to determine the site's ability to handle load at peak periods. The desired output is an empirical distribution of the access response

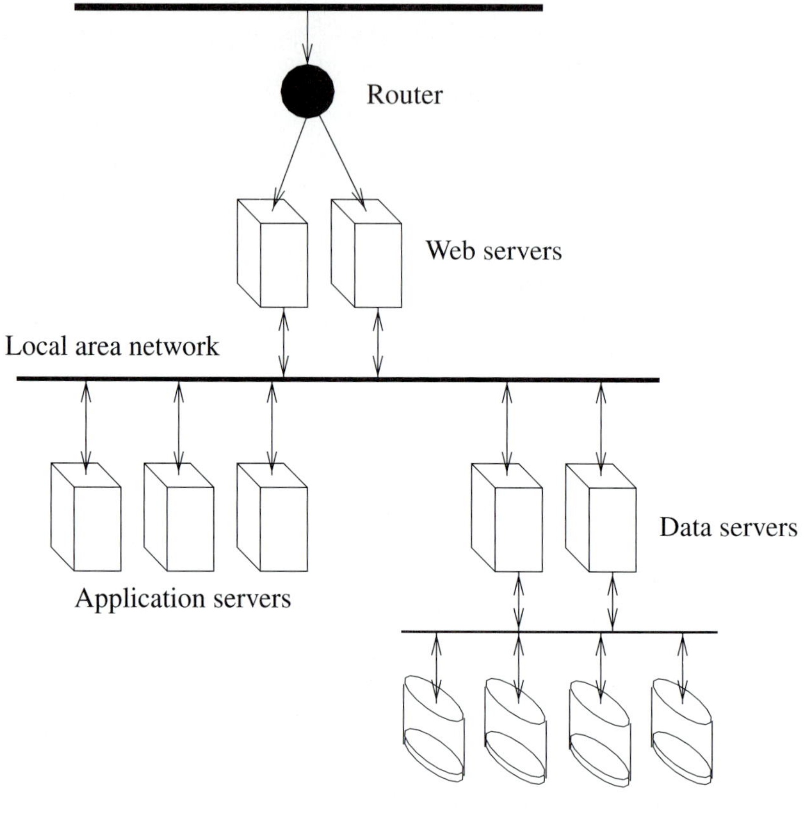

Figure 14.11. Web-site server system.

time. Thus, the high-level simulation model should focus on the impact of timing at each level that is used, system factors that affect that timing, and the effects of timing on contention for resources. To understand where those delays occur, let us consider the processing associated with a typical query.

All entries into the system are through a dedicated router, which examines the request and forwards it to some web server. Time is required to exercise the logic of looking at the request, determining whether it is a new request (requiring load balancing) or part of an ongoing session. It is reasonable to assume one switching time for a preexisting request and a different time for a new request. The result of the first step is selection of a web server, and enqueueing there of a request for service. A web server can be thought of as having a queue of threads of new requests, a queue of threads that are suspended awaiting a response from an application server, and a queue of threads "ready" to process responses from application servers. An accepted request from the router creates a new request thread. We may assume the web server has adequate memory to deal with all requests. It has a queueing policy that manages access to the CPU; the distinction between new requests and responses from application servers is maintained for the sake of scheduling and for the sake of assigning service times, the distributions of which depend on the type. The servicing of a new request amounts to identification of an application and associated application server. A request for service is formatted and forwarded to an application server, and the requesting thread joins the suspended queue. At an application server, requests for service are organized along application types. A new request creates a thread that joins a new request queue associated with the identified application. An application request is modeled as a sequence of sets of requests from data servers, interspersed with computational bursts — e.g.,

```
burst 1
request data from D1, D3 and D5
burst 2
request data from D1 and D2
burst 3
```

In this model we assume that all data requests from a set must be satisfied before the subsequent computational burst can begin. Query search on a database is an example of an application that could generate a long sequence of bursts and data requests, with large numbers of data requests in each set. We need not assume that every execution of an application is identical in terms of data requests or execution bursts, these can be generated stochastically. An application thread's state will include a description of its location in its sequence, and a list of data requests still outstanding before the thread can execute again. Thus, for each application we will maintain a list of threads that are ready to execute, and a list of threads that are suspended awaiting responses from data servers. An application server will implement a scheduling policy over sets of ready application threads. A data server creates a new thread to respond to a data request and places it in a queue of ready threads. A data server may implement memory-management policies, and may require further coordination with the application server to know when to release used memory. Upon receiving service the thread requests data from a disk, and suspends until the the disk operation completes, at which point the thread is moved from the suspended list to the ready list, and when executed again reports back to the application server associated with the request. The thread suspended at the application server responds; eventually the application thread finishes and reports its completion back to the web-server thread that initiated it, which in turn communicates the results back over the Internet.

Stepping back from the details, we see that a simulation model of this system must specify a number of features, listed in Table 14.2. All of these affect timing in some way. The query-response-time distribution can be estimated by measuring, for each query, the time between when a request first hits the router, to when the web-server thread communicates the results. From the set of simulated queries one can build up a histogram. As should be evident, a

Table 14.2. Required Specification for Web System Model

Subsystem	*Specifications*
Router	Load-balancing policy, execution times
Web server	Server count, queueing policy, execution times
Application server	Server count, queueing policy, behavior model
Data server	Server count, disk count, queueing policy, Memory policy, disk timing

response time reflects a great many different factors related to execution bursts, scheduling policies, and disk access times. Deeper understanding of the system is obtained by measuring behavior at each server of each type. One would look especially for evidence of bottlenecks. CPU bottlenecks would be reflected at servers with high CPU utilization; IO bottlenecks would be reflected at disks with high utilization. To assess system capacity at peak loads we would simulate to identify bottlenecks, and then look to see how to reduce load at bottleneck devices by changes in scheduling policies, binding of applications to servers, or increasing the number of CPUs or disks in the system. Normally one must resimulate a reconfigured system under the same load as before to assess the effects of the changes.

The web-site model is an excellent candidate for a threaded (process-oriented) approach to modeling. There are two natural approaches for defining a process. One emulates our CSIM example, defining each request as a process. The model would be expressed from the point of view of a request going through the router, to a web server, and so on. To support this view, the simulator would need to allow the modeler some control over process priorities at simulated servers; from the model narrative we see how web servers, application servers, and data servers might schedule processes differently. A modeling system that allowed user manipulation of process priorities might serve this need. One complication with equating a query with a process is that it is not an exact fit to what happens in this model. A query passed from application server to data server may actually cause multiple concurrent requests to the data disks—it is insufficient to "push" a query process from router to disks and back. A query process can spawn concurrent supporting processes, which implies encoding a fork-and-join synchronization mechanism (a process spawns new processes, and waits for them to return). The CSIM example illustrated this by having the main thread suspend, once all job arrival processes had been generated, until the last one finished. The clear advantage to a query-centered modeling approach is clarity of expression, and the ease with which query-oriented statistics can be gathered. A given query process can simply measure its delays at every step along the way, and upon departing the system include its observations in the statistical record the simulation maintains for the system.

One downside to equating a query to a process is that the approach can lead to many many concurrent processes! Depending on the implementation of the threading system, this can place severe memory demands on the simulator. Some thread implementations place an upper bound on the allowable number of concurrent threads. An alternative process-oriented approach is to associate processes with *servers*. The simulation model is expressed from an abstracted point of view of the servers' operating system. Individual queries become messages that are passed between server processes. In addition to limiting the number of processes, an advantage of this approach is that it explicitly exposes the scheduling of query processing at the user level. The modeler has both the opportunity and the responsibility to provide the logic of scheduling actions that model processing done on behalf of a query. It is a modeling viewpoint that simplifies analysis of server behavior—an overloaded server is easily identified by the (modeler-observable) length of its queue of runnable queries. However, it is a modeling viewpoint that is a bit lower in abstraction than the first one, and requires more modeling and coding on the part of the user.

An event-oriented model of this system need not look a great deal different than the second of our process-oriented models. Queries passed as messages between servers have an obvious event-oriented expression. A modeler would have to add to the logic, events, and event handlers that describe the way a CPU passes through simulation time. For example, consider a call to *hold(qt)* in a process-oriented model to express the CPU allocating *qt* units of service to a query, during which time it does nothing else. In an event-oriented model one would need to define events that reflect "starting" and "stopping" the processing of a query, with some scheduling logic interspersed. Additional events and handlers need to be defined for any "signaling" that might be done between servers in a process-oriented model, for example when a data-server process awaits completion of modeled IO requests sent to its disks. A process-oriented approach, even one focused on servers rather than queries, lifts model expression to a higher level of abstraction and reduces the amount of code that must be written. In a system as complex as the web site, one must factor complexity of expression into the overall model-development process. ◀

14.5 CPU Simulation

Next we consider a lower level of abstraction and look at the simulation of a central processing unit. Whereas the high-level simulation of the previous example treated execution time of a program as a constant, at the lower level we do the simulation to determine what the execution time is. The input driving this simulation is a stream of instructions. The simulation works through the mechanics of the CPU's logical design to determine what happens in response to that stream, how long it takes to execute the program, and where bottlenecks exist in the CPU design. Our discussion illustrates some of the functionality of a modern CPU, and the model characteristics that such a simulation seeks to determine. Examples of such simulations include those described in (Cmelik and Keppel [1994], Bedicheck [1995], and Witchel and Rosenblum [1996]). The view of the CPU taken in our discussion is similar to that taken by the RSIM system (Pai et al. [1997]).

The main challenge to making effective use of a CPU is to avoid stalling it, which happens whenever the CPU commits to executing an instruction whose inputs are not all present. A leading cause of stalls is the latency delay between CPU and main memory, which can be tens of CPU cycles. One instruction may initiate a read, e.g.,

```
load $2, 4($3)
```

which is an assembly-language statement that instructs the CPU to use the data in register 3 (after adding value 4 to it) as a memory address, and to put the data found at that address in register 2. If the CPU insisted on waiting for that data to appear in register 2 before further execution, the instruction could stall the CPU for a long time if the referenced address was not found in the cache. High-performance CPUs avoid this by recognizing that additional instructions *can* be executed, up to the point where the CPU attempts to execute an instruction that reads the contents of register 2, e.g.,

```
add $4,$2,$5
```

This instruction adds the contents of registers 2 and 5 and places the result in register 4. If the data expected in register 2 is not yet present, the CPU will stall. So we see that to allow the CPU to continue past a memory load it is necessary to (1) mark the target register as being unready, (2) allow the memory system to asynchronously load the target register while the CPU continues on in the instruction stream, (3) stall the CPU if it attempts to read a register marked as unready, and (4) clear the unready status when the memory operation completes.

The sort of arrangement described above was first used in the earliest supercomputers, designed in the 1960s. Modern microprocessors add some additional capabilities to exploit *instruction-level parallelism* (ILP). We outline some of the current architecture ideas in use to illustrate what a simulation model of an ILP CPU involves.

The technique of pipelining has long been recognized as a way of accelerating the execution of computer instructions (see Hennessy and Patterson [1994]). Pipelining exploits the fact that each instruction goes sequentially through several stages in the course of being processed; separate hardware resources are dedicated to each stage, permitting multiple instructions to be in various stages of processing concurrently. A typical sequence of stages in an ILP CPU are as follows:

1. *Instruction fetch:* the instruction is fetched from the memory.
2. *Instruction decode:* the memory word holding the instruction is interpreted to determine what operation is specified; the registers involved are identified.

3. *Instruction issue:* an instruction is "issued" when there are no constraints holding it back from being executed. Constraints that keep an instruction from being issued include data not yet being ready in an input register, and unavailability of a functional unit (e.g., arithmetic-logical-unit) needed to execute the instruction.
4. *Instruction execute:* the instruction is performed.
5. *Instruction complete:* results of the instruction are stored in the destination register.
6. *Instruction graduate:* executed instructions are graduated in the order that they appear in the instruction stream.

Ordinary pipelines permit at most one instruction to be represented in each stage; the degree of parallelism (number of concurrent instructions) is limited to the number of stages. ILP designs allow multiple instructions to be represented in some stages. This necessarily implies the possibility of executing some stages of successively fetched instructions out-of-order. For example, it is entirely possible for the nth instruction, I_n, to be constrained from being issued for several clock cycles, while the next instruction I_{n+1} is not so constrained. An ILP processor will push the evaluation of I_{n+1} along as far as it can without waiting on I_n. However, the instruction graduate stage will reimpose order and insist on graduating I_n before I_{n+1}.

ILP CPUs use architectural sleight of hand with respect to register useage to accelerate performance. An ILP machine typically has more registers available than appear in the instruction set. Registers named in instructions need not precisely be the registers actually used in the implementation of those instructions. This is acceptable, of course, so long as the effect of the instructions is the same in the end. One factor motivating this design is the possibility of having multiple instructions, involving the same logical registers (those named by the instructions themselves), actively being processed concurrently. By providing each instruction with its own "copy" of a register we eliminate one source of stalls. Another factor involves branches — that is, instructions that interrupt the sequential flow of control. An ILP, encountering a branch instruction, will predict whether the branch is taken or not and possibly alter the instruction stream as a result. Various methods exist to predict branching, but any of them will occasionally predict incorrectly. When an incorrect prediction is made, the register state computed as a result of speculating on branch outcome needs to be discarded and execution resumed at the branch point. Thus, another use of additional registers is to store the "speculative register state." With dedicated hardware resources to track register usage following speculative branch decision, speculative state can be discarded in a single cycle and control resumed at the mis-predicted branch point. In all of these cases the hardware implements techniques for renaming the logical registers that appear in the instructions to physical registers, for maintaining the mapping of logical to physical registers, and for managing physical register usage.

A simulation model of an ILP CPU will model the logic of each stage and coordinate the movement of instructions from stage to stage. We consider each stage in turn.

An instruction fetch stage may interact with the simulated memory system, if that is present. However, if the CPU simulation is driven by a direct-execution simulation or a trace file, there is little for a model of this stage to do but get the next instruction in the stream. If a memory system is present, this stage may look into an instruction cache for the next referenced instruction, stalling if a miss is suffered.

Following an instruction fetch, an instruction will be the CPU's list of active instructions, until it exits the pipeline altogether. The instruction decode stage places an instruction in this list; a logical register that appears as the target of an operation is assigned a physical register—registers used as operand sources will have been assigned physical registers in instructions that defined their values (sequencing issues associated with having multiple representations of the same register are dealt with at a later stage in the pipeline). Branch instructions are identified in this stage, predictions of branch outcomes are made, and resources for tracking speculative execution are committed here.

Decoded instructions pass into the instruction issue stage. The logic here is complex and very timing dependent. An instruction cannot be issued until values in its input registers are available and a functional unit needed to perform the instruction is available. An input value may not yet be in a register, for instance, if that value is loaded from memory by a previous instruction and has not yet appeared. A functional unit may not be available because all appropriate ones may be busy with multicycle operations initiated by other instructions. Implementation of the issue stage model (and hardware) depends on marking registers and functional units as busy or pending, and making sure that when the state of a register or functional unit changes, any instruction that cannot yet issue because of that register or functional unit is reconsidered for issue.

Simulation of the instruction execute stage is a matter of computing the result specified by the instruction, e.g., an addition. At this point the action of depositing the result into a register or memory is scheduled for the instruction complete stage. This latter stage also cleans up the status bits associated with registers and functional units involved in the instruction, and resolves the final outcome of a predicted branch. If a branch was mispredicted, the speculatively fetched and processed instructions that follow it are removed from other pipeline stages, the hardware that tracks speculative instruction is released, and the instruction stream is reset to follow the branch's other decision direction.

Between the instruction issue and instruction complete stages instructions may be processed in an order that does not correspond to the original instruction stream. The last stage, graduation, reorders them. Architecturally this permits an ILP CPU to identify exceptions (e.g., a page fault or division-by-zero) with the precise instruction with which it is associated. Simulation of this stage is a matter of knowing the sequence number of the next instruction to be graduated, and graduating it when it appears.

EXAMPLE 14.2

An example helps to show what goes on. Consider the sequence of assembly-language instructions below for a hypothetical computer.

```
load $2, 0($6) ; I1-- load $2 from memory
mult $5, 2     ; I2-- multiply $5 with constant 2
add  $4, 12    ; I3-- add constant 12 to $4
add  $5, $2    ; I4-- $5 <- $5 + $2
add  $5, $4    ; I5-- $5 <- $5 + $4
```

Let us suppose that the register load misses the first-level cache but hits in the second-level cache, resulting in a delay of 4 cycles before the register gets the value. Suppose further that separate hardware exists for addition and multiplication, and that addition takes one cycle and multiplication 2 cycles to complete. Time is assumed to advance in units of a single clock tick.

Table 14.3 shows a timeline of when each instruction is in each stage. Cycles when an instruction cannot proceed through the pipeline are marked as "stall" cycles. Processing is most easily understood by tracing individual instructions through.

I1 After being fetched in cycle 1, the decode of I1 assigns physical register \$p1 as the target of the load operation, and marks \$p1 as unready. No constraints prohibit I1 from being issued in cycle 3 or executed in cycle 4. Because the memory operation takes 4 cycles to finish, I1 is stalled in cycles 5–8. Cycle 9 commits the data from memory to physical register \$p1 and clears its unready flag; the instruction is graduated in cycle 10.

I2 Instruction I2 is fetched in cycle 2 and has physical register \$p2 allocated to receive the results of the multiplication in the cycle 3 decode stage; \$p2's unready flag is raised. No constraints keep I2 from being issued in cycle 4 or executed in cycle 5, but the 2-cycle delay of the multiplier means the result is not committed to register \$p2 until cycle 7, at which point the \$p2 unready flag

Table 14.3. Pipeline Stages, ILP CPU Simulation

Inst./Cycle	1	2	3	4	5	6	7
I1	fetch	decode	issue	exec.	stall	stall	stall
I2		fetch	decode	issue	exec.	stall	comp.
I3			fetch	decode	issue	exec.	comp.
I4				fetch	decode	stall	stall
I5					fetch	decode	stall
Inst./Cycle	**8**	**9**	**10**	**11**	**12**	**13**	**14**
I1	stall	comp.	grad.				
I2	stall	stall	stall	grad.			
I3	stall	stall	stall	stall	grad.		
I4	stall	stall	issue	exec.	comp.	grad.	
I5	stall	stall	stall	stall	stall	issue	comp.

is cleared. The instruction remains stalled through cycles 8–10, awaiting the graduation of I1.

I3 Instruction I3 is fetched in cycle 3 and has physical register $p3 allocated to receive the results of its addition in the cycle 4 decode stage. The $p3 unready flag is raised. There are no constraints keeping I3 from being issued in cycle 5 and executed in cycle 6, with results written into $p3 in cycle 7, when $p3's unready flag is cleared. I3 must stall, however, during cycles 8–11 awaiting the graduation of I2.

I4 Instruction I4 is fetched in cycle 4 and has physical register $p4 allocated to receive the results of the addition during the cycle 5 decode stage. $p4's unready flag is raised at that point. Physical registers $p1 and $p2 are operands to the addition; I4 stalls in cycles 6–9 waiting for their unready flags to clear. It then passes the remaining stages without further delay, clearing the $p4 unready flag in cycle 12.

I5 Instruction I5 is fetched in cycle 5 and has physical register $p5 allocated to receive the results of its addition in the cycle 6 decode stage, at which point the $p5 unready flag is set. Physical registers $p3 and $p4 contain the addition's operands; I5 stalls through cycles 7–12 waiting for their unready flags to both clear. From that point forward I5 passes through the remaining stages without further delay.

The performance benefit of pipelining and ILP can be appreciated if we compare the execution time of this sequence on a nonpipelined, non-ILP machine. Assuming that each stage must be performed for each instruction but that one instruction is processed in its entirety before another one begins, 51 cycles are needed to execute I1 through I5. With the advanced architectural features only 15 cycles are needed. The example illustrates both the parallelism that pipelining exposes, and the latency tolerance that the ILP design supports. Even though I1 stalls for four cycles awaiting a result from memory, the pipeline keeps moving other instructions through to some extent. The bottom line for someone using a model like this is the rate at which instructions are graduated, as this reflects the effectiveness of the CPU design. Secondary statistics would try to pinpoint where in the design stalls occur which might be alleviated, e.g., if many stalls occur waiting for the multiplier (no such stalls occur in the example), then one could consider including an additional multiplier in the CPU design.

Our explanation of the model's workings was decidedly process oriented, taking the view of an instruction. However, the computational demands of a model like this are enormous, owing to the very large number of instructions that must be simulated to assess the CPU design on, say, a single program run. The relatively high cost of context switching would deter use of a normal process-oriented language. One could implement what is essentially a process-oriented view using events—each time an instruction passes through a stage, an event is scheduled to take that instruction through the next stage, accounting for stalls. The amount of simulation work accomplished per event is thus the amount of work done on behalf of one instruction in one stage. An alternative approach is to eschew explicit events altogether and simply use a cycle-by-cycle activity scan. At each cycle, one would examine each active instruction to see if any activity associated with that instruction can be done. An instruction that was at one stage at cycle j will, at cycle $j+1$, be examined for constraints that would keep it at cycle j. Finding none, that instruction would be advanced to the next stage. An activity scanning approach has the attractiveness of eliminating event-list overhead, but the disadvantage of expending computational effort checking the status of a stalled instruction, on every cycle during which it is stalled. Implementation details and model behavior largely determine whether an activity-scanning approach is faster than an event-oriented approach (with the nod going to activity scanning when few instructions stall). ◀

14.6 Memory Simulation

One of the great challenges of computer architecture is finding ways to effectively deal with the increasing gap in operation speed between CPUs and main memory. A factor of 100 in speed is not far from the mark. The main technique that has evolved is to build hierarchies of memories. A relatively small memory—the L1 cache—operates at CPU speed. A larger memory—the L2 cache—is larger, and operates more slowly. The main memory is larger still, and slower still. The smaller memories hold data that was referenced recently, and nearby data that one hopes will also be referenced soon. Data moves up the hierarchy on demand, and ages out as it becomes disused, to make room for the data in current use. For instance, when the CPU wishes to read memory location 100000, hardware will look for it in the L1 cache, and failing to find it there, look for it in the L2 cache. If found there, an entire block containing that reference is moved from the L2 cache into the L1 cache. If not found in the L2 cache, a (larger) block of data containing location 10000 is copied from the main memory to the L2 cache, and part of that block (containing location 10000, of course) is copied into the L1 cache. It may take 50 cycles or more to accomplish this. After this cost has been suffered, the hope and expectation is that the CPU will continue to make references to data in the block brought in, because accesses to L1 data are made at CPU speeds. Fortunately most programs exhibit locality of reference at this scale (as well as at the paging scale discussed earlier in the chapter), so the strategy works. However, after a block ceases to be referenced for a time, it is ejected from the L1 cache. It may remain in the L2 cache for a while, and be brought back into the L1 cache if any element of the block is referenced again. Eventually a block remains unreferenced long enough so that it is ejected also from the L2 cache.

The astute reader will realize that data that is written into an L1 cache by the CPU creates a consistency problem, in that a memory address has different values associated with it at different levels of the memory hierarchy. One way of dealing with this is to write-through all cache levels every time there is a write—the new value is asynchronously pushed from L1 through L2 to the main memory. An alternative method copies back a block from one memory level to the lower level, at the point the block is being ejected from the faster level. The write-through strategy avoids writing back blocks when they are ejected, whereas the write-back strategy requires that an entire block be written back when ejected, even if only one word of the block was modified, once. One of the roles simulation plays is to compare performance of these two write-back strategies, taking into consideration all costs and contention for the resources needed to support writing back modifications.

Like paging systems, the principal measure of the quality of a memory hierarchy is its hit ratio at each level. As with CPU models, to evaluate a memory hierarchy design one must study the design in response to a very long string of memory references. Direct-execution simulation can provide such a reference stream, as can long traces of measured reference traffic. Nearly every caching system is a demand system, which means that a new block is not brought into a cache before a reference is made to a word in that block. Decisions left still to the designer include whether to write-through or write-back modifications, the replacement policy, and the "set associativity."

The concept of set associativity arises in response to the cost of the mechanism used to look for a match. Imagine we have an L2 cache with 2 million memory words (an actual figure from an actual machine). The CPU references location 10000—the main memory has, say, 2^{12} words, so the L2 cache holds but a minute fraction of the main memory. How does the hardware determine whether location 10000 is in the L2 cache? It uses what is called an associative memory, one that associates search keys with data. One queries an associative memory by providing some search key. If the key is found in the memory, then the data associated with the key is returned, otherwise indication of failure is given. In the caching context the search key is derived from the reference address, and the return data is the data stored at that address. Caches must be very very fast, which means that the search process has to be abbreviated. This is accomplished by dedicating comparison hardware with every location in the associative memory. Presented with a search key, every comparator looks for a match with the key at its location. At most one comparator will see a match and return the data, it is possible that none will.

A *fully associative* cache is one where any address can appear anywhere in the cache. This means building the cache to have a unique comparator associated with every address in the cache, which is prohibitively expensive. Tricks are played with memory addresses in order to greatly reduce the costs. The idea is to partition the address space into sets. Figure 14.12 illustrates how a 48-bit memory address might be partitioned in key, set id, and block offset. Any given memory address is mapped to the set identified by its set id address bits. This scheme assigns the first block of 2^b addresses to set 0, the second block of 2^b addresses to set 1, and so on, wrapping around back to set 0 after 2^s blocks have been assigned. Each set is given a small portion of the cache—the set size—which typically is 2 or 4 or 8 words. Only those

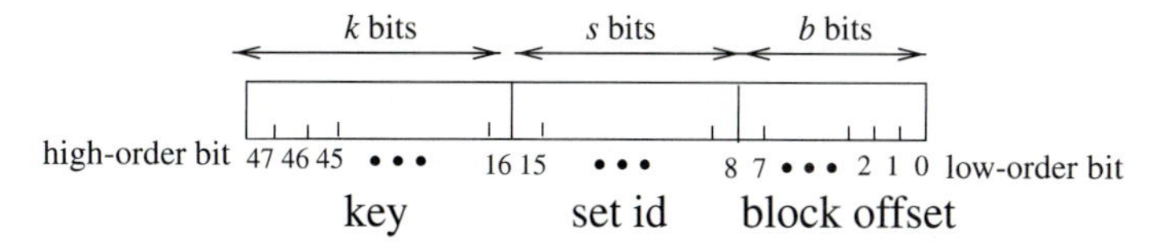

Figure 14.12. 48-bit address partitioned for cache.

addresses mapped to the same set compete for storage in that space. Only as many comparators as words in the set are needed. Given an address, the hardware uses the set id bits to identify the set number, and the key bits to identify the key. The hardware matches the keys of the blocks already in the identified set to comparator inputs, and also provides the the key of the sought address as input to all the comparators. Comparisons are made in parallel; in the case of a match, the block offset bits are used to index into the identified block to select the particular address being referenced.

The overall size of this cache is seen to be the total number of sets times the set size. One role of simulation is to determine, for a given cache size, how the space ought to be partitioned into sets. This is largely a cost consideration, for increasing the set size (thereby reducing the number of sets) typically increases the hit ratio. However, if a set size of 4 yields a sufficiently large hit ratio, then there is little point to increasing the set size (and cost).

Least-recently-used (LRU) is the replacement policy most typically used. When a reference is made and is not found in a set, some block in the set is ejected to make room for the one containing the new reference. Under LRU the block selected for ejection is the one which, among all blocks in the set, was last referenced most distantly in the past.

LRU is one of several replacement policies known as *stack* policies (see Stone [1990]). These are characterized by the behavior that for any reference in any reference string, if that reference misses in a cache of size n, then it also misses in every cache of size $m < n$, and if it hits in a cache of size m, then it hits in every cache of size $n > m$. Simulations can exploit this fact to compute the miss ratio of many different set sizes, in just one pass of the reference string! Suppose that we do not wish to consider any set size larger than 64. Now we conduct the simulation with set sizes of 64. Every block in the cached set is marked with a priority, the temporal index of the last reference made to it; e.g., the block containing the first reference in the string is marked with 1, the block containing the second reference is marked with a 2 (overwriting the 1, if the same as the previous block), etc. When a block must be replaced, the one with smallest index is selected.

Imagine that the simulation organizes and maintains the contents of a cached set in LRU order, with the most recently referenced block first in the order. The *stack distance* of a block in this list is its distance from the front; the most recently referenced block has stack distance 1, the next most recently referenced block has stack distance 2, the LRU block has stack distance 64. Presented with a reference, the simulation searches the list of cache blocks for

Reference trace	A	B	C	A	D	B	A	D	C	D	F	C	B	F	E	Hits array
Stack distance 1	A	B	C	A	D	B	A	D	C	D	F	C	B	F	E	0
Stack distance 2		A	B	C	A	D	B	A	D	C	D	F	C	B	F	1
Stack distance 3			A	B	C	A	D	B	A	A	C	D	F	C	B	5

Figure 14.13. LRU stack evolution.

a match. If no match is found, then by the stack property no match will be found in any cache of a size smaller than 64, on this reference, for this reference string. If a match is found and the block has stack distance k, then no match will be found in any cache smaller than size k, and a match will always be found in a cache of size larger than k. Rather than record a hit or miss, one increments the kth element of a 64-element array that records the number of matches at each LRU level. To determine how many hits occured in a cache of size n, one sums up the counts of the first n elements of the array. Thus, with a little arithmetic at the end of the run, one can determine (for each set cache) the number of hits for every set of every size between 1 and 64.

Figure 14.13 illustrates the evolution of an LRU list in response to a reference string. Under each reference (given as an alphabetic symbol rather than actual memory address) is the state of the LRU stack *after* the reference is processed. The horizontal direction from left to right symbolizes the trace, reading from left to right. A hit is illustrated by a circle, with an arrow showing the migration of the symbol to the top of the heap. The "hits" array counts the number of hits found at each stack distance. Thus we see that a cache of size 1 will have a hit ratio of 0/15, a cache of size 1 will have a hit ratio of 1/15, and a cache of size 3 will have a hit ratio of 6/15.

In the context of a set-associative cache simulation, each set must be managed separately as shown. In one pass one can get hit ratios for varying set sizes, but it is important to note that each change in set size corresponds to a change in the overall size of the entire cache. This technique alone does not let us in one pass determine the hit ratios for all the difference ways one might partition a cache of a given capacity (e.g., 256 sets with set size 1 versus 128 sets with set size 2 versus 64 sets with set size 4). It actually is possible to evaluate all these possibilities in one pass, but the technique is beyond the scope of this discussion.

14.7 Summary

This chapter looked at the broad area of simulating computer systems. It emphasized that computer-system simulations are performed at a number of levels of abstraction. Inevitably, it discussed a good deal of computer science along with the simulation aspects, for in computer-systems simulation the two are inseparable.

The chapter outlined fundamental implementation issues behind computer-systems simulators — principally, how process orientation is implemented and how object-oriented concepts such as inheritence are fruitfully employed. Next it considered model input, ranging from stochastically generated traffic to stochastically generated memory-referencing patterns, to measured traces and direct-execution techniques. The chapter concluded by looking at examples of simulation at different levels of abstraction: a WWW site server system, an instruction-level CPU simulation, and simulation of set-associative memory systems.

The main point is that computer-systems simulators are tailored to the tasks at hand. Appropriate levels of abstraction need to be chosen, as must appropriate simulation techniques.

REFERENCES

ASHENDEN, P. J. [1996], *The Designer's Guide to VHDL*, Morgan Kaufmann, San Fransisco, CA.

BEDICHECK, R. C. [1995], "Talisman: Fast and Accurate Multicomputer Simulation," *Proceedings of the 1995 ACM SIGMETRICS Conference*, pp. 14–24, Ottawa, Canada, May.

CMELIK, B., AND D. KEPPLE [1994], "A Fast Instruction-Set Simulator for Execution Profiling," *Proceedings of the 1994 ACM SIGMETRICS Conference*, pp. 128–137, Nashvill, TN, May.

COVINGTON, R., S. DWARKADAS, J. JUMP, S. MADALA, AND J. SINCLAIR [1991], "Efficient Simulation of Parallel Computer Systems," *International Journal on Computer Simulation*, Vol. 1, No. 1, pp. 31–58.

COWIE, J., A. OGIELSKI, AND D. NICOL [1999], "Modeling the Global Internet," *Computing in Science and Engineering*, Vol. 1, No. 1, pp. 42–50.

DICKENS, P., P. HEIDELBERGER, AND D. NICOL [1996], "Parallelized Direct Execution Simulation of Message Passing Programs," *IEEE Transactions on Parallel and Distributed Systems*, Vol. 7, No. 10, pp. 1090–1105.

FISCHER, W., AND K. MEIER-HELLSTERN [1993], "The Markov-Modulated Poisson (MMPP) Cookbook," *Performance Evaluation*, Vol. 18, No. 2, pp. 149–171.

FISHWICK, P. [1992], "Simpack: Getting Started with Simulation Programming in C and C++," *Proceedings of the 1992 Winder Simulation Conference*, pp. 154–162, Washington, D.C.

GRAND, M. [1997], *Java Language Reference*. O'Reilly, Cambridge, MA.

GRUNWALD, D. [1995], *User's Guide to Awesime-II*, Department of Computer Science, University of Colorado.

HENNESSY, J. L., AND D. A. PATTERSON [1994], *Computer Organization and Design, The Hardware/Software Interface*. Morgan Kaufmann, Palo Alto, CA.

LEBECK, A., AND D. WOOD [1997], "Active Memory: A New Abstraction for Memory System Simulation," *ACM Transactions on Modeling and Computer Simulation*, Vol. 7, No. 1, pp. 42–77.

LITTE, M. C., AND D. L. MCCUE [1994], "Construction and Use of a Simulation Package in C++," *C User's Journal*, Vol. 3, No. 12. Also available as Technical Report 437, Dept. of Computing Science, University of Newcastle on Tyne, 1993.

MANO, M. [1993], *Computer Systems Architecture*, 3d ed., Prentice Hall, Upper Saddle River, NJ.

NICHOLS, B., D. BUTTLAR, and J. PROULX FARRELL [1996], *Pthreads Programming*. O'Reilly, Cambridge, MA.

NUTT, G. [2000], *Operating Systems, A Modern Perspective*, 2d ed., Addison Wesley, Reading, MA.

PAI, V., P. RANGANATHAN, AND S. ADVE [1997], *Rsim Reference Manual*, Vol. 1.0, Technical Report 9705, Rice University.

SCHWETMAN, H. [1986], "CSIM: A C-Based, Process Oriented Simulation Language," *Proceedings of the 1986 Winter Simulation Conference*, pp. 387–396.

STONE, H. [1990], *High Performance Computer Architecture*, Addison Wesley, Reading, MA.

WITCHEL, E., AND M. ROSENBLUM [1996], "Embra: Fast and Flexible Machine Simulation," *Proceedings of the 1996 ACM SIGMETRICS Conference*, pp. 68–79, Philadelphia, PA, May.

EXERCISES

1. Sketch the logic of an event-oriented model of an $M/M/1$ queue. Estimate the number of events executed when processing the arrival of 5000 jobs. How many context switches does the process-oriented model in Figure 14.2 undergo?

2. For each of the systems below, sketch the logic of a process-oriented model and of an event-oriented model. For both approaches develop and simulate the model in any language.

- a central server queueing model; when a job leaves the CPU queue it joins the IO queue with shortest length.
- a queueing model of a database system, which implements fork-join: a job receives service in two parts. When it first enters the server it spends a small amount of simulation time generating a random number of requests to disks. It then suspends (freeing the server) until such time as *all* the requests it made have finished, and then enqueues for its second phase of service where it spends a larger amount of simulation time, before finally exiting. Disks may serve requests from various jobs concurrently, but serve them using FCFS ordering. Your model should report on the statistics of a job in service—how long (on average) it waited for phase 1, how long it waits on average for its IO requests to complete, how long it waits on average for service after its IO requests complete.

3. Consider a Markov modulated process with three states, OFF, ON, BURSTY, with the following characteristics:

- the MMP is in the OFF state for 25% of the time, on average.
- the MMP is in the BURSTY state for 5% of the time, on average.
- the MMP transitions from OFF to BURSTY with probability 0.05, and from OFF to ON with probability 0.95.

- the MMP transitions from ON to OFF with probability 0.9, and from ON to BURSTY with probability 0.1.
- the MMP transitions from BURSTY to ON with probability 0.5, and from BURSTY to OFF with probability 0.5.

If the time spent in the OFF state is an exponential with mean 0.25, what are the means of the exponential times spent in states ON and BURSTY? Simulate and plot the results.

4. Recall the pseudocode for generating reference traces (Figure 14.9). Write routines `new_wrkset`, `from_wrkset`, and `not_from_wrkset` to model the following types of programs:

  (a) A scientific program with a large working set during initialization, a small working set for the bulk of the computation, and a different working set to complete the computation. (You will need to modify the control code in the figure slightly to force phase transitions in desired places.)

  (b) A program whose working set always contains a core set of pages present in every phase, with the rest of the pages clustered elsewhere in the address space.

5. Using any simulator or language you like, model some aspect of your local computing-system environment — e.g., the print queue for a network printer. Carefully describe your workload model.

6. Using any simulator or language you like, model the router-to-web-server logic of the system described in Section 14.4. Pay special attention to the load-balancing mechanism that the router employs.

7. Using any simulator or language you like, model the interaction between application server and data server described in Section 14.4. Pay special attention to the logic of requesting multiple data services, and waiting until all are completed until advancing to the next burst.

8. From our web-site fetch Object-Oriented Example 1, which uses base classes `SimObj`, `SimScheduler`, and `SimEvt` to implement a simple simulation in which a single message is passed between simulated computer nodes. In this example, when one node schedules the arrival of a message at another node, it passes the memory address of the recipient node to the scheduler.

- Define a new simulation object, called `Link`, that simulates a computer link between two nodes. Define its interface so that nodes attach themselves to both ends, and when a node sends a message, it passes it to the link, not the recipient, directly. The link ought to impose a deterministic latency delay d and have associated bandwidth b. The arrival time of the last bit of an L-bit message sent at time t is $t + d + L/b$. The link ought to limit the number of messages it can concurrently carry to M, and ought to schedule the arrival time of the message at the target node at the instant the last bit arrives. Implement and test this new object.
- Modify the link object defined above so that its fixed bandwidth b is shared equally among all messages in transit. That is, if there are n messages on the link, then each receives b/n bandwidth (bits per second).

9. From our web-site fetch Object-Oriented Example 1, as before. This time develop new classes (derived from the abstract base classes) to model the systems described in Exercise 2.

10. Consider the simple language below for describing CPU instructions.

```
op  r1  r2
```

describes an operation, where

```
op=1  means add, op=2 means subtract.  Each require 1 cycle.
op=3  means mult.   r1 receives the result r1 op r2. A multiplication
      requires 2 cycles.
op=4  means a load from memory, into r1, using the value
      in r2 as the memory address.   Every $10^{th}$ load requires
      4 cycles, the remaining loads require 1.
op=5  means a store to memory, storing the data found in r1,
      using the value in r2 as the memory address.  Each store
      requires 1 cycle.
```

Write a CPU simulation along the lines of that described in Section 14.5 that accepts a stream of instructions in the format just described. Your simulator should use a logical-to-physical-register mapping, use the timing information sketched above, and stall instructions as described in the example.

11. Write a one-pass simulator of a fully associative cache of size N that accepts as input reference traces from the trace generator in Exercise 4. At the end of the simulation, report on the hit ratio for all cache sizes that are powers of 2 no greater than N.

12. Integrate the trace generator created in Exercise 4 with the one-pass simulator written in Exercise 11, to effectively create a pseudo "direct-execution simulator."

13. Analyze the log of WWW requests to your site's server, produce a stochastic model of the request stream, and simulate it.

Appendix

Tables

Table A.1. Random Digits

94737	08225	35614	24826	88319	05595	58701	57365	74759
87259	85982	13296	89326	74863	99986	68558	06391	50248
63856	14016	18527	11634	96908	52146	53496	51730	03500
66612	54714	46783	61934	30258	61674	07471	67566	31635
30712	58582	05704	23172	86689	94834	99057	55832	21012
69607	24145	43886	86477	05317	30445	33456	34029	09603
37792	27282	94107	41967	21425	04743	42822	28111	09757
01488	56680	73847	64930	11108	44834	45390	86043	23973
66248	97697	38244	50918	55441	51217	54786	04940	50807
51453	03462	61157	65366	61130	26204	15016	85665	97714
92168	82530	19271	86999	96499	12765	20926	25282	39119
36463	07331	54590	00546	03337	41583	46439	40173	46455
47097	78780	04210	87084	44484	75377	57753	41415	09890
80400	45972	44111	99708	45935	03694	81421	60170	58457
94554	13863	88239	91624	00022	40471	78462	96265	55360
31567	53597	08490	73544	72573	30961	12282	97033	13676
07821	24759	47266	21747	72496	77755	50391	59554	31177
09056	10709	69314	11449	40531	02917	95878	74587	60906
19922	37025	80731	26179	16039	01518	82697	73227	13160
29923	02570	80164	36108	73689	26342	35712	49137	13482
29602	29464	99219	20308	82109	03898	82072	85199	13103
94135	94661	87724	88187	62191	70607	63099	40494	49069
87926	34092	34334	55064	43152	01610	03126	47312	59578
85039	19212	59160	83537	54414	19856	90527	21756	64783
66070	38480	74636	45095	86576	79337	39578	40851	53503
78166	82521	79261	12570	10930	47564	77869	16480	43972
94672	07912	26153	10531	12715	63142	88937	94466	31388
56406	70023	27734	22254	27685	67518	63966	33203	70803
67726	57805	94264	77009	08682	18784	47554	59869	66320
07516	45979	76735	46509	17696	67177	92600	55572	17245
43070	22671	00152	81326	89428	16368	57659	79424	57604
36917	60370	80812	87225	02850	47118	23790	55043	75117
03919	82922	02312	31106	44335	05573	17470	25900	91080
46724	22558	64303	78804	05762	70650	56117	06707	90035
16108	61281	86823	20286	14025	24909	38391	12183	89393
74541	75808	89669	87680	72758	60851	55292	95663	88326
82919	31285	01850	72550	42986	57518	01159	01786	98145
31388	26809	77258	99360	92362	21979	41319	75739	98082
17190	75522	15687	07161	99745	48767	03121	20046	28013
00466	88068	68631	98745	97810	35886	14497	90230	69264

Table A.2. Random Normal Numbers

0.23	−0.17	0.43	2.18	2.13	0.49	2.72	−0.18	0.42
0.24	−1.17	0.02	0.67	−0.59	−0.13	−0.15	−0.46	1.64
−1.16	−0.17	0.36	−1.26	0.91	0.71	−1.00	−1.09	−0.02
−0.02	−0.19	−0.04	1.92	0.71	−0.90	−0.21	−1.40	−0.38
0.39	0.55	0.13	2.55	−0.33	−0.05	−0.34	−1.95	−0.44
0.64	−0.36	0.98	−0.21	−0.52	−0.02	−0.15	−0.43	0.62
−1.90	0.48	−0.54	0.60	−0.35	−1.29	−0.57	0.23	1.41
−1.04	−0.70	−1.69	1.76	0.47	−0.52	−0.73	0.94	−1.63
−.78	0.11	−0.91	−1.13	0.07	0.45	−0.94	1.42	0.75
0.68	1.77	−0.82	−1.68	−2.60	1.59	−0.72	−0.80	0.61
−0.02	0.92	1.76	−0.66	0.18	−1.32	1.26	0.61	0.83
−0.47	1.04	0.83	−2.05	1.00	−0.70	1.12	0.82	0.08
−0.40	1.40	1.20	0.00	0.21	−2.13	−0.22	1.79	0.87
−0.75	0.09	−1.50	0.14	−2.99	−0.41	−0.99	−0.70	0.51
−0.66	−1.97	0.15	−1.16	−0.60	0.50	1.36	1.94	0.11
−0.44	−0.09	−0.59	1.37	0.18	1.44	−0.80	2.11	−1.37
1.41	−2.71	−0.67	1.83	0.97	0.06	−0.28	0.04	−0.21
1.21	−0.52	−0.20	−0.88	−0.78	0.84	−1.08	−0.25	0.17
0.07	0.66	−0.51	−0.04	−0.84	0.04	1.60	−0.92	1.14
−0.08	0.79	−0.09	−1.12	−1.13	0.77	0.40	0.69	−0.12
0.53	−0.36	−2.64	0.22	−0.78	1.92	−0.26	1.04	−1.61
−1.56	1.82	−1.03	1.14	−0.12	−0.78	−0.12	1.42	−0.52
0.03	−1.29	−0.33	2.60	−0.64	1.19	−0.13	0.91	0.78
1.49	1.55	−0.79	1.37	0.97	0.17	0.58	1.43	−1.29
−1.19	1.35	0.16	1.06	−0.17	0.32	−0.28	0.68	0.54
−1.19	−1.03	−0.12	1.07	0.87	−1.40	−0.24	−0.81	0.31
0.11	−1.95	−0.44	−0.39	−0.15	−1.20	−1.98	0.32	2.91
−1.86	0.06	0.19	−1.29	0.33	1.51	−0.36	−0.80	−0.99
0.16	0.28	0.60	−0.78	0.67	0.13	−0.47	−0.18	−0.89
1.21	−1.19	−0.60	−1.22	0.07	−1.13	1.45	0.94	0.54
−0.82	0.54	−0.98	−0.13	1.52	0.77	0.95	−0.84	2.40
0.75	−0.80	−0.28	1.77	−0.16	−0.33	2.43	−1.11	1.63
0.42	0.31	1.56	0.56	0.64	−0.78	0.04	1.34	−0.01
−1.50	−1.78	−0.59	0.16	0.36	1.89	−1.19	0.53	−0.97
−0.89	0.08	0.95	−0.73	1.25	−1.04	−0.47	−0.68	−0.87
0.19	0.85	1.68	−0.57	0.37	−0.48	−0.17	2.36	−0.53
0.49	0.32	−2.08	−1.02	2.59	−0.53	0.15	0.11	0.05
−1.44	0.07	−0.22	−0.93	−1.40	0.54	−1.28	−0.15	0.67
−0.21	−0.48	1.21	0.67	−1.10	−0.75	−0.37	0.68	−0.02
−0.65	−0.12	0.94	−0.44	−1.21	−0.06	−1.28	−1.51	1.39
0.24	−0.83	1.55	0.33	−0.59	−1.24	0.70	0.01	0.15
−0.73	1.24	0.40	−0.61	0.68	0.69	0.07	−0.23	−0.66
−1.93	0.75	−0.32	0.95	1.35	1.51	−0.88	0.10	−1.19
0.08	0.16	0.38	−0.96	1.99	−0.20	0.98	0.16	0.26
−0.47	−1.25	0.32	0.51	−1.04	0.97	2.60	−0.08	1.19

Table A.3. Cumulative Normal Distribution

$$\phi(z_\alpha) = \int_{-\infty}^{z_\alpha} \frac{1}{\sqrt{2\pi}} e^{-u^2/2}\, du = 1 - \alpha$$

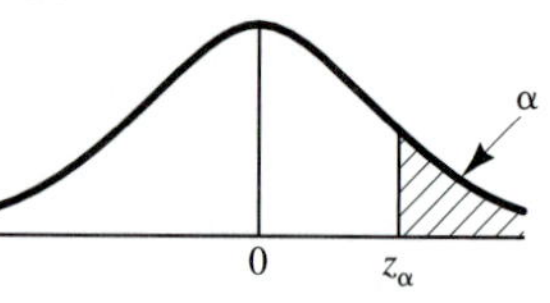

z_α	0.00	0.01	0.02	0.03	0.04	z_α
0.0	0.500 00	0.503 99	0.507 98	0.511 97	0.515 95	**0.0**
0.1	0.539 83	0.543 79	0.547 76	0.551 72	0.555 67	**0.1**
0.2	0.579 26	0.583 17	0.587 06	0.590 95	0.594 83	**0.2**
0.3	0.617 91	0.621 72	0.625 51	0.629 30	0.633 07	**0.3**
0.4	0.655 42	0.659 10	0.662 76	0.666 40	0.670 03	**0.4**
0.5	0.691 46	0.694 97	0.698 47	0.701 94	0.705 40	**0.5**
0.6	0.725 75	0.729 07	0.732 37	0.735 65	0.738 91	**0.6**
0.7	0.758 03	0.761 15	0.764 24	0.767 30	0.770 35	**0.7**
0.8	0.788 14	0.791 03	0.793 89	0.796 73	0.799 54	**0.8**
0.9	0.815 94	0.818 59	0.821 21	0.823 81	0.826 39	**0.9**
1.0	0.841 34	0.843 75	0.846 13	0.848 49	0.850 83	**1.0**
1.1	0.864 33	0.866 50	0.868 64	0.870 76	0.872 85	**1.1**
1.2	0.884 93	0.886 86	0.888 77	0.890 65	0.892 51	**1.2**
1.3	0.903 20	0.904 90	0.906 58	0.908 24	0.909 88	**1.3**
1.4	0.919 24	0.920 73	0.922 19	0.923 64	0.925 06	**1.4**
1.5	0.933 19	0.934 48	0.935 74	0.936 99	0.938 22	**1.5**
1.6	0.945 20	0.946 30	0.947 38	0.948 45	0.949 50	**1.6**
1.7	0.955 43	0.956 37	0.957 28	0.958 18	0.959 07	**1.7**
1.8	0.964 07	0.964 85	0.965 62	0.966 37	0.967 11	**1.8**
1.9	0.971 28	0.971 93	0.972 57	0.973 20	0.973 81	**1.9**
2.0	0.977 25	0.977 78	0.978 31	0.978 82	0.979 32	**2.0**
2.1	0.982 14	0.982 57	0.983 00	0.983 41	0.983 82	**2.1**
2.2	0.986 10	0.986 45	0.986 79	0.987 13	0.987 45	**2.2**
2.3	0.989 28	0.989 56	0.989 83	0.990 10	0.990 36	**2.3**
2.4	0.991 80	0.992 02	0.992 24	0.992 45	0.992 66	**2.4**
2.5	0.993 79	0.993 96	0.994 13	0.994 30	0.994 46	**2.5**
2.6	0.995 34	0.995 47	0.995 60	0.995 73	0.995 85	**2.6**
2.7	0.996 53	0.996 64	0.996 74	0.996 83	0.996 93	**2.7**
2.8	0.997 44	0.997 52	0.997 60	0.997 67	0.997 74	**2.8**
2.9	0.998 13	0.998 19	0.998 25	0.998 31	0.998 36	**2.9**
3.0	0.998 65	0.998 69	0.998 74	0.998 78	0.998 82	**3.0**
3.1	0.999 03	0.999 06	0.999 10	0.999 13	0.999 16	**3.1**
3.2	0.999 31	0.999 34	0.999 36	0.999 38	0.999 40	**3.2**
3.3	0.999 52	0.999 53	0.999 55	0.999 57	0.999 58	**3.3**
3.4	0.999 66	0.999 68	0.999 69	0.999 70	0.999 71	**3.4**
3.5	0.999 77	0.999 78	0.999 78	0.999 79	0.999 80	**3.5**
3.6	0.999 84	0.999 85	0.999 85	0.999 86	0.999 86	**3.6**
3.7	0.999 89	0.999 90	0.999 90	0.999 90	0.999 91	**3.7**
3.8	0.999 93	0.999 93	0.999 93	0.999 94	0.999 94	**3.8**
3.9	0.999 95	0.999 95	0.999 96	0.999 96	0.999 96	**3.9**

Table A.3. (Continued)

z_α	0.05	0.06	0.07	0.08	0.09	z_α
0.0	0.519 94	0.523 92	0.527 90	0.531 88	0.535 86	**0.0**
0.1	0.559 62	0.563 56	0.567 49	0.571 42	0.575 34	**0.1**
0.2	0.598 71	0.602 57	0.606 42	0.610 26	0.614 09	**0.2**
0.3	0.636 83	0.640 58	0.644 31	0.648 03	0.651 73	**0.3**
0.4	0.673 64	0.677 24	0.680 82	0.684 38	0.687 93	**0.4**
0.5	0.708 84	0.712 26	0.715 66	0.719 04	0.722 40	**0.5**
0.6	0.742 15	0.745 37	0.748 57	0.751 75	0.754 90	**0.6**
0.7	0.773 37	0.776 37	0.779 35	0.782 30	0.785 23	**0.7**
0.8	0.802 34	0.805 10	0.807 85	0.810 57	0.813 27	**0.8**
0.9	0.824 94	0.831 47	0.833 97	0.836 46	0.838 91	**0.9**
1.0	0.853 14	0.855 43	0.857 69	0.859 93	0.862 14	**1.0**
1.1	0.874 93	0.876 97	0.879 00	0.881 00	0.882 97	**1.1**
1.2	0.894 35	0.896 16	0.897 96	0.899 73	0.901 47	**1.2**
1.3	0.911 49	0.913 08	0.914 65	0.916 21	0.917 73	**1.3**
1.4	0.926 47	0.927 85	0.929 22	0.930 56	0.931 89	**1.4**
1.5	0.939 43	0.940 62	0.941 79	0.942 95	0.944 08	**1.5**
1.6	0.950 53	0.951 54	0.952 54	0.953 52	0.954 48	**1.6**
1.7	0.959 94	0.960 80	0.961 64	0.962 46	0.963 27	**1.7**
1.8	0.967 84	0.968 56	0.969 26	0.969 95	0.970 62	**1.8**
1.9	0.974 41	0.975 00	0.975 58	0.976 15	0.976 70	**1.9**
2.0	0.979 82	0.980 30	0.980 77	0.981 24	0.981 69	**2.0**
2.1	0.984 22	0.984 61	0.985 00	0.985 37	0.985 74	**2.1**
2.2	0.987 78	0.988 09	0.988 40	0.988 70	0.988 99	**2.2**
2.3	0.990 61	0.990 86	0.991 11	0.991 34	0.991 58	**2.3**
2.4	0.992 86	0.993 05	0.993 24	0.993 43	0.993 61	**2.4**
2.5	0.994 61	0.994 77	0.994 92	0.995 06	0.995 20	**2.5**
2.6	0.995 98	0.996 09	0.996 21	0.996 32	0.996 43	**2.6**
2.7	0.997 02	0.997 11	0.997 20	0.997 28	0.997 36	**2.7**
2.8	0.997 81	0.997 88	0.997 95	0.998 01	0.998 07	**2.8**
2.9	0.998 41	0.998 46	0.998 51	0.998 56	0.998 61	**2.9**
3.0	0.998 86	0.998 89	0.998 93	0.998 97	0.999 00	**3.0**
3.1	0.999 18	0.999 21	0.999 24	0.999 26	0.999 29	**3.1**
3.2	0.999 42	0.999 44	0.999 46	0.999 48	0.999 50	**3.2**
3.3	0.999 60	0.999 61	0.999 62	0.999 64	0.999 65	**3.3**
3.4	0.999 72	0.999 73	0.999 74	0.999 75	0.999 76	**3.4**
3.5	0.999 81	0.999 81	0.999 82	0.999 83	0.999 83	**3.5**
3.6	0.999 87	0.999 87	0.999 88	0.999 88	0.999 89	**3.6**
3.7	0.999 91	0.999 92	0.999 92	0.999 92	0.999 92	**3.7**
3.8	0.999 94	0.999 94	0.999 95	0.999 95	0.999 95	**3.8**
3.9	0.999 96	0.999 96	0.999 96	0.999 97	0.999 97	**3.9**

Source: W. W. Hines and D. C. Montgomery, *Probability and Statistics in Engineering and Management Science*, 2d ed., ©1980, pp. 592–3. Reprinted by permission of John Wiley & Sons, Inc., New York.

Table A.4. Cumulative Poisson Distribution

	α = Mean											
x	.01	.05	.1	.2	.3	.4	.5	.6	.7	.8	.9	x
0	.990	.951	.905	.819	.741	.670	.607	.549	.497	.449	.407	0
1	1.000	.999	.995	.982	.963	.938	.910	.878	.844	.809	.772	1
2		1.000	1.000	.999	.996	.992	.986	.977	.966	.953	.937	2
3				1.000	1.000	.999	.998	.997	.994	.991	.987	3
4						1.000	1.000	1.000	.999	.999	.998	4
5									1.000	1.000	1.000	5

	α = Mean											
x	1.0	1.1	1.2	1.3	1.4	1.5	1.6	1.7	1.8	1.9	2.0	x
0	.368	.333	.301	.273	.247	.223	.202	.183	.165	.150	.135	0
1	.736	.699	.663	.627	.592	.558	.525	.493	.463	.434	.406	1
2	.920	.900	.879	.857	.833	.809	.783	.757	.731	.704	.677	2
3	.981	.974	.966	.957	.946	.934	.921	.907	.891	.875	.857	3
4	.996	.995	.992	.989	.986	.981	.976	.970	.964	.956	.947	4
5	.999	.999	.998	.998	.997	.996	.994	.992	.990	.987	.983	5
6	1.000	1.000	1.000	1.000	.999	.999	.999	.998	.997	.997	.995	6
7					1.000	1.000	1.000	1.000	.999	.999	.999	7
8									1.000	1.000	1.000	8

Table A.4. (Continued)

x	α = Mean											x
	2.2	2.4	2.6	2.8	3.0	3.5	4.0	4.5	5.0	5.5	6.0	
0	.111	.091	.074	.061	.050	.030	.018	.011	.007	.004	.002	0
1	.355	.308	.267	.231	.199	.136	.092	.061	.040	.027	.017	1
2	.623	.570	.518	.469	.423	.321	.238	.174	.125	.088	.062	2
3	.819	.779	.736	.692	.647	.537	.433	.342	.265	.202	.151	3
4	.928	.904	.877	.848	.815	.725	.629	.532	.440	.358	.285	4
5	.975	.964	.951	.935	.916	.858	.785	.703	.616	.529	.446	5
6	.993	.988	.983	.976	.966	.935	.889	.831	.762	.686	.606	6
7	.998	.997	.995	.992	.988	.973	.949	.913	.867	.809	.744	7
8	1.000	.999	.999	.998	.996	.990	.979	.960	.932	.894	.847	8
9		1.000	1.000	.999	.999	.997	.992	.983	.968	.946	.916	9
10				1.000	1.000	.999	.997	.993	.986	.975	.957	10
11						1.000	.999	.998	.995	.989	.980	11
12							1.000	.999	.998	.996	.991	12
13								1.000	.999	.998	.996	13
14									1.000	.999	.999	14
15										1.000	.999	15
16											1.000	16

Table A.4. (Continued)

x	α = Mean											x
	6.5	7.0	7.5	8.0	9.0	10.0	12.0	14.0	16.0	18.0	20.0	
0	.002	.001	.001									0
1	.011	.007	.005	.003	.001							1
2	.043	.030	.020	.014	.006	.003	.001					2
3	.112	.082	.059	.042	.021	.010	.002					3
4	.224	.173	.132	.100	.055	.029	.008	.002				4
5	.369	.301	.241	.191	.116	.067	.020	.006	.001			5
6	.527	.450	.378	.313	.207	.130	.046	.014	.004	.001		6
7	.673	.599	.525	.453	.324	.220	.090	.032	.010	.003	.001	7
8	.792	.729	.662	.593	.456	.333	.155	.062	.022	.007	.002	8
9	.877	.830	.776	.717	.587	.458	.242	.109	.043	.015	.005	9
10	.933	.901	.862	.816	.706	.583	.347	.176	.077	.030	.011	10
11	.966	.947	.921	.888	.803	.697	.462	.260	.127	.055	.021	11
12	.984	.973	.957	.936	.876	.792	.576	.358	.193	.092	.039	12
13	.993	.987	.978	.966	.926	.864	.682	.464	.275	.143	.066	13
14	.997	.994	.990	.983	.959	.917	.772	.570	.368	.208	.105	14
15	.999	.998	.995	.992	.978	.951	.844	.669	.467	.287	.157	15
16	1.000	.999	.998	.996	.989	.973	.899	.756	.566	.375	.221	16
17		1.000	.999	.998	.995	.986	.937	.827	.659	.469	.297	17

Table A.4. (Continued)

x	α = Mean											x
	6.5	7.0	7.5	8.0	9.0	10.0	12.0	14.0	16.0	18.0	20.0	
18			1.000	.999	.998	.993	.963	.883	.742	.562	.381	18
19				1.000	.999	.997	.979	.923	.812	.651	.470	19
20					1.000	.998	.988	.952	.868	.731	.559	20
21						.999	.994	.971	.911	.799	.644	21
22						1.000	.997	.983	.942	.855	.721	22
23							.999	.991	.963	.899	.787	23
24							.999	.995	.978	.932	.843	24
25							1.000	.997	.987	.955	.888	25
26								.999	.993	.972	.922	26
27								.999	.996	.983	.948	27
28								1.000	.998	.990	.966	28
29									.999	.994	.978	29
30									.999	.997	.987	30
31									1.000	.998	.992	31
32										.999	.995	32
33										1.000	.997	33
34											.999	34
35											.999	35
36											1.000	36

Source: J. Banks and R. G. Heikes, *Handbook of Tables and Graphs for the Industrial Engineer and Manager*, ©1984, pp. 34–35. Reprinted by permission of John Wiley & Sons, Inc., New York.

Table A.5. Percentage Points of the Students t Distribution with v Degrees of Freedom

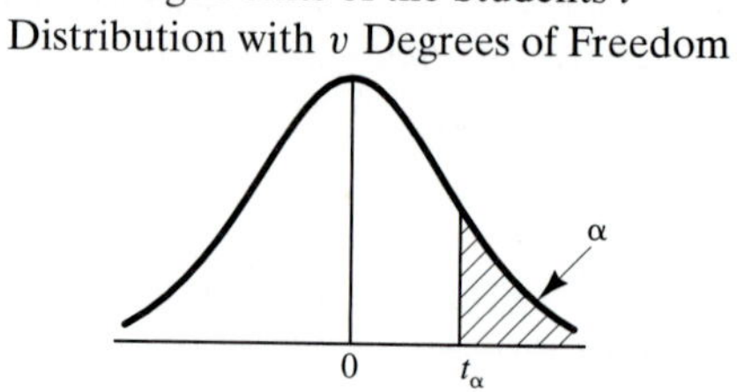

v	$t_{0.005}$	$t_{0.01}$	$t_{0.025}$	$t_{0.05}$	$t_{0.10}$
1	63.66	31.82	12.71	6.31	3.08
2	9.92	6.92	4.30	2.92	1.89
3	5.84	4.54	3.18	2.35	1.64
4	4.60	3.75	2.78	2.13	1.53
5	4.03	3.36	2.57	2.02	1.48
6	3.71	3.14	2.45	1.94	1.44
7	3.50	3.00	2.36	1.90	1.42
8	3.36	2.90	2.31	1.86	1.40
9	3.25	2.82	2.26	1.83	1.38
10	3.17	2.76	2.23	1.81	1.37
11	3.11	2.72	2.20	1.80	1.36
12	3.06	2.68	2.18	1.78	1.36
13	3.01	2.65	2.16	1.77	1.35
14	2.98	2.62	2.14	1.76	1.34
15	2.95	2.60	2.13	1.75	1.34
16	2.92	2.58	2.12	1.75	1.34
17	2.90	2.57	2.11	1.74	1.33
18	2.88	2.55	2.10	1.73	1.33
19	2.86	2.54	2.09	1.73	1.33
20	2.84	2.53	2.09	1.72	1.32
21	2.83	2.52	2.08	1.72	1.32
22	2.82	2.51	2.07	1.72	1.32
23	2.81	2.50	2.07	1.71	1.32
24	2.80	2.49	2.06	1.71	1.32
25	2.79	2.48	2.06	1.71	1.32
26	2.78	2.48	2.06	1.71	1.32
27	2.77	2.47	2.05	1.70	1.31
28	2.76	2.47	2.05	1.70	1.31
29	2.76	2.46	2.04	1.70	1.31
30	2.75	2.46	2.04	1.70	1.31
40	2.70	2.42	2.02	1.68	1.30
60	2.66	2.39	2.00	1.67	1.30
120	2.62	2.36	1.98	1.66	1.29
∞	2.58	2.33	1.96	1.645	1.28

Source: Robert E. Shannon, *Systems Simulation: The Art and Science*, ©1975, p. 375. Reprinted by permission of Prentice Hall, Upper Saddle River, NJ.

Table A.6. Percentage Points of the Chi-Square Distribution with v Degrees of Freedom

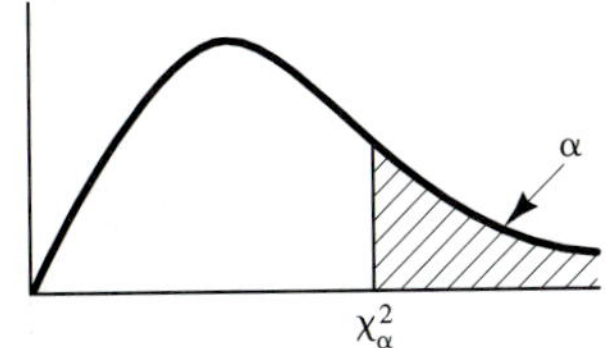

v	$\chi^2_{0.005}$	$\chi^2_{0.01}$	$\chi^2_{0.025}$	$\chi^2_{0.05}$	$\chi^2_{0.10}$
1	7.88	6.63	5.02	3.84	2.71
2	10.60	9.21	7.38	5.99	4.61
3	12.84	11.34	9.35	7.81	6.25
4	14.96	13.28	11.14	9.49	7.78
5	16.7	15.1	12.8	11.1	9.2
6	18.5	16.8	14.4	12.6	10.6
7	20.3	18.5	16.0	14.1	12.0
8	22.0	20.1	17.5	15.5	13.4
9	23.6	21.7	19.0	16.9	14.7
10	25.2	23.2	20.5	18.3	16.0
11	26.8	24.7	21.9	19.7	17.3
12	28.3	26.2	23.3	21.0	18.5
13	29.8	27.7	24.7	22.4	19.8
14	31.3	29.1	26.1	23.7	21.1
15	32.8	30.6	27.5	25.0	22.3
16	34.3	32.0	28.8	26.3	23.5
17	35.7	33.4	30.2	27.6	24.8
18	37.2	34.8	31.5	28.9	26.0
19	38.6	36.2	32.9	30.1	27.2
20	40.0	37.6	34.2	31.4	28.4
21	41.4	38.9	35.5	32.7	29.6
22	42.8	40.3	36.8	33.9	30.8
23	44.2	41.6	38.1	35.2	32.0
24	45.6	43.0	39.4	36.4	33.2
25	49.6	44.3	40.6	37.7	34.4
26	48.3	45.6	41.9	38.9	35.6
27	49.6	47.0	43.2	40.1	36.7
28	51.0	48.3	44.5	41.3	37.9
29	52.3	49.6	45.7	42.6	39.1
30	53.7	50.9	47.0	43.8	40.3
40	66.8	63.7	59.3	55.8	51.8
50	79.5	76.2	71.4	67.5	63.2
60	92.0	88.4	83.3	79.1	74.4
70	104.2	100.4	95.0	90.5	85.5
80	116.3	112.3	106.6	101.9	96.6
90	128.3	124.1	118.1	113.1	107.6
100	140.2	135.8	129.6	124.3	118.5

Source: Robert E. Shannon, *Systems Simulation: The Art and Science*, ©1975, p. 372. Reprinted by permission of Prentice Hall, Upper Saddle River, NJ.

Table A.7. Percentage Points of the F Distribution with $\alpha = 0.05$

Degrees of Freedom for Denominator (v)	*Degrees of Freedom for the Numerator* (v_1) 1	2	3	4	5	6	7	8	9	10	12	15	20	24	30	40	60	120	∞
1	161.4	199.5	215.7	224.6	230.2	234.0	236.8	238.9	240.5	241.9	243.9	245.9	248.0	249.1	250.1	251.1	252.2	253.3	254.3
2	18.51	19.00	19.16	19.25	19.30	19.33	19.35	19.37	19.38	19.40	19.41	19.43	19.45	19.45	19.46	19.47	19.48	19.49	19.50
3	10.13	9.55	9.28	9.12	9.01	8.94	8.89	8.85	8.81	8.79	8.74	8.70	8.66	8.64	8.62	8.59	8.57	8.55	8.53
4	7.71	6.94	6.59	6.39	6.26	6.16	6.09	6.04	6.00	5.96	5.91	5.86	5.80	5.77	5.75	5.72	5.69	5.66	5.63
5	6.61	5.79	5.41	5.19	5.05	4.95	4.88	4.82	4.77	4.74	4.68	4.62	4.56	4.53	4.50	4.46	4.43	4.40	4.36
6	5.99	5.14	4.76	4.53	4.39	4.28	4.21	4.15	4.10	4.06	4.00	3.94	3.87	3.84	3.81	3.77	3.74	3.70	3.67
7	5.59	4.74	4.35	4.12	3.97	3.87	3.79	3.73	3.68	3.64	3.57	3.51	3.44	3.41	3.38	3.34	3.30	3.27	3.23
8	5.32	4.46	4.07	3.84	3.69	3.58	3.50	3.44	3.39	3.35	3.28	3.22	3.15	3.12	3.08	3.04	3.01	2.97	2.93
9	5.12	4.26	3.86	3.63	3.48	3.37	3.29	3.23	3.18	3.14	3.07	3.01	2.94	2.90	2.86	2.83	2.79	2.75	2.71
10	4.96	4.10	3.71	3.48	3.33	3.22	3.14	3.07	3.02	2.98	2.91	2.85	2.77	2.74	2.70	2.66	2.62	2.58	2.54
11	4.84	3.98	3.59	3.36	3.20	3.09	3.01	2.95	2.90	2.85	2.79	2.72	2.65	2.61	2.57	2.53	2.49	2.45	2.40
12	4.75	3.89	3.49	3.26	3.11	3.00	2.91	2.85	2.80	2.75	2.69	2.62	2.54	2.51	2.47	2.43	2.38	2.34	2.30
13	4.67	3.81	3.41	3.18	3.03	2.92	2.83	2.77	2.71	2.67	2.60	2.53	2.46	2.42	2.38	2.34	2.30	2.25	2.21
14	4.60	3.74	3.34	3.11	2.96	2.85	2.76	2.70	2.65	2.60	2.53	2.46	2.39	2.35	2.31	2.27	2.22	2.18	2.13
15	4.54	3.68	3.29	3.06	2.90	2.79	2.71	2.64	2.59	2.54	2.48	2.40	2.33	2.29	2.25	2.20	2.16	2.11	2.07
16	4.49	3.63	3.24	3.01	2.85	2.74	2.66	2.59	2.54	2.49	2.42	2.35	2.28	2.24	2.19	2.15	2.11	2.06	2.01
17	4.45	3.59	3.20	2.96	2.81	2.70	2.61	2.55	2.49	2.45	2.38	2.31	2.23	2.19	2.15	2.10	2.06	2.01	1.96
18	4.41	3.55	3.16	2.93	2.77	2.66	2.58	2.51	2.46	2.41	2.34	2.27	2.19	2.15	2.11	2.06	2.02	1.97	1.92
19	4.38	3.52	3.13	2.90	2.74	2.63	2.54	2.48	2.42	2.38	2.31	2.23	2.16	2.11	2.07	2.03	1.98	1.93	1.88
20	4.35	3.49	3.10	2.87	2.71	2.60	2.51	2.45	2.39	2.35	2.28	2.20	2.12	2.08	2.04	1.99	1.95	1.90	1.84
21	4.32	3.47	3.07	2.84	2.68	2.57	2.49	2.42	2.37	2.32	2.25	2.18	2.10	2.05	2.01	1.96	1.92	1.87	1.81
22	4.30	3.44	3.05	2.82	2.66	2.55	2.46	2.40	2.34	2.30	2.23	2.15	2.07	2.03	1.98	1.94	1.89	1.84	1.78
23	4.28	3.42	3.03	2.80	2.64	2.53	2.44	2.37	2.32	2.27	2.20	2.13	2.05	2.01	1.96	1.91	1.86	1.81	1.76
24	4.26	3.40	3.01	2.78	2.62	2.51	2.42	2.36	2.30	2.25	2.18	2.11	2.03	1.98	1.94	1.89	1.84	1.79	1.73
25	4.24	3.39	2.99	2.76	2.60	2.49	2.40	2.34	2.28	2.24	2.16	2.09	2.01	1.96	1.92	1.87	1.82	1.77	1.71
26	4.23	3.37	2.98	2.74	2.59	2.47	2.39	2.32	2.27	2.22	2.15	2.07	1.99	1.95	1.90	1.85	1.80	1.75	1.69
27	4.21	3.35	2.96	2.73	2.57	2.46	2.37	2.31	2.25	2.20	2.13	2.06	1.97	1.93	1.88	1.84	1.79	1.73	1.67
28	4.20	3.34	2.95	2.71	2.56	2.45	2.36	2.29	2.24	2.19	2.12	2.04	1.96	1.91	1.87	1.82	1.77	1.71	1.65
29	4.18	3.33	2.93	2.70	2.55	2.43	2.35	2.28	2.22	2.18	2.10	2.03	1.94	1.90	1.85	1.81	1.75	1.70	1.64
30	4.17	3.32	2.92	2.69	2.53	2.42	2.33	2.27	2.21	2.16	2.09	2.01	1.93	1.89	1.84	1.79	1.74	1.68	1.62
40	4.08	3.23	2.84	2.61	2.45	2.34	2.25	2.18	2.12	2.08	2.00	1.92	1.84	1.79	1.74	1.69	1.64	1.58	1.51
60	4.00	3.15	2.76	2.53	2.37	2.25	2.17	2.10	2.04	1.99	1.92	1.84	1.75	1.70	1.65	1.59	1.53	1.47	1.39
120	3.92	3.07	2.68	2.45	2.29	2.17	2.09	2.02	1.96	1.91	1.83	1.75	1.66	1.61	1.55	1.55	1.43	1.35	1.25
∞	3.84	3.00	2.60	2.37	2.21	2.10	2.01	1.94	1.88	1.83	1.75	1.67	1.57	1.52	1.46	1.39	1.32	1.22	1.00

Source: W. W. Hines and D. C. Montgomery, *Probability and Statistics in Engineering and Management Science*, 2d ed., ©1980, p. 599. Reprinted by permission of John Wiley & Sons, Inc., New York.

Table A.8. Kolmogorov–Smirnov Critical Values

Degrees of Freedom (N)	$D_{0.10}$	$D_{0.05}$	$D_{0.01}$
1	0.950	0.975	0.995
2	0.776	0.842	0.929
3	0.642	0.708	0.828
4	0.564	0.624	0.733
5	0.510	0.565	0.669
6	0.470	0.521	0.618
7	0.438	0.486	0.577
8	0.411	0.457	0.543
9	0.388	0.432	0.514
10	0.368	0.410	0.490
11	0.352	0.391	0.468
12	0.338	0.375	0.450
13	0.325	0.361	0.433
14	0.314	0.349	0.418
15	0.304	0.338	0.404
16	0.295	0.328	0.392
17	0.286	0.318	0.381
18	0.278	0.309	0.371
19	0.272	0.301	0.363
20	0.264	0.294	0.356
25	0.24	0.27	0.32
30	0.22	0.24	0.29
35	0.21	0.23	0.27
Over 35	$\frac{1.22}{\sqrt{N}}$	$\frac{1.36}{\sqrt{N}}$	$\frac{1.63}{\sqrt{N}}$

Source: F. J. Massey, "The Kolmogorov–Smirnov Test for Goodness of Fit," *The Journal of the American Statistical Association*, Vol. 46, ©1951, p. 70. Adapted with permission of the American Statistical Association.

Table A.9. Maximum-Likelihood Estimates of the Gamma Distribution

$1/M$	β	$1/M$	β
0.020	0.0187	5.200	2.755
0.030	0.0275	5.400	2.855
0.040	0.0360	5.600	2.956
0.050	0.0442	5.800	3.056
0.060	0.0523	6.000	3.156
0.070	0.0602	6.200	3.257
0.080	0.0679	6.400	3.357
0.090	0.0756	6.600	3.457
0.100	0.0831	6.800	3.558
0.200	0.1532	7.000	3.658
0.300	0.2178	7.300	3.808
0.400	0.2790	7.600	3.958
0.500	0.3381	7.900	4.109
0.600	0.3955	8.200	4.259
0.700	0.4517	8.500	4.409
0.800	0.5070	8.800	4.560
0.900	0.5615	9.100	4.710
1.000	0.6155	9.400	4.860
1.100	0.6690	9.700	5.010
1.200	0.7220	10.000	5.160
1.300	0.7748	10.300	5.311
1.400	0.8272	10.600	5.461
1.500	0.8794	10.900	5.611
1.600	0.9314	11.200	5.761
1.700	0.9832	11.500	5.911
1.800	1.034	11.800	6.061
1.900	1.086	12.100	6.211
2.000	1.137	12.400	6.362
2.100	1.188	12.700	6.512
2.200	1.240	13.000	6.662
2.300	1.291	13.300	6.812
2.400	1.342	13.600	6.962
2.500	1.393	13.900	7.112
2.600	1.444	14.200	7.262
2.700	1.494	14.500	7.412
2.800	1.545	14.800	7.562
2.900	1.596	15.100	7.712
3.000	1.646	15.400	7.862
3.200	1.748	15.700	8.013
3.400	1.849	16.000	8.163
3.600	1.950	16.300	8.313
3.800	2.051	16.600	8.463
4.000	2.151	16.900	8.613
4.200	2.252	17.200	8.763
4.400	2.353	17.500	8.913
4.600	2.453	17.800	9.063
4.800	2.554	18.100	9.213
5.000	2.654	18.400	9.363
		18.700	9.513
		19.000	9.663
		19.300	9.813
		19.600	9.963
		20.000	10.16

Source: S. C. Choi and R. Wette, "Maximum Likelihood Estimates of the Gamma Distribution and Their Bias," *Technometrics*, Vol. 11, No. 4, ©1969, pp. 688–9. Adapted with permission of the American Statistical Association.

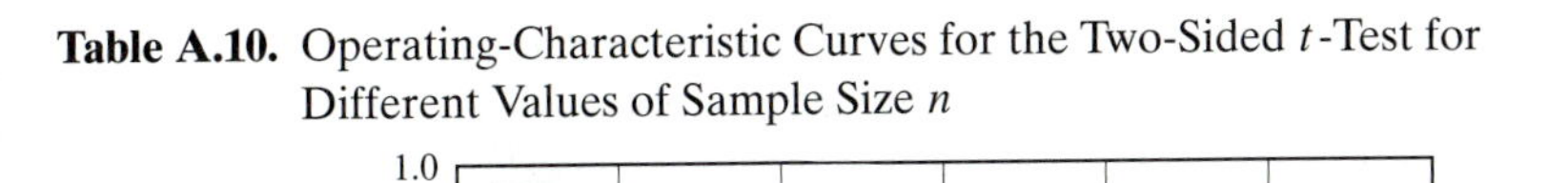

Table A.10. Operating-Characteristic Curves for the Two-Sided t-Test for Different Values of Sample Size n

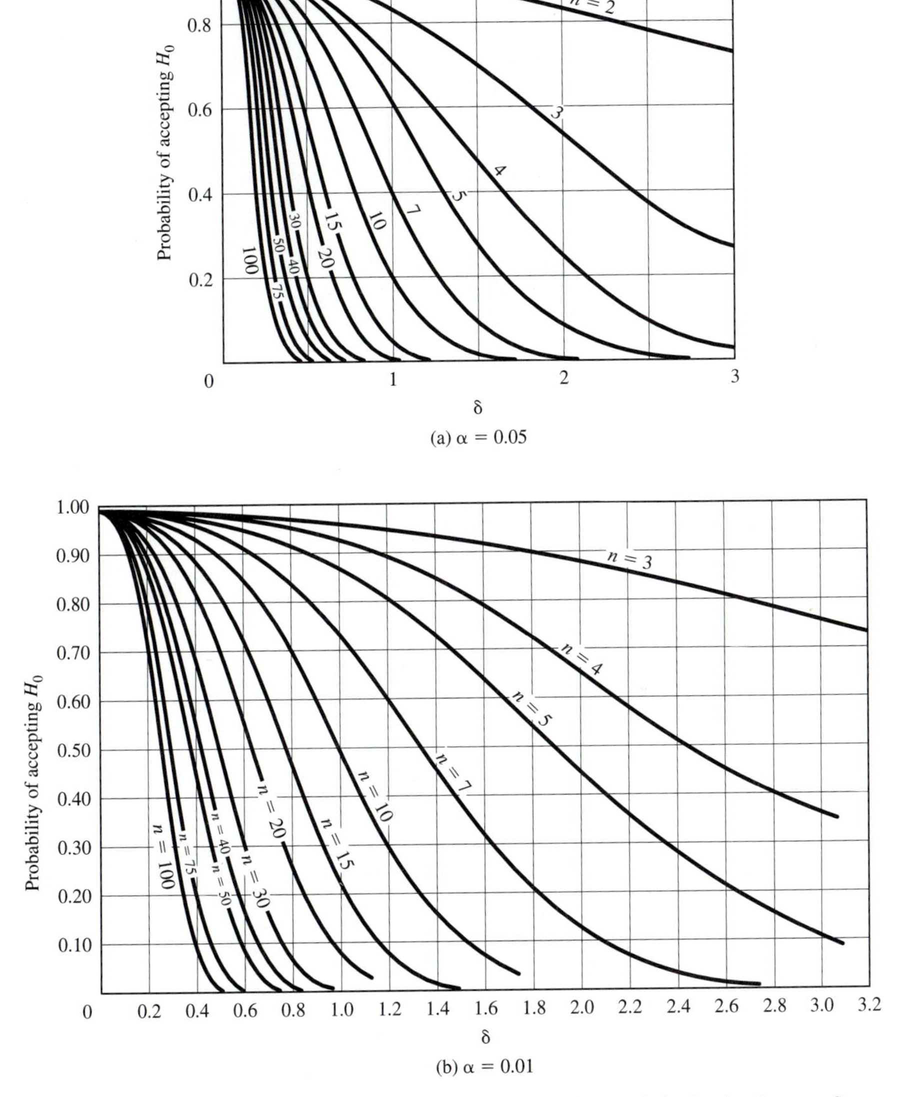

(a) $\alpha = 0.05$

(b) $\alpha = 0.01$

Source: C. L. Ferris, F. E. Grubbs, and C. L. Weaver, "Operating Characteristics for the Common Statistical Tests of Significance," *Annals of Mathematical Statistics*, ©1946. Reproduced with permission of The Institute of Mathematical Statistics.

Table A.11. Operating-Characteristic Curves for the One-Sided t-Test for Different Values of Sample Size n

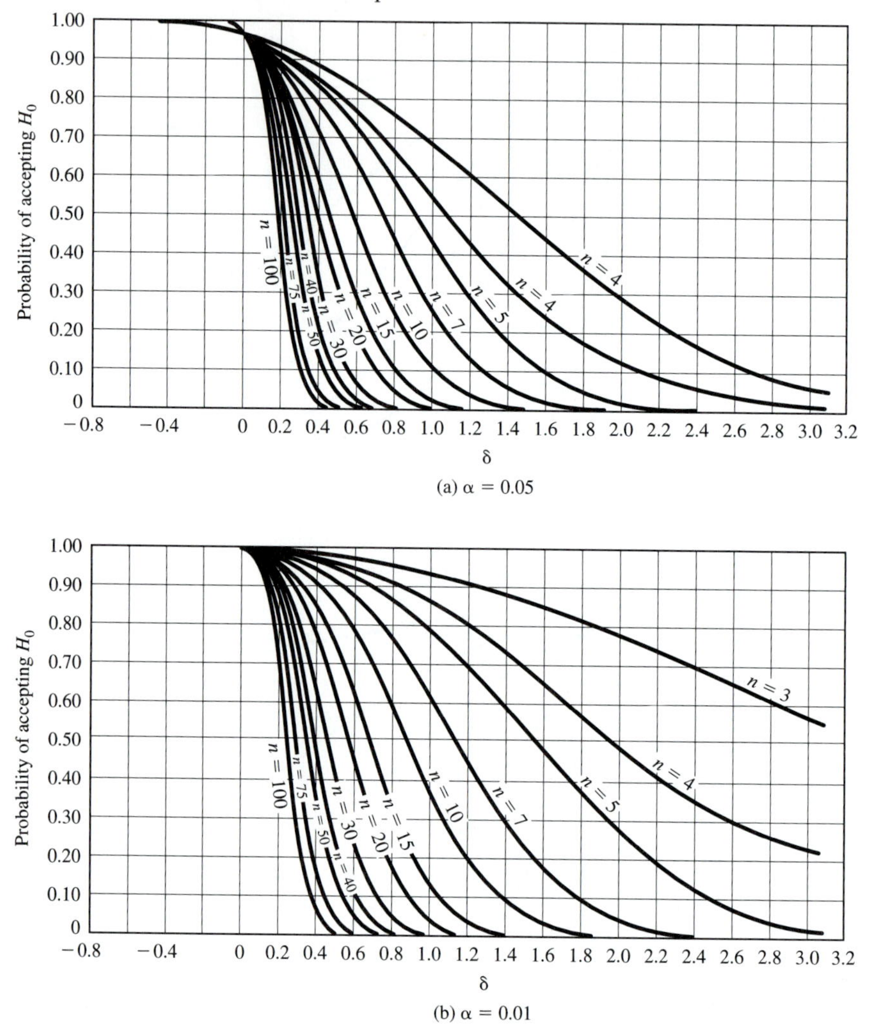

(a) $\alpha = 0.05$

(b) $\alpha = 0.01$

Source: A. H. Bowker and G. J. Lieberman, *Engineering Statistics*, 2d ed., ©1972, p. 203. Reprinted by permission of Prentice Hall, Upper Saddle River, NJ.

Index

A

Abstraction levels
  computer systems, 529, 542
Acceptance-rejection technique
  random-variate generation, 310–315
Activity, 10
  deterministic, 65
  discrete-event simulation concept and examples, 64, 66–85
  duration types, 65
  examples, 11t
  state variable, 65
  statistical, 65
Analysis features
  simulation software, 103t
Analytical methods
  model solving, 14
Animation
  simulation software, 102t
Approximation
  multiserver queue, 235–236
Arena, 123–124, 129–131
Arrays
  list processing, 87–88
Arrival process, 207–209
Attribute, 10
  discrete-event simulation concept and examples, 64, 66–85
  examples, 11t
Autocorrelation
  batch size, 437
  steady-state simulation, 427–429
Autocorrelation test
  for random numbers, 264, 278–280
Automated guided vehicles (AGVs), 506
AutoMod, 124–125
AutoSimulations, 124–125
AutoStat, 130
AutoView, 124–125

B

Batch means, 422
  steady-state simulation, 436–440
Batch size
  steady-state simulation, 437–438
Bernoulli distribution, 165
Bernoulli trials, 165
Beta distribution, 165
  selection, 332
Binomial distribution, 165–166
  selection, 332
Bonferroni approach
  multiple system design comparison, 468–473
  selection procedure, 473–476
Bootstrapping, 71
Bottom
  list, 86
Boundary, 9
Breakdowns
  simulation, 510–513

C

C++, 104–106
  examples, 106–114
Cache, 564–566
Calibration
  models, 374–393
Calling population, 206–207
Capacity
  system, 207
Chi-square percentage points
  distribution, 581t
Chi-square test
  input modeling, 343–345
Clock
  discrete-event simulation concept and examples, 65, 66–85
Clock-time breakdowns, 510
Closed-form inverse
  continuous distributions, 300–301
Common random numbers. *See also* Correlated sampling
  two system design comparison, 456–459
Comparison
  two system design, 451–467
Component libraries, 133
Computer system
  high-level simulation, 553–557
  simulation, 528–567

Conditional wait, 65
Confidence interval estimation
  quantiles, 416–417, 440
  terminating simulation, 411–412, 414–416
  two system design comparison, 451, 466–467
Conservation equation, 216
Continuous data
  examples, 329–330
Continuous distributions, 170–190
  inverse transform technique, 300–301
Continuous empirical distributions, 193–195
Continuous random variables, 155–157
Continuous system, 12
Continuous-time data, 407
Continuous-time Markov chain (CTMC), 543–547
Continuous uniform distributions
  selection, 333, 337–338
Control sampling variability
  optimization via simulation, 491–492
Conventional limitations
  input modeling selection, 352
Conveyors
  material handling systems, 506–507
Convolution method
  random-variate generation, 309–310
Correlated sampling, 450
  standard errors comparison, 473t
  two system design comparison, 456–459, 464
Correlation, 354
Covariance, 354
Covariance stationary, 354
  steady-state simulation, 426
CPU
  level, 529
  simulation, 528–567
CSIM, 119–120, 533–537
  examples, 121–123
Cumulative average sample mean, 423, 425
Cumulative distribution function, 157–158
Cumulative normal distribution, 574t–575t
Cumulative Poisson distribution, 576t–579t
Current contents, 371
`Customer` threads, 535
Customers
  queueing system, 205
Cycle breakdowns, 510

D

Data collection, 324–327
  steady-state simulation, 420
Data distribution identification, 327–336
Data exchange standards, 132
Delay
  discrete-event simulation concept and examples, 64, 66–85
Deletion point, 424
Deneb, 125–126
Deneb Robotics, 125–126
Deterministic activity, 65
Deterministic models, 13
Deterministic queue, 221f
Direct execution simulation
  input generation, 552
Direct transformation
  normal and lognormal distribution, 307–309
Discrete data
  examples, 328–329
Discrete distributions, 165–170
  inverse transform technique, 301–307
Discrete empirical distributions, 193–195
Discrete-random variables, 154, 165
Discrete system, 12
Discrete-time data, 407
Discrete uniform distributions
  examples, 304
  selection, 333
Discrete-event system simulation, 14
  concepts and methodologies, 63–93
Distribution. *See also* individual types
  Chi-square percentage points, 581t
  cumulative normal, 574t
  cumulative Poisson, 576t
  identification, 327–336
  F percentage points, 582t
  selection, 331–333
  Student percentage points, 580t
Documentation, 18
  simulation software, 103t
Downtime model
  manufacturing and material handling systems, 508–511
Dynamic allocation
  linked lists, 90
Dynamic simulation, 13

E

Embedded simulation, 134

Empirical discrete distribution
examples, 302–303
selection, 333
Empirical distributions, 193–195
inverse transform technique, 296–300
End-service event, 73
Endogenous event, 10
Engineering Animation Incorporated, Inc., 132
Engineering data
input modeling selection, 352
Ensemble averages, 422, 424, 425
Entity, 10
discrete-event simulation concept and examples, 64, 66–85
examples, 11t
Equal probabilities
Chi-square test, 346–347
Equal variances
two system design comparison, 454–456
Equipment
material handling systems, 506–507
Ergodic, 544
Erlang distribution, 176–178
selection, 332
Event, 10
discrete-event simulation concept and examples, 64, 66–85
examples, 11t
primary, 65
queueing system, 25
Event list
discrete-event simulation concept and examples, 64, 66–85
future, 67–72
Event notice
discrete-event simulation concept and examples, 64, 66–85
Event orientation, 537–542
Event scheduling
algorithm, 67–72
approach, 86–87
manual simulation, 75–85
Exogenous event, 10
Expectation, 158–160
Experimentation features, 129–131
simulation software, 103t
Expert option
input modeling selection, 352
Exponential distribution, 172–174
Chi-square test, 347–348
inverse transform technique, 290–294
Kolmogorov-Smirnov goodness-of-fit tests, 349–350
selection, 332, 340
Extend, 126–127

F

F&H Simulations, 128
Face validity
validation process, 376
Failures
manufacturing and material handling systems, 508–511
Family distribution selection
data distribution identification, 331–333
Fixed total sample size, 437, 467
Frequency test
for random numbers, 264, 266–269
Fully associative cache, 564
Future event list (FEL), 67–72

G

Gamma distribution, 174–175
acceptance-rejection technique, 314–315
maximum likelihood estimates, 584t
selection, 332, 339–340
Gap test
for random numbers, 264, 281–282
Garbage in, garbage out (GIGO), 324
Gate level, 529
gcc compiler
scatter-plotted referencing pattern, 550f
General-purpose programming languages
simulation model development, 95
Genetic algorithms (GA), 489
robust heuristics, 490–491
Geometric distribution, 167
examples, 306–307
Goodness-of-fit tests, 343–351
Chi-square, 343–348
Kolmogorov-Smirnov, 348–349
GPSS, 114–115
examples, 115–119

H

Head
list, 86
headptr, 88
High-fidelity simulation, 131–132
High Level Architecture (HLA), 133–134
High-level computer system
simulation, 553–557

Histograms
  data distribution identification, 327–331
Historical input data
  validation process, 388–393
Historical output data
  validation process, 388–393
Hit ratio, 547

I

IIE Magazine, 515
Imagine That Inc., 126–127
Independent sampling, 450
  standard errors comparison, 473t
  two system design comparison, 454–456, 456
Individual batch means
  steady-state simulation, 421t
Initial conditions
  terminating simulation, 399
Initialization bias
  steady-state simulations, 419–426
Input modeling, 323–359
  selection without data, 351–353
Input transformations
  validation process, 378–388
Instruction complete, 559
Instruction decode, 558
Instruction execute, 559
Instruction fetch, 558
Instruction graduate, 559
Instruction issue, 559
Instruction-level parallelism (ILP), 558–560
Intelligent initialization. *See* Point estimation
Internet, 132–133
Interval estimation
  performance parameter, 407–409
  steady-state simulation, 436–440
Inventory system
  examples, 42–47
  simulation table, 46t
  statistical models, 163–164
Inverse transform technique
  random-variate generation, 290–307

K

Kolmogorov-Smirnov
  critical values, 583t
  goodness-of-fit tests, 348–349

L

Lag, 426
Lag-*h* autocorrelation, 354
Lag-*h* autocovariance, 354
Lanner Group, 128–129
Layout features
  simulation software, 102t
Lead-time demand
  examples, 53–55
Least recently used (LRU), 565–566
Length of runs
  runs test, 274–278
Limited data
  statistical models, 164–165
Linear regression
  metamodeling, 477–481
  multiple, 484
Linked lists
  dynamic allocation, 90
List
  discrete-event simulation concept and examples, 64, 85–92
  processing, 85–88
  properties and operations, 86–88
Lognormal distribution, 189–190
  direct transformation, 307–309
  selection, 332, 338–339
Long-run measures of performance
  queueing systems, 212–224
Long-run simulation. *See* Steady-state simulations

M

Maintainability
  statistical models, 164
Manual simulation
  event scheduling, 75–85
Manufacturing systems
  case studies, 515–517
  goals and performance measures, 507–508
  models, 503–504
  simulation, 502–517
Markov chain, 530
Markovian models
  steady-state behavior of finite population models, 239–243
  steady-state behavior of infinite population, 224–239
Material handling systems
  case studies, 515–517
  equipment, 506–507

Material handling systems (*continued*)
 goals and performance measures, 507–508
 models, 505–507
 simulation, 502–517
Measures of performance
 estimation, 407–410
 of system, 3
Memory simulation, 563–566
Metamodeling, 476–485
Methods
 analytical, 14
 convolution, 309–310
 numerical, 14
 replication, 430, 451
Micro Analysis and Design, Inc., 127
Micro Saint, 127
Mode, 160
Model
 adequacy testing, 481–484
 calibration and validation, 374–393
 complex system principles, 63–93
 definition, 3
 deterministic, 13
 discrete-event simulation concept and examples, 64, 66–85
 downtime, 508–511
 downtime and failures, 508–511
 input, 323–359, 343–345, 542–552
 manufacturing and material handling systems, 502–517
 manufacturing systems, 503–504
 Markovian, 224–239
 material handling systems, 505–507
 multivariate input, 353–358
 queueing, 204–246, 425–426, 531–533
 simulation study steps, 15–20
 solving, 14
 statistical, 163–164, 164–165
 of system, 13
 time-series input, 353–358
 trace-driven, 513–515
 types, 13–14
 validation, 374–393
Model-building features
 simulation software, 101t
 verification and validation, 368–369
Modern Material Handling, 515
ModL, 127
Modulated Poisson process (MPP), 543–547
Monotonicity, 459
Multiple linear regression, 484
Multiserver queue, 231–235
 approximation, 235–236
 Poisson arrivals, 237–238
Multivariate input models, 353–358

N

Negative binomial distribution
 selection, 332
Networks
 queues, 243–245
Nonstationary simulation. *See* Transient simulation
Nonterminating simulation
 output analysis, 401
Normal distribution, 178–184
 direct transformation, 307–309
 selection, 332, 339
Numerical methods
 model solving, 14

O

Operating characteristic curves
 one-sided test, 586t
 two-sided test, 585t
Optimization, 134
Optimization via simulation, 485–486
 complications, 488–489
 definition, 487–488
OptQuest, 130
Output analysis
 comparison and evaluation, 450–496
 single model, 398–441
 steady-state simulations, 418–440
 terminating simulations, 410–417
Output analyzer, 129–130
Output transformations
 validation process, 378–388

P

p-values
 goodness-of-fit tests, 350–351
Page fault, 547
Page frames, 547
Paging systems, 564
Parameter
 scale and shape, 175
Parameter estimation
 input modeling, 336–343
Performance measures
 manufacturing systems, 507–508
 material handling systems, 507–508

Performance parameter
  estimation, 407–410
Petri nets, 531–533
Physical limitation
  input modeling selection, 352
Pipeline stages
  ILP CPU simulation model, 561–562
Pipelining, 558
Point estimation
  performance parameter, 407–409
  steady-state simulations, 419–426
Poisson arrivals
  multiserver queue, 237–238
Poisson distribution, 168–170
  acceptance-rejection technique, 311–314
  Chi-square test, 345
  selection, 332, 338
Poisson process, 190–193, 224
  single-channel queueing system, 225–227
Poker test
  for random numbers, 264, 283–284
Pooled process, 193
Practically significant
  versus statistically significant, 454
Preliminary statistics, 336–337
Primary event, 65
Probabilistic nature. *See* Stochastic simulation
Process analyzer, 129–130
Process orientation, 533–537, 556
Processor level, 529
Product documentation
  simulation software, 103t
ProModel, 127–128
PROMODEL Corporation, 127–128
Proof Animation, 126–127
Pseudocode
  generating reference trace, 551f
Pseudorandom numbers
  generation, 256–258

Q

Quantile-quantile plots, 333–334
Quantiles
  confidence interval estimation, 416–417, 440
QUEST, 125–126
Queue behavior, 209
Queue discipline, 209
Queueing models, 204–246, 531–533
  simulation data, 425–426
Queueing notation, 211–212
Queueing situations
  costs, 223–224
Queueing system, 24–30
  characteristics, 205–211
  examples, 206t
  long-run measures of performance, 212–224
  simulation table, 34t
  statistical models, 161–163
Queues
  networks, 243–245

R

Random digits, 572t
Random normal numbers, 573t
Random-number assignment
  regression analysis, 484–485
Random numbers, 31
  examples, 51–53
  generating techniques, 258–264
  generation, 255–284
  properties, 255–256
  pseudo, 256–258
  tests for, 264–284
Random-search algorithm
  optimization via simulation, 492–495
Random splitting, 193
Random variables
  continuous, 155–157
  discrete, 154, 165
Random-variate generation, 289–315
Reference trace
  generating pseudocode, 551f
Register-transfer language, 530
Regression analysis, 481–484
  multiple, 484
  random-number assignment, 484–485
  significance testing, 481–484
Reliability
  examples, 47–51
  statistical models, 164
Replication method
  steady-state simulation, 430
  two system design comparison, 451
Restarting
  optimization via simulation, 492
Robust heuristics, 489–491
Runs above and below the mean
  runs test, 272–274
Runs test
  for random numbers, 264, 270–278

Runs up and runs down
  runs test, 270–272
Runtime environment
  simulation software, 102t

S

Sample mean, 336–337
Sample size, 437
  steady-state simulation, 434–436
Sample variance, 336–337
Scale parameter, 175
Scatter-plotted referencing pattern
  gcc compiler, 550f
Schedule threads, 535
SDX, 132
Sensitivity analysis
  validation process, 376
Sequential total sample size, 437–438, 467
Server utilization, 218–223
Servers
  infinite, 236–237
  queueing system, 205
Service channel
  multiple examples, 37–42
Service mechanism, 209
Service rate, 218
Service times, 209
Service variable
  utilization, 230–231
Set associativity
  concept, 564
Shape parameter, 175
Significance of regression
  testing, 481–484
SIMAN simulation language, 124
Simple linear regression
  metamodeling, 477–481
SimRunner, 131
Simulation
  advantages, 6
  application, 7–9
  breakdowns, 510–513
  clock, 25
  computer system, 528–567
  definition, 3
  disadvantages, 6–7
  examples, 23–55
  high-fidelity, 131–132
  high-level computer system, 553–557
  manufacturing and material handling systems, 502–517
  model (*See also* Model)
  optimization, 485–486
  steps, 15–20
  tools, 531–542
  when to use, 4–5
Simulation data
  queueing models, 425–426
Simulation Data Exchange, 132
Simulation environments
  simulation model development, 95
Simulation kernel library, 552
Simulation model
  ILP CPU, 560
  verification and validation, 367–394
Simulation model-building process, 16f
Simulation programming languages
  simulation model development, 95
Simulation software, 95–134
  history, 96–99
  selection, 100–103
Simulation table
  inventory system, 46t
  queueing problems, 34t
Single-channel queueing system, 28
  example(s), 30–37, 104
  Poisson process, 225–227
Software
  simulation, 95–134
Stack distance, 565–566
Standard errors
  comparison, 473t
State variable activity, 65
State variables, 10
  discrete-event simulation concept and examples, 64, 66–85
  examples, 11t
  queueing system, 25
Static simulation, 13
Stationary probability, 544
Statistical activity, 65
Statistical analysis tools, 129–131
Statistical background
  steady-state simulation, 426–430
  terminating simulation, 410–411
Statistical models, 153–196
  terminology and concepts, 154–160
  useful, 160–165
Statistical theorems, 61
Statistically significant
  versus practically significant, 454
Steady-state behavior of finite population Markovian models, 239–243

Steady-state behavior of infinite population Markovian models, 224–239
Steady-state simulation(s)
 output analysis, 399, 401–402, 418–440
Stochastic simulation, 14
 output analysis, 402–407
Stopping time
 terminating simulation, 399
Support
 simulation software, 103t
System
 components, 10–11
 definition, 9
 discrete and continuous, 12
 discrete-event simulation concept and examples, 64, 66–85
 examples, 11t
System capacity, 207
System design
 comparison and evaluation, 450–496
System environment, 9
System performance
 server utilization, 221
System state
 discrete-event simulation concept and examples, 64, 66–85
Systems Modeling Corporation, 123–124

T

Tabu search (TS), 489
 robust heuristics, 491
Tail
 list, 86
tailptr, 88
Taylor Enterprise Dynamics, 128
Terminating simulation(s)
 output analysis, 399–401, 410–417
Test vectors, 530
Theorems
 statistical, 61
Threaded approach, 556
Time-advance algorithm, 67–72
Time-average number, 213
Time-integrated average, 214
Time-series input models, 353–358
Top
 list, 86
Total counts, 371
Total sample size
 fixed, 437
 sequentially, 437–438
Trace-driven models, 513–515
Transient simulation
 output analysis, 399, 401
Triangular distribution, 165, 187–189
 inverse transform technique, 295–296
 selection, 333
Turing test
 input-output validation, 392–393
Two-stage Bonferroni approach, 474–476
Two system design
 comparison and evaluation, 450–496

U

Unconditional wait, 65
Unequal variances
 two system design comparison, 456
Uniform distribution, 164, 170–172
 inverse transform technique, 294
User threads, 533
Utilization
 service variable, 230–231

V

Validate model assumption
 validation process, 377
Validation
 models, 374–393
 simulation models, 367–394
Verification
 simulation models, 367–394
VHDL, 531–533
Virtual memory referencing, 547–552

W

Website
 server system, 553–557, 554f
 specifications, 556f
 WSC, 8
Weibull distribution, 185–186
 inverse transform technique, 294–295
 selection, 332, 341–343
Winter Simulation Conference (WSC), 7
 sponsored by, 7
 website, 8
Winter Simulation Conference Proceedings, 515
Witness, 128–129
Work in process (WIP), 505
Working set, 548